ACTIVATION AND CATALYTIC REACTIONS OF SATURATED HYDROCARBONS IN THE PRESENCE OF METAL COMPLEXES

Catalysis by Metal Complexes

Volume 21

The titles published in this series are listed at the end of this volume.

ACTIVATION AND CATALYTIC REACTIONS OF SATURATED HYDROCARBONS IN THE PRESENCE OF METAL COMPLEXES

by

ALEXANDER E. SHILOV

Institute of Biochemical Physics,
Moscow, Russia
and
Semenov Institute of Chemical Physics,
Russian Academy of Sciences,
Moscow, Russia

and

Georgiy B. Shul'pin

Semenov Institute of Chemical Physics,
Russian Academy of Sciences,
Moscow, Russia

KLUWER ACADEMIC PUBLISHERS
DORDRECHT / BOSTON / LONDON

A C.I.P. Catalogue record for this book is available from the Library of Congress.

ISBN 0-7923-6101-6

Published by Kluwer Academic Publishers,
P.O. Box 17, 3300 AA Dordrecht, The Netherlands.

Sold and distributed in North, Central and South America
by Kluwer Academic Publishers,
101 Philip Drive, Norwell, MA 02061, U.S.A.

In all other countries, sold and distributed
by Kluwer Academic Publishers,
P.O. Box 322, 3300 AH Dordrecht, The Netherlands.

Printed on acid-free paper

Printed in the Netherlands.

CONTENTS

PREFACE

Chemistry is the science about breaking and forming of bonds between atoms. One of the most important processes for organic chemistry is breaking bonds C–H, as well as C–C in various compounds, and primarily, in hydrocarbons. Among hydrocarbons, saturated hydrocarbons, alkanes (methane, ethane, propane, hexane etc.), are especially attractive as substrates for chemical transformations. This is because, on the one hand, alkanes are the main constituents of oil and natural gas, and consequently are the principal feedstocks for chemical industry. On the other hand, these substances are known to be the less reactive organic compounds. Saturated hydrocarbons may be called the "noble gases of organic chemistry" and, if so, the first representative of their family – methane – may be compared with extremely inert helium. As in all comparisons, this parallel between noble gases and alkanes is not fully accurate. Indeed the transformations of alkanes, including methane, have been known for a long time. These reactions involve the interaction with molecular oxygen from air (burning – the main source of energy!), as well as some mutual interconversions of saturated and unsaturated hydrocarbons. However, all these transformations occur at elevated temperatures (higher than 300–500 °C) and are usually characterized by a lack of selectivity. The conversion of alkanes into carbon dioxide and water during burning is an extremely valuable process – but not from a chemist viewpoint.

The chemical inertness of alkanes can be overcome if the transformations are carried out at high temperatures. However, the low selectivity of such processes motivates chemists into searching principally for new routes of alkane conversion which could transform them into very valuable products (hydroperoxides, alcohols, aldehydes, ketones, carboxylic acids, olefins, aromatic compounds etc.) under mild conditions and selectively. This is also connected with the necessity for the development of intensive technologies and for solving the problems of ecology. Finally, one more very important problem is the complete and efficient chemical processing of oil and gas components, which becomes pertinent because of gradual depletion of hydrocarbon natural resources.

In the last decades, new reactions of saturated hydrocarbons under mild conditions have been discovered. For example, new reactions include alkane transformations in superacid media, interactions with metal atoms and ions, and reactions with some radicals and carbenes. In the same period, the development of coordination metal-complex catalysis led to the discovery of the ability of various types of molecules, including molecular hydrogen, carbon monoxide, oxygen, nitrogen, olefins, acetylenes, aromatic compounds, to take part in catalytic reactions in homogeneous solutions. In such processes, a molecule or its fragment entering the coordination sphere of the metal complex, as a ligand, is chemically activated. It means that a molecule or its fragment attains the ability to enter into reactions that either do not proceed in the absence of a metal complex or occur at very slow rates. At last, the list of compounds capable of being activated by metal complexes has been enriched with alkanes.

This monograph is devoted to the activation and various transformations of saturated hydrocarbons, i.e., reactions accompanied by the C–H and C–C bond cleavage. A special attention is paid to the recently found reactions with the alkane activation in the presence of metal complexes being described in more detail. In addition to the reactions of saturated hydrocarbons which are the main topic of this book, the activation of C–H bonds in arenes and even olefins and acetylenes is considered. In some cases, this activation exhibits similarities for all types of compounds, and sometimes they proceed by different mechanistic pathways.

Chapter I discusses some general questions relevant to the chemistry of alkanes and especially their reactions with metal compounds. Transformations of saturated hydrocarbons in the absence of metal derivatives and in the presence of solid metal and metal oxide surfaces are described in Chapters II and III (Figure 1). Since these reactions are not the main topic of the monograph their consideration here is far from comprehensiveness but the knowledge of such processes is very important for understanding the peculiarities and mechanisms of the reactions with metal complexes. Chapters IV–X are the main chapters of this book because they describe the activation of hydrocarbons in the presence of metal complexes. Finally, Chapter XI is devoted to a brief description of the hydrocarbon reactions with enzymes, which mainly contain metal ions and are true metal complexes.

We clearly understand that this monograph does not cover all references that have appeared on the reactions of alkanes and other hydrocarbons with metal

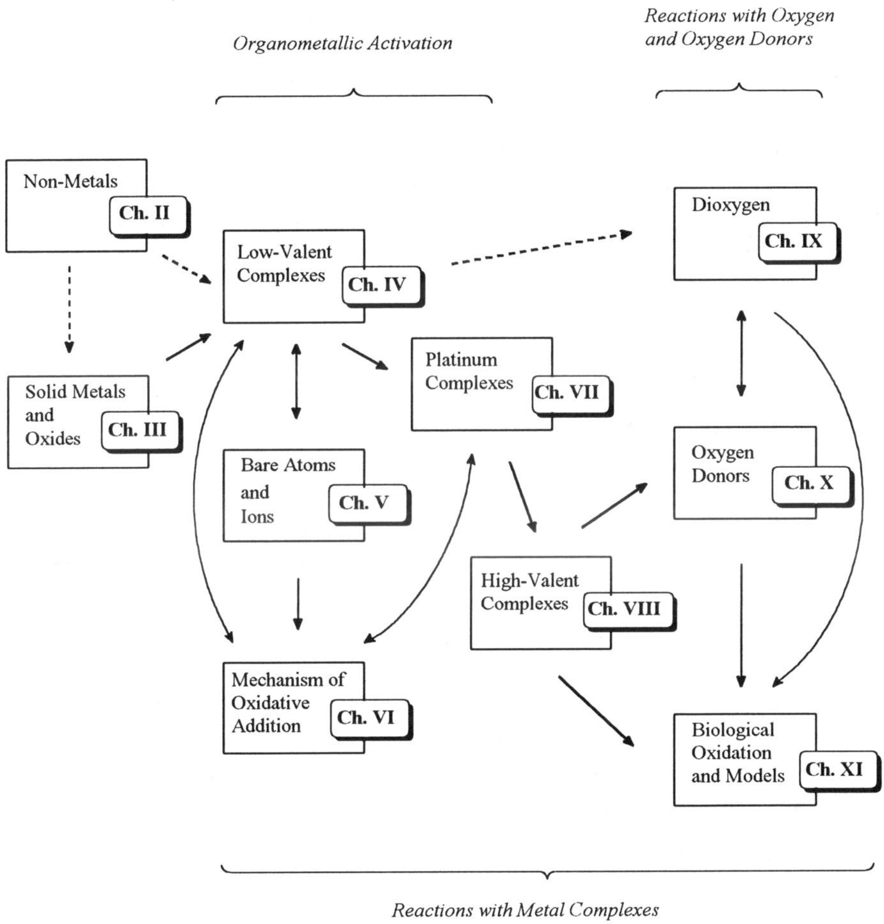

Figure 1. Structure of this monograph.

complexes (and especially with various reagents that are not metal complexes). Moreover, we suspect that not all interesting works on alkane activation will be described here in proper detail and some important findings will not be referenced in this edition. We wish to apologize in advance to all scientists who decide that their works are covered too briefly. The subjective factor here is very great. In searching and selecting the references for various chapters, we gave the preferences to recent publications assuming that the reader will be able to find many publications on a certain topic having only one very recent paper. In some cases, we restricted our citation by a review and a few recent original publications (this is especially necessary for citation of works on heterogeneous activation on solid catalysts where the total number of papers is enormous). We tried also to give more detailed descriptions of some hard to obtain works (e.g., published in Russian.). The material of our previous reviews and books published either in Russian or English have been partially used in this monograph.

The authors hope that this book will be useful not only for those who are interested in activation of alkanes and other hydrocarbons by metal complexes, but also for the specialists in homogeneous and heterogeneous catalysis, petrochemistry, and organometallic chemistry. Some parts of the monograph will be interesting for the specialists in inorganic and organic chemistry, theoretical chemistry, biochemistry and even biology, and also for those who work in petrochemical industry and industrial organic synthesis. This book covers studies which appeared up to early 1999.

We are grateful to the scientists who have helped to create this book, who discussed with us certain problems of alkane activation, and also provided us with reprints and manuscripts: D. M. Camaioni, B. Chaudret, E. G. Derouane, R. H. Fish, Y. Fujiwara, A. S. Goldman, T. Higuchi, C. L. Hill, Y. Ishii, B. R. James, G. V. Nizova, A. Kitaygorodskiy, the late R. S. Drago, D. R. Ketchum, J. A. Labinger, J. R. Lindsay Smith, J. M. Mayer, J. Muzart, L. Nice, R. A. Periana, E. S. Rudakov, S. Sakaguchi, U. Schuchardt, H. Schwarz, A. Sen, A. A. Shteinman, G. Süss-Fink and many others.

Aleksandr Evgenievich SHILOV

Georgiy Borisovich SHUL'PIN

INTRODUCTION

Hydrocarbons occurring in oil and natural gas are of great significance for the contemporary civilization, due to their being the most commonly used fuel, on the one hand, and the source of row materials for chemical industry, on the other hand [1]. Oil contains (see examples in Figure 2) a considerable amount of alkanes, as well as, aromatic and other hydrocarbons. Methane is usually a major component of natural gas (Figure 3) [2]. The distribution of total natural gas reserves (4,933 trillion cubic feet or 1.4×10^{11} m^3) is the following: Eastern Europe (40.1%), Africa/Middle East (39.2%), Asia-Pacific (6.6%), North America (6.1%), Latin America (4.1%), and Western Europe (3.9%) [1c].

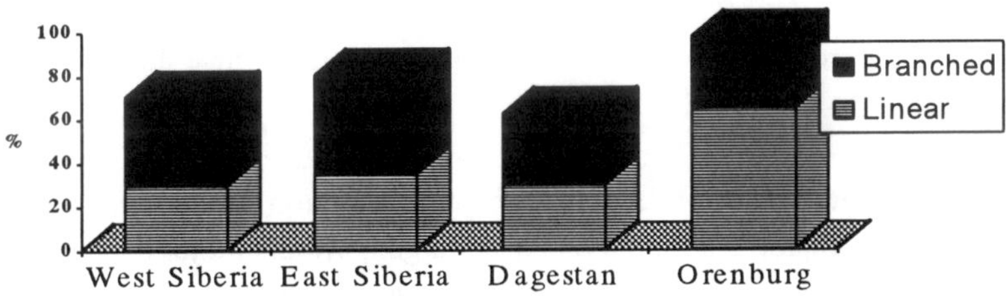

Figure 2. Alkane content (%) in petroleum fraction 28-200 °C from certain Russian oil deposits.

Oil cracking gives additional amounts of lower alkanes and olefins, the latter being even more valuable products. Methane and other alkanes are also contained in gases evolved in coal mines; saturated hydrocarbons are obtained by hydrogenation and dry distillation of coal and peat [3a–c]. Paraffins may be produced synthetically, i.e., an alkane mixture is formed of carbon monoxide and hydrogen (Fischer-Tropsch synthesis); multiring aromatics can be converted to isoparaffins and cycloparaffins [3d]. Polymeric materials including natural products are also the source of alkanes [4]. Finally, alkanes are involved during some biochemical processes [5].

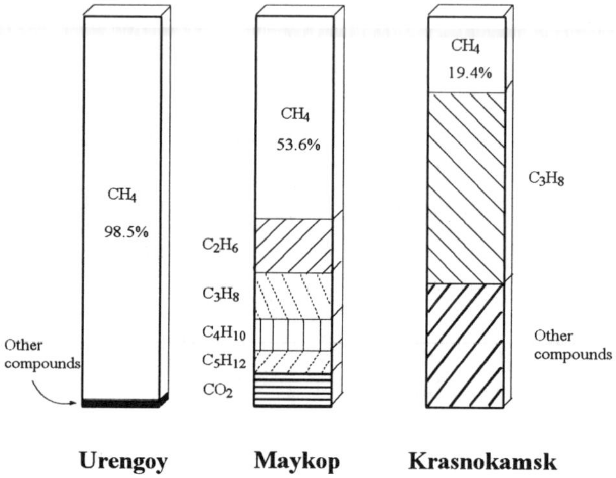

Figure 3. Composition of natural gas (volume %) from different Russian deposits.

The major field of hydrocarbon consumption is power supply. Energy evolved from alkane combustion is used in gas, diesel and jet engines. Nowadays, chemical processing of hydrocarbon raw materials, in particular alkanes, requires usually participation of heterogeneous catalysts and elevated temperatures (above 200–300 °C) [6]. Natural gas is used mainly in the production of synthesis gas or hydrogen [6e]. Liquefiable components of natural gas find more extensive application. In the USA, the gas condensate and other liquefied components account for 18% of the overall production of liquid hydrocarbons and 70% of the raw material for the production of ethene and other valuable products.

It should be noted that some processes that proceed at relatively low temperatures are well known – chain radical and microbiological oxidation. Biological transformations of alkanes and other hydrocarbons are extremely important because they give convenient routes to valuable chemical products [7]. For example, the bacterial enzyme methane monooxygenase converts about 10^9 tons per year of methane to methanol. Besides, they are the basis for the microbiological remediation of soil and waters of seas, rivers and lakes polluted with oil and products of petrochemical industry [8].

The interest in new reactions of alkanes is prompted mainly by the need for selective and efficient industrial transformations of hydrocarbons from oil, gas and coal. Development of this area is necessary because fundamentally new routes from hydrocarbons to valuable products, for example, alcohols, ketones, acids and peroxides, may be discovered. In addition, important environmental problems might be solved using such types of transformations, for instance, the removal of petroleum pollution. However, well-known chemical inertness of alkanes causes great difficulties in their activation especially under mild conditions. Thus, efficient reactions of saturated hydrocarbons with various reagents and particularly with metal complexes make it an extremely difficult, but also excitedly interesting and important problem both for industry and academic theoretical science. Only in the past decades, the vigorous development of metal-complex catalysis allowed the beginning of an essentially new chemistry of alkanes and enriched the knowledge about transformations of unsaturated hydrocarbons. Transformations of hydrocarbons (both saturated and unsaturated) under the action of metal complexes, particularly when these complexes play a role of catalysts, seems to be a very promising field. Indeed, in contrast to almost all presently employed processes, reactions with metal complexes occur at low temperatures and can be selective.

Several monographs [9] and many reviews [10], wholly or partly devoted to the metal complex activation of C–H and C–C bonds in hydrocarbons, appeared in recent decades. Reviews and books devoted to some more narrow topics will be cited later throughout this book.

References for Introduction

1. (a) Waddams, A. L. *Chemicals from Petroleum*; Gulf Publ. Co: Houston, 1980. (b) Absi-Halabi, M.; Stanislaus, A.; Qabazard, H. *Hydrocarbon Processing*, Feb. **1997**, p. 45. (c) Weirauch, W. *Hydrocarbon Processing*, Apr. **1997**, p. 23. (d) Weirauch, W. *Hydrocarbon Processing*, May **1997**, p. 27. (e) Manning, T. J. *Hydrocarbon Processing*, May **1997**, p. 85. (f) *Industrial Gases in Petrochemical Processing*; Gunardson, H., Ed.; Dekker: New York, 1997. (g) Morse, P. M. *Chem. & Eng. News* **1998**, *March 23*, p. 17. (h) *Chem. Engineering* May **1998**, 99. (i) *Natural Gas Conversion V*; Parmaliana, A.; Sanfilippo, D.; Frusteri, F.; Vaccari, A.; Arena, F., Eds.; Elsevier: Amsterdam, 1998. (j) Tominaga, H.; Takehi, N. *Petrotech.* **1998**, *21*, 11 (in Japanese). (k) Sato, S.; Morita, K.; Sugioka, M.; Takita, Y.; Fujita, K. *Petrotech.* **1998**, *21*, 188 (in Japanese). (l) Shimizu, K.; Iida, W.; Shintani, O.; Nakamura, M.; Iwai, R. *Petrotech.* **1998**, *21*, 388 (in Japanese). (m) Sedriks, W. *CHEMTECH*, Feb. **1998**, p. 47. (n) Wiltshire, J. *The Chemical Engineer*, 12 March **1998**, p. 20. (o) Molenda, J. *Gaz, Woda Tech. Sanit.* **1998**, *72*, 11 (in Polish). (p) Kaneko, H. *Nensho Kenkyu* **1998**, *111*, 39 (in Japanese).
2. (a) Petrov Al. A. *Hydrocarbons of Petroleum*; Nauka: Moscow, 1984 (in Russian). (b) Adel'son, S. V.; Vishnyakova, T. P.; Paushkin, Ya. M. *Technology of Petrochemical Synthesis*; Khimiya: Moscow, 1985 (in Russian). (c) *Chemistry of Oil and Gas*; Proskuryakov, V. A.; Drabkin, A. E., Eds.; Khimiya: Moscow, 1989 (in Russian). (d) Vyakhirev, R. I. *Perspectives in Energy*, **1997**, *1*, 4. (e) Philp, R. P.; Mansuy, L. *Energy & Fuels*, **1997**, *11*, 753. (f) Berkowitz, N. *Fossil Hydrocarbons: Chemistry and Technology*; Academic Press: San Diego, 1997. (g) Zhuze, N. G.; Kruglyakov, N. M. *Geol. Nefti Gaza* **1998**, No 3, 2 (in Russian). (h) Wang, P.; Zhu, J.; Fang, X.; Zhao, H.; Zhu, C. *Shiyou Xuebao* **1998**, *19*, 24 (in Chinese).
3. (a) Nelson, C. R.; Li, W.; Lazar, I. M.; Larson, K. H.; Malik, A.; Lee, M. L. *Energy & Fuels* **1998**, *12*, 277. (b) Bonfanti, L.; Comellas, L.; Liberia, J.; Vallhonrat-Matalonga, R.; Pich-Santacana, M.; Lopez-Pinol, D. *J. Anal. Appl. Pyrolysis* **1997**, *44*, 89. (c) Lapidus, A. L.; Krylova, A. L.; Eliseev, O. L.; Khudyakov, D. S. *Khim. Tverd. Topl.* 1998, No. 1, 3 (in Russian). (d) Demirel, B.; Wiser, W. H. *Fuel Process. Technol.* **1998**, *55*, 83.
4. (a) Ding, W.; Liang, J.; Anderson, L. L. *Fuel Process. Technol.* **1997**, *51*, 47. (b) Dufaud, V.; Basset, J.-M. *Angew. Chem., Int. Ed. Engl.* **1998**, *37*, 806. (c)

Idem, R. O.; Katikaneni, S. P. R.; Bakhshi, N. N. *Fuel Process. Technol.* **1997**, *51*, 101. (d) Joo, H. K.; Curtis, C. W. *Fuel Process. Technol.* **1998**, *53*, 197. (e) Elliott, D. C.; Sealock, L. J.; Baker, E. G.; Richland, W. A. *Pat. US 5, 616,154* (to Battelle Memorial Institute) [*J. Mol. Catal. A: Chem.* **1998**, *129*, 112].

5. (a) *Fuels and Chemicals from Biomass*; Saha, B. C.; Woodward, J., Eds.; ACS: Washington, DC, 1997. (b) Panow, A.; FitzGerald, J. M. P.; Mainwaring, D. E. *Fuel Process. Technol.* **1997**, *52*, 115. (c) Premuzic, E. T.; Lin, M. S.; Lian, H.; Zhou, W. M.; Yablon, J. *Fuel Process. Technol.* **1997**, *52*, 207. (d) Prinzhofer, A.; Pernaton, E. *Chem. Geol.* **1997**, *142*, 193. (e) Gazso, L. G. *Fuel Process. Technol.* **1997**, *52*, 239. (f) Morikawa, M.; Iwasa, T.; Yanagida, S.; Imanaka, T. *J. Ferment. Bioeng.* **1998**, *85*, 243. (g) Tokuda, M.; Ohta, N.; Morimura, S.; Kida, K. *J. Ferment. Bioeng.* **1998**, *85*, 495. (h) Behns, W.; Friedrich, K.; Haida, H. *Chem. Eng. Technol.* **1998**, *21*, 4. (i) Morikawa, M.; Jin, S.; Shigenori, K.; Imanaka, T. *Nippon Nogei Kagaku Kaishi* **1998**, *72*, 532 (in Japanese).
6. (a) Petrov, Al. A. *Chemistry of Alkanes*, Nauka: Moscow, 1974 (in Russian). (b) Ponec, V. *J. Mol. Catal. A: Chem.* **1998**, *133*, 221. (c) *Catalysis in Petroleum Refining and Petrochemical Industries*; Absi-Halabi, M.; Beshara, J.; Qabazard, H.; Stanislaus, A. Eds.; Elsevier: Amsterdam, 1996. (d) Maxwell, I. E. *Cattech.* **1997**, *1*, 5. (e) Muradov, N. Z. *Energy & Fuels* **1998**, *12*, 41. (f) Hadjigeorge, G. A. *Pat. US 5,622,677* (to Shell Oil Company) [*J. Mol. Catal. A: Chem.* **1998**, *129*, 114]. (g) Krishna, A. S.; Skocpol, R. C.; Fredrickson, L. A. *Pat. US 5,616,237* (to Chevron Research and Development Company) [*J. Mol. Catal. A: Chem.* **1998**, *129*, 112].
7. (a) Hanson, R. S.; Hanson, T. E. *Microbiol. Rev.* **1996**, *60*, 439. (b) King, G. M.; Schnell, S. *Appl. Environ. Microbiol.* **1998**, *64*, 253. (c) Grosskopf, R.; Janssen, P. H.; Liesack, W. *Appl. Environ. Microbiol.* **1998**, *64*, 960. (d) Benstead, J.; King, G. M.; Williams, H. G. *Appl. Environ. Microbiol.* **1998**, *64*, 1091. (e) Calhoun, A.; King, G. M. *Appl. Environ. Microbiol.* **1998**, *64*, 1099. (f) Jensen, S.; Prieme, A.; Bakken, L. *Appl. Environ. Microbiol.* **1998**, *64*, 1143.
8. (a) Atlas, R. M. *Microbiol. Rev.* **1981**, *45*, 180. (b) King, G. M. *Adv. Microbiol. Ecol.* **1992**, *12*, 431. (c) Alexander, M. *Biodegradation and Bioremediation*; Academic Press: San Diego, 1994. (d) Conrad, R. *Adv. Microbiol. Ecol.* **1995**, *14*, 207. (e) Mohn, W. W. *Biodegadation* **1997**, *8*, 15. (f) Xiao, W.; Clarkson, W. W. *Biodegradation* **1997**, *8*, 61. (g) Holden, P. A.; Halverson, L. J.; Firestone, M. K. *Biodegradation* **1997**, *8*, 143. (h) Zhang, W.;

Bouwer, E. J. *Biodegradation* **1997**, *8*, 167. (i) Uraizee, F. A.; Venosa, A. D.; Suidan, M. T. *Biodegradation* **1998**, *8*, 287. (j) Bucke, C. *J. Chem. Technol. Biotechnol.* **1998**, *71*, 356. (k) Hofrichter, M.; Scheibner, K.; Schneegass, I.; Fritsche, W. *Appl. Environ. Microbiol.* **1998**, *64*, 399. (l) .Shen, Y.; Stehmeier, L. G.; Voordouw, G. *Appl. Environ. Microbiol.* **1998**, *64*, 637. (m) Willardson, B. M.; Wilkins, J. F.; Rand, T. A.; Schupp, J. M.; Hill, K. K.; Keim, P.; Jackson, P. J. *Appl. Environ. Microbiol.* **1998**, *64*, 1006. (n) Aislabie, J.; McLeod, M.; Fraser, R. *Appl. Microbiol. Biotechnol.* **1998**, *49*, 210. (o) Yerushalmi, L.; Guiot, S. R. *Appl. Microbiol. Biotechnol.* **1998**, *49*, 475. (p) Margesin, R.; Schinner, F. *Appl. Microbiol. Biotechnol.* **1998**, *49*, 482. (q) Willmann, G.; Fakoussa, R. M. *Fuel Process. Technol.* **1997**, *52*, 27.

9. (a) Sheldon, R. A.; Kochi, J. K. *Metal-Catalysed Oxidations of Organic Compounds*; Academic Press: New York, 1981. (b) Shilov, A. E. *Activation of Saturated Hydrocarbons by Transition Metal Complexes*; D. Reidel: Dordrecht, 1984. (c) Gubin, S. P.; Shul'pin, G. B. *The Chemistry of Complexes with Metal–Carbon Bonds*; Nauka: Novosibirsk, 1984 (in Russian). (d) Hines, A. H. *Methods for the Oxidation of Organic Compounds*; Academic Press: London, 1985. (e) Rudakov, E. S. *The Reactions of Alkanes with Oxidants, Metal Complexes, and Radicals in Solutions*; Naukova Dumka: Kiev, 1985 (in Russian). (f) Omae, I. *Organometallic Intramolecular-Coordination Compounds*; Elsevier: Amsterdam, 1986. (g) Chipperfield, J. R.; Webster, D. E. In *The Chemistry of the Metal–Carbon Bond*; Hartley, F. R., Ed.; J. Wiley: Chichester, 1987, Vol. 4, p. 1073. (h) Shul'pin, G. B. *Organic Reactions Catalyzed by Metal Complexes*; Nauka: Moscow, 1988 (in Russian). (i) Ephritikhine, M. In *Industrial Applications of Homogeneous Catalysis*; Mortreux, A; Petit, F., Eds.; D Reidel: Dordrecht, 1988. (j) *Perspectives in the Selective Activation of C–H and C–C Bonds in Saturated Hydrocarbons*; Meunier, B., Chaudret, B., Eds.; Sci Affaires Division – NATO: Brussels, 1988. (k) *Activation and Functionalization of Alkanes*; Hill, C. L. , Ed.; Wiley: New York, 1989. (l) *Selective Hydrocarbon Activation*; Davies, J. A.; Watson, P. L.; Liebman, J. F.; Greenberg, A., Eds.; VCH Publishers: New York, 1990. (m) Shilov, A. E.; Shul'pin, G. B. *The Activation and Catalytic Reactions of Hydrocarbons*; Nauka: Moscow, 1995 (in Russian). (n) Olah, G. A.; Molnár, A. *Hydrocarbon Chemistry*; Wiley: New York, 1995. (o) *Theoretical Aspects of Homogeneous Catalysis*; van Leeuwen, W. N. M.; Morokuma, K.; van Lenthe, J. H., Eds.; Kluwer: Dordrecht, 1995. (p) *Applied Homogeneous Catalysis with Organometallic Compounds*; Cornils, B.; Herrmann, W. A., Eds.; VCH:

Weinheim, 1996. (q) *Catalytic Activation and Functionalisation of Light Alkanes*; Derouane, E. G.; Haber, J.; Lemos, F.; Ribeiro, F. R.; Guisnet, M., Eds.; Kluwer Acad. Publ.: Dordrecht, 1998.

10. (a) Parshall, G. W. *Acc. Chem. Res.* **1970**, *3*, 139. (b) Parshall, G. W. *Acc. Chem. Res.* **1975**, *8*, 113. (c) Shilov, A. E.; Shteinman, A. A. *Kinetika i Kataliz* **1977**, *18*, 1129 (in Russian). (d) Shilov, A. E.; Shteinman, A. A. *Coord. Chem. Rev.* **1977**, *24*, 97. (e) Webster, D. E. *Adv. Organomet. Chem.* **1977**, *15*, 147. (f) Bruce, M. I. *Angew. Chem.* **1977**, *89*, 75. (g) Muetterties, E. L. *Chem. Soc. Rev.* **1983**, *12*, 283. (h) Grigoryan, E. A. *Usp. Khim.* **1984**, *53*, 347 (in Russian). (i) Crabtree, R. H. *Chem. Rev.* **1985**, *85*, 245. (j) Schwartz, J. *Acc. Chem. Res.* **1985**, *18*, 302. (k) Rothwell, I. P. *Polyhedron* **1985**, *4*, 177. (l) Green, M. L. H.; O'Hare, D. *Pure Appl. Chem.* **1985**, *57*, 1897. (m) Watson, P. L.; Parshall, G. W. *Acc. Chem. Res.* **1985**, *18*, 51. (o) Deem, M. L. *Coord. Chem. Rev.* **1986**, *74*, 101. (p) Artamkina, G. A.; Beletskaya I. P. *Zh. Vses. Khim. Obsh. im. D. I. Mendeleeva* **1986**, *31*, 196 (in Russian). (q) Shilov, A. E.; Shul'pin, G. B. *Russ. Chem. Rev.* **1987**, *56*, 442. (r) Mimoun, H. *Nouv. J. Chim.* **1987**, *11*, 513. (s) Soloveichik, G. L. *Metalloorg. Khim.* **1988**, *1*, 729 (in Russian). (t) Rothwell, I. P. *Acc. Chem. Res.* **1988**, *21*, 153. (u) Bagriy, E. I.; Nekhaev, A. I. *Zh. Vses. Khim. Obsh. im. D. I. Mendeleeva* **1989**, *34*, 634 (in Russian). (v) Jones, W. D.; Feher, F. J. *Acc. Chem. Res.* **1989**, *22*, 91. (w) Moiseev, I. I. *Usp. Khim.* **1989**, *58*, 1175 (in Russian). (x) Shilov, A. E.; Shul'pin, G. B. *Russ. Chem. Rev.* **1990**, *59*, 853. (y) Crabtree, R. H. *Chem. Rev.* **1995**, *95*, 987. (z) Arndten, B. A.; Bergman, R. G.; Mobley, T. A.; Peterson, T. H. *Acc. Chem. Res.* **1995**, *28*, 154. (aa) Schneider, J. J. *Angew. Chem., Int. Ed. Engl.* **1996**, *35*, 1069. (ab) Lohrenz, J. C. W.; Jacobsen, H. *Angew. Chem., Int. Ed. Engl.* **1996**, *35*, 1305. (ac) Herrmann, W. A.; Cornils, B. *Angew. Chem., Int. Ed. Engl.* **1997**, *36*, 1049. (ad) Shilov, A. E.; Shul'pin, G. B. *Chem. Rev.* **1997**, *97*, 2879.

CHAPTER I

PROCESSES OF C–H BOND ACTIVATION

This chapter is devoted to some general problems which should be discussed before consideration of the reactions of alkanes and other hydrocarbons with non-metal and metal-containing substances. First of all we will consider general chemical properties of hydrocarbons and principle mechanisms of reactions with participation of these compounds.

I.1. CHEMICAL REACTIVITY OF HYDROCARBONS

Chemical inertness of alkanes is reflected in one of their old names, "paraffins", from the Latin *parum affinis* (without affinity). However, saturated hydrocarbons can be involved very easily in a total oxidation with air (simply speaking, burning) to produce thermodynamically very stable products: water and carbon dioxide. It should be emphasized that at room temperature alkanes are absolutely inert toward air, if a catalyst is absent. At the same time, some active reagents, e.g., atoms, free radicals, and carbenes, can react with saturated hydrocarbons at room and lower temperatures. These compounds are easily transformed into various products under elevated (above 1000 °C) temperatures, in the absence of other reagents.

Some important reactions of alkanes have been developed, e.g., autoxidation by molecular oxygen at elevated temperatures, which proceeds via a radical chain mechanism. The main feature of this and many other reactions is a lack of selectivity. Reactions with radicals give rise to the formation of many products; all possible isomers may be obtained. As far as burning is concerned, this process can be very selective, producing solely carbon dioxide, but apart from being an important source of energy, is useless from the viewpoint of the synthesis of valuable organic products. Chemical inertness of alkanes is due to very high values of C–H bond energies and ionization potentials. Proton affinities are far lower than for unsaturated hydrocarbons (see Table I.1). Alkane acidities are much smaller than those of other molecules listed in Table I.1.

Table I.1. Characteristics of some hydrocarbons.

RH	→ R•	*D*(R–H) [a] kcal mol^{-1}	I.P. [b] eV	E.A. [c] eV	P.A. [d] eV	pK_a [e]
CH_4	CH_3•	104	12.7		5.3	40
C_2H_6	C_2H_5•	98	11.5		5.6	42
C_3H_8	*n*-C3H_7•	97	11.1		6.1	
	iso-C_3H_7•	94				44
C_4H_{10}	*tert*-C_4H_9•		10.6			
cyclo-C_6H_{12}	*cyclo*-C_6H_{11}•	94	9.9			45
C_6H_6	C_6H_5•	109	9.2	–1.1	7.5	37
CH_2=CH_2	CH_2=CH•	106	10.5	–1.8	6.9	36
CH≡CH	CH≡C•	120	11.4			25
$C_6H_5CH_3$	$C_6H_5CH_2$•	85	8.8	–1.3	7.3	35
CH_3CN	$NCCH_2$•	79	12.1		8.1	24
H_2	H•	104	15.4	–3.6	4.4	25
H_2O	HO•	118	12.6	1.8	7.1	16

[a] Energy of C–H bond.
[b] Ionization potential.
[c] Electron affinity.
[d] Proton affinity.
[e] Acidity.

Thus, we should conclude, that alkanes are extremely inert toward "normal" (i.e., not very reactive) reagents in reactions that proceed more or less selectively. In many respects, alkanes, especially lower ones (methane, ethane) are similar to molecular hydrogen. Indeed, like alkanes, dihydrogen while being inert towards dioxygen at ambient temperatures can be burned in air to produce thermodynamically stable water. The values of the C–H and H–H dissociation energy for methane and dihydrogen molecules are almost exactly equal (104 kcal mol^{-1}). Like methane, dihydrogen is relatively unreactive with respect to many reagents. Ethylene, acetylene and benzene, compounds with stronger C–H bonds (106, 120 and 109 kcal mol^{-1}) in contrast to methane, are known to exhibit much higher reactivity. This is due to the fact that both methane and dihydrogen are completely saturated compounds, i.e., contain neither π- nor n-electrons. Therefore, it is not surprising that most reactions with unsaturated hydrocarbons proceed as addition, followed in some cases by elimination. In the last decades, the reactions involving metal complex activation of C–H bonds in unsaturated hydrocarbons were described that do not involve the addition to a double or triple bond. These reactions proceed via oxidative addition of C–H bonds to metal centers. Using the term "the activation" of a molecule, we mean that the reactivity of this molecule increases due to some action. In contrast to saturated compounds (and saturated bonds), the activation of the unsaturated species (or their fragments) may be induced by coordination of a particle followed by the addition to this bond or by the rupture of the unsaturated bond. For example, for olefins and arenes such activation can be caused by π-complexation. It is known that π-coordination of the olefinic double bond with some metal ion gives rise to the enhanced reactivity of the organic fragment in its interaction with nucleophiles [1a,b]. Carbonyl group, CO, when coordinated to a metal, becomes reactive with nucleophilic reagent [1a, c–f].

However, what is "the activation of ordinary σ-bond"? It is reasonable to propose that the activation of, for example, the C–H bond, is the increasing of the reactivity of this bond toward a reagent. As a consequence, such a bond is capable of splitting, thus producing two particles instead of one initial species. In many cases such a rupture of a saturated bond is implied when the term "activation" is used. However, strictly speaking, the splitting of the bond is in fact a consequence of its activation. It seems that in some respects the term "splitting of C–H bond" would be more correct. It is noteworthy that in the last years examples of coordination between some particles (and metal complexes also) and saturated hydrocarbons or their fragments were demonstrated [2]. In the present monograph, we will consider all processes of splitting C–H bonds in hydrocarbons by metal complexes as well as the problem of coordination of alkanes or alkyl groups (from various organic compounds) with metal complexes. To concern this problem is important when we discuss the possible mechanisms of the C–H bond splitting, because such adducts of alkanes with metal complexes can lie on the reaction coordinate. It should be noted that the term "the activation of C–H bond" is noticeably more narrow than that of "the activation of hydrocarbons". Indeed, the activation of alkanes may also involve the splitting of C–C bonds. As for unsaturated hydrocarbons, their activation is outside the content of this book. Nevertheless, some reactions of formal substitution at aromatic C–H bond which proceed as addition followed by elimination (e.g., electrophilic metalation) will be surveyed. Metal complex activation of C–H bond in unsaturated hydrocarbons which does not involve the addition to double or triple bond (proceeding usually as oxidative addition) will also be a topic of this monograph.

I.2. CLEAVAGE OF THE C–H BOND PROMOTED BY METAL COMPLEXES

Bearing in mind a mechanistic consideration, we propose to divide all the C–H bond splitting reactions promoted by metal complexes into three groups. This formal classification is based on the reaction mechanisms.

I.2.A. THREE TYPES OF PROCESSES

In the previous section, we discussed the term "activation" when applied to saturated compounds and concluded that the cleavage of an ordinary bond (e.g., C–H) can be a result of such activation, and in many cases, we might consider the activation and splitting as synonymous. We wish to describe here the classification that is based on types of interaction between the alkane and metal complex.

First Type: "True" (Organometallic) Activation

Processes where organometallic derivatives, i.e., compounds containing an M–C σ-bond (M = metal), are formed as an intermediate or as the final product, can be conveniently assigned to the first type. The σ-ligand in the resulting compound is an organyl group, i.e., alkyl, aryl, vinyl, acyl, etc. (all these groups are bound to M via a carbon atom). Subsequently, the M–C σ-bond can be cleaved; naturally, in catalytic processes the dissociation of this bond is inevitable. The cleavage of the C–H bond with direct participation of a transition metal ion proceeds via the oxidative addition mechanism:

$$\mathrm{RH} + \mathrm{M}^{n+} \longrightarrow \mathrm{R{-}M}^{(n+2)+}\mathrm{{-}H} \tag{I.1}$$

or an electrophilic substitution:

$$\mathrm{RH} + \mathrm{M}^{n+} \longrightarrow \mathrm{R{-}M}^{n+} + \mathrm{H}^{+} \tag{I.2}$$

Conventionally speaking, metals in low and high oxidation state respectively take part in these reactions. In the case of electrophilic metalation of the aromatic nucleus, the reaction proceeds in two stages, the electrophilic species adding to the arene in the first step with the formation of a Wheland intermediate [3]. An analogous intermediate, which might be formed in the interaction of a saturated hydrocarbon with an electrophilic metal-containing species, should be much less stable. Much more probable is that this structure is a transition state corresponding to a maximum on the energy diagram (Figure I.1).

It is therefore not surprising that the reactivities of arenes and alkanes in electrophilic substitution reactions are very different, with the former being much more active. At the same time, the mechanism of the interaction (oxidative addition) of both saturated and aromatic hydrocarbons with complexes of metals in a low oxidation state is in principle the same. The reactivities of arenes and alkanes in oxidative addition reactions with respect to low-valent metal complexes therefore usually differ insignificantly. Furthermore, a metal complex via the oxidative addition mechanism can easily cleave the C–H bond in olefin or acetylene.

Thus according to our classification, the first group includes reactions involving "true", "organometallic" metal complex activation of the C–H bonds. We call this type of activation "true", because only in this case, the closest contact between metal ion and the C–H bond (i.e., normal σ-bond between M and C) is realized. A molecule of C–H compound enters in the form of an organyl σ-ligand into the coordination sphere of the metal complex.

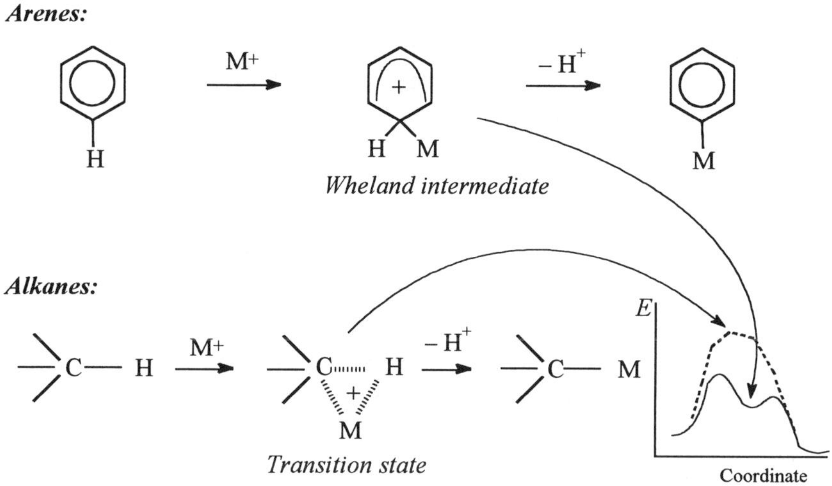

Figure I.1. Schematic representation of routes for the interactions of arenes and alkanes with electrophilic transition metal complex, M.

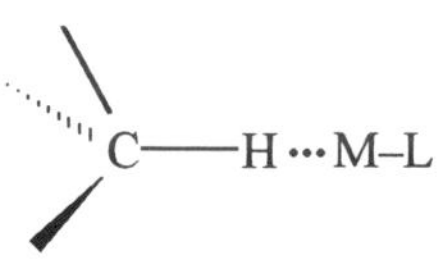

I-1

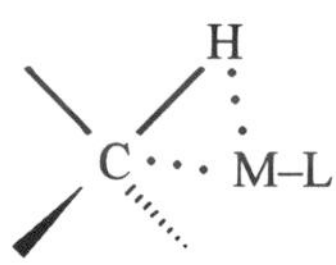

I-2

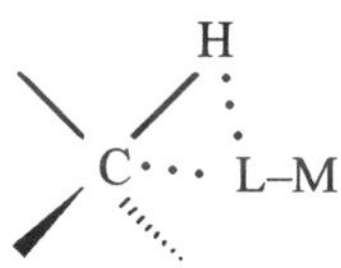

I-3

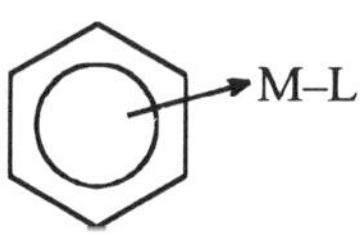

I-4

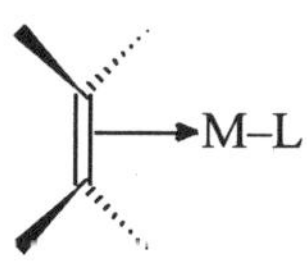

I-5

Very weak adducts of metal complexes with alkanes (or alkyl groups), in which the C–H bond is directly coordinated to the metal (**I-1, I-2**) or its ligands (**I-3**), do not necessarily lead to subsequent cleavage of the C–H bond. However, in some cases, compounds like **I-1, I-2** or **I-3** can be intermediate species lying on the reaction coordinate, which leads to σ-organyl products. These intermediates are analogous in some respects to much more stable π-complexes, (e.g., **I-4** and **I-5**) which are known to be formed by unsaturated hydrocarbons. This coordination of the metal complex to the C–H bond may be referred to as preactivation of the compound's C–H bond.

Second Type: Interaction of the Complex and the C–H Bond only via the Ligand

In the second group, we include reactions in which the contact between the complex and the C–H bond is only via a complex' ligand during the process of the C–H bond cleavage and the σ-C–M bond is not generated directly at any stage. The function of the metal complex usually consists under these conditions in abstracting an electron or a hydrogen atom from the hydrocarbon, RH. The radical-ions RH^{+}• or radicals R• which are formed, then interact with other species present in solution, for example, with molecular oxygen. One of the ligands of the metal complex can also serve as a species of this sort. An example is provided by the hydroxylation of an alkane by an oxo complex of a high-valent metal:

$$RH + O{=}M^{n+} \longrightarrow R\bullet + HO{-}M^{(n-1)+} \longrightarrow ROH + M^{(n-2)+} \quad (I.5)$$

In this reaction, the oxo complex is an oxidant of the type CrO_4^{2-} or MnO_4^{-} and also, for example, one of the states of the cytochrome P450 enzyme – an oxoferryl species containing the $Porph^{+\bullet}Fe^{V}{=}O$ species.

It is noteworthy that the reaction with the intermediate participation of radicals can result in the formation of alkyl σ-complexes, for example via the following mechanism

$$RH + M^{n+} \longrightarrow R\bullet + M^{(n-1)+} + H^{+} \longrightarrow \sigma\text{-}R{-}M^{n+} + H^{+} \quad (I.6)$$

and the reaction should then be assigned to the first type. Since the alkyl σ-derivative is usually unstable, it is difficult to demonstrate its intermediate formation. The mediated (i.e., without contact with metal center of the metal complex) splitting of the C–H bond in the hydrocarbon, RH, by a complex can be apparently effected also via a molecular mechanism, where RH is in direct contact only with a ligand of the complex. In some cases, the metal complex is capable of activating not only as a hydrocarbon but also as another reactant. Thus, for example, the active center of cytochrome P450 initially causes the transition of the oxygen molecule to a reactive state (one or the two atoms of dioxygen is coordinated to the iron ion, forming an oxo ligand). After this, the same active center activates the hydrocarbon molecule (with the participation of the oxo ligand).

Third Type: Metal Complex Generates an Independent Reactive Species Which Then Attacks the C–H Bond

Whereas the reactions included in the second group require direct contact between a molecule of the C–H compound and the metal complex (albeit via the ligand). In the processes belonging to the third type, the complex activates initially not the hydrocarbon but the other reactant (e.g., H_2O_2 or O_2). The reactive species formed (a carbene or radical, e.g., hydroxyl radical) attacks then the hydrocarbon molecule without any participation in the latter process of the metal complex which has generated this species. Oxidation of alkanes by Fenton's reagent [4a–c] is an example of a such type process:

$$H_2O_2 + Fe^{2+} \longrightarrow HO\bullet + HO^- + Fe^{3+}$$

$$HO\bullet + RH \longrightarrow H_2O + R\bullet$$

$$R\bullet + O_2 \longrightarrow ROO\bullet$$

$$ROO\bullet + Fe^{2+} \longrightarrow ROO^- + Fe^{3+}$$

$$ROO^- + H^+ \longrightarrow ROOH$$

It should be mentioned that this is only a simplified scheme. Actual mechanism of this reaction is much more complex.

The Ishii oxidation reaction uses a combination of *N*-hydroxyphthalimide (NHPI, **I-7**) and cobalt derivative, e.g., $Co(acac)_2$, as a catalyst in the transformation of alkanes, RH, and other compounds into oxygenates under the action of molecular oxygen [4d]. The reaction proceeds, though much less efficiently, in the absence of a metal compound. It is assumed that the role of metal complex is to facilitate conversions between NHPI and phthalimide *N*-oxyl (PINO, **I-6**) under the action of molecular oxygen.

I-6 **I-7**

It is evidently that the metal catalyst does not take part in the direct "activation" of the alkane C–H bond by the radical.

I.2.B. MECHANISMS OF THE C–H BOND CLEAVAGE

The classification described above is an approximate subdivision of all reactions known in accordance with their mechanisms. One example was given in eq. (I.6). Such process can proceed with participation of ligands of metal complex. Photochemical reaction between, for example, alkane, RH, and $PtCl_6^{2-}$ [5], depicted by eq. (I.7) and initiated via mechanism of the third type can lead to the formation of an σ-organyl derivative of the metal and the entire process then belongs to the first type.

$$RH + [M\text{–}Cl] \xrightarrow{h\nu} R\bullet + M + H^+ + Cl^- \longrightarrow \sigma\text{-}R\text{–}M + H^+ + Cl^- \quad (I.7)$$

Evidently the unambiguous assignment of a process to a particular type requires a detailed knowledge of the reaction mechanism. However, at the present time the mechanisms of many processes have not been elucidated even in a broad outline. For example, organomagnesium compounds formed by co-condensation of magnesium and aryl halides from the gas phase are capable of catalyzing alkane transformations at room temperature [6]. Thus *n*-hexane can be converted into a mixture of saturated and unsaturated hydrocarbons C_4–C_8, with the conversion reaching 75%. The mechanism of this very interesting reaction seems to be unclear.

We may note that the mechanisms of reactions included in the last two types are, in general, not the same for paraffins, on the one hand, and aromatic hydrocarbons, on the other hand, even if the products of these reactions are of the same type. For example, alcohols and phenols may be obtained from alkanes and arenes respectively by the reaction in air with hydroxyl radicals generated by the action of a metal complex. However, in the case of alkane, an alcohol can be formed by the reduction of alkyl peroxide, whereas hydroxyl is added to an arene with subsequent oxidation of a radical formed. Hence follows the possibility that arenes and alkanes may exhibit different reactivities in each specific reaction.

I.3. Brief History of Metal-Complex Activation of C–H Bonds

Although first metal-containing systems capable of reacting with hydrocarbons and other C–H compounds such as Fenton's reagent (hydroxylation) and mercury salts (direct mercuration) were discovered as early as the end of nineteenth century, the 1930s should apparently be regarded as the start of investigations into activation of C–H compounds with participation of transition metal complexes. During this period, the reaction involving the electrophilic auration of arenes was described [7a], the radical chain autoxidation of hydrocarbons initiated by metal derivatives was developed [7b], and the method was proposed for the oxidation of alkenes and arenes by hydrogen peroxide promoted by oxo-complexes [7c].

A second spurt in pioneering research into this field occurred in the 1960s. Reactions involving the cyclometalation (i.e., the cleavage of a C–H bond in the ligand linked to the metal via an atom of nitrogen, phosphorus etc.) of the aromatic nucleus [8a] and of a sp^3-hybridized carbon atom [8b] were found. It was demonstrated that palladium(II) derivatives induce the oxidative coupling of arenes [9a] and also the arylation of alkenes [9b], while platinum(II) salts catalyze the H–D exchange between benzene and D_2O [9c]. In 1969 the first reactions involving activation of C–H bond in alkanes were discovered [10]: it was found that platinum(II) salts catalyze the H–D exchange between methane or its analogs and D_2O at 100 °C and the complex $CoH_3(PPh_3)_3$ induces the deuteration of methane by D_2 at room temperature. It has now become evident that, as expected, organometallic derivatives are formed as intermediates in all these instances. It was shown in the 1970s that alkanes are oxidized by platinum(IV) [11a], palladium(II) [11b], ruthenium(IV) [11c], and cobalt(III) [11d,e] compounds, and that complexes of iridium(III) [11f] and titanium(II) [11g] catalyze the H–D exchange.

The next decade was marked by a vigorous development of studies on the activation of alkanes and arenes by low-valent metal complexes via an oxidative addition mechanism with the formation of alkyl and aryl derivatives of metals (or alkenes) [12]. The number of known examples of the activation of the C–H bond by complexes of metals in a high oxidation state with formation of organometallic compounds is so far much less. Thus, the methyllutetium σ-complex, which enters into an exchange reaction with $^{13}CH_4$ presents another case of C–H bond activation by high-valent metal complex [13]. Ion $PtCl_6^{2-}$ metalates arenes similarly to palladium(II). However, unlike palladium(II) σ-aryl complexes of Pt(IV) are stable compounds and have been isolated [14]. The $PtCl_6^{2-}$ ion easily platinates arenes under the action of light [14] (or γ-irradiation [14]) giving the first example of photo-electrophilic substitution of arenes. Under the same conditions, alkanes are dehydrogenated to afford π-olefin complexes of platinum(II).

At the end of the 1980s and during 1990s, the intensity of investigations of the C–H bond activation by low-valent metal complexes began to diminish somewhat and interest gradually shifted to the field involving the oxidation of hydrocarbons by high-valent metal oxo-compounds and oxygen (e.g., [4d, 15]). Attention is being especially concentrated nowadays on biological oxidation and its chemical models. Studies on the modeling of cytochrome P450 were stimulated by the use of iodosyl benzene and some other compounds as the donors of an oxygen atom in catalytic oxidation reactions and of metalloporphyrin as a model of the active center of the enzyme (e.g., [16]).

References for Chapter I

1. (a) Shul'pin, G. B. *Organic Reactions Catalyzed by Metal Complexes*; Nauka: Moscow, 1988, p. 69 (in Russian). (b) Eisenstein, O.; Hoffmann, R. *J. Am. Chem. Soc.* **1981**, *103*, 4308. (c) Doxsee, K. M.; Grubbs, R. H. *J. Am. Chem. Soc.* **1981**, *103*, 7696. (d) Trautman, R. J.; Gross, D. C.; Ford, P. C. *J. Am. Chem. Soc.* **1985**, *107*, 2355. (e) Nakamura, S.; Dedieu, A. *Theor. Chim. Acta* **1982**, *61*, 587. (f) Dedieu, A.; Nakamura, S. *Nouv. J. Chim.* **1984**, *8*, 317.
2. (a) Brookhart, M.; Green, M. L. H. *J. Organomet. Chem.* **1983**, *250*, 395. (b) Crabtree, R. H.; Hamilton, D. G. *Adv. Organomet. Chem.* **1988**, *28*, 299. (c) Ginzburg, A. G.; Bagatur'yants, A. A. *Metalloorg. Khim.* **1989**, *2*, 249 (in Russian). (d) Crabtree, R. H. *Acc. Chem. Res.* **1990**, *23*, 95. (e) Crabtree, R. H. *Angew. Chem., Int. Ed. Engl.* **1993**, *32*, 789. (f) Heinekey, D. M.; Oldham, W. J., Jr. *Chem. Rev.* **1993**, *93*, 913. (g) Hall, C.; Perutz, R. N. *Chem. Rev.* **1996**, *96*, 3125.
3. (a) Sykes, P. *A Guidebook to Mechanism in Organic Chemistry*; Longman: London, 1971, Ch. 6. (b) Dneprovskiy, A. S.; Temnikova, T. I. *Theoretical Principles of Organic Chemistry*, Khimiya: Leningrad, 1979, Chpts. 14, 15 (in Russian).
4. (a) Taqui Khan, M. M.; Martell, A. E. Homogeneous Catalysis by Metal Complexes; Vol. 1, Academic Press: New York, 1974, p. 143. (b) Walling, C. *Acc. Chem. Res.* **1975**, *8*, 125. (c) Karakhanov, E. A.; Narin, S. Yu.; Dedov, A. G. *Appl. Organomet. Chem.* **1990**, *5*, 445. (d) Ishii, Y. *J. Mol. Catal. A: Chem.* **1997**, *117*, 123.
5. Shul'pin, G. B.; Nizova, G. V.; Shilov, A. E. *J. Chem. Soc., Chem. Commun.* **1983**, 671.
6. Tyurina, L. A.; Kombarova, S. V.; Smirnov, V. V. *Dokl. Akad. Nauk SSSR* **1989**, *309*, 122 (in Russian).
7. (a) Kharash, M. S.; Isbell, H. C. *J. Am. Chem. Soc.* **1931**, *53*, 3053. (b) Haber, F.; Willstätter, R. *Ber.* **1931**, *64B*, 2844. (c) Milas, N. A.; Sussman, S. *J. Am. Chem. Soc.* **1936**, *58*, 1302.
8. (a) Kleiman, J. P.; Dubeck, M.; *J. Am. Chem. Soc.* **1963**, *85*, 1544. (b) Chatt, J.; Davidson, J. M. *J. Chem. Soc..* **1965**, 843.
9. (a) van Helden, R.; Verberg, G. *Recl. Trav. Chim. Pays-Bas* **1965**, *84*, 1263. (b) Fujiwara, Y.; Moritani, I.; Danno, S. *et al. J. Am. Chem. Soc.* **1969**, *91*, 7166. (c) Garnett, J. L.; Hodges, R .J. *J. Am. Chem. Soc.* **1967**, *89*, 4546.

10. Gol'dshleger, N. F.; Tyabin, M. B.; Shilov, A. E.; Shteinman, A. A. *Zh. Fiz. Khim.* **1969**, *43*, 2174 (in Russian).

11. (a) Gol'dshleger, N. F.; Es'kova, V. V.; Shilov, A. E.; Shteinman, A. A. *Zh. Fiz. Khim.* **1972**, *46*, 1353 (in Russian). (b) Rudakov, E. S.; Zamashchikov, V. V.; Belyaeva, N. P.; Rudakova, R. I. *Zh. Fiz. Khim.* **1973**, *47*, 2732 (in Russian). (c) Tret'akov, V. P.; Arzamaskova, L. N.; Ermakov, Yu. I. *Kinet. Katal.* **1974**, *15*, 538 (in Russian). (d) Cooper, T. A.; Waters, W. A. *J. Chem. Soc. B* **1967**, 687. (e) Hanotier, J.; Camerman, P.; Hanotier-Bridoux, M.; De Radzitsky, P. *J. Chem. Soc.., Perkin Trans. 2.* **1972**, 2247. (f) Garnett, J. L.; Long M. A.; Peterson, K. B. *Aust. J. Chem.* **1974**, *27*, 1823. (g) Grigoryan E. A.; D'ychkovskiy, F. S.; Mullagaliev, I. R. *Dokl. Akad. Nauk SSSR* **1975**, *224*, 859 (in Russian).

12. (a) Baudry, D.; Ephritikhine, M.; Felkin, H. *J. Chem. Soc., Chem. Commun.* **1980**, 1243. (b) Crabtree, R. H.; Mihelcic, J. M.; Quirk, J. M. *J. Am. Chem. Soc.* **1979**, *101*, 7738. (c) Janowicz, A. H.; Bergman, R. G. *J. Am. Chem. Soc.* **1982**, *104*, 352. (d) Hoyano, J. K.; Graham, W. A. G., *J. Am. Chem. Soc.* **1982**, *104*, 3723. (e) Fendrick, C. M.; Marks, T. J. *J. Am. Chem. Soc.* **1984**, *106*, 2214. (f) Jones, W. D.; Feher, F. J. *Organometallics* **1983**, *2*, 562. (g) Green, M. L. H. *J. Chem. Soc., Dalton Trans.* **1986**, 2469.

13. Watson, P. L. *J. Am. Chem. Soc.* **1983**, *105*, 6491.

14. Shul'pin, G. B.; Nizova, G. V.; Nikitaev, A. T. *J. Organomet. Chem.* **1984**, *276*, 115.

15. (a) Renneke, R. F.; Hill, C. L. *Angew. Chem.* **1988**, *100*, 1583. (b) Periana, R. A.; Taube, D. J.; Evitt, E. R.; Löffer, D. G.; Wentrecek, P. R.; Voss, G.; Masuda, T. *Science,* **1993,** *259*, 340. (c) Lin, M.; Hogan, T.E.; Sen, A. *J. Am. Chem. Soc.* **1996**, *118*, 4574.

16. (a) Groves, J.T.; Nemo, T.E.; Myers, R.C. *J. Am. Chem. Soc.* **1979**, *101*, 1032. (b) Ohtake, H.; Higuchi, T.; Hirobe, M. *Heterocycles* **1995**, *40*, 867.

CHAPTER II

HYDROCARBON TRANSFORMATIONS THAT DO NOT INVOLVE METALS OR THEIR COMPOUNDS

In this chapter we will briefly survey the main types of hydrocarbon transformations that occur without the participation of solid metals or oxides and metal complexes. The alkane transformations described here should be taken into consideration for comparison when discussing the metal complex activation.

II.1. Transformations Under the Action of Heat or Irradiation

Under the action of heat or irradiation, alkanes decompose to produce free radicals which further form stable products. Such processes, especially pyrolysis, are of practical importance.

II.1.A. PYROLYSIS

Heating alkanes [1a–c] at temperature 900–2000 °C gives rise to the formation of radicals:

$$CH_4 \rightleftharpoons CH_3\bullet + H\bullet$$

$$CH_3\bullet + CH_4 \rightleftharpoons C_2H_6 + H\bullet$$

$$H\bullet + CH_4 \rightleftharpoons CH_3\bullet + H_2$$

$$2\,H\bullet \rightleftharpoons H_2$$

$$2\,CH_3\bullet \rightleftharpoons C_2H_6$$

$$H\bullet + CH_3\bullet \rightleftharpoons CH_4$$

and carbenes:

$$CH_4 \rightleftharpoons :CH_2 + H_2$$
$$:CH_2 + CH_4 \longrightarrow C_2H_6$$
$$C_2H_6 \longrightarrow CH_2{=}CH_2 + H_2$$
$$CH_2{=}CH_2 \longrightarrow CH{\equiv}CH + H_2$$
$$CH{\equiv}CH \longrightarrow 2\,C + H_2$$

In accordance with these equations, ethane, ethylene, acetylene and elemental carbon are produced by the cracking of methane. Activation energy of this process is 76 kcal mol^{-1}. The formation of acetylene starting from methane or ethane are endothermic processes:

$$2\,CH_4 \rightleftharpoons CH{\equiv}CH + 3\,H_2 \qquad -\Delta H_{298}{}^0 = -90 \text{ kcal mol}^{-1}$$
$$C_2H_6 \rightleftharpoons CH{\equiv}CH + 2\,H_2 \qquad -\Delta H_{298}{}^0 = -74 \text{ kcal mol}^{-1}$$

The equilibrium in these reactions is shifted to the right at temperatures >900 °C for ethane and > 1100 °C for methane. Higher alkanes are formed from methane via a recombination of methyl radicals [1d].

High-temperature pyrolysis of methane is very important since it is a main source of raw material (e.g., acetylene) for chemical industry. Cracking of higher alkanes gives a wider set of products. For example, pyrolysis of propane (activation energy is 64 kcal mol^{-1}) yielding ethylene, methane, propene, hydrogen and ethane consists of the following stages:

$$C_3H_8 \longrightarrow CH_3\bullet + C_2H_5\bullet$$
$$CH_3\bullet + C_3H_8 \longrightarrow CH_4 + n\text{-}C_3H_7\bullet$$
$$CH_3\bullet + C_3H_8 \longrightarrow CH_4 + iso\text{-}C_3H_7\bullet$$
$$n\text{-}C_3H_7\bullet \longrightarrow CH_2{=}CH_2 + CH_3\bullet$$

$iso\text{-}C_3H_7\bullet \longrightarrow CH_3CH{=}CH_2 + H\bullet$

$H\bullet + C_3H_8 \longrightarrow H_2 + n\text{-}C_3H_7\bullet$

$H\bullet + C_3H_8 \longrightarrow H_2 + iso\text{-}C_3H_7\bullet$

$2\,H\bullet \longrightarrow H_2$

$H\bullet + CH_3\bullet \longrightarrow CH_4$

$H\bullet + n\text{-}C_3H_7\bullet \longrightarrow C_3H_8$

$H\bullet + iso\text{-}C_3H_7\bullet \longrightarrow H_2 + CH_3CH{=}CH_2$

$H\bullet + iso\text{-}C_3H_7\bullet \longrightarrow C_3H_8$

$CH_3\bullet + n\text{-}C_3H_7\bullet \longrightarrow CH_4 + CH_3CH{=}CH_2$

$CH_3\bullet + n\text{-}C_3H_7\bullet \longrightarrow C_4H_{10}$

$CH_3\bullet + iso\text{-}C_3H_7\bullet \longrightarrow CH_4 + CH_3CH{=}CH_2$

$CH_3\bullet + iso\text{-}C_3H_7\bullet \longrightarrow C_4H_{10}$

$2\,CH_3\bullet \longrightarrow C_2H_6$

$2\,n\text{-}C_3H_7\bullet \longrightarrow C_3H_8 + CH_3CH{=}CH_2$

$2\,n\text{-}C_3H_7\bullet \longrightarrow C_6H_{14}$

$2\,iso\text{-}C_3H_7\bullet \longrightarrow C_3H_8 + CH_3CH{=}CH_2$

$2\,iso\text{-}C_3H_7\bullet \longrightarrow C_6H_{14}$

$n\text{-}C_3H_7\bullet + iso\text{-}C_3H_7\bullet \longrightarrow CH_3CH{=}CH_2 + C_3H_8$

$n\text{-}C_3H_7\bullet + iso\text{-}C_3H_7\bullet \longrightarrow C_6H_{14}$

Plasma reforming of methane can be efficiently used to produce hydrogen-rich gas (50-75% H_2, 25-50% CO) [1e]. Methane has also been converted to higher hydrocarbons (including ethylene and acetylene) by a microwave plasma [1f]. Heating induces important transformations not only in alkanes, but also in aromatic hydrocarbons, for example [1g]:

II.1.B. PHOTOLYSIS

Analogous transformations of alkanes may be induced by light irradiation under ambient temperature [2]. When irradiated with light λ = 121.6 nm, methane (which absorbs light λ < 143 nm) decomposes to generate the following species (Φ is quantum yield):

$$CH_4 \longrightarrow {}^1{:}CH_2 + H_2 \qquad \Phi = 0.4$$

$$CH_4 \longrightarrow {}^1{:}CH_2 \text{ (or } {}^3{:}CH_2) + H\bullet + H\bullet \qquad \Phi = 0.5$$

$$CH_4 \longrightarrow CH + H\bullet + H_2 \qquad \Phi = 0.08$$

Then the following stable products are formed:

$$:CH_2 + :CH_2 \longrightarrow HC{\equiv}CH + H\bullet + H\bullet$$

$$CH + CH_4 \longrightarrow CH_2{=}CH_2 + H\bullet$$

$$:CH_2 + H_2 \longrightarrow CH_3\bullet + H\bullet$$

$$CH + H_2 \longrightarrow CH_3\bullet$$

$$:CH_2 + CH_3\bullet \longrightarrow CH_2{=}CH_2 + H\bullet$$

$$CH_3\bullet + CH_3\bullet \longrightarrow C_2H_6$$

$$CH_2 + HC{\equiv}CH \longrightarrow CH_3C{\equiv}CH$$

II.1.C. RADIOLYSIS

The interaction of high-energy irradiation with alkanes leads, at the first stage of the process, to the excitation of the hydrocarbon molecule [3a]. Furthermore, the excited molecule decomposes to generate free radicals and carbenes. Radiolysis of methane produces ethane, ethylene, and higher hydrocarbons:

$$CH_4 \rightsquigarrow CH_4^{+} + e\text{–} \quad CH_4 \rightsquigarrow CH_4^{+*} + e^{-}$$

$$CH_4 \rightsquigarrow CH_4^* \qquad CH_4^* \longrightarrow CH_3^+ + H\bullet$$
$$CH_4^{+*} \longrightarrow CH_2^+ + H_2 \qquad CH_4^{+*} \longrightarrow CH^+, C^+, H_2, H$$
$$CH_4^* \longrightarrow CH_3\bullet + H\bullet$$
$$CH_4^* \longrightarrow CH_2 + H_2$$
$$CH_4^+ + CH_4 \longrightarrow CH_5^+ + CH_3\bullet$$
$$CH_3^+ + CH_4 \longrightarrow C_2H_7^{+*}$$
$$C_2H_7^{+*} \longrightarrow C_2H_5+* + H_2$$
$$CH_2^+ + CH_4 \longrightarrow C_2H_6^{+*} \qquad C_2H_6^{+*} \longrightarrow C_2H_4^+ + H_2$$
$$CH_2 + CH_4 \longrightarrow C_2H_6^*$$
$$C_2H_6^* \longrightarrow 2\ CH_3\bullet$$
$$C_2H_6^* + M \longrightarrow C_2H_6 + M$$
$$CH_5^+ + C_2H_6 \longrightarrow CH_4 + C_2H_7^{+*} \qquad C_2H_7^{+*} \longrightarrow C_2H_5^+ + H_2$$

Pulse radiolysis of N_2O-saturated methylcyclohexane (MCH) gives rise to the formation of the solvent radical cation, MCH^+, but in argon-saturated MCH, the olefinic fragment cation (methylcyclohexene)$^+$ is obtained [3b].

II.2. REACTIONS WITH ATOMS, FREE RADICALS AND CARBENES

Atoms and free radicals are very reactive species toward all organic substances including hydrocarbons, even saturated ones. Reactions of alkanes with atoms

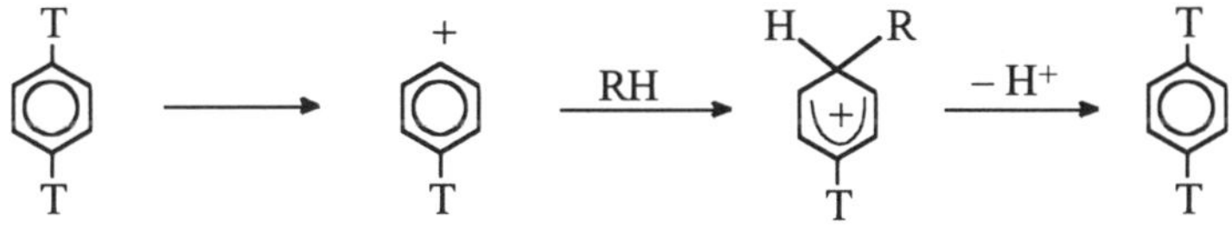

$R = CH_3, C_2H_5, C_3H_7$

Scheme II.1. The interaction of 1,4-T_2-benzene with light alkanes.

and radicals (see, for example, [4]) are stages of various chain-radical processes, occurring both in the gas phase [5] and in solution [6]. Interactions of hydrocarbons with ions in the gas phase are also known; for example, phenylium ion formed from 1,4-T_2-benzene reacts with methane and other lower alkanes to give tritiated alkylbenzene (Scheme II.1) [7].

A carbon atom inserts into C–H bond of methane to produce a species which is transformed into ethylene and some other hydrocarbons [8a].

$$C + CH_4 \longrightarrow C_2H_2 + C_2H_4 + C_2H_6 + CH_2{=}C{=}CH_2 + HC{\equiv}C{-}C{\equiv}CH$$

In one of the stages of the reaction between carbon atoms and *tert*-butyl-benzene, a carbon atom inserts into a methyl C–H bond to give a carbene followed by a 1,2-hydrogen shift (Scheme II.2) [8b].

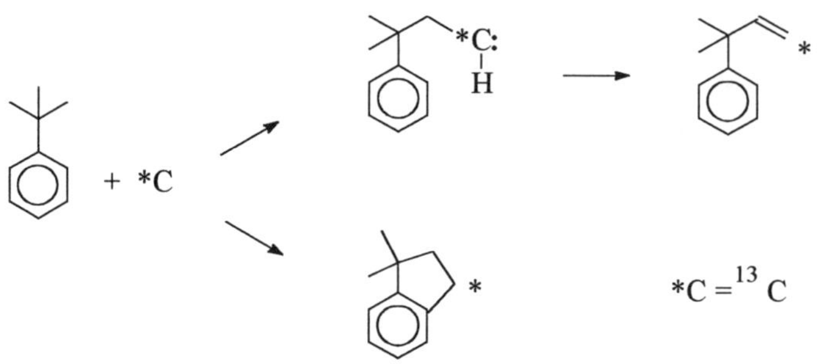

Scheme II.2. Reaction of arc generated carbon atoms with *tert*-butylbenzene.

The electronically excited oxygen atom reacts with alkanes, the dominant reaction mechanism being the insertion of this atom into the C–H bond, thus yielding chemically activated alcohol followed by fragmentation [8c,d].

$$O(s) + CH_4 \longrightarrow CH_3OH$$

The same species in triplet state reacts with alkanes similarly to free radicals since the insertion is a spin-forbidden reaction for a triplet to form a singlet state molecule. Investigation of the singlet state oxygen $O(^1D_2)$ reaction with alkanes in the gas phase has revealed that two distinct channels exist for these reactions. They were interpreted as insertion into C–H bond and H atom abstraction. The insertion component dominates for small molecules (CH_4, C_2H_6) with strong C–H bonds, while the abstraction component is the dominant mechanism for formation of OH radical in the cases of larger molecules [C_3H_8, $C(CH_3)_4$] with stronger steric hindrances and weaker C–H bonds. An insertion mechanism of the interaction of singlet state oxygen with methane was confirmed in *ab initio* theoretical calculations [8e]. The minimum energy path of the reaction was found to be an almost collinear approach of $O(^1D)$ to one of the hydrogen atoms of methane to the distance of 1.66 Å, and then oxygen atom migrates off-axis.

We will see that similar conclusions with respect to the mechanism (possibility of insertion as well as hydrogen atom abstraction) may be made for reactions of oxygen atoms bound to metal ("oxenes") with alkanes in enzymatic and model chemical reactions. The excited nitrogen atom, $N(^2D)$, reacts analogously [8f,g].

The most usual reaction of alkanes and other C–H compounds with atoms and free radicals is the hydrogen atom abstraction [9]:

$$\text{Rad}\bullet + \text{R'H} \longrightarrow \text{RadH} + \text{R'}\bullet$$

The activation energy of this reaction is usually low. To estimate the activation energy (E) as the function of the reaction heat (Q), one may use a simple approximate equation (so called Polanyi-Semenov rule) [10]:

$$E = A - \alpha Q$$

Here A and α are constants approximately equal to 11.5 and 0.25, respectively. Modern theoretical analysis gives more complicated and more precise, but also approximate formulae. According to Denisov [11], the dependence of the activation energy on enthalpy (ΔH) is expressed by the formula:

$$\Delta E_H = [\alpha'/(1 + \alpha')]\, \Delta H_e + (\alpha' \Delta H_e / 2br_e)^2.$$

Here α' is a coefficient which is constant for each pair of fixed bonds, the one being cleaved, and the other being formed; br_e is a parameter also constant for a given series of reactions and not very different for different classes of reactions. The latter formula gives non-linear dependence of activation energy on the reaction's enthalpy; whereas, less precise Polanyi-Semenov rule predicts a linear dependence:

$$\Delta E = \alpha\Delta\Delta H.$$

Nevertheless, the thermodynamics of these reactions remain an important basis for their kinetic characteristics. Polar factors, while they should not be completely neglected, are not of great importance in reactions of neutral free radicals with molecules. If we have a series of substrates reacting via C–H homolytic cleavage, we can then expect that as a rule, the stronger the C–H bond, the smaller the rate constant.

When a branched alkane is used as the substrate, a radical usually attacks the C–H bonds in accordance with "normal" selectivity, i.e., the reactivity of C–H bonds decreases in the following order:

tertiary > *secondary* > *primary* (or in another notation 3° > 2° > 1°).

The hydrogen abstraction process by photo-excited ketones and some other organic compounds have been thoroughly investigated [12a–f]. The rates of intramolecular hydrogen atom abstraction from methylene groups of anthraquinone-2-carboxylic acid derivatives have been measured [12a]. For n = 9, the abstraction constant is $k = 10^4\ s^{-1}$. Each methylene group added to the alkyl chain increases k by approximately $1.5 \times 10^4\ s^{-1}$.

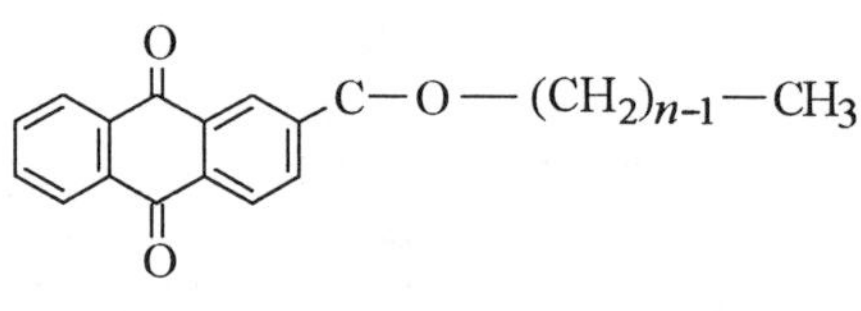

Alkyl radicals formed via abstraction of the hydrogen atom may react with molecular oxygen present in the solution; if dioxygen is replaced with carbon monoxide, carbonylation of the alkane occurs to produce an aldehyde or a ketone

[12d]. A theoretical study of H-abstraction from aliphatic alcohols was carried out [12f]. The atomic oxygen anion, O^- [12g], reacts with alkanes via the mechanism of hydrogen atom transfer [12h]:

$$O^- + C_nH_{2n+2} \longrightarrow HO^- + C_nH_{2n+1}^\bullet$$

$$O^- + C_nH_{2n+2} \longrightarrow H_2O + C_nH_{2n} + e^-$$

Table II.1 presents total rate constants and HO^- branching percentages for the reactions of O^- with various straight-chain and branched alkanes [12h].

Table II.1. Reactions of the oxide anion O^- with alkanes at 300 K

Alkane	$k \times 10^{-9}$, $cm^3\ s^{-1}$	% HO^-
Methane	0.088	100
Ethane	0.59	60
Propane	1.2	50
n-Butane	2.0	27
2-Methylpropane	1.9	32
n-Pentane	2.1	21
2-Methylbutane	2.2	25
2,2-Dimethylpropane	2.1	97
n-Hexane	2.3	17
n-Heptane	2.4	10
n–Octane	2.6	9
n-Decane	2.8	4

Unsaturated hydrocarbons and even alkanes are capable of formation of weak complexes with various radicals [13]. A theoretical consideration of a model system $H_2NO^{\cdot\cdot}HCH_3$ (by method INDO UHF) showed that the stabilization energy of this complex at the equilibrium distance $r(O^{\cdot\cdot}H) \cong 2$ Å is <1 kcal mol^{-1}. One may suggest that complexes of such type are intermediate species in alkane reactions with radicals.

II.2.A. HALOGENATION

Halogenation and especially chlorination of alkanes is a very important process as it gives halogenated alkyls [14]. If the temperature of chain radical chlorination of methane is high, and time of contact between hydrocarbon and chlorine is short, in addition to chlorinated hydrocarbons, large amounts of ethane (at ~1000 K) or ethylene and hydrogen (at ~2000 K) are formed (Benson's process [15]). Thus the content of ethylene, propylene and acetylene is up to 77% at 1273 K. The radical chain process includes the following stages:

I:

$$CH_4 \longrightarrow CH_3\bullet + H\bullet$$

$$Cl_2 \longrightarrow 2\,Cl\bullet$$

$$CH_4 + H\bullet \longrightarrow CH_3\bullet + H_2$$

$$Cl_2 + H\bullet \longrightarrow Cl\bullet + HCl$$

$$CH_4 + Cl\bullet \longrightarrow CH_3\bullet + HCl$$

$$CH_3\bullet + CH_3\bullet \longrightarrow C_2H_6$$

$$CH_3\bullet + Cl\bullet \longrightarrow CH_3Cl$$

$$CH_3\bullet + H\bullet \longrightarrow CH_4$$

II:

$$\bullet CH_2Cl + RH \rightleftharpoons CH_3Cl + R\bullet$$

$$\bullet CH_2Cl + Cl_2 \longrightarrow CH_2Cl_2 + Cl\bullet$$

$$C_2H_5\bullet \longrightarrow CH_2{=}CH_2 + H\bullet$$

$$C_2H_5\bullet + RH \rightleftharpoons C_2H_6 + R\bullet$$

$$C_2H_5\bullet + Cl_2 \longrightarrow C_2H_5Cl + Cl\bullet$$

$$CH_3\bullet + \bullet CH_2Cl \longrightarrow C_2H_5Cl$$

$$\bullet CH_2Cl + \bullet CH_2Cl \longrightarrow ClCH_2CH_2Cl$$

III:

$$CH_3\bullet + CH_2{=}CH_2 \rightleftharpoons CH_3CH_2CH_2\bullet$$

$$C_2H_5\bullet + CH_2{=}CH_2 \rightleftharpoons CH_3CH_2CH_2CH_2\bullet$$

$$CH_3CH_2CH_2\bullet \longrightarrow CH_3CH{=}CH_2 + H\bullet$$

$$CH_3CH_2CH_2CH_2\bullet \longrightarrow CH_3CH_2CH{=}CH_2 + H\bullet$$

$$R\bullet + CH_2{=}CH_2 \longrightarrow RH + C_2H_3\bullet$$

$$C_2H_3\bullet \longrightarrow HC{\equiv}CH + H\bullet$$

IV:

$$CH_3\bullet + HC{\equiv}CH \rightleftharpoons CH_3CH{=}CH\bullet$$

$$R\bullet + CH_3CH{=}CH_2 \longrightarrow RH + \bullet CH_2CH{=}CH_2$$

$$R\bullet + HC{\equiv}CH \longrightarrow RH + HC{\equiv}C\bullet$$

$$CH_3C{=}CH\bullet \longrightarrow CH_3C{\equiv}CH + H\bullet$$

$$HC{\equiv}C\bullet \longrightarrow 2\,C + H\bullet$$

$$\bullet CH_2CH{=}CH_2 \longrightarrow CH_2{=}C{=}CH_2 + H\bullet$$

Chlorine atoms can also be generated at ambient temperature by light irradiation of molecular chlorine mixed with hydrocarbon [14a, 16]. The following scheme describes the process in the liquid phase:

$$Cl_2 \overset{h\nu}{\rightleftharpoons} [2\,Cl\,]_S \longrightarrow 2\,Cl\bullet$$

$$Cl\bullet + RH \xrightarrow{k_1} R\bullet + HCl$$

$$R\bullet + Cl_2 \longrightarrow RCl + Cl\bullet$$

$$2\,Cl\bullet \longrightarrow Cl_2$$

$$R\bullet + Cl\bullet \longrightarrow RCl$$

$$2\,R\bullet \longrightarrow R_2$$

For ethane at 293 K lg k_1 = 7.6 dm^3 mol^{-1} s^{-1}. Chlorination of alkanes also occurs if solutions of metal chlorides (for example, $AuCl_4^-$ or $PtCl_6^{2-}$; see [17]) are irradiated in the presence of an alkane. Bromination of alkylaromatic hydrocarbons with anion Br^- in the presence of cerium ammonium nitrate proceeds as a radical reaction [18a]:

$$Ce^{IV}NO_3 + Br^- \longrightarrow Ce^{IV}Br + NO_3^-$$

$$Ce^{IV}Br \longrightarrow Ce^{III} + Br\bullet$$

$$Br\bullet + C_6H_5CH_3 \longrightarrow HBr + C_6H_5CH_2\bullet$$

$$C_6H_5CH_2\bullet + Ce^{IV}Br \longrightarrow C_6H_5Br + Ce^{III}$$

Bromination of toluene by the system $NaBrO_3$ – $NaHSO_3$ gives three main products:

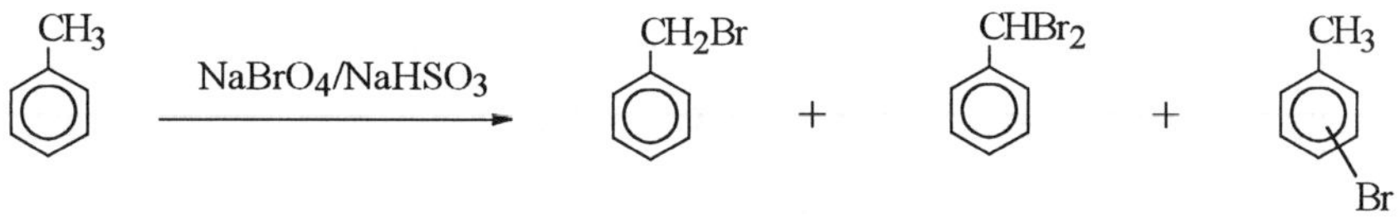

The free-radical bromination by Br• occurs in ethyl acetate yielding α-bromotoluene and α,α′-dibromotoluene, while the ring-bromination by Br^+ occurs in aqueous acetonitrile [18b].

Photochemical reaction of 4-*tert*-butyltoluene with HBr/H_2O_2 gives benzyl bromide easily hydrolyzed to 4-*tert*-butylbenzaldehyde [18c]. Thermal and photochemical bromination of decalin gives mainly tetrabromooctalin [18d]:

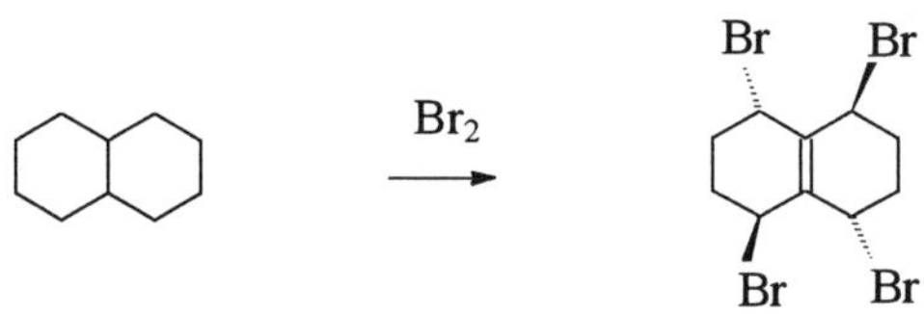

Adamantane has been brominated with the system CBr_4/NaOH under phase transfer catalysis conditions [18e]:

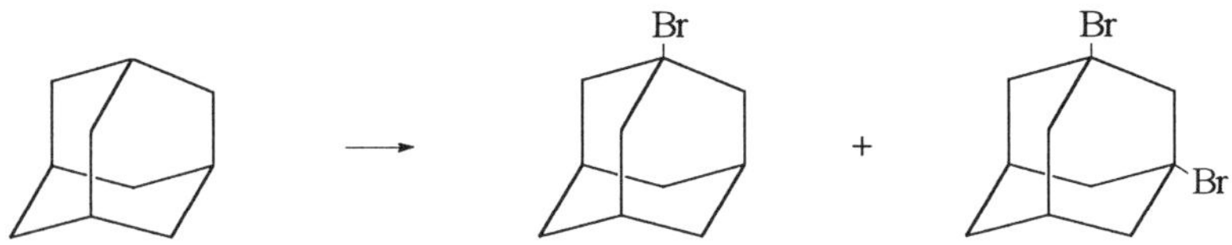

The reaction is believed to be initiated by single-electron oxidation of HO^- by CBr_4:

$$CBr_4 + HO^- \rightleftharpoons CBr_4{\bullet}^- + HO\bullet$$
$$CBr_4{\bullet}^- \longrightarrow {\bullet}CBr_3 + Br^-$$
$$RH + {\bullet}CBr_3 \longrightarrow R\bullet + HCBr_3$$
$$R\bullet + CBr_4 \longrightarrow RBr + Br_3C\bullet$$

The behavior of 1-adamantyl radical in chlorine atom abstraction has been investigated recently [18f]. Initiators of radical reactions and some metal complexes induce dark chlorination of saturated hydrocarbons with Cl_2 or CCl_4 [19].

II.2.B. REACTIONS WITH OXYGEN- AND NITROGEN-CONTAINING RADICALS

Reactions of alkanes with hydroxyl radicals, which take place in the atmosphere under the action of irradiation and occur as crucial stages in oxidation of hydrocarbons are of great importance and have been investigated in detail [20]. The data for reactions of HO• with some alkanes are summarized in Table II.2 [20f]. Reaction of alkanes with nitric acid occurs as a radical chain process with participation of radicals HO• and $NO_2\bullet$ [21]:

$$HNO_3 \longrightarrow HO\bullet + NO_2\bullet$$
$$RH + HO\bullet \longrightarrow R\bullet + H_2O$$
$$R\bullet + NO_2\bullet \longrightarrow RNO_2$$
$$R\bullet + HNO_3 \longrightarrow RNO_2 + HO\bullet$$

Table II.2. Rate constants, $k \times 10^{-13}$, $cm^3\ s^{-1}$, for the reaction of alkanes with radicals HO• at different temperatures

Alkane	300 K	340 K	390 K
Ethane	2.6	3.9	6.0
Propane	10.9	14.6	18.3
n-Butane	24.3	28.7	35.4
2-Methylpropane	20.9	24.7	27.2
n-Pentane	39.8	45.9	64.4
n-Hexane	54.5	62.3	72.9
Cyclopentane	58.9	61.8	75.3
Cyclohexane	76.6	78.4	103
Cycloheptane	120	133	164
Cyclooctane	134	149	191

Radical nitration of a hydrocarbon with cerium ammonium nitrate begins from the photohomolysis [22]:

$$Ce^{IV}NO_3 \overset{h\nu}{\rightleftharpoons} Ce^{III} + NO_3\bullet$$

$$RH + NO_3\bullet \longrightarrow R\bullet + HNO_3$$

The photochemically excited nitroxide **II-1** reacts with such alkanes as cyclohexane, isobutane and *n*-butane abstracting hydrogens and generating carbon-centered radicals which are trapped by ground state **II-1** [23]:

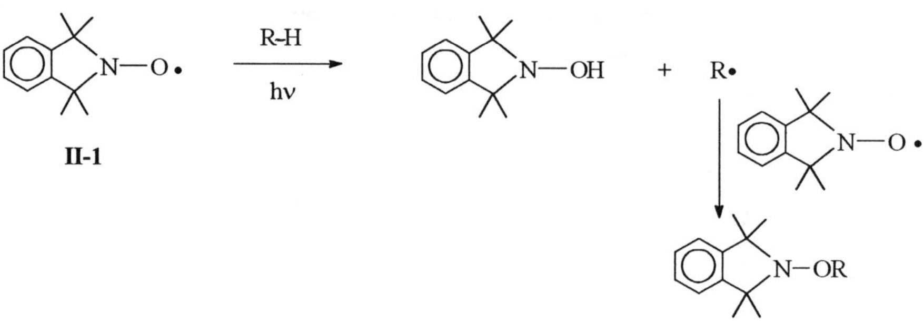

II-1

II.2.C. REACTIONS WITH CARBENES

Singlet state carbenes produced in photo- or thermal decomposition of ketene or diazomethane

$$CH_2{=}C{=}O \longrightarrow CH_2 + CO$$

$$CH_2NN \longrightarrow CH_2 + N_2$$

are known to insert into the C–H bond of an alkane:

$$CH_2(s) + CH_4 \longrightarrow C_2H_6$$

Usually insertion of a simple carbene into the C–H bond of an alkane proceeds with retention of configuration. Some other carbenes react much more selectively than simple CH_2. For example, phenylcarbene reacts approximately eight times faster and species CHCl twenty times faster with a secondary C–H bond of *n*-pentane than with a primary one. Diclorocarbene reacts even more selectively and inserts only into secondary and tertiary C–H bonds. Even secondary bonds are not affected if a tertiary bond is present. Calculations demonstrated that in the methylene insertion into the C–H bond, the hydrogen atom is first attacked to produce a complex similar to that which is formed in the recombination of two radicals [24]. Low stereoselectivities and a large deuterium isotope effect (k_H/k_D = 4–8) for the intramolecular insertion reaction of 2-alkoxyphenylcarbenes and 2-alkylphenylcarbenes led to the conclusion that a triplet state carbene was involved in the reaction [25a]. Even greater kinetic isotope effects (k_H/k_D = 14.7) have been measured for the intramolecular C–H insertion reaction of 2-alkylphenylnitrenes [25b]:

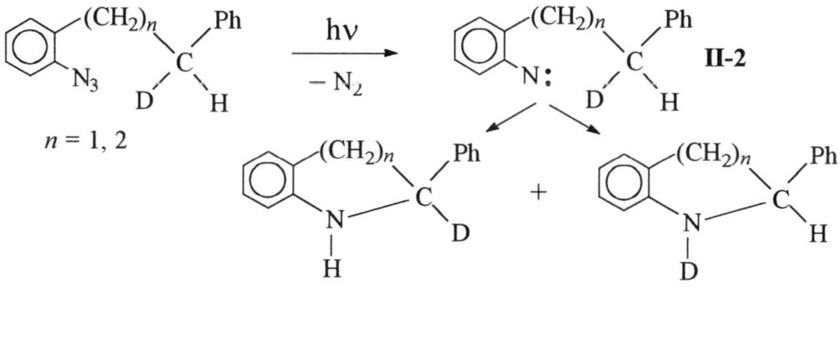

II-2

A large isotope effect supports the hydrogen abstraction–recombination mechanism involving the triplet state nitrene **II-2**.

The reaction of phosphorylnitrene species with alkanes gives a set of products [25c]:

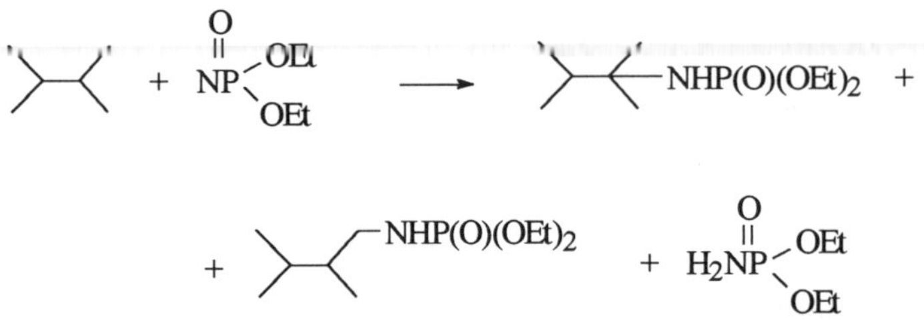

Methylidyne and silylidyne insert into methane and silane (see [26]):

$$CH + CH_4 \longrightarrow CH_2CH_3$$
$$CH + SiH_4 \longrightarrow CH_2SiH_3$$
$$SiH + CH_4 \longrightarrow SiH_2CH_3$$
$$SiH + SiH_4 \longrightarrow SiH_3SiH_3$$

Processes with participation of methylidyne take place in combustion and planetary atmosphere chemistry, while silylidyne is an important intermediate in the chemical vapor deposition of amorphous silicon film from SiH_4.

II.2.D. REACTIONS WITH PARTICIPATION OF ION RADICALS

The interaction of arenes or alkylarenes, $ArCH_2R$, with some oxidants, Ox, [27] for example, with cerium(IV) salts [27c] or photoactivated quinones [27d], gives rise to the formation of the ion-radical pair:

$$ArCH_2R + Ox \rightleftharpoons [ArCH_2R^{+}\bullet, Ox^{-}\bullet]$$

which can subsequently undergo proton transfer to generate radical species:

$$[ArCH_2R^{+}\bullet,\ Ox^{-}\bullet] \rightleftharpoons ArC\bullet H_2R + Ox\bullet$$

Carbon-centered arylalkyl radicals can then enter into various transformations, for example, can react with molecular oxygen (see next section).

It is interesting that in a similar reaction [27e] of perfluoroalkanes with the organometallic photosensitizer, decamethylferrocene, electron transfer leads to the formation of anion radical which is transformed further to a radical:

$$R^{f}_{2}CF\text{–}CF_2R^{f} + e^{-} \longrightarrow [R^{f}_{2}CF\text{–}CF_2R^{f}]\bullet^{-} \longrightarrow R^{f}_{2}C\bullet\text{–}CF_2R^{f} + F^{-}$$

Under the action of a reductant, Red, the radical produces an olefin

$$R^{f}_{2}C\bullet\text{–}CF_2R^{f} + Red \longrightarrow R^{f}_{2}C{=}CFR^{f} + RedF$$

II.3. Oxidation by Molecular Oxygen

The importance of chemical processes based on the oxidation of hydrocarbons by molecular oxygen is connected with the necessity of rational use of hydrocarbons from natural gas, petroleum and coal. Despite the great amount of work devoted to this problem, it is far from being completely solved. At present its significance, in terms of the urgent necessity of more economical consumption of natural resources, is increasing with time. Thermodynamically the formation of oxygen-containing products from saturated hydrocarbons and molecular oxygen is always favorable because oxidation reactions are highly exothermic. Nevertheless, it is this very fact that hinders the creation of selective processes, with the difficulty usually lying in the prevention of different parallel and secondary oxidation reactions.

II.3.A. HIGH-TEMPERATURE OXIDATION IN THE GAS PHASE

The complete oxidation of alkanes by air (burning) to produce water and carbon dioxide is a very important source of energy [28a–c]. There can also be partial oxidation of saturated hydrocarbons producing various valuable organic substances, e.g., alkyl hydroperoxides, alcohols and ketones or aldehydes [28a,b,d,e]. Oxidation of benzene and alkylbenzene under UV irradiation takes place in the atmosphere [29].

The oxidation of methane and other light alkanes in the gas phase at high temperature gives alcohols, aldehydes and ketones and occurs via a chain radical mechanism (for certain stages of the process, see recent publications [30]). The methane oxidation produces methanol and formaldehyde as main oxygenates. Due to the fact that both products are much more reactive than hydrocarbon, the yield of alcohol and aldehyde is only a few percents. Usually the reactions were

Table II.3. Partial non-catalyzed oxidation of methane with molecular oxygen.

Pressure, atm	Tempe-, rature, °C	$[O_2]$, %	Reaction time, s	Conversion, of CH_4, %	S_{MeOH} [a]	S_{HCHO} [b]	Yield of MeOH, %
141	475	4.5	48	2.51	45.0	3.66	1.13
141	475	8.7	48	5.30	29.6	2.48	1.57
3500	264	7.9	600	6.29	40.1	0.64	2.52
40	455	2.04	~ 0.08	2.10	59.6	17.04	1.25
40	461	3.2	~ 0.08	2.90	54.4	17.0	1.58
40	483	7.2	~ 0.08	6.10	34.6	13.83	2.11
65	310	3.85	2–1000	7.5	95		6
65	410	7	2–1000	21	80		17
100	410–500	2.8	~ 1		46.0	9.1	2.8
65.3	456	5.1	~ 150	8.0	83.0		6.6
40	474–498	2.35–3.14	1.2–2.1	1.3–2.2	43.6	6.4	0.73–1.0
70	448	3.3	~ 2	3.9	67.5	1.8	2.56
20	400	5.0	10–100	>4.6	~ 32		
30	447	9.5		9.0	23.5		1.9–2.1
50	350–550	1.5	>10		44		1.1

[a] S_{MeOH} is the selectivity of the formation of methanol.

[b] S_{HCHO} is the selectivity of the formation of formaldehyde.

carried out in broad pressure (1–230 atm) and temperature (300–500 °C and higher) ranges. The reaction time varied from fractions of a second to tens of minutes. Some data on partial non-catalyzed methane oxidation are summarized in Table II.3 [28d].

The oxidation of ethane yields not only expected ethanol (and also less amount of acetaldehyde) but additionally large amounts of the C–C bond splitting products: methanol, formaldehyde, formic acid, as well as CO and CO_2 [28d, 31a] (Table II.4).

Table II.4. The selectivity (products, % based on the combusted ethane) of the partial non-catalyzed ethane oxidation with molecular oxygen.

Pressure, atm	Tempe-, rature,°C	EtOH	MeOH	MeCHO	HCHO	AcOH	HCOOH	CO	CO_2
15 [a]	315	16.0	19.4	1.9	4.5	0	0	–	–
50 [b]	286	24.4	14.1	8.3	2.0	1.7	0.9	34.8	10.0
75 [c]	279	18.0	16.6	6.8	0.4	3.6	0.6	–	–
100 [d]	266.5	23.7	11.2	6.0	0.05	23.8	0.6	9.3	14.2

[a] Initial mixture composition: 88.2% ethane, 11.8% O_2; reaction time 3.0 min.

[b] Initial mixture composition: 88.4% ethane, 11.6% O_2; reaction time 10.7 min.

[c] Initial mixture composition: 88.2% ethane, 11.8% O_2; reaction time 2.5 min.

[d] Initial mixture composition: 88.4% ethane, 11.6% O_2; reaction time 34.5 min.

The composition of oxygenate mixture in the propane oxidation depends on the conditions (pressure and temperature of the process) [28d, 31b,c] (Table II.5).

Table II.5. The selectivity (products, %) of the partial non-catalyzed propane oxidation with air (propane : air = 1 : 3.6).

Products	1 atm, 373 °C	20 atm, 281 °C	60 atm, 252 °C	100 atm, 250 °C
The sum of aldehydes	20.5	21.8	13.5	13.7
The sum of normal alcohols	19.7	21.0	17.5	15.2
Isopropanol	1.3	2.8	6.2	16.0
Acetone	0.5	4.3	12.5	7.9
The sum of acids	4.3	17.0	19.0	18.9
CO_2	7.3	17.1	21.4	20.6
CO	21.3	16.0	9.9	7.7
Propene	25.1	0	0	0

Table II.6 demonstrates that the efficiency of oxidation of alkanes into useful products is much higher for propane and *n*-butane in comparison with light alkanes (ethane and especially methane) [28d].

Table II.6. The optimal parameters of the alkane oxidation.

Alkane	Pressure, atm	Tempera-ture °C	Ratio alkane : O_2	Degree of conversion into useful products (%)
Methane	100	450	30	2
Ethane	50	360	10	5
Propane	20	280	1	20
n-Butane	9	260	1	20

Table II.7. The mechanism of the methane oxidation and the parameters of the equation $k = AT^n\exp(-E/RT)$.

No	Reaction	A [a]	n	E [b]
1	$CH_4 + O_2 = CH_3• + HOO•$	1.00×10^{-10}	0	56
2	$CH_3• + HOO• = CH_4 + O_2$	1.00×10^{-12}	0	0
3	$CH_3• + O_2 = CH_3OO•$	2.00×10^{-12}	0	0
4	$CH_3OO• = CH_3• + O_2$	8.90×10^{13}	0	31
5	$HO• + CH_4 = H_2O + CH_3•$	1.32×10^{-17}	1.9	2.7
6	$CH_3OO• + CH_3OO• = CH_3O• + CH_3O• + O_2$	1.71×10^{-13}	0	0
7	$CH_3OO• + CH_3OO• = HCHO + CH_3OH + O_2$	0.74×10^{-13}	0	0
8	$CH_3OO• + CH_3• = CH_3O• + CH_3O•$	4.50×10^{-11}	0	0
9	$CH_3OO• + HOO• = CH_3OOH + O_2$	7.70×10^{-14}	0	−2.6
10	$CH_3• + HOO• = CH_3O• + HO•$	3.00×10^{-11}	0	0
11	$HOO• + HOO• = H_2O_2 + O_2$	2.20×10^{-13}	0	−1.2
12	$CH_3• + CH_3• = C_2H_6$	4.00×10^{-11}	0	0
13	$CH_3O• + HOO• = CH_3OH + O_2$	1.70×10^{-11}	0	0
14	$CH_3O• + CH_4 = CH_3OH + CH_3•$	1.00×10^{-12}	0	11
15	$CH_3O• + O_2 = HCHO + HOO•$	1.00×10^{-13}	0	2.6
16	$CH_3O• = HCHO + H•$	1.00×10^{14}	0	30
17	$CH_3OO• + CH_4 = CH_3OOH + CH_3•$	1.00×10^{-12}	0	21.5
18	$CH_3OO• + HCHO = CH_3OOH + CHO•$	4.70×10^{-13}	0	12
19	$HOO• + HCHO = H_2O_2 + CHO•$	2.00×10^{-12}	0	11
20	$CH_3• + HCHO = CH_4 + CHO•$	1.40×10^{-12}	0	7
21	$CH_3O• + HCHO = CH_3OH + CHO•$	1.00×10^{-12}	0	3.6
22	$HO• + HCHO = H_2O + CHO•$	1.25×10^{-11}	0	0.17
23	$H• + CH_4 = H_2 + CH_3•$	1.30×10^{-10}	0	12
24	$H• + O_2 + M = HOO• + M$	1.00×10^{-32}	0	−1.0
25	$H• + HCHO = H_2 + CHO•$	3.27×10^{-11}	0	3.7
26	$CHO• + O_2 = CO + HOO•$	5.50×10^{-11}	−0.4	0
27	$CHO• + M = CO + H• + M$	4.00×10^{-10}	0	17
28	$CH_3OOH = CH_3O• + HO•$	4.00×10^{15}	0	43
29	$H_2O_2 = HO• + HO•$	3.00×10^{14}	0	50
30	$HO• + H_2 = H_2O + H•$	1.06×10^{-17}	2.0	1.5
31	$HOO• + H_2 = H_2O_2 + H•$	5.60×10^{-12}	0	22
32	$CH_3OO• + H_2 = + H•$	3.60×10^{-12}	0	22
33	$CH_3O• + H_2 = CH_3OH + H•$	3.60×10^{-12}	0	10

34 $CH_3\bullet + H_2 = CH_4 + H\bullet$	3.60×10^{-12}	0	11
35 $CH_3OO\bullet + CH_3O\bullet = CH_3OOH + HCHO$	1.50×10^{-12}	0	0
36 $CH_4 + HOO\bullet = CH_3\bullet + H_2O_2$	3.00×10^{-12}	0	21.5
37 $CH_3\bullet + CH_3OH = CH_4 + \bullet CH_2OH$	3.30×10^{-13}	0	10
38 $CH_3OO\bullet + CH_3OH = CH_3OOH + \bullet CH_2OH$	5.00×10^{-13}	0	14.5
39 $\bullet CH_2OH + O_2 = HCHO + HOO\bullet$	2.00×10^{-12}	0	0
40 $HOO\bullet + CH_3OH = H_2O_2 + \bullet CH_2OH$	1.50×10^{-12}	0	14.5
41 $HO\bullet + CH_3OH = H_2O + \bullet CH_2OH$	5.70×10^{-12}	0	1.4
42 $HO\bullet + CH_3OH = H_2O + CH_3O\bullet$	1.70×10^{-11}	0	3.3
43 $CH_3O\bullet + CH_3OH = CH_3OH + \bullet CH_2OH$	6.60×10^{-13}	0	5.3
44 $H\bullet + CH_3OH = H_2 + \bullet CH_2OH$	2.16×10^{-11}	0	5.3
45 $H\bullet + H_2O_2 = HO\bullet + H_2O$	3.00×10^{-10}	0	6.3
46 $OH\bullet + H_2O_2 = H_2O + HOO\bullet$	3.70×10^{-12}	0	0.5
47 $H\bullet + H_2O_2 = H_2 + HOO\bullet$	1.70×10^{-11}	0	5
48 $CH_3OO\bullet + H_2O_2 = CH_3OOH + HOO\bullet$	2.50×10^{-13}	0	13
49 $CH_3O\bullet + H_2O_2 = CH_3OH + HOO\bullet$	2.50×10^{-13}	0	4
50 $CH_3\bullet + H_2O_2 = CH_4 + HOO\bullet$	2.50×10^{-13}	0	2.8
51 $HO\bullet + CO = H\bullet + CO_2$	2.50×10^{-17}	1.3	–0.8
52 $HOO\bullet + CO = HO\bullet + CO_2$	1.70×10^{-10}	0	23
53 $CH_3O\bullet + CO = CH_3\bullet + CO_2$	2.60×10^{-11}	0	12
54 $CH_3O\bullet + CH_3O\bullet = CH_3OH + HCHO$	3.00×10^{-11}	0	0
55 $CH_3\bullet + CH_3O\bullet = CH_4 + HCHO$	2.80×10^{-11}	0	0
56 $CH_3\bullet \xrightarrow{\text{heter.}}$ radical destruction	5.00×10^{-3}	0	0
57 $CH_3OO\bullet \xrightarrow{\text{heter.}}$ radical destruction	5.00×10^{-3}	0	0
58 $HOO\bullet \xrightarrow{\text{heter.}}$ radical destruction	5.00×10^{-3}	0	0
59 $CH_3OOH \xrightarrow{\text{heter.}}$ radical destruction	5.00×10^{-3}	0	0
60 $H_2O_2 \xrightarrow{\text{heter.}} H_2O$	1.50×10^{-3}	0	0
60 $H_2O_2 \xrightarrow{\text{heter.}} H_2O + O_2$	1.50×10^{-3}	0	0

[a] Preexponential factors are expressed in s^{-1} for unimolecular reactions, cm^3 mol^{-1} s^{-1} for bimolecular reactions, or cm^6 mol^{-2} s^{-1} for the termolecular reaction 24.

[b] Activation energy, kcal mol^{-1}.

The base model of the direct oxidation of methane to methanol, proposed in [28d, 31d], takes into account all the main reliably established elementary reactions that play a significant role at high pressures and moderate temperatures, including heterogeneous destruction of radicals and peroxides on the reactor surface. This mechanism and parameters of its stages are presented in Table II.7.

When propane is oxidized, the dehydrogenation and C–C bond splitting occurs to yield propylene and acetaldehyde in considerable amounts. The oxidation of the latter is accompanied by the appearance of so-called cool flames [32]. The mechanism proposed for the propane oxidation is presented in Table II.8 [32c].

Table II.8. The mechanism of the propane oxidation.

No	Reaction	E^{a}
1	$C_3H_8 + O_2 = C_3H_7{\bullet} + HOO{\bullet}$	47.4
–1	$C_3H_7{\bullet} + HOO{\bullet} = C_3H_8 + O_2$	0
2	$C_3H_7{\bullet} + O_2 = C_3H_7OO{\bullet}$	0
–2	$C_3H_7OO{\bullet} = C_3H_7{\bullet} + O_2$	31.5
3	$C_3H_7{\bullet} + O_2 = C_3H_6 + HOO{\bullet}$	2.2
4	$C_3H_7OO{\bullet} + C_3H_8 = C_3H_7OOH + C_3H_7{\bullet}$	15.0
5	$C_3H_7OOH = C_3H_7O{\bullet} + HO{\bullet}$	43.0
6	$C_3H_7O{\bullet} + C_3H_8 = C_3H_7OH + C_3H_7{\bullet}$	7.0
7	$HO{\bullet} + C_3H_8 = H_2O + C_3H_7{\bullet}$	0.9
8	$HOO{\bullet} + C_3H_8 = H_2O_2 + C_3H_7{\bullet}$	15.0
9	$H_2O_2 + M = HO{\bullet} + HO{\bullet} + M$	45.8
10	$C_3H_7OO{\bullet} + C_3H_7OO{\bullet} = C_3H_7O{\bullet} + C_3H_7O{\bullet} + O_2$	0
11	$C_3H_7OO{\bullet} + C_3H_7OO{\bullet} = C_3H_7OH + C_2H_5CHO + O_2$	0
12	$HOO{\bullet} + HOO{\bullet} = H_2O_2 + O_2$	1.23
13	$HOO{\bullet} + C_3H_7OO{\bullet} = C_3H_7O{\bullet} + HO{\bullet} + O_2$	0
14	$C_3H_7{\bullet} + C_3H_7{\bullet} = C_3H_8 + C_3H_6$	0
15	$C_3H_7{\bullet} + C_3H_7OO{\bullet} = C_3H_7O{\bullet} + C_3H_7O{\bullet}$	0
16	$C_3H_7{\bullet} + O_2 = C_2H_5CHO + HOO{\bullet}$	2.6
17	$C_3H_7{\bullet} + HOO{\bullet} = C_3H_7O{\bullet} + HO{\bullet}$	0

18	$C_3H_7O\bullet = C_2H_5\bullet + HCHO$	21.1
19	$C_2H_5\bullet + O_2 = C_2H_5OO\bullet$	0
–19	$C_2H_5OO\bullet = C_2H_5\bullet + O_2$	31.5
20	$C_2H_5OO\bullet + C_3H_8 = C_2H_5OOH + C_3H_7\bullet$	15.0
21	$C_2H_5OOH = C_2H_5O\bullet + HO\bullet$	43.0
22	$C_2H_5O\bullet + O_2 = CH_3CHO + HOO\bullet$	2.6
23	$C_2H_5\bullet + O_2 = C_2H_4 + HOO\bullet$	2.2
24	$C_2H_5O\bullet + C_3H_8 = C_2H_5OH + C_3H_7\bullet$	7.0
25	$C_2H_5OO\bullet + C_2H_5OO\bullet = C_2H_5O\bullet + C_2H_5O\bullet + O_2$	0
26	$C_2H_5OO\bullet + C_2H_5OO\bullet = C_2H_5OH + CH_3CHO + O_2$	0
27	$C_2H_5OO\bullet + HOO\bullet = C_2H_5OOH + O_2$	0
28	$C_2H_5OO\bullet + C_3H_7OO\bullet = C_2H_5O\bullet + C_3H_7O\bullet + O_2$	0
29	$C_2H_5OO\bullet + C_3H_7OO\bullet = C_2H_5OH + C_2H_5CHO + O_2$	0
30	$C_2H_5\bullet + C_2H_5OO\bullet = C_2H_5O\bullet + C_2H_5O\bullet$	0
31	$C_2H_5\bullet + C_3H_7OO\bullet = C_2H_5O\bullet + C_3H_7O\bullet$	0
32	$C_2H_5\bullet + HOO\bullet = C_2H_6 + O_2$	0
33	$C_2H_5\bullet + HOO\bullet = C_2H_5O\bullet + HO\bullet$	0
34	$C_2H_5\bullet + C_2H_5\bullet = C_2H_4 + C_2H_6$	0
35	$C_2H_5\bullet + C_3H_7\bullet = C_2H_6 + C_3H_8$	0
36	$C_2H_5O\bullet = CH_3\bullet + HCHO$	21.1
37	$CH_3\bullet + O_2 = CH_3OO\bullet$	0
–37	$CH_3OO\bullet = CH_3\bullet + O_2$	31.5
38	$CH_3OO\bullet + C_3H_8 = CH_3OOH + C_3H_7\bullet$	15.0
39	$CH_3OOH = CH_3O\bullet + HO\bullet$	43.0
40	$CH_3O\bullet + C_3H_8 = CH_3OH + C_3H_7\bullet$	7.0
41	$CH_3O\bullet + O_2 = HCHO + HOO\bullet$	2.6
42	$CH_3OO\bullet + CH_3OO\bullet = CH_3O\bullet + CH_3O\bullet + O_2$	0
43	$CH_3OO\bullet + CH_3OO\bullet = CH_3OH + HCHO + O_2$	0
44	$CH_3OO\bullet + C_2H_5OO\bullet = CH_3O\bullet + C_2H_5O\bullet\ O_2$	0
45	$CH_3OO\bullet + C_2H_5OO\bullet = CH_3OH + CH_3CHO + O_2$	0
46	$CH_3OO\bullet + C_3H_7OO\bullet = CH_3O\bullet + C_3H_7O\bullet + O_2$	0
47	$CH_3OO\bullet + C_3H_7OO\bullet = CH_3OH + C_2H_5CHO + O_2$	0
48	$CH_3\bullet + CH_3\bullet = C_2H_6$	0
49	$CH_3\bullet + C_2H_5\bullet = C_3H_8$	0
50	$CH_3\bullet + C_3H_7\bullet = CH_4 + C_3H_6$	0
51	$HO\bullet + CH_3OO\bullet = CH_3OOH + O_2$	0
52	$HOO\bullet + CH_3\bullet = CH_3O\bullet + HO\bullet$	0
53	$CH_3OO\bullet + CH_3\bullet = CH_3O\bullet + CH_3O\bullet$	0
54	$CH_3OO\bullet \longrightarrow$ termination	0
55	$C_2H_5OO\bullet \longrightarrow$ termination	0

56	$C_3H_7OO\bullet \longrightarrow$	termination	0
57	$CH_3\bullet \longrightarrow$	termination	0
58	$C_2H_5\bullet \longrightarrow$	termination	0
59	$C_3H_7\bullet \longrightarrow$	termination	0

[a] Activation energy, kcal mol^{-1}.

The oxidation of methane can be stimulated by addition of tetrafluorohydrazine [33a]. The reaction proceeds via a chain radical mechanism:

$$N_2F_4 + M \longrightarrow 2\,\bullet NF_2 + M$$
$$CH_4 + \bullet NF_2 \longrightarrow CH_3\bullet + HNF_2$$
$$CH_3\bullet + O_2 + (M) \longrightarrow CH_3OO\bullet + (M)$$
$$CH_3OO\bullet + \bullet NF_2 \longrightarrow CH_3O\bullet + F\bullet + FNO\bullet$$
$$CH_4 + F\bullet \longrightarrow CH_3\bullet + HF$$
$$CH_3O\bullet + CH_4 \longrightarrow CH_3OH + CH_3\bullet$$

The chain branching stage in this process is the interaction of $CH_3OO\bullet$ and $\bullet NF_2$.

It is very important that if molecular oxygen is introduced into the oxidation reaction only in small amounts (and the role of dioxygen is to initiate the radical chain process), methane homologues can be converted not to oxygenated products but to olefins. For example, laser heating transforms propane into monoolefins [33b]:

$$C_3H_8 + O_2 \longrightarrow C_3H_7\bullet + HOO\bullet$$
$$C_3H_7\bullet \rightleftharpoons CH_3CH{=}CH_2 + H\bullet$$
$$C_3H_7\bullet \rightleftharpoons CH_2{=}CH_2 + CH_3\bullet$$
$$H\bullet + C_3H_8 \longrightarrow CH_4 + C_3H_7\bullet$$
$$CH_3\bullet + C_3H_8 \longrightarrow CH_4 + C_3H_7\bullet$$
$$CH_3\bullet + CH_3\bullet \longrightarrow C_2H_6$$

$$CH_3\bullet + C_3H_7\bullet \longrightarrow CH_4 + CH_3CH{=}CH_2$$
$$C_3H_7\bullet + C_3H_7\bullet \longrightarrow C_3H_8 + CH_3CH{=}CH_2$$
$$H\bullet + C_3H_7\bullet \longrightarrow CH_3CH{=}CH_2 + H_2$$
$$H\bullet + CH_3\bullet \longrightarrow CH_4$$

On the contrary, an excess of oxygen leads to deep oxidation producing mainly carbon dioxide and water. Mechanism of methane burning includes the following stages [28b]:

$$CH_4 + HO\bullet \longrightarrow CH_3\bullet + H_2O$$
$$CH_4 + O \longrightarrow CH_3\bullet + HO\bullet$$
$$CH_3\bullet + O_2 \longrightarrow HCHO + HO\bullet$$
$$HCHO + HO\bullet \longrightarrow HCO\bullet + H_2O$$
$$CO + HO\bullet \longrightarrow CO_2 + H\bullet$$
$$H\bullet + H_2O \longrightarrow HO\bullet + H_2$$
$$O + H_2 \longrightarrow HO\bullet + H\bullet$$
$$HO\bullet + HO\bullet \longrightarrow H_2O + O$$
$$O + O + M \longrightarrow O_2 + M$$

II.3.B. NON-CATALYZED AUTOXIDATION IN THE LIQUID PHASE

Autoxidation of saturated hydrocarbons in the liquid phase is usually a branched chain process with a so-called 'degenerate' chain branching. This means that branching of each chain happens much later than its termination (as distinct from non-degenerate branching which occurs virtually simultaneously with chain propagation) and is caused by the formation of a rather stable intermediate. This intermediate is nevertheless chemically more active than the initial hydrocarbon and can form free radicals at a greater rate than that of the chain initiation process. Autoxidation of C–H compounds is of great importance for living organisms.

Hydroperoxides are the intermediates in liquid phase oxidation. The comparatively low energy of the O–O bond in hydroperoxides brings about its cleavage with formation of free radicals. The study of hydroperoxides, ROOH, produced from hydrocarbons, RH, has shown the structure of R in ROOH to be the same as in the initial RH, which confirms that a hydroperoxide is formed in the interaction of ROO• radical with the molecule of the initial hydrocarbon

$$ROO\bullet + RH \longrightarrow ROOH + R\bullet$$

The evidence for the chain mechanism of hydrocarbon oxidation is primarily based on the observation of the enhancing effect of light, ionizing radiation and small additions of various initiators easily decomposing into free radicals, as well as the inhibiting effect of such compounds as phenols or aromatic amines that readily react with free radicals. For some oxidation reactions ROO• radicals involved in the chain propagation were directly observed by ESR. The following classical scheme [6a] presents a typical mechanism of liquid-phase hydrocarbon oxidation for the early stages when the effect of reaction products may be neglected. The numbering of this radical-chain reaction scheme is one conventionally accepted in the literature.

Chain initiation:

$$(0) \qquad RH + O_2 \xrightarrow{k_0} R\bullet + HOO\bullet$$

Chain propagation:

$$(1) \qquad R\bullet + O_2 \xrightarrow{k_1} ROO\bullet$$

$$(2) \qquad ROO\bullet + RH \xrightarrow{k_2} ROOH + R\bullet$$

Chain branching:

$$(3) \qquad ROOH \xrightarrow{k_3} RO\bullet + HO\bullet$$

or

$$(3') \qquad 2\,ROOH \longrightarrow RO\bullet + ROO\bullet + H_2O$$

Chain termination:

$$(4) \qquad R\bullet + R\bullet \longrightarrow R\text{–}R$$

(5) $ROO\bullet + R\bullet \longrightarrow ROOR$

(6) $ROO\bullet + ROO\bullet \xrightarrow{k_6} ROH + R'COR'' + O_2$

The branching chain mechanism has some general features. The rate constants of reactions 1–6 of R• and ROO• radicals are high and their concentrations quickly reach stationary values. At sufficiently high dioxygen concentration [ROO•] >> [R•] and the termination proceeds only by reaction 6. For the steady-state conditions

$$d[R\bullet]/dt = v_i - k_1 [O_2] [R\bullet] + k_2 [RH] [ROO\bullet] = 0$$
$$d[ROO\bullet]/dt = k_1 [O_2] [R\bullet] - k_2 [RH] [ROO\bullet] - k_6 [ROO\bullet]^2 = 0$$

where v_i is the overall rate of radicals formation in chain initiation. Summing up both the equations we get

$$v_i = k_6 [ROO\bullet]^2$$
$$[ROO\bullet] = (v_i/k_6)^{1/2}$$

and the hydrocarbon oxidation rate at early stages where the rate of branching may be neglected will be

$$v = d[RH]/dt = k_2 [RH] [ROO\bullet] = k_2 [RH] (v_i/k_6)^{1/2}$$

The low O–O bond energy causes the rate of radical formation in reactions 3 and 3' in the process of hydroperoxide accumulation, to exceed the chain initiation rate, which is the reason for the total reaction rate increase during the process of hydroperoxide accumulation. The reaction rate will stop increasing when the hydroperoxide accumulation rate via reaction 2 becomes equal to the rate of its decomposition into radicals via reactions 3 and 3'. The latter, in its turn, must be equal to the rate of radical termination via reaction 6. Thus, the reaction rate will reach its maximum when

$$nk_2 [ROO\bullet] [RH] = 2k_6 [ROO\bullet]^2$$

Table II.9. The mechanism of the cyclohexane oxidation.

No	Reaction	A [a]	E [b]	k_i [c]
1	$RH + O_2 = R\bullet + HOO\bullet$	7.94×10^{12}	40.0	1.8×10^{-8}
2	$R\bullet + O_2 = ROO\bullet$	10^9	0	10^9
3	$RH + ROO\bullet = ROOH + R\bullet$	1.6×10^8	13.8	4.65
4	$RH + RO\bullet = ROH + R\bullet$	2.95×10^8	3.5	7.65×10^7
5	$ROOH = RO\bullet + HO\bullet$	6.3×10^{11}	32.4	5.07×10^{-5}
6	$ROOH + ROH = 2\,RO\bullet + H_2O$	4.5×10^8	22.0	1.91×10^{-3}
7	$ROOH + R'{=}O = 0.7\,R\bullet + 0.3\,P + RO\bullet + HO\bullet$	5.6×10^{-1}	7.3	9.47×10^{-4}
8	$ROOH + RH = RO\bullet + R\bullet + H_2O$	6.8×10^{10}	32.0	4.52×10^{-6}
9	$ROOH + RO\bullet = ROH + ROO\bullet$	2.9×10^9	3.1	7.65×10^7
10	$ROOH + ROO\bullet = R'{=}O + HO\bullet + ROO\bullet$	9.3×10^6	9.1	7.44
11	$ROH + ROO\bullet = R'{=}O + ROO\bullet$	2.69×10^7	12.0	4.56
12	$R'{=}O + ROO\bullet = P + ROO\bullet$	5.69×10^7	12.0	8.32
13	$ROOH + R\bullet = ROH + RO\bullet$	2.0×10^6	3.9	4.59×10^4
14	$ROOH + R\bullet = ROO\bullet + RH$	6.31×10^8	5.0	1.64×10^6
15	$2\,ROO\bullet = R'{=}O + ROH + O_2$	1.03×10^7	28.0	1.55×10^6
16	$ROOH + P = R'{=}O + H_2O + P$	6.35×10^6	18.3	2.18×10^{-3}
17	$ROH + P = P'$	3.1×10^5	15.6	1.67×10^{-3}
18	$2\,ROO\bullet = P'$	8.77×10^6	25.4	1.55×10^6
19	$RH + HO\bullet = R\bullet + H_2O$	2.9×10^{10}	3.1	7.65×10^8

[a] Preexponential factors are expressed in $cm^3\ mol^{-1}\ s^{-1}$.

[b] Activation energy, $kcal\ mol^{-1}$.

[c] Calculated rate constants, k_i, are expressed in $cm^3\ mol^{-1}\ s^{-1}$; for stage No. 5 in s^{-1}.

and according to Walling [34c] the maximum reaction rate will be the following

$$v_{max} = -\,d[RH]/dt = f^{-1}k_2\,[ROO\bullet]\,[RH] = (nk_2^2/2fk_6)\,[RH]^2$$

Here n is the average number of radicals produced per ROOH decomposed and f is the fraction of RH consumed which disappears by one ROO• attack. If the branching proceeds via reaction 3 ($n = 2, f = {}^1/_3$), then

$$v_{max} = -\,d[RH]/dt = (3k_2^2/k_6)\,[RH]^2$$

After the decomposition of ROOH molecule both RO• and HO• radicals react with RH in fast reactions, e.g.,

$$HO\bullet + RH \longrightarrow H_2O + R\bullet,$$

hence per each reaction 2, three RH molecules disappear. If radicals are formed in bimolecular reaction 3' ($n = 1, f = {}^2/_3$), then

$$v_{max} = {}^3/_4(k_2^2/k_6)\,[RH]^2$$

This maximum rate in non-catalytic oxidation (if [RH] is the initial hydrocarbon concentration) is never really reached due to a slow increase of hydroperoxide concentration and complications resulting from the accumulation of products.

The stages of non-catalyzed oxidation of cyclohexane, RH, are summarized in Table II.9 [34d].

Although liquid-phase oxidations of alkanes can be carried out even in the absence of any metal derivative (the role of an initiator of chain radical process can be played by a non-metal compound), derivatives of transition metals are often used in these reactions. Metal-catalyzed autoxidation will be considered in Chapter IX.

Liquid-phase oxidation of butane (in acetic acid as a solvent) is a commercial method of acetic acid production. The mechanism of the initial stage of the process initiated by $(CH_3)_3COOC(CH_3)_3$ is the following [35]:

$$(CH_3)_3COOC(CH_3)_3 \longrightarrow 2\,(CH_3)_3CO\bullet$$

$$(CH_3)_3CO\bullet + RH \longrightarrow (CH_3)_3COH + R\bullet$$

(where $R = CH_3CH_2CH\bullet CH_3$)

$$R\bullet + O_2 \longrightarrow ROO\bullet$$

$$ROO\bullet + RH \longrightarrow ROOH + R\bullet$$
$$2\,ROO\bullet \longrightarrow 2\,RO\bullet + O_2$$
$$RO\bullet + RH \longrightarrow ROH + R\bullet$$
$$RO\bullet \longrightarrow CH_3COOH + CH_3CH_2\bullet$$
$$2\,ROO\bullet \longrightarrow CH_3CH_2COCH_3 + ROH + O_2$$
$$CH_3CH_2\bullet + O_2 \longrightarrow CH_3CH_2OO\bullet$$
$$CH_3CH_2OO\bullet + RH \longrightarrow CH_3CH_2OOH + R\bullet$$
$$2\,CH_3CH_2OOH \longrightarrow 2\,CH_3CH_2O\bullet + O_2$$

Alkylarenes can be oxidized with molecular oxygen at 70–140° C if ion Br^- and tertiary ammonium salts are present in the solution [36]. The oxidation of *p*-xylene has been proposed to be initiated by arylalkyl hydroperoxide:

$$RCH_2OOH + Br^- \longrightarrow RCH_2\bullet + HO^- + Br\bullet$$
$$Br\bullet + RCH_3 \longrightarrow HBr + RCH_2\bullet$$
$$RCH_2\bullet + O_2 \longrightarrow RCH_2OO\bullet$$
$$RCH_2OO\bullet + RCH_3 \longrightarrow RCH_2OOH + RCH_2\bullet$$
$$2\,RCH_2OO\bullet \longrightarrow \text{products} + O_2$$

II.3.C. PHOTOCHEMICAL OXIDATION IN THE LIQUID PHASE

Photolysis of alkanes in air gives rise to the formation of oxygenated products in very low yields. Thus irradiated with high-pressure mercury arc, cyclohexane saturated with air ($[O_2] = 1.9 \times 10^{-3}$ mol dm^{-3}) yields cyclohexanol, cyclohexanone and cyclohexyl hydroperoxide (ratio 1.0 : 0.6 : 0.9) [37a]. Under analogous conditions, toluene is transformed into the mixture of benzaldehyde and benzyl alcohol (after 24 h the total yield 0.15 %, ratio aldehyde/alcohol is 3 : 1) [37b]. Benzyl hydroperoxide is the initial product of this reaction which decomposes further to yield more stable oxygenates [37c].

Photosensitized oxidation of hydrocarbons by oxygen or air gives alkyl hydroperoxides, ketones and alcohols in much higher yields. Anthraquinone, cyanonaphthalenes and some other compounds are widely used as sensitizers [38, 39]. An electron of a hydrogen atom is transferred from the hydrocarbon, RH, to 1,4-dicyanonaphthalene (S) at the first stage of the reaction to generate ion radicals or radicals (Scheme II.1) [38b].

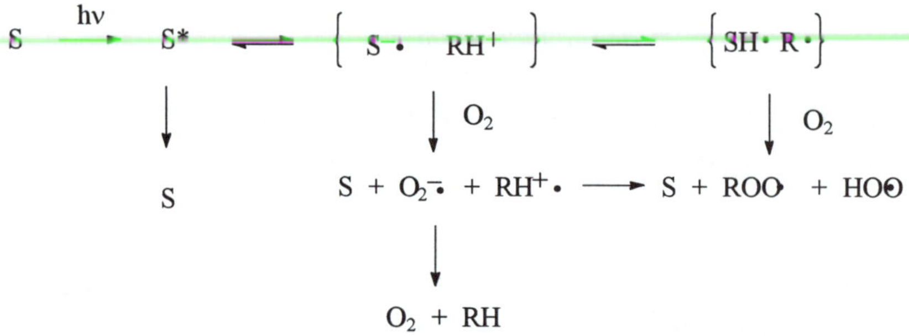

Scheme II.1. The mechanism of photosensitized by compound S oxygenation of hydrocarbon, RH.

Irradiation ($\lambda > 310$ nm) of an air-saturated solution of alkylbenzene (or even cyclohexane) in acetonitrile [38c] or acetic acid [38d] in the presence of catalytic amounts of *o*-phenanthroline or 2,2′-bipyridine and sulfuric acid gives oxygenated products (benzaldehyde from toluene). It is noteworthy that activity of 4,4′-bipyridine and methylviologen, as well as, pyridine, pyrazole, pycolinic acid in the photo-oxygenation is very low. An explanation of the fact that only biphenyls with nitrogen atoms in the α-positions show appreciable activity is as follows [38d]. In the photo-oxidation, a form of *o*-phenanthroline that is protonated only on one nitrogen atom is active. The second nitrogen atom may then act as a base and take up the proton eliminated from the toluene radical cation. Since the electron from toluene is transferred to the ring containing the protonated nitrogen atom, the second nitrogen atom, which acts as a base, must be situated a fairy short distance from the protonated nitrogen atom. This requirement is probably not satisfied by the nitrogen atoms in 4,4′-bipyridine, which are at a considerable distance from one another. The acceptance of a proton by the basic nitrogen atom favors a shift in the equilibrium (*a*) on Scheme II.2 to the right. We can also suppose the possibility of simultaneous activation at the two nitrogen atoms of *o*-phenanthroline or 2,2′-bipyridine for molecules of toluene and dioxygen (Scheme II.2, equation *b*). The intermediately formed benzyl radical and superoxide radical anion, which are close in space, recombine giving a peroxide anion which can accept a proton from the solvent.

Irradiation of a solution containing cyclohexane (0.46 mol dm^{-3}) and pyrazine-2-carboxylic acid (4×10^{-4} mol dm^{-3}) in acetonitrile for 4 h gave cyclohexanol, cyclohexanone and cyclohexyl hydroperoxide (total concentration

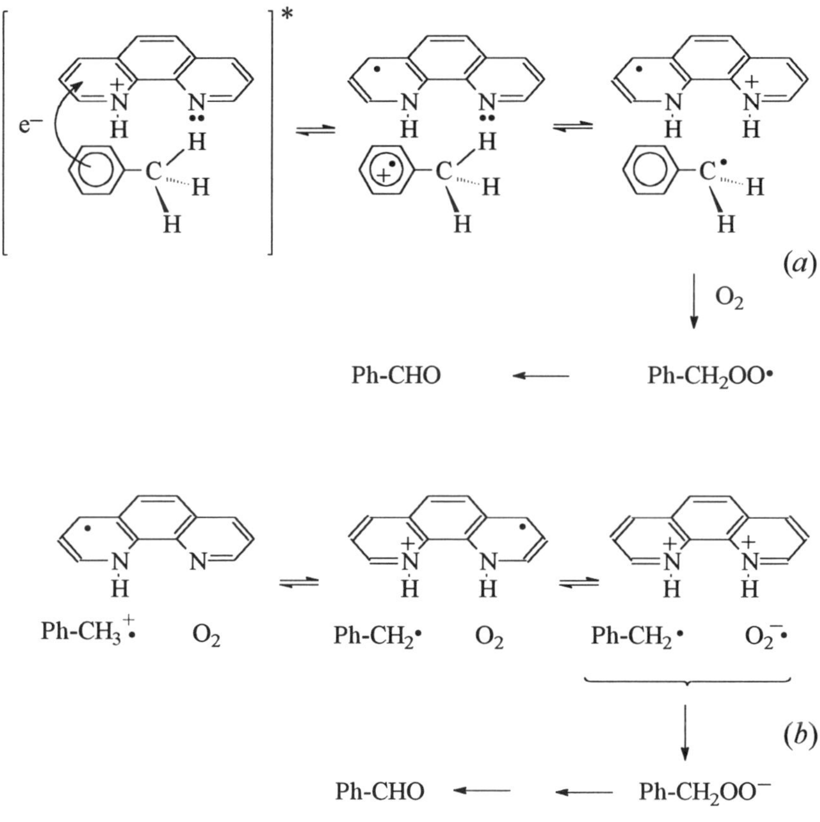

Scheme II.2. Proposed mechanism of the toluene photo-oxygenation sensitized by protonated *o*-phenanthroline.

2.2×10^{-3} mol dm^{-3}; the ratio is 1 : 1 : 1.5) [39b]. Methyl pyrazine-2-carboxylate exhibits higher activity after 4 h irradiation: 3.7×10^{-3} mol dm^{-3}; the ratio is 1 : 1 : 1.5 , and unsubstituted pyrazine (1.8×10^{-3} mol dm^{-3}) and pyrimidine (0.5×10^{-3} mol dm^{-3}) are less active. Some other heterocyclic compounds (imidazole, picolinic acid, phenanthroline) or anthranilic acid do not sensitize oxygenation at all.

Recently, for the photochemical oxidation of indan, tetralin, 2-ethylnaphthalene and cyclohexane by nitropyridinium salts, the mechanism has been proposed [39d], which also involves the stages of electron and hydrogen atom transfer (Scheme II.3).

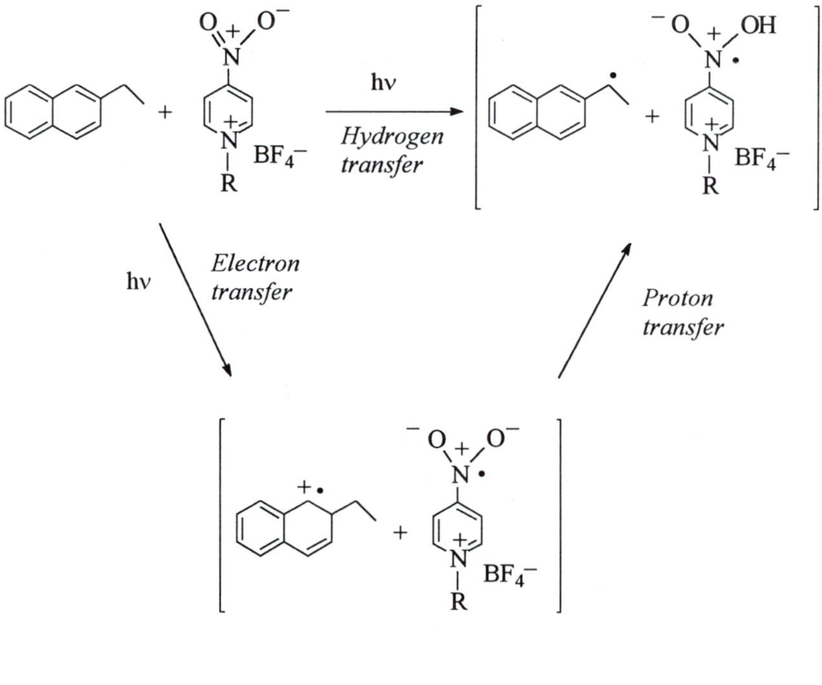

Scheme II.3. Proposed initial stages in the photo-oxygenation of 2-ethylnaphthalene by nitropyridinium salt.

Photolysis of tetrachlorodiazocyclopentadiene in oxygen-saturated hydrocarbon solvents results in the transient formation of tetrachlorocyclopentadienone *O*-oxide, which decomposing reacts with the hydrocarbon (Scheme II.4) [39e].

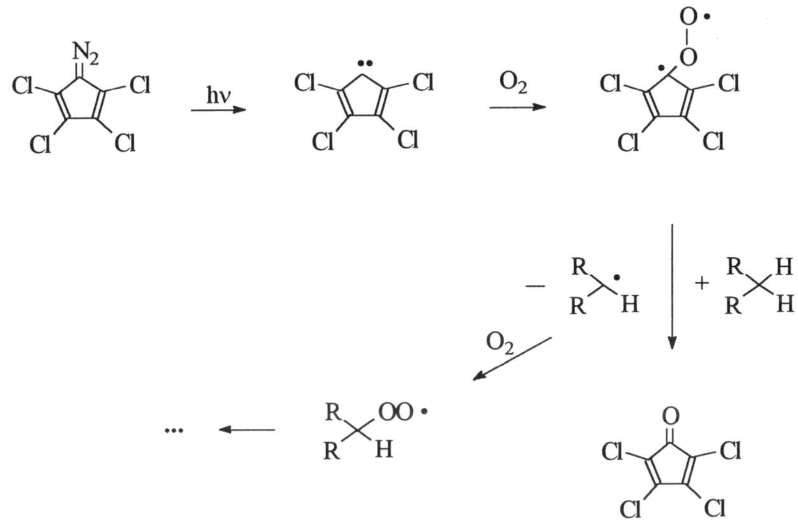

Scheme II.4. Initial stages of the alkane photo-oxygenation by tetrachlorodiazocyclopentadiene.

II.3.D. OTHER REACTIONS INITIATED BY RADICALS

Using the combination of molecular oxygen, acetaldehyde and *N*-hydroxyphthalimide (the last compounds is a component – in addition to a metal complex – of the catalyst in the Ishii oxidation reaction), it is possible to oxidize alkylaromatic hydrocarbons and cycloalkanes in acetonitrile at room temperature [40a]. The reaction occurs via a radical mechanism depicted in Scheme II.5.

Radical ion $SO_4^{-}\bullet$ (formed from $S_2O_8^{2-}$) in aqueous solution at 100° C abstracts a hydrogen atom from methane and ethane (AlkH) to generate radicals

Alk•. These radicals react with SO_4^{-}• affording $AlkOSO_3^{-}$. If the reaction is carried out in a carbon monoxide atmosphere, AlkCOOH may be obtained [40b].

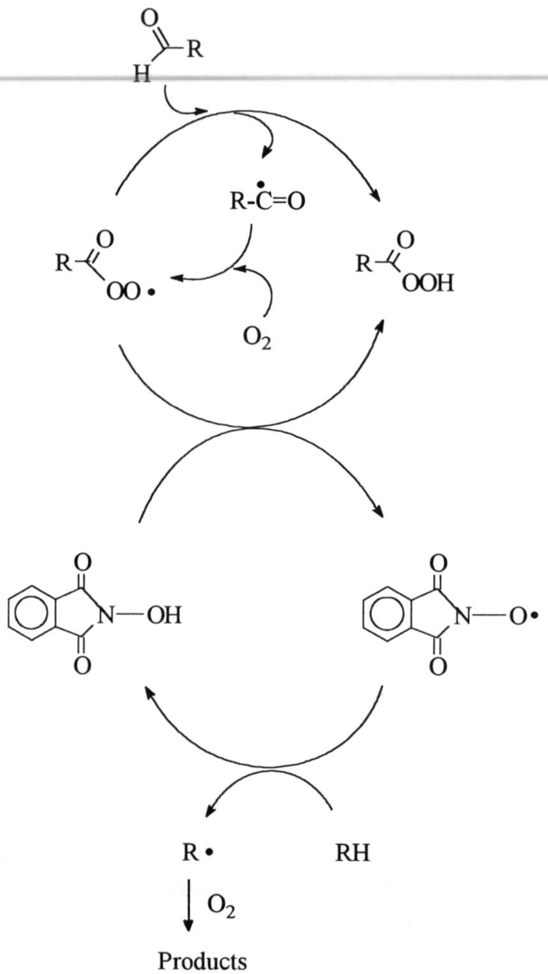

Scheme II.5. The mechanism of oxygenation of alkylaromatic and cyclic saturated hydrocarbons by the "O_2 – CH_3CHO – NHPI" system.

II.4. Oxidation with Oxygen-containing Compounds

Careful investigation of reaction of ozone with C–H compounds (including alkanes) showed that a hydride abstraction and the concerted insertion of ozone into a C–H bond are most consistent with the kinetic data [40a] (see also [40b]). Table II.10 contains the relative reactivities of some alkanes toward ozone as well as three radical species [40a].

Table II.10. Relative reactivities of alkanes and arylalkanes toward ozone and some radicals.

C–H Bond	Br•	Me_3COO•	Me_3CO•	Ozone
Primary (1°)	0.02	0.024	0.08	0.003
Secondary (2°)	1.0	1.0	1.0	1.0
Tertiary (3°)	40.0	10.0	3.7	87.0
Ar–CH_3 (value ρ)	–1.88	–0.76	–0.34	–2.07

Ozone is also a promoter of arene nitration with nitrogen oxides [40c]. In the presence of ozone at –78 °C, nitrogen dioxide reacts with adamantane at a bridgehead position to yield the corresponding nitro derivative. The proposed mechanism involves the initial hydrogen abstraction by nitrogen trioxide, followed by rapid trapping of the resulting adamantyl radical with nitrogen dioxide [40d]:

$$NO_2 + O_3 \longrightarrow NO_3$$

$$NO_3 + NO_2 \rightleftharpoons N_2O_5$$

$$RH + NO_3 \longrightarrow R\bullet + HNO_3$$

$$R\bullet + NO_2 \longrightarrow RNO_2 + RONO$$

In this reaction, ozone does not interact with the hydrocarbon directly. However, dry ozonation (low-temperature ozonation of adsorbed on silica surface substrate) of adamantane produces adamantylhydrotrioxide [40e]:

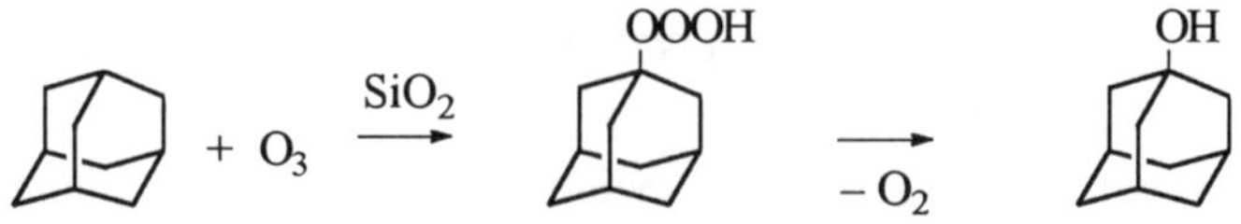

Although hydroxyl radicals are apparently the active species in oxygenation of benzene [40a], cyclohexane [40a] and even methane [40b] by oxygen-containing derivatives of xenon (which may be generated by dissolution of XeO_3 or XeF_2 in water of aqueous acetonitrile), the mechanism of direct arene epoxidation as the first stage is not excluded [40a]:

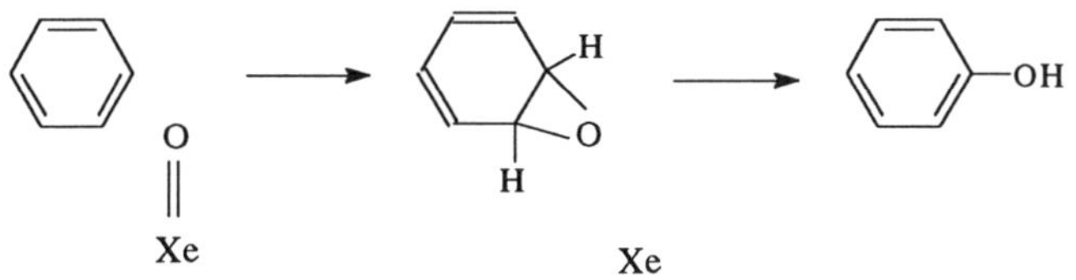

This stage is analogous to the crucial step in arene hydroxylation in living cells under the action of cytochrome P450.

On the basis of MO calculations of alkane oxidation by a model singlet atom donor (water oxide, H_2O–O), the following mechanism is suggested [43]. The electrophilic oxygen approaches the hydrocarbon to form a carbon-oxygen σ bond. A concerted hydrogen migration to an adjacent oxygen lone pair of electrons gives the alcohol insertion product. The activation barriers for oxygen atom insertion into the weakest C–H bond were calculated for methane (10.7), ethane (8.2), propane (3.9), butane (4.8), isobutane (4.5), and methanol (3.3 kcal mol^{-1}).

II.4.A. PEROXIDES

Although alkyl peroxides are usually intermediates and products in autoxidation of alkanes, the reactions between alkanes and alkyl hydroperoxides need relatively high temperature or a metal complex as a catalyst. Sometimes other compounds can induce such reactions. For example, *n*-pentane and *n*-hexane are oxidized by *tert*-butyl hydroperoxide in benzene solution if $(t\text{-Bu})_3Al$ is present in the system [44].

Peroxyacids are more reactive than alkyl peroxides [45]. The mechanism of alkane hydroxylation by peroxyacids is believed to involve an electrophilic attack of the oxygen atom, for example, in the reaction with trifluoroperoxyacetic acid [45c]:

Indeed, the hydroxylation of *cis*- and *trans*-1,2-dimethylcyclohexane occurs with retention of configuration. The electrophilic nature of this reaction is also confirmed by the normal selectivity sequence in the attack of the C–H bonds: 3° > 2° > 1°. See, however, [45b]. Alkyl sulfoperoxyacids also hydroxylate alkanes [45d].

II.4.B. DIOXIRANES

Dioxiranes, $RR'CO_2$ [usually 3,3-dimethyldioxirane and 3-(trifluoromethyl)-3-methyldioxirane], as efficient reagents for oxyfunctionalization of C–H compounds have been described and investigated (see, for example, reviews [46] and recent original publications [47]). Stereoselective hydroxylation of a steroid compound with 3-(trifluoromethyl)-3-methyldioxirane is shown below [47g]:

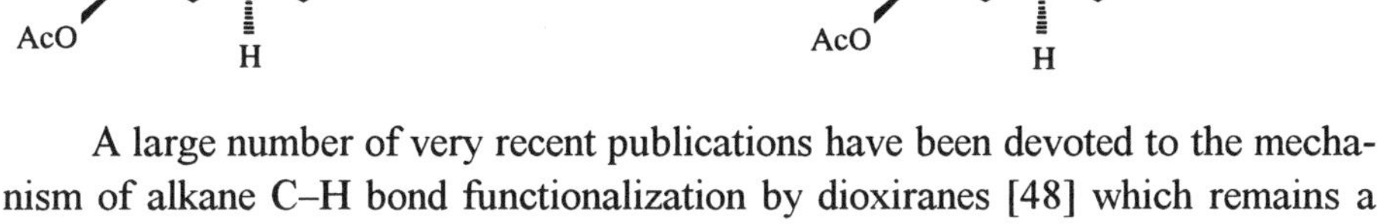

A large number of very recent publications have been devoted to the mechanism of alkane C–H bond functionalization by dioxiranes [48] which remains a matter of debate. Two alternative mechanisms have been proposed:

Concerted Electrophilic Oxygen Insertion:

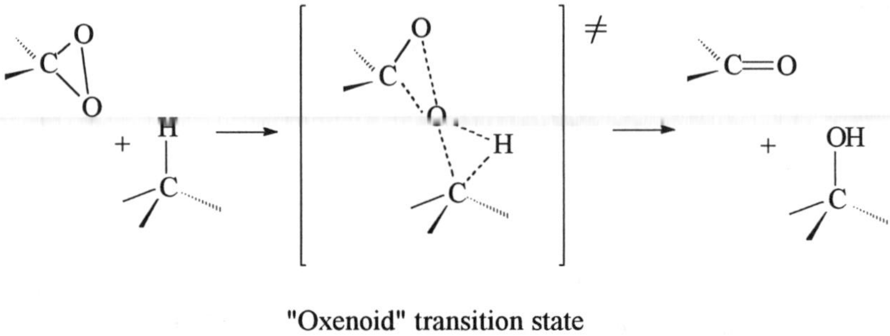

Radical Mechanism:

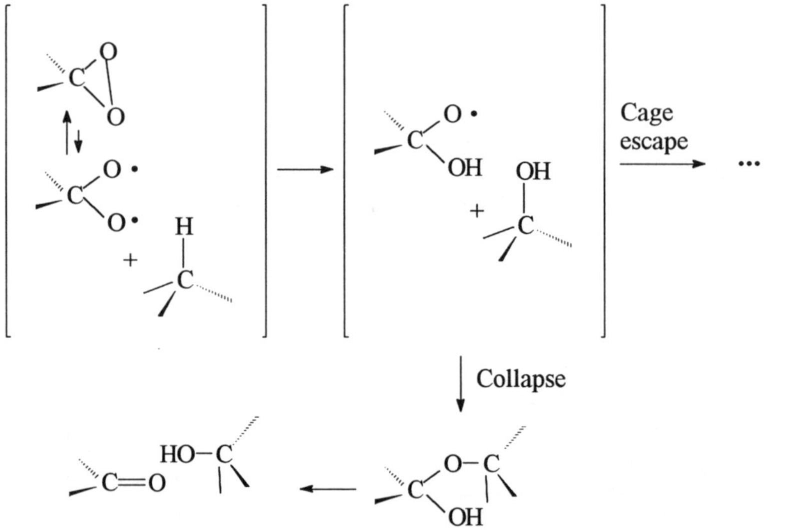

Scheme II.6. Mechanisms proposed for the oxygenation of saturated hydrocarbons by dioxirane.

(i) a concerted electrophilic oxygen insertion involving an "oxenoid" transition state, and

(ii) a radical mechanism involving hydrogen abstraction from the alkane by the ring-opened dioxymethane biradical, followed by collapse of the radical pair in the solvent cage (Scheme II.6).

It has been shown by *ab initio* calculations that the alkane oxidation proceeds via a highly polar asynchronous transition state, which is common for either concerted oxygen insertion into the C–H bond or formation of a radical pair [48c,d]. The transition state having considerable diradical character (and also is polarized) is shown below:

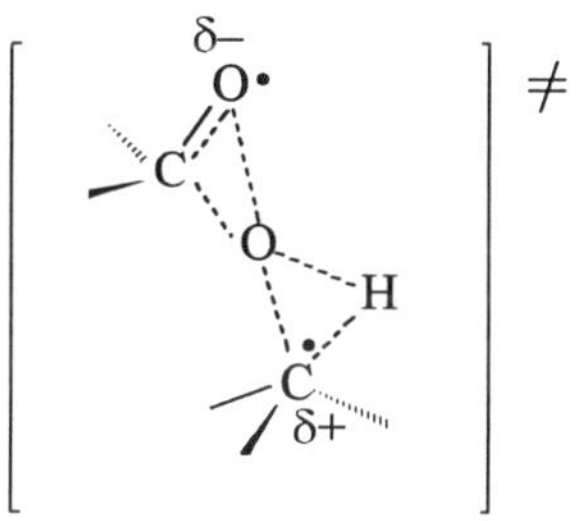

A theoretical study on the oxidation of methane, propane and isobutane with dioxirane, dimethyldioxirane, difluorodioxirane and methyl(trifluoromethyl)dioxirane has provided a rational for the formation of radical intermediates when dioxygen is rigorously excluded and supported the generally accepted, highly exothermic, concerted oxygen insertion mechanism for the oxidation under typical preparative conditions. The activation barriers for the oxidation of methane (44.2), propane (30.3) and isobutane (22.4 kcal mol^{-1}) with dimethyldioxirane have been evaluated [48e]. Perfluorodialkyloxaziridines are also mild and selective reagents for the hydroxylation of alkanes [49]:

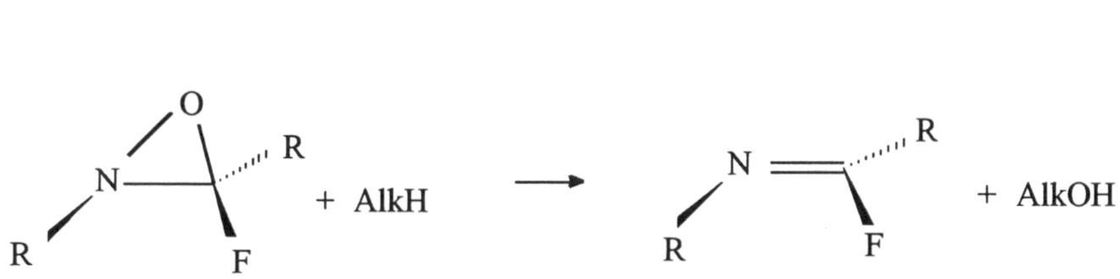

R = C_4F_9, C_6F_{13}, C_3F_7, C_5H_{11}

II.5. CARBOXYLATION

Alkyl radicals that formed from alkanes under the action of various radical-like species can react rapidly not only with molecular oxygen, but can also be trapped by carbon monoxide. In this case, carboxylic acids can be obtained. Thus adamantane has been recently converted into 1-adamantanecarboxylic acid as the main product (Scheme II.7) [50]. *N*-Hydroxyphthalimide was used for this transformation as an efficient radical catalyst. The reaction occurs under mild conditions (CO pressure up to 15 atm and temperature below 100 °C) and gives 1-adamantanecarboxylic acid with selectivity 56% and conversion 75%.

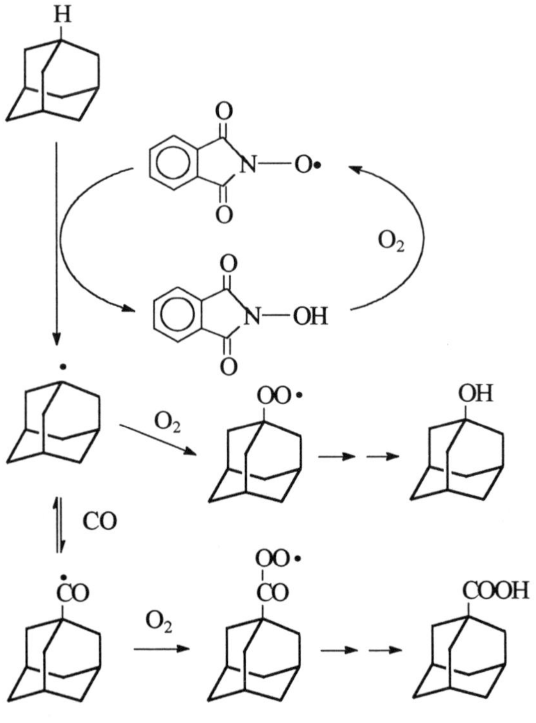

Scheme II.7. Carboxylation of adamantane with CO catalyzed by *N*-hydroxyphthalimide.

It is interesting that methane and carbon monoxide are converted to light hydrocarbons by microwave plasma [50b]. Acetylene and ethane are the dominant organic products; a small amount of ethylene and hydrocarbons containing three carbon atoms have been also found. Higher hydrocarbons and oxygenated organic products were not detected.

II.6. ELECTROPHILIC SUBSTITUTION OF HYDROGEN IN ALKANES

In contrast to olefins and aromatic hydrocarbons, alkanes are too weak as bases and nucleophiles to react with the majority of electrophilic reagents. In reactions with electrophiles, tertiary C–H bonds are much more reactive in comparison with secondary and especially with primary C–H bonds. Strained, polycyclic and cage saturated hydrocarbons react rather easily with electrophilic reagents, and in these cases not only C–H but also C–C bonds can be activated. A single-electron transfer (SET) mechanism must be considered as an important potential mechanism for the activation of strained aliphatic hydrocarbons [51a]. MO theory has been applied to investigate the reactions of electrophiles NO^+ and NOH^{2+} with methane [51b].

II.6.A. TRANSFORMATIONS IN THE PRESENCE OF SUPERACIDS

When alkanes are dissolved in superacids, they are transformed into various products [52]. In protic media, the first step of the reaction is the protonation to form in the case of methane, the methonium ion, CH_5^+. For the theoretical description of such a structure see refs. [53]. The ground state of this ion has C_s symmetry which may be regarded as a complex of CH_3^+ cation with H_2 molecule (H–H distance has been found to be 0.936 Å which suggests formation of a bond). The ground state is only insignificantly different (ca. 1 kcal mol^{-1}) from the excited state of C_{4v} symmetry. Even closer are two transition states of other symmetries [53c]. This finding corresponds to extremely fast hydrogen atom movements in CH_5^+ and other similar carbonium ions; therefore, fast isomerization, e.g., epimerization of alkane molecules, is expected to proceed in superacid solutions. More recently the structures and stabilities of cations CH_5^{3+} and CH_6^{4+} were calculated [53d]. The planar D_{5h} symmetric structure of the first ion in which the five hydrogens are bonded to the carbon atom by sharing the six valence electrons is in a minimum. The cation CH_6^{4+} does not correspond to a minimum. However, *ab initio* calculations have shown that triprotonated methane, CH_7^{3+}, resides at a minimum on its potential energy surface, although its deprotonation is highly exothermic [53e]. The potential energy surface of the $C_3H_9^+$ cations was recently calculated [53f,g].

Formed in superacid media species exhibiting electrophilic properties are able to attack alkanes primarily via electrophilic addition to C–H bond followed by other reactions. In particular, Olah *et al.* observed O atom insertion in hydrogen peroxide reaction with methane in Magic Acid above 0 °C to produce methanol with very high (>95%) selectivity [54a]. The particle $(OH)^+$, which may be considered to be a protonated oxygen atom in the singlet state is apparently the active species in the reaction. Methyl alcohol formed is immediately protonated to methyloxonium ion, $CH_3OH_2^+$, and this prevents further oxidation.

Superacid DF–SbF_5 induces protium–deuterium exchange in isobutane [54b]. Strong acids (BF_3, BF_3–H_2O, HF–BF_3, CF_3SO_3H) catalyze carbonylation of alkanes including methane by carbon monoxide [54c], whereas sulfuric acid can induce carbonylation of iso- and cycloalkanes [54e,d]. In both cases, carboxylic acids are obtained. The elimination of molecular hydrogen from alkyl can occur (see a recent theoretical study of the H_2 elimination from $C_2H_5^+$ [54f]).

Fluorination in superacid of a non-activated C–H bond in a complex molecule has been described [55a]:

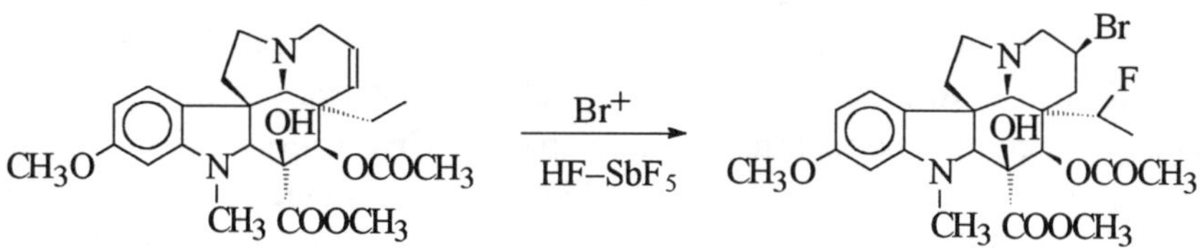

Strong acids promote electrophilic hydroxylation of aromatics. Thus, sodium perborate – trifluoromethanesulfonic acid has been found to be a versatile reagent for the monohydroxylation of arenes to phenols [55b].

II.6.B. REACTIONS WITH NOVEL ELECTROPHILIC REAGENTS

Novel aprotic superelectrophilic reagents ($RCOX \cdot 2AlX_3$, where R = Alkyl, Aryl; X = Br, Cl) have been described by Vol'pin and co-workers. These reagents are highly active toward alkanes under mild conditions (see reviews [56] and recent publications [57]).

The reactions of alkanes with a superelectrophilic reagent have a common stage, namely, the generation of a carbenium ion from the alkane and the reduction of the acyl halide. The difference in the activity of systems $RCOX \cdot 2AlX_3$ and $RCOX \cdot AlX_3$ is explained by the assumption that it is only systems with an excess of an AlX_3 that can generate acylium ions in solution. Active systems apparently exist in equilibrium, which is strongly shifted to the left, between acylium salts and more electrophilic complexes, in which an acylium cation, RCO^+, is coordinated to a Lewis acid:

$$RCO^+Al_2Br_7^- \rightleftharpoons$$
$$\rightleftharpoons RC^+C{=}O \rightarrow AlBr_3\ AlBr_4^- \longleftrightarrow [RCO^{2+}{-}OAlBr_3^-]AlBr_4^-$$

Such highly electrophilic cations are assumed to be present in very small concentrations and cannot be detected experimentally.

Some solid acids, that have been suggested as superacids capable of catalyzing various alkane reactions are listed in Table II.11 [58].

The conversion of propane and cycloalkanes into dialkyl (dicycloalkyl) sulfides under the action of elemental sulfur has been recently reported [59]. Sulfur is converted completely into dicyclopentylsulfide in the reaction with cyclopentane in the presence of $CX_4 \cdot nAlBr_3$ (X = Cl, Br; n = 2 or 3) at –20 °C for 20 min. Propane forms i-Pr_2S in 60% at 20 °C after 2 h. Transformations of alkanes and other hydrocarbons under the action of solid acids (metal oxides) occurring at relatively high temperature will be considered in the next chapter.

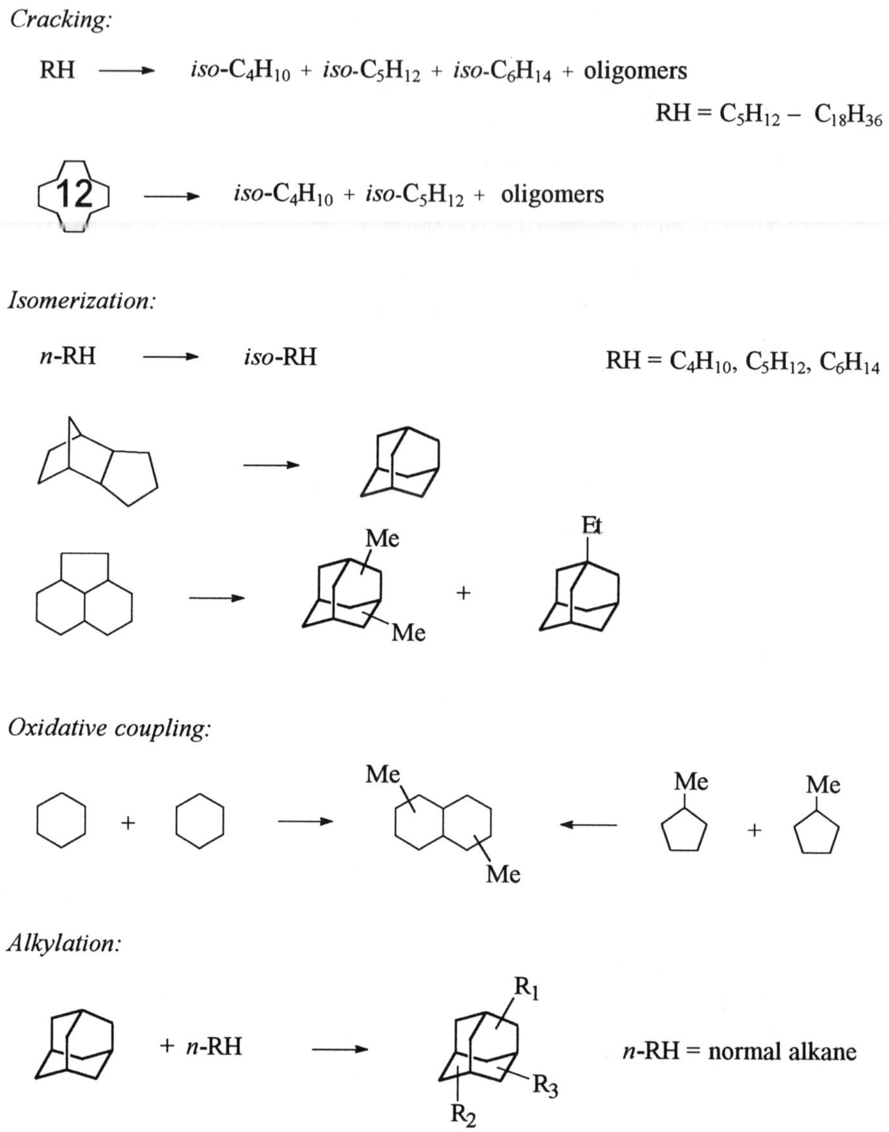

Scheme II.8. Cracking and some other transformations of alkanes induced and catalyzed by superelectrophilic reagents $RCOX \cdot 2AlX_3$.

Acylation and dehydroacylation:

$$n\text{-}C_nH_{2n+2} + RCO^+ \longrightarrow iso\text{-}C_nH_{2n+1}COR + iso\text{-}C_nH_{2n-1}COR$$

R = Me, Pr
$n = 4 - 6$

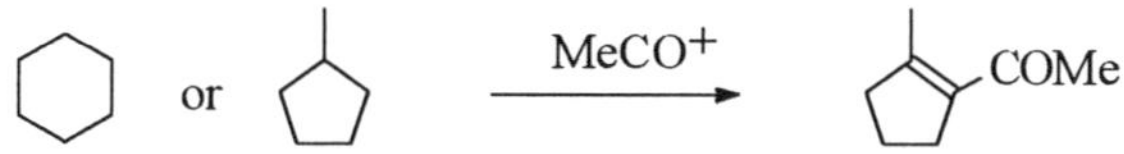

Alkylacylation of arenes:

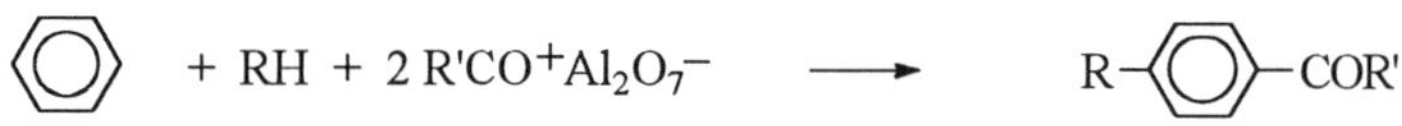

Halogenation:

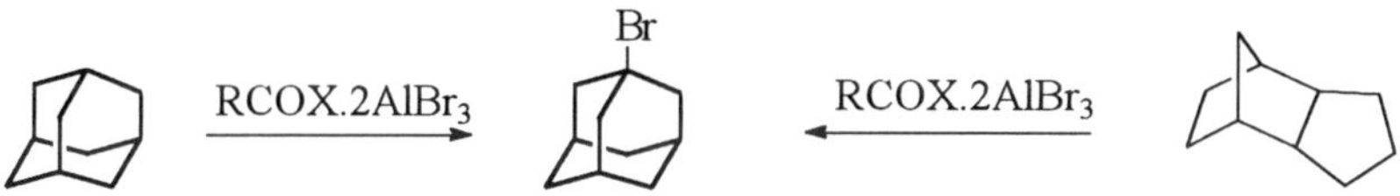

Ionic bromination:

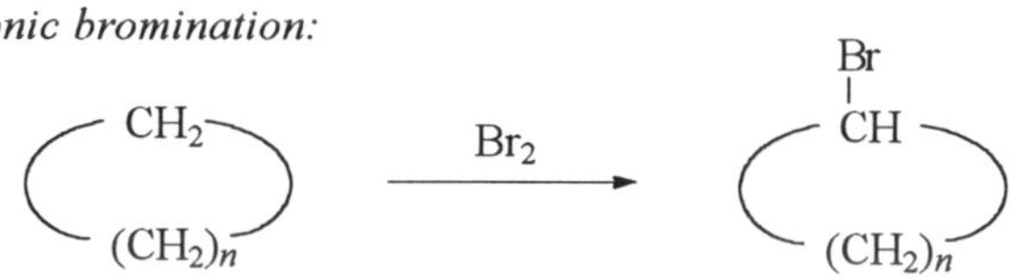

Thioacylation:

$$RH + S + 2\,R'CO^+Al_2O_7^- \longrightarrow RSCOR'$$

Carbonylation:

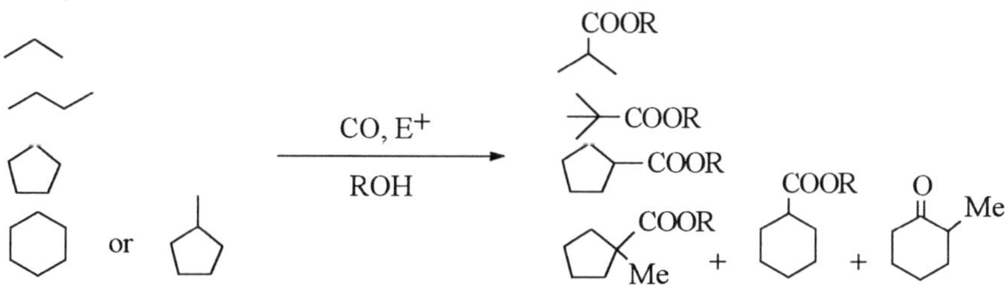

Scheme II.9. Functionalization of alkanes under the action of superelectrophilic reagents $RCOX{\cdot}2AlX_3$.

Table II.11. Solid superacids used as catalysts.

Class of catalyst	Example	Evidence of acid strength [a]	Rate of isomerization [b]
Metal halide supported on sulfonic acid resin	$AlCl_3$ – sulfonic acid resin	Catalysis of *n*-butane isomerization	2×10^{-9}
Metal halide supported on inorganic salt	$AlCl_3$ – $CuSO_4$	$H_0 = -14$	
Metal halide supported on solid	SbF_6 – SiO_2 – Al_2O_3; SbF_5 – sulfonic acid resin	$H_0 = -14$; catalysis of *n*-butane isomerization	
Heteropoly compound	$Cs_{2.5}H_{0.5}PW_{12}O_{40}$	$H_0 = -13$; catalysis of *n*-butane isomerization	
Sulfated metal oxide	SO_4^{2-} – ZrO_2	$H_0 = -16$; catalysis of *n*-butane isomerization	5×10^{-9}
Promoted sulfated zirconia	Fe-, Mn-promoted SO_4^{2-} – ZrO_2	Catalysis of *n*-butane isomerization	7×10^{-6}

[a] H_0 is Hammett acidity function. Reaction conditions for isomerization were 60 °C and 0.55 atm *n*-butane partial pressure.

[b] Rate of *n*-butane isomerization, mol s^{-1} g^{-1}.

References for Chapter II

1. (a) *Pyrolysis: Theory and Industrial Practice*; Albright, L.F.; Crynes, B.L.; Corcoran, W.H., Eds.; Acad. Press: New York, 1983. (b) Arutyunov, V.S.; Vedeneev, V.I. *Usp. Khim.* **1991**, *60*, 2663 (in Russian). (c) Bauer, S. H.; Javanovic, S.; Yu, C.-L.; Cheng, H.-Z. *Energy & Fuels* **1997**, *11*, 1204. (d) Naroznik, M. *J. Chem. Soc., Faraday Trans.* **1998**, *94*, 2531. (e) Bromberg, L.; Cohn, D. R.; Rabinovich, A.; O'Brien, C.; Hochgreb, S. *Energy & Fuels* **1998**, *12*, 11. (f) Cho, W.; Baek, Y.; Park, D.; Kim, Y. C.; Anpo, M. *Res. Chem. Intermed.* **1998**, *24*, 55. (g) Cioslowski, J.; Piskorz, P.; Moncrieff, D. *J. Org. Chem.* **1998**, *63*, 4051.
2. Dorofeev, Yu. I.; Skurat, V. E. *Usp. Khim.* **1982**, *51*, 925 (in Russian).
3. (a) Sarayeva, V. V. *Radiolysis of Hydrocarbons in the Liquid Phase*; MGU: Moscow, 1986 (in Russian). (b) Bühler, R. E.; Katsumura, Y. *J. Phys. Chem. A* **1998**, *102*, 111.
4. (a) Johnston, G. W.; Park, J.; Satyapal, S.; Shafer, N.; Tsukiyama, K.; Bersohn, R.; Katz, B. *Acc. Chem. Res.* **1990**, *23*, 232. (b) Page, M.; Brenner, D. W. *J. Am. Chem. Soc.* **1991**, *113*, 3270. (c) Lu, D.; Maurice, D.; Truhlar, D. G. *J. Am. Chem. Soc.* **1990**, *112*, 6206. (d) Suzuki, H.; Nonoyama, N., *Chem. Comm.* **1996**, 1783. (e) Tsuji, M.; Komatsu, T.; Tanaka, M.; Nakamura, M.; Nishimura, Y.; Obase, H. *Chemistry Lett.* **1997**, 359. (f) Liguori, L.; Bjorsvik, H.-R.; Bravo, A.; Fontana, F.; Minischi, F. *Chem. Comm.* **1997**, 1501. (g) Mella, M.; Fagnoni, M.; Freccero, M.; Fasani, E.; Albini, A. *Chem. Soc. Rev.* **1998**, *27*, 81.
5. Kondratiev, V. N.; Nikitin, E. E. *Kinetics and Mechanism of Reactions in the Gase Phase*; Nauka: Moscow, 1975 (in Russian).
6. (a) Emanuel, N. M.; Denisov, E. T.; Maizus, Z. K. *Liquid Phase Oxidation of Hydrocarbons*; Plenum: New York, 1967. (b) Rudakov, E. S. *The Reactions of Alkanes with Oxidants, Metal Complexes, and Radicals in Solutions*; Naukova Dumka: Kiev, 1985 (in Russian).
7. Angelini, G.; Sparapani, C.; Speranza, M. *J. Chem. Soc., Perkin Trans. 2* **1988**, 1393.
8. (a) Gi, H. J.; Klabunde, K. J.; Pan, O.-G. *et al.* *J. Am. Chem. Soc.* **1989**, *111*, 8784. (b) Armstrong, B. M.; Zheng, F.; Shevlin, P. B. *J. Am. Chem. Soc.* **1998**, *120*, 6007. (c) Wada, S.; Obi, K. *J. Phys. Chem. A* **1998**, *102*, 3481. (d)

Brownsword, R. A.; Hillenkamp, M.; Schmiechen, P.; Volpp, H.-R.; Upadhyaya, H. P. *J. Phys. Chem. A* **1998**, *102*, 4438. (e) Arai, H.; Kato, S.; Koda, S. *J. Phys. Chem.* **1994**, *98*, 12. (f) Umemoto, H.; Kimura, Y.; Asai, T. *Bull. Chem. Soc. Jpn.* **1997**, *70*, 2951. (g) Kurosaki, Y.; Takayanagi, T.; Sato, K.; Tsunashima, S. *J. Phys. Chem. A* **1998**, *102*, 254.

9. (a) Roberts, B. P. *J. Chem. Soc., Perkin Trans. 2* **1996**, 2719. (b) Denisov, E. T. *Russ. Chem. Rev.* **1997**, *66*, 859. (c) Zavitsas, A. A. *J. Chem. Soc., Perkin Trans. 2* **1998**, 499. (d) Lee, W. T.; Masel, R. I. *J. Phys. Chem. A* **1998**, *102*, 2332. (e) Kamagai, J.; Kumada, T.; Kitagawa, N.; Morishita, N.; Miyazaki, T. *J. Phys. Chem. A* **1998**, *102*, 2842. (f) Rothenberg, G.; Sasson, Y. *Tetrahedron*, **1998**, *54*, 5417.

10. Semenov, N. N. *Some Problems in Chemical Kinetics and Reactivity*; Princeton Univ. Press: Princeton, New York, Vol 1, 1958, Vol. 2, 1959.

11. (a) Denisov, E. T. *Mendeleev Commun.* **1992**, 1. (b) Denisov, E. T. *Kinet. Katal.* **1994**, *35*, 671 (in Russian).

12. (a) Tanimoto, Y.; Uehara, M.; Takashima, M.; Itoh, M. *Bull. Chem. Soc. Jpn.* **1988**, *61*, 3121. (b) Nekhaev, A. I.; Bagriy, E. I.; Kuzmichev, A. V.; Ageev, V. P.; Konov, V. I.; Mikaya, A. I.; Zaikin, V. G. *Neftekhimiya* **1990**, *30*, 728 (in Russian). (c) Nekhaev, A. I.; Bagriy, E. I.; Kuzmichev, A. V.; Ageev, V. P.; Konov, V. I.; Mikaya, A. I.; Zaikin, V. G. *Mendeleev Commun.* **1991**, 18. (d) Boese, W. T.; Goldman, A. S. *Tetrahedron Lett.* **1992**, *33*, 2119.(e) Julliard, M. *New J. Chem.* **1994**, *18*, 243. (f) Pardo, L.; Banfelder, J. R.; Osman, R. *J. Am. Chem. Soc.* **1992**, *114*, 2382. (g) Lee, J.; Grabowski, J. J. *Chem. Rev.* **1992**, *92*, 1611. (h) Arnold, S. T.; Morris, R. A.; Viggiano, A. A. *J. Phys. Chem. A* **1998**, *102*, 1345.

13. Buchachenko, A. L. *Complexes of Radicals and Molecular Oxygen with Organic Molecules*; Nauka: Moscow, 1984 (in Russian).

14. (a) Treger, Yu. A.; Rozanov, V. N. *Usp. Khim.* **1989**, *58*, 138 (in Russian). (b) Lebedev, D. D. *Kinet. Katal.* **1988**, *29*, 1310 (in Russian). (b) Tanko, J. M.; Anderson, F. E., III. *J. Am. Chem. Soc.* **1988**, *110*, 3525. (c) Tanko, J. M.; Mas, R. H. *J. Org. Chem.* **1990**, *55*, 5145. (d) Seetula, J. A. *J. Chem. Soc., Faraday Trans.* **1998**, *94*, 891. (e) Kanderl, S. A.; Rakitzis, T. P.; Lev-On, T.; Zare, R. N. *J. Phys. Chem. A* **1998**, *102*, 2270. (f) Roberto-Neto, O.; Coitiño, E. L.; Truhlar, D. G. *J. Phys. Chem. A* **1998**, *102*, 4568.

15. (a) Benson S. W. *J. Am. Chem. Soc.* **1978**, *100*, 3214. (b) Benson S. W. *Chem. Eng.* **1982**, *89* No. 4, 19. (c) Weissmann, M.; Benson, S. W. *Int. J. Chem.*

Kinet. **1984**, *16*, 307. (d) Chambon, M.; Marquaire, P.-M.; Come, G.-M. *C_1 Mol. Chem.* **1987**, *2*, 47.

16. Averyanov, V. A.; Treger, Yu. A. *Khimich. Promysh.* **1988**, 266 (in Russian).

17. (a) Shul'pin, G. B.; Lederer, P.; Geletiy Yu. V. *J. Gen. Chem. USSR* **1987**, *57*, 543. (b) Shul'pin, G. B.; Nizova, G. V.; Geletiy Yu. V. *J. Gen. Chem. USSR* **1987**, *57*, 548.

18. (a) Baciocchi, E.; Crescenzi, M. *Tetrahedron*, **1988**, *44*, 6525. (b) Kikuchi, D.; Sakaguchi, S.; Ishii, Y. *J. Org. Chem.* **1998**, *63*, 6023. (c) Jones, C. W.; Carter, N. G.; Oakes, S. C.; Wilson, S. L.; Johnstone, A. *J. Chem. Technol. Biotechnol.* **1998**, *71*, 111. (d) Dastan, A.; Tahir, M. N.; Ülkü, D.; Shevlin, P. B.; Balci, M. *J. Org. Chem.* **1997**, *62*, 4018. (e) Schreiner, P. R.; Lauenstein, O.; Kolomitsyn, I. V.; Nadi, S.; Fokin, A. A. *Angew. Chem., Int. Ed. Engl.* **1998**, *37*, 1895. (f) Recupero, F.; Bravo, A.; Bjørsvik, H.-R.; Fontana, F.; Minisci, F.; Pirreda, M. *J. Chem. Soc., Perkin Trans. 2* **1997**, 2399.

19. (a) Davis, R.; Durrant, J. L. A.; Rowland, C. C. *J. Organometal. Chem.* **1986**, *316*, 147. (b) Golubeva, E. N.; Nevskaya, S. M.; Vorontsov, V. V.; Abdrashitov, Ya. M. *Russ. Chem. Bull.* **1997**, *46*, 1741.

20. (a) Atkinson, R. *Chem. Rev.* **1986**, *86*, 69. (b) Butkovskaya, N. I.; Setser, D. W. *J. Chem. Phys.* **1998**, *108*, 2434. (c) Aliagas, I.; Gronert, S. *J. Phys. Chem. A* **1998**, *102*, 2609. (d) Kramp, F.; Paulson, S. E. *J. Phys. Chem. A* **1998**, *102*, 2685. (e) Smith, I. W. M.; Sims, I. R.; Rowe, B. R. *Chem. Eur. J.* **1997**, *3*, 1925. (f) Donahue, N. M.; Anderson, J. G.; Demerjian, K. L. *J. Phys. Chem. A* **1998**, *102*, 3121.

21. (a) Ballod, A. P.; Shtern, V. Ya. *Usp. Khim.* **1976**, *45*, 1428 (in Russian). (b) Ballod, A. P.; Molchanova, S. I.; Prokhortseva, N. G.; Shtern, V. Ya. *Kinetika i Kataliz* **1986**, *27*, 7 (in Russian). (c) Kazakov, A. I.; Rubtsov, Yu. I.; Andrienko, L. P.; Manelis, G. B. *Russ. Chem. Bull.* **1997**, *46*, 1694. (d) Rubtsov, Yu. I.; Kazakov, A. I.; Rubtsova, E. Yu.; Andrienko, L. P.; Manelis, G. B. *Russ. Chem. Bull.* **1998**, *47*, 32. (e) Kazakov, A. I.; Rubtsov, Yu. I.; Kirpichev, E. P.; Manelis, G. B. *Russ. Chem. Bull.* **1998**, *47*, 1084.

22. (a) Baciocchi, E.; Del Giacco, T.; Sebastiani, G. V. *Tetrahedron Lett.* **1987**, *28*, 1941. (b) Ito, O.; Akiho, S.; Iino, M. *J. Org. Chem.* **1989**, *54*, 2436.

23. Bottle, S. E.; Chand, U.; Micallef, A. S. *Chemistry Lett.* **1997**, 857.

24. (a) Dobson, P. C.; Hayes, D. M.; Hoffman, *J. Am. Chem. Soc.* **1971**, *93,* 6188. (b) Bodor, N.; Dewar, M. J. S.; Wasson, J. S. *J. Am. Chem. Soc.* **1972**, *94*, 9095.

25. (a) Kirmse, W.; Schnitzler, D. *Tetrahedron Lett.* **1994**, *35*, 1699. (b) Murata, S.; Tsubone, Y.; Tomioka, H. *Chemistry Lett.* **1998**, 549. (c) Maslak, P. *J. Am. Chem. Soc.* **1989**, *111*, 8201.

26. Wang, Z.-X.; Huang, M.-B. *J. Chem. Soc., Faraday Trans.* **1998**, *94*, 635.

27. (a) *Photoinduced Electron Transfer*; Fox, M. A., Chanon, M., Eds.; Elsevier: New York, 1988. (b) Astruc, D. Electron Transfer and Radical Processes in Transition-Metal Chemistry; VCH: Cambridge, UK, 1995. (c) Freccero, M.; Pratt, A.; Albini, A.; Long, C. *J. Am. Chem. Soc.* **1998**, *120*, 284. (d) Bockman, T. M.; Hubib, S. M.; Kochi, J. K. *J. Am. Chem. Soc.* **1998**, *120*, 2826. (e) Burdeniuc, J.; Crabtree, R. H. *Organometallics* **1998**, *17*, 1582.

28. (a) Arutyunov, V. S.; Krylov, O. V. *Oxidative Conversion of Methane*; Nauka: Moscow, 1998 (in Russian). (b) Basevich, V. Ya. *Usp. Khim.* **1987**, *56*, 705 (in Russian). (c) Kim, H. M.; Enomoto, H.; Kato, H.; Tsue, M.; Kono,. M. *Combust. Sci. Technol.* **1997**, *128*, 197. (d) Gesser, H. D.; Hunter, N. R.; Prakash, C. B. *Chem. Rev.* **1985**, 85, 235. (e) Arutyunov, V. S.; Basevich, V. Ya.; Vedeneev, V. I *Usp. Khim.* **1996**, *65*, 211 (in Russian).

29. (a) *Free Radicals in Biology and Environment*; Minisci, F., Ed.; Kluwer Academic Publishers: Dordrecht, 1997. (b) *Active Oxygen in Chemistry*; Valentine, J. S., Foote, C. S.; Greenberg, A., Liebman, J. F., Eds.; Blakie Academic and Professional (Chapmann & Hall): New York, 1995. (c) Ghigo, G.; Tonachini, G. *J. Am. Chem. Soc.* **1998**, *120*, 6753.

30. (a) Heck, S. M.; Pritchard, H. O.; Griffiths, J. F. *J. Chem. Soc., Faraday Trans.* **1998**, *94*, 1725. (b) Chan, W.-T.; Hamilton, I. P.; Pritchard, H. O. *J. Chem. Soc., Faraday Trans.* **1998**, *94*, 2303. (c) Wilson, C.; Balint-Kurti, G. G. *J. Phys. Chem. A* **1998**, *102*, 1625. (d) Tyndall, G. S.; Wallington, T. J.; Ball, J. C. *J. Phys. Chem. A* **1998**, *102*, 2547. (e) Kaiser, E. W. *J. Phys. Chem. A* **1998**, *102*, 5903. (f) Abdel-Aal, H. K.; Shalabi, M. A.; Al-Harbi, D. K.; Hakeem, T. *Int. J. Hydrogen Energy* **1998**, *23*, 457.

31. (a) Newitt, D. M.; Bloch, A. M. *Proc. R. Soc. London, A* **1933**, *140*, 426. (b) Newitt, D. M. *Chem. Rev.* **1937**, *21*, 299. (c) Newitt, D. M.; Schmidt, W. G. *J. Chem. Soc.* **1937**, 1665. (d) Vedeneev, V. I.; Gol'denberg, M. Ya.; Gorban', N. I.; Teitel'boim, M. A. *Kinetika i Kataliz* **1988**, *29*, 7 (in Russian).

32. (a) Stern, V. Ya. *Gas Phase Oxidation of Hydrocarbons*; Pergamon Press: Oxford, 1964. (b) Skrumeda, L. L.; Ross, J. *J. Phys. Chem. A* **1995**, *99*, 12835. (c) Vedeneev, V. I.; Romanovich, L. B.; Basevich, V. Ya.; Arutyunov, V. S.; Sokolov, O. V.; Parfenov, Yu. V. *Izvest. Akad. Nauk, Ser. Khim.* **1997**, 2120 (in Russian).

33. (a) Bedzhanyan, Yu. P.; Gershenzon, Yu. M.; Minasyan, V. T.; Rozenshtein, V. B. *Kinetika i Kataliz* **1989**, *30*, 247 (in Russian). (b) Sukiasyan, A. A.; Khachatryan, L. A.; Mantashyan, V. T. *Kinetika i Kataliz* **1989**, *30*, 277 (in Russian).

34. (a) Haliwell, B.; Gutteridge, J. M. C. *Free Radicals in Biology and Medicine*; Clarendon Press: Oxford, 1989. (b) Lucarini, M.; Pedulli, G. F.; Valgimigli, L. *J. Org. Chem.* **1998**, *63*, 4497. (c) Walling, C. *J. Am. Chem. Soc.* **1969**, *91*, 7590. (d) Khar'kova, T. V.; Arest-Yakubovich, I. L.; Lipes, V. V. *Kinetika i Kataliz* **1989**, *30*, 954 (in Russian).

35. Korablev, L. I. *Neftekhimiya* **1989**, *29*, 684 (in Russian).

36. (a) Harustiak, M.; Hronec, M.; Ilavsky, J. *React. Kinet. Catal. Lett.* **1988**, *37*, 215. (b) Harustiak, M.; Hronec, M.; Ilavsky, J. *J. Mol. Catal.* **1988**, *48*, 335. (c) Opeyda, I. A.; Zalevskaya, N. M. *Neftekhimiya* **1989**, *29*, 244 (in Russian).

37. (a) Galimova, L. G.; Maslennikov, S. I.; Nikolaev, A. I. *Izvest. Akad. Nauk SSSR, Ser. Khim.* **1980**, 2464 (in Russian). (b) Pasternak, M.; Morduchowitz, A. *Tetrahedron Lett.* **1983**, *24*, 4275. (c) Sydnes, L. K.; Burkow, I. C.; Hansen, S. H. *Acta Chem. Scand.* **1985**, *39*, 829.

38. (a) Kryukov, A. I.; Sherstyuk, V. P.; Dilung, I. I. *Electron Photo-transfer and Its Applied Aspects*; Naukova Dumka: Kiev, 1982 (in Russian). (b) Albini, A; Spreti, S. *J. Chem. Soc.., Perkin Trans. 2.* **1987**, 1175. (c) Shul'pin, G. B.; Kats, M. M.; Nizova, G. V. *Bull. Acad. Sci. USSR, Div. Chem. Sci.* **1988**, 2395. (d) Shul'pin, G. B.; Kats, M. M.; Lederer, P. *J. Gen. Chem. USSR* **1989**, *59*, 2450. (e) Rontani J.-F.; Giusti, G. *J. Photochem. Photobiol. A: Chem.* **1990**, *53*, 69.

39. (a) Mella, M.; Freccero M.; Albini, A. *J. Chem. Soc., Chem. Commun.* **1995**, 41. (b) Nizova, G. V.; Shul'pin, G. B. *Russ. Chem. Bull.* **1995**, *44*, 1982. (c) D'Auria, M.; Ferri, T.; Mauriello, G. *Gazz. Chim. Ital.* **1996**, *126*, 313 (d) Negele, S.; Wieser, K.; Severin, T. *J. Org. Chem.* **1998**, *63*, 1138. (e) Dunkin, I. R.; McCluskey, A. *Bull. Chem. Soc. Jpn.* **1998**, *71*, 1397.

40. (a) Einhorn, C.; Einhorn, J.; Marcadal, C.; Pierre, J.-L. *Chem. Comm.* **1997**, 447. (b) Lin, M.; Sen, A. *J. Chem. Soc., Chem. Commun.* **1992**, 892.

41. (a) Giamalva, D. H.; Church D. F., Pryor, W. A. *J. Am. Chem. Soc.* **1986**, *108*, 7678. (b) Denisov, E. T.; Denisova, T. G. *Kinetika i Kataliz* **1996**, *37*, 51 (in Russian). (c) Suzuki, H.; Murashima, T.; Shimizu, K.; Tsukamoto, K. *Chem. Lett.* **1991**, 817. (d) Suzuki, H.; Nonoyama, N. *J. Chem. Soc., Perkin Trans. 1* **1997**, 2965. (e) Avzyanova, E. V.; Timergazin, Q. K.; Khalizov, A. F.; Khursan, S. L.; Spirikhin, L. V.; Shereshovets, V. V. *Mendeleev Commun.* **1997**, 227.

42. (a) Kats, M. M.; Kozlov, Yu. N.; Shul'pin, G. B. *J. Gen. Chem. USSR.* **1991**, *61*, 1694. (b) Kozlov, Yu. N.; Moravskiy, A. P.; Uskov, A. M.; Shilov, A. E. *Kinetika i Kataliz.* **1990**, *31*, 251 (in Russian).

43. Bach, R. D.; Andrés, J. L.; Su, M.-D.; McDouall, J .J. W. *J. Am. Chem. Soc.* **1993**, *115*, 5768.

44. (a) Dodonov, V. A.; Stepovik, L. P.; Soskova, A. S.; Zaburdaeva, E. A. *Zhurn. Obsch. Khim.* **1994**, *64*, 1713 (in Russian). (b) Stepovik, L. P., Zaburdaeva, E. A.; Razuvayeva, E. A. *Zhurn. Obsch. Khim.* **1996**, *66*, 1242 (in Russian). (c) Stepovik, L. P.; Dodonov, V. A.; Zaburdaeva, E. A. *Zhurn. Obsch. Khim.* **1997**, *67*, 116 (in Russian). (d) Dodonov, V. A.; Zaburdaeva, E. A.; Dolganova, N. V.; Stepovik, L. P *Zhurn. Obsch. Khim.* **1997**, *67*, 988 (in Russian).

45. (a) Fossey, J.; Lefort, D.; Massoudi, M.; Nedelec, J.-Y.; Sobra, J. *Can. J. Chem.* **1985**, *63*, 678. (b) Bravo, A.; Bjorsvik, H.-R.; Fontana, F.; Minisci, F.; Serri, A. *J. Org. Chem.* **1996**, *61*, 9409. (c) Deno, N. C.; Jedziniak, E. J.; Messer, L. A.; Meyer, M. D.; Stroud, S. G.; Tomezsko, E. S. *Tetrahedron*, **1977**, *33*, 2503. (d) Safiullin, R. L.; Zaripov, R. N.; Komissarov, V. D.; Tolstikov, G. A. *Izvest. Akad. Nauk SSSR, Ser. Khim.* **1989**, 973 (in Russian).

46. (a) Adam, W.; Curci, R.; Edwards, J. O. *Acc. Chem. Res.* **1989**, *22*, 205. (b) Murray, R. W. *Chem. Rev.* **1989**, *89*, 1187. (c) Curci, R.; Dinoi, A.; Rubino. M. F. *Pure Appl. Chem.* **1995**, *67*, 811.

47. (a) Bravo, A.; Fontana, F.; Fronza, G.; Mele, A.; Minisci, F. *J. Chem. Soc., Chem. Commun.* **1995**, 1573. (b) Asensio, G.; Mello, R.; González-Núñez, M. E.; Castellano, G.; Corral, J. *Angew. Chem., Int. Ed. Engl.* **1996**, *35*, 217. (c) Curci, R.; Dinoi, A.; Fusco, C.; Lillo, M. A. *Tetrahedron Lett.* **1996**, *37*, 249. (d) Asensio, G.; Mello, R.; González-Núñez, M. E.; Boix, C.; Royo, J. *Tetrahedron Lett.* **1997**, *38*, 2373. (e) Adam, W. Curci, R.; D'Accolti, L. *et al.* *Chem. Eur. J.* **1997**, *3*, 105. (f) Yang, D.; Wong, M.-K.; Wang, X.-C.; Tang, Y.-C. *J. Am. Chem. Soc.* **1998**, *120*, 6611. (g) Bovicelli, P.; Lupattelli, P.; Fiorini, V.; Mincione, E. *Tetrahedron Lett.* **1993**, *38*, 6103.

48. (a) Bravo, A.; Fontana, F.; Fronza, G.; Minisci, F.; Zhao, L. *J. Org. Chem.* **1998**, *63*, 254. (b) Simakov, P. A.; Choi, S.-Y.; Newcomb, M. *Tetrahedron Lett.* **1998**, *39*, 8187. (c) Du, X.; Houk, K. N. *J. Org. Chem.* **1998**, *63*, 6480. (d) Shustov, G. V.; Rauk, A. *J. Org. Chem.* **1998**, *63*, 5413. (e) Glukhovtsev, M. N.; Canepa, C.; Bach, R. D. *J. Am. Chem. Soc.* **1998**, *120*, 10528.

49. DesMarteau, D. D.; Donadelli, A.; Montanari, V.; Petrov, V. A., Resnati, G. *J. Am. Chem. Soc.* **1993**, *115*, 4897.

50. (a) Kato, S.; Iwahama, T.; Sakaguchi, S.; Ishii, Y. *J. Org. Chem.* **1998**, *63*, 222. (b) Chen, D.-L.; Lei, Z.-L.; Liu, W.-Y.; Wang, Z.; Hong, P.-J.; Dai, S.-S. *J. Nat. Gas Chem.* **1998**, *7*, 80.

51. (a) Fokin, A. A.; Schreiner, P. R.; Schleyer, P. von R.; Gunchenko, P. A. *J. Org. Chem.* **1998**, *63*, 6494. (b) Olah, G. A.; Hartz, N.; Rasul, G.; Prakash G. K. S. *J. Am. Chem. Soc.* **1995**, *117*, 1336.

52. (a) Sommer, J.; Bukala, J. *Acc. Chem. Res.* **1993**, *26*, 370. (b) Olah, G. A.; Molnar, A. *Hydrocarbon Chemistry*; Wiley: Chichester, 1995. (c) Olah, G. A. *Angew. Chem., Int. Ed. Engl.* **1995**, *34*, 1393.

53. (a) Schleyer, P. von R.; Carnieiro, J. de M. *J. Comput. Chem.* **1992**, *13*, 997. (b) Marx, D.; Savin, A. *Angew. Chem., Int. Ed. Engl.* **1997**, *36*, 2077. (c) Schreiner, P. R.; Kim, S.-J.; Shaefer, H. F.; Schleyer, P. von R. *J. Chem. Phys.* **1993**, *99*, 3716. (d) Olah, G. A.; Rasul, G. *J. Am. Chem. Soc.* **1996**, *118*, 12922. (e) Olah, G. A.; Rasul, G. *J. Am. Chem. Soc.* **1996**, *118*, 8503. (f) Esteves, P. M.; Mota, C. J. A.; Ramirez-Solis, A.; Hernandez-Lamoneda, R. *J. Am. Chem. Soc.* **1998**, *120*, 3213. (g) Esteves, P. M.; Mota, C. J. A.; Ramirez-Solis, A.; Hernandez-Lamoneda, R. *Topics Catal.* **1998**, *6*, 163.

54. (a) Olah, G. A. In *Activation and Functionalization of Alkanes*, Hill, C. L., Ed., Wiley: New York, 1989, Chap. 3. (b) Sommer, J.; Bukala, J.; Rouba, S.; Graff, R. *J. Am. Chem. Soc.* **1992**, *114*, 5884. (c) Bagno, A.; Bukala, J.; Olah, G. A. *J. Org. Chem.* **1990**, *55*, 4284. (d) Lazzeri, V.; Jalal, R.; Poinas, R.; Gallo, R. *New J. Chem.* **1992**, *16*, 521. (e) Garni, M.; Lazzeri, V.; Deschamps, P.; Gallo, R. *J. Mol. Catal. A: Chem.* **1997**, *125*, 33. (f) del Rio, E.; Lopez, R.; Sordo, T. L. *J. Phys. Chem. A* **1998**, *102*, 6831.

55. (a) Berrier, C.; Jacquesy, J.-C.; Jouannetaud, M.-P.; Lafitte, C.; Vidal, Y.; Zunino, F.; Fahy, J.; Duflos, A. *Tetrahedron* **1998**, *54*, 13761. (b) Prakash, G. K. S.; Krass, N.; Wang, Q.; Olah, G. A. *SYNLETT* **1991**, 39.

56. (a) Vol'pin, M. E.; Akhrem, I. S.; Orlinkov, A. V. *New J. Chem.* **1989**, *13*, 771. (b) Akhrem, I. S.; Orlinkov, A. V.; Vol'pin, M. E. *Usp. Khim.* **1996**, *65*, 1120 (in Russian). (c) Akhrem, I. S.; Orlinkov, A. V. *Izvest. Akad. Nauk, Ser. Khim.* **1998**, 740 (in Russian). (d) Akhrem, I. *Topics Catal.* **1998**, *6*, 27.

57. (a) Orlinkov, A. V.; Akhrem, I. S.; Afanas'eva, L. V.; Mysov, E. I.; Vol'pin, M. E. *Mendeleev Commun.* **1997**, 61. (b) Akhrem, I. S.; Chistyakov, A. L.; Gambaryan, N. P.; Stankevich, I. V.; Vol'pin, M. E. *J. Organometal. Chem.* **1997**, *536-537*, 489. (c) Akhrem, I. S.; Churilova, I. M.; Orlinkov, A. V.; Afanas'eva, L. V.; Vitt, S. V.; Petrovskii, P. V. *Izvest. Akad. Nauk, Ser. Khim.* **1998**, 918 (in Russian).

58. Cheung, T.-K.; Gates, B. C. *CHEMTECH* September **1997**, 28.

59. Akhrem, I.; Orlinkov, A.; Vitt, S. *Inorg. Chim. Acta* **1998**, *280*, 355.

CHAPTER III

HETEROGENEOUS HYDROCARBON REACTIONS WITH PARTICIPATION OF SOLID METALS AND METAL OXIDES

Transformations of hydrocarbons promoted by solid metals and their oxides play very important roles in chemical industry [1]. Heterogeneous metal-containing catalysts [2] are widely employed for cracking (see, e.g., [3]) of oil fractions, oxidation, dehydrogenation, isomerization and many other processes of saturated as well as alkylaromatic hydrocarbons.

Examples of these processes are depicted in Scheme III.1. Usually such reactions occur only at high temperature (> 200 °C); microwave irradiation [4a], and electrons [4b] can induce catalyzed processes.

It is necessary to note that many hydrocarbon transformations on heterogeneous catalysts proceed with intermediate formation of the same species; although, they give rise to different products depending on the nature of the catalyst or conditions of the process. For example, dehydrogenation and isomerization of alkanes involve stages of the formation of coordinated to the metal surfaces species in the case of catalysis by metals (see below, Figure III.3). Carbenium ions are formed if a metal oxide is used as a catalyst.

Further transformations of these intermediates lead to either olefins (dehydrogenation) or branched alkanes (isomerization). However, usually heterogeneous processes are not highly selective and afford simultaneously products of dehydrogenation and isomerization (as well as of other reactions, e.g., cracking and condensation).

In this chapter, we will consider only the general characteristics of these reactions, which are especially important from the point of view of their comparison with metal complex activation.

Cracking:

$$n\text{-}C_7H_{16} \longrightarrow CH_4 + C_2H_6 + \ldots + C_6H_6 + C_6H_5CH_3 + H_2$$

Hydrogenolysis (hydrocracking):

$$n\text{-}C_7H_{16} + H_2 \longrightarrow CH_4 + C_2H_6 + \ldots + C_6H_6 + C_6H_5CH_3$$

Isomerization:

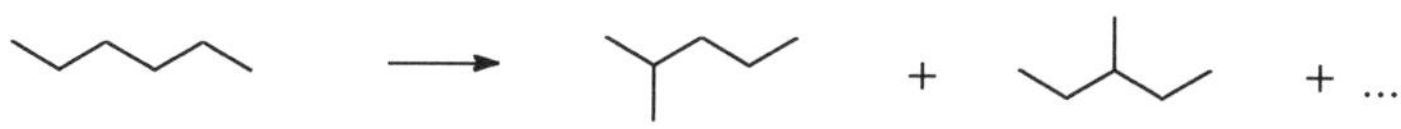

Dehydrogenation:

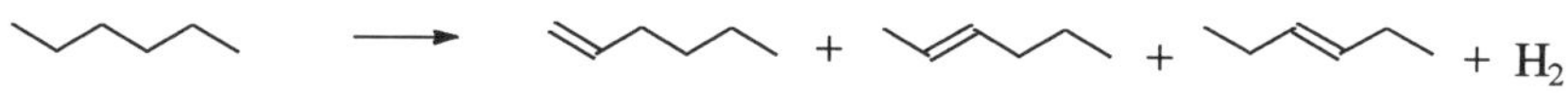

Dehydrocyclization:

$$n\text{-}C_7H_{16} \longrightarrow C_6H_5CH_3 + H_2 \qquad CH_4 \longrightarrow C_6H_6$$

Oxidative coupling:

$$CH_4 \longrightarrow CH_2{=}CH_2$$

Condensation:

$$CH_4 \longrightarrow C_2H_6 + C_3H_8 + H_2$$

Oxidation:

$$CH_4 \xrightarrow{O_2} CO_2 + H_2O$$

$$CH_4 \xrightarrow{O_2} CH_3OH + HCHO + HCOOH$$

Scheme III.1. Some examples of processes catalyzed by metals, metal oxides and other heterogeneous catalysts.

III.1. Mechanisms of the Interaction between Alkanes and Catalyst Surfaces

The mechanism of heterogeneous catalytic reactions is in general analogous to the reactions in homogeneous solutions, and that metal complexes behave in a similar way on surfaces and in solution. Metal atoms or ions on surfaces can be looked upon as analogous to metal atoms or ions surrounded by certain ligands. In the case of metal surfaces, ligands are other metal atoms surrounding a given atom. Chemisorption may be looked upon as a complex formation on the surface. In those cases when a combined participation of several metal atoms is essential for the process, clusters can serve as homogeneous analogs [5]. The advantage of metals and their simple oxides over complexes in solutions for the special case of alkanes consists of the possibility of using elevated temperatures, which are essential for those processes having comparatively high activation energy. In this case there is no such problem as C–H bonds contained in solvents or ligands, which can be effective competitors with alkanes.

The first stages of any heterogeneous catalytic process is chemisorption of the substrate on the catalyst surface followed by splitting the C–H bonds. A large number of experimental and theoretical works have been devoted to the inves-

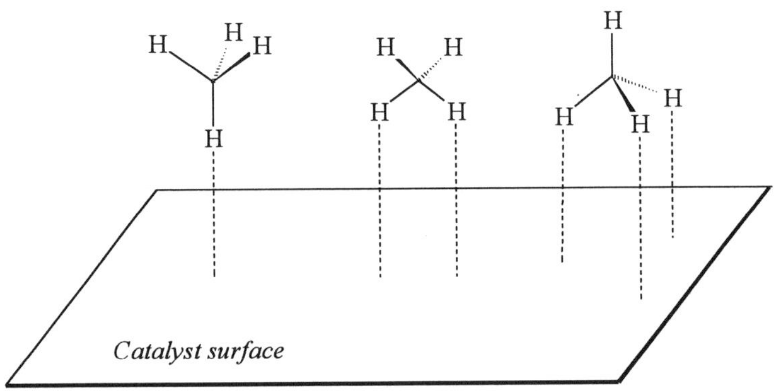

Figure III.1. Modes of coordination of methane molecule with the metal catalyst surface.

tigation of the interaction between metal surfaces with alkanes, often with the simplest alkane, methane [6]. Thus methane interacts with the metal surface with participation of weak three-center bonds C–H–M via one, two or three hydrogen atoms involved in the coordination (Figure III.1).

In the calculations of methane interaction with nickel or titanium surfaces, five different modes of coordination have been considered (Figure III.2) [6c]. It has been concluded that acts of methane activation by a separate molecule of a metal complex on the one hand and the metal surface on the another hand are very similar. In both processes the C–H bonds of methane are ruptured and new M–H bonds are formed. Intermediate species formed during alkane chemisorption on oxide catalysts have been investigated both by MO calculations and spectroscopic methods [7].

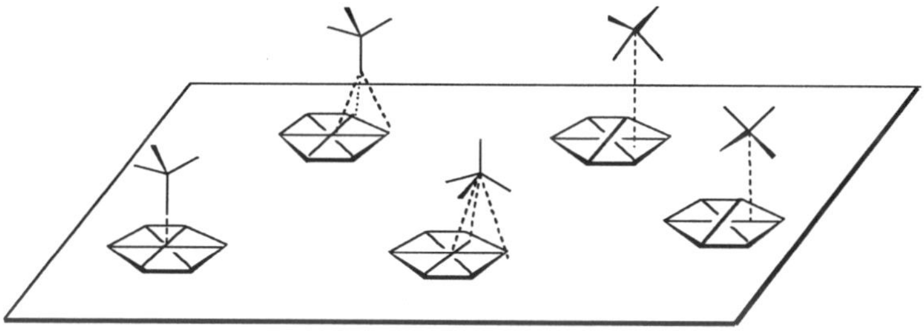

Figure III.2. Possible modes of coordination of methane molecule with the surface of nickel or titanium.

III.2. ISOTOPE EXCHANGE

Metal surfaces are capable of inducing isotope exchange [8a], with aromatic and alkylaromatic hydrocarbons being more reactive than alkanes. Metal films and wires were used as catalysts. The isotope exchange occurs even at room temperature, though it is generally studied under heating conditions. The highest reactivity is exhibited by the arene hydrogens in *meta* and *para* positions to a substituent. Low reactivity of *ortho*-hydrogen is due to the steric restrictions. The most reactive atoms in the side chain of alkylaromatics are benzylic hydrogens. The most active among the alkanes are cycloalkanes, cyclopentane being usually somewhat more reactive than cyclohexane. The open-chain alkanes, starting from ethane onwards, have similar reactivities, higher hydrocarbons being slightly more reactive than ethane. Neopentane and particularly methane are the least reactive. For the hydrogen–deuterium exchange of propane with D_2, the following order of metal reactivities was found:

$$Pt > Rh > Ir > Ru > Pd$$

which coincides with that found for hydrocracking. However, depending on the method of preparation and reaction conditions, catalyst activity and order of reactivity for various metals may vary significantly.

The exchange reaction generally proceeds via the cleavage of the C–H bond (dissociative mechanism) and the formation of an alkyl group bound to the surface metal atom, C–H cleavage being probably the most difficult stage. Possible types of surface intermediates [8b] are shown in Figure III.3. The dissociation of the hydrogen molecule normally occurs much more easily than that of hydrocarbons. This assertion is supported by the isotope effect (the usual rate of exchange between RH and D_2 is markedly higher than between RD and H_2), and the largest isotope effect among hydrocarbons is usually observed in the case of alkanes. A very important feature of the heterogeneous isotope exchange of hydrocarbons with molecular deuterium of deuterated water on metal surfaces is the ability of many catalysts to carry out the exchange of several hydrogen atoms in a hydrocarbon molecule within one contact with the catalyst center (a so-called multiple exchange). This type of exchange is characterized by a multiple exchange factor (M) denoting the average number of hydrogen atoms exchanged within one contact. To differentiate between multiple exchange and isotope exchange where $M = 1$ the latter is called stepwise exchange.

Multiple exchange proceeds most readily in the presence of β-C–H bonds, which can be explained by β-elimination leading to an olefin-hydrido complex either with a single metal atom or with the participation of a neighboring metal atom. The importance of intermediate complex formation with an olefin is confirmed by the absence of exchange via a quaternary carbon atom. Cyclopentane in the isotope exchange with molecular deuterium catalyzed by metals of the Group VIII elements provides a large amount of d_5-molecules with low d_2-content. That means that hydrogen atoms of one side of the ring are primarily exchanged in the chemisorption of this molecule.

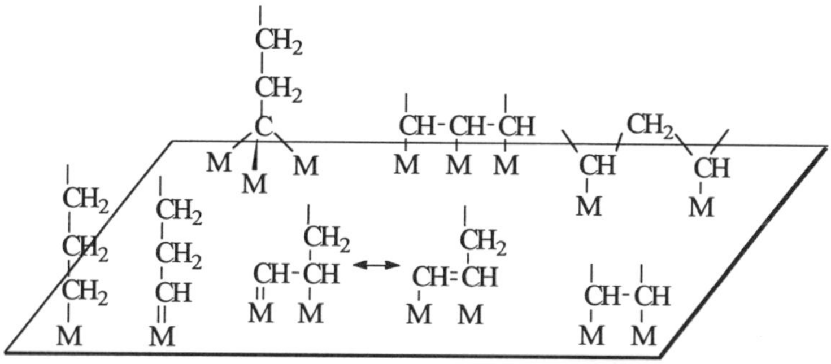

Figure III.3. Types of intermediates on the metal surface.

For various metals the factor *M* may vary greatly. For example, only small quantities of heavily exchanged cycloalkanes are produced on rhodium whereas, in the exchange on palladium, a great amount of cycloalkanes with strong multiple exchange is formed. Here along with the maximum on cyclopentane-d_5, the second maximum for completely exchanged cyclopentane-d_{10} is also observed. When increasing the temperature, this second maximum increases and becomes prevalent according to which the cyclopentane molecule turns over, without losing the link with the surface catalyst atoms [8c]. The formation of π-bound olefin is not the only path leading to the multiple exchange of hydrocarbon, as is clear in view of methane multiple exchange, where such a path is impossible. The stepwise exchange (resulting in methane-d_1) and multiple exchange have been shown to proceed via different mechanisms, supported by some kinetic data. Methane exchange on transition metals is characterized by negative order with respect to molecular deuterium, the multiple exchange being more strongly inhibited by D_2 than the stepwise one. It is assumed that the subsequent dissociation of the surface methyl group to form a chemisorbed carbene is necessary for the stepwise exchange to take place. The activation energy for the multiple exchange is higher than for the stepwise one, i.e., methyl dissociation evidently requires greater energy than does methyl formation from methane. Upon the formation of adsorbed carbene $(CH_2)_{ads}$, a rapid exchange to produce $(CD_2)_{ads}$ occurs leading to the appearance of $(CD_3)_{ads}$ and free CD_4, which explains the maximum formation of completely deuterated methane.

Metal oxides (Cr_2O_3, Al_2O_3, NiO, Co_3O_4, ZnO, Fe_2O_3 etc.) catalyze H–D exchange between alkanes and D_2, as well as between alkanes and deuterated alkanes. Zeolites are also appropriate catalysts of isotope exchange (see, e.g., [9]). Cycloalkanes exchange more readily than linear alkanes. Thus, cyclopropane exchanges even at room temperature. Molecules containing one deuterium atom prevail in the initial stages of the reaction. Benzene is also exchanged with D_2, the reaction proceeds readily at 80 °C.

Active centers which catalyze the exchange of various types of hydrocarbons and other reactions on the surface of chromium oxide can be different: the hydrogen-deuterium exchange of benzene and hydrogenation of cyclohexene do not inhibit each other, hence these reactions might be thought to be catalyzed by different centers. When toluene reacts with molecular deuterium on chromium oxide, the hydrogen atoms of the ring are exchanged about one order of magnitude faster than those of the methyl group. For alkanes, the H–D exchange with D_2 was observed to proceed somewhat faster for primary than for secondary carbon-hydrogen bonds, which can be explained by a carbanion character of the intermediate species formed on the oxide surface. The possibility of steric hindrance should be also taken into account, being of minimum significance for primary bonds.

The system naphthalene–sodium is able to catalyze the hydrogen isotope exchange at room temperature in aromatic hydrocarbons and methane in tetrahydrofuran [10a]. The authors assumed that atomic sodium can aggregate to form clusters:

$$Na_{met} + ArH\bullet^{-}Na^{+} \rightleftharpoons Na_{at} + Na^{+}ArH\bullet^{-}$$

$$Na^{+}ArH\bullet^{-} + ArH\bullet^{-}Na^{+} \rightleftharpoons ArH + Na^{+}ArH\bullet^{-} + Na_{at}$$

$$Na^{+}ArH\bullet^{-} \rightleftharpoons ArH + Na_{at}$$

These clusters are stabilized by the formation of surface complexes with naphthalene. The hydrogen-deuterium exchange in hydrocarbon, RH, takes place apparently by reversible cleavage of the hydrocarbon C–H bonds with formation of surface groups Na_{surf}–Ar, Na_{surf}–H, Na_{surf}–R, and Na_{surf}–D. The hydrogen–deuterium exchange reactions at room temperature can be also catalyzed by the adducts of active carbons with metallic potassium [10b]. The redistribution of hydrogen isotopes in CD_3H, $C_6H_5CD_3$ and $C_6D_5CH_3$ as well as the similar isotope exchange reactions in the systems C_2H_4–C_2D_4, C_6H_6–C_6D_6, n-C_6H_{14}–n-C_6D_{14} and CD_4–n-C_5H_{12} has been investigated.

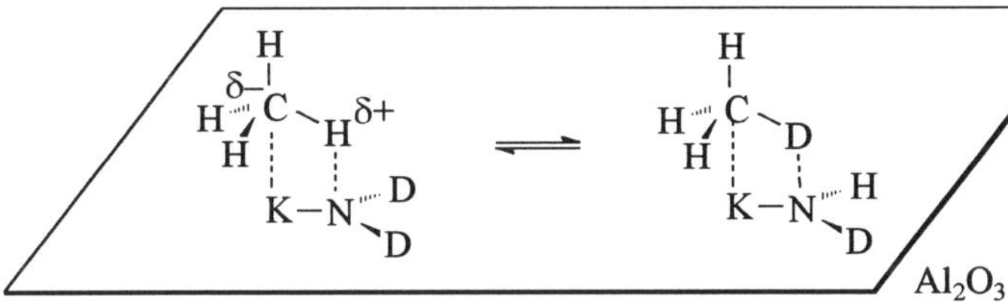

Figure III.4. The schematic mechanism of the H–D exchange between CH_4 and KND_2/Al_2O_3.

It has been also demonstrated recently that the H–D exchange reaction between methane and deuterated potassium amide supported on alumina (KND_2/Al_2O_3) proceeds at room temperature [10c]. The proposed mechanism is shown schematically in Figure III.4.

III.3. ISOMERIZATION

Heterogeneous catalysts can under heating induce skeletal isomerization [11] of saturated hydrocarbons which proceeds with rupture of the C–C bonds in the hydrocarbon chain. Metals [12], metal oxides [13], and metal oxides promoted with metals (see certain recent papers [5, 14] and patents [15]) are used as catalysts for such transformations. The process can be carried out in the presence of molecular hydrogen (hydroisomerization).

Isomerization of alkanes on metal surfaces occurs via the intermediate formation of α,γ-adsorbed, metallacyclobutane, carbene and π-bounded species (Scheme III.2).

The reaction mechanism of isomerization on bifunctional zeolite-based catalysts involves the activation of hydrocarbon molecule via dehydrogenation followed by protonation of the double bond to generate alkylcarbenium ions. The mechanism of the hydroisomerization of *n*-hexane consists of the following stages (Scheme III.3) [5]. The platinum metal serves to equilibrate hexane and hexene (stage *a*). The olefin is protonated by the zeolitic protons (step *b*) and the carbenium ion formed can isomerize (stage *c*). Dehydrogenation of isomerized carbenium ion gives rise to the formation of isomerized olefin (stage *d*) which is further hydrogenated over platinum to branched hexane (stage *e*).

The following expression for the rate of hexane isomerization has been deduced, taking into account that at high H_2/hexane ratio and large Pt/H^+ ratio, equilibration by platinum is fast and the rate-limiting step of the reaction is the rate of isomerization of protonated hexene:

$$\frac{d\,[\text{isohexane}]}{d\,t} = k_{iso} \times \frac{K_{ads}{}^{p} \cdot K_{eq} \cdot ([\text{hexane}]/[H_2])}{1 + K_{ads} \cdot K_{eq} \cdot ([\text{hexane}]/[H_2])} =$$

$$= \left\{ k_{iso} \cdot K_{ads}{}^{p} \cdot K_{eq} \cdot \frac{[\text{hexane}]}{[H_2]} \right\}^{n} \qquad 0 < n < 1$$

Here k_{iso} is the rate of isomerization of protonated hexane, $K_{ads}{}^{p}$ is the equilibrium constant of protonation of hexene and K_{eq} is the equilibrium constant of the hexane/hexene equlibrium.

Examples of mechanisms of skeletal isomerization of alicyclic alkylcarbenium ions are presented in Scheme III.4 [14k].

Heterogeneous metal oxides also catalyze the isomerization of alkylbenzenes, for example, under the action of H-ZSM-5 zeolite at 723 K 1,2,4-trimethylbenzene is transformed into 1,2,3- and 1,3,5-trymethylbenzenes [16].

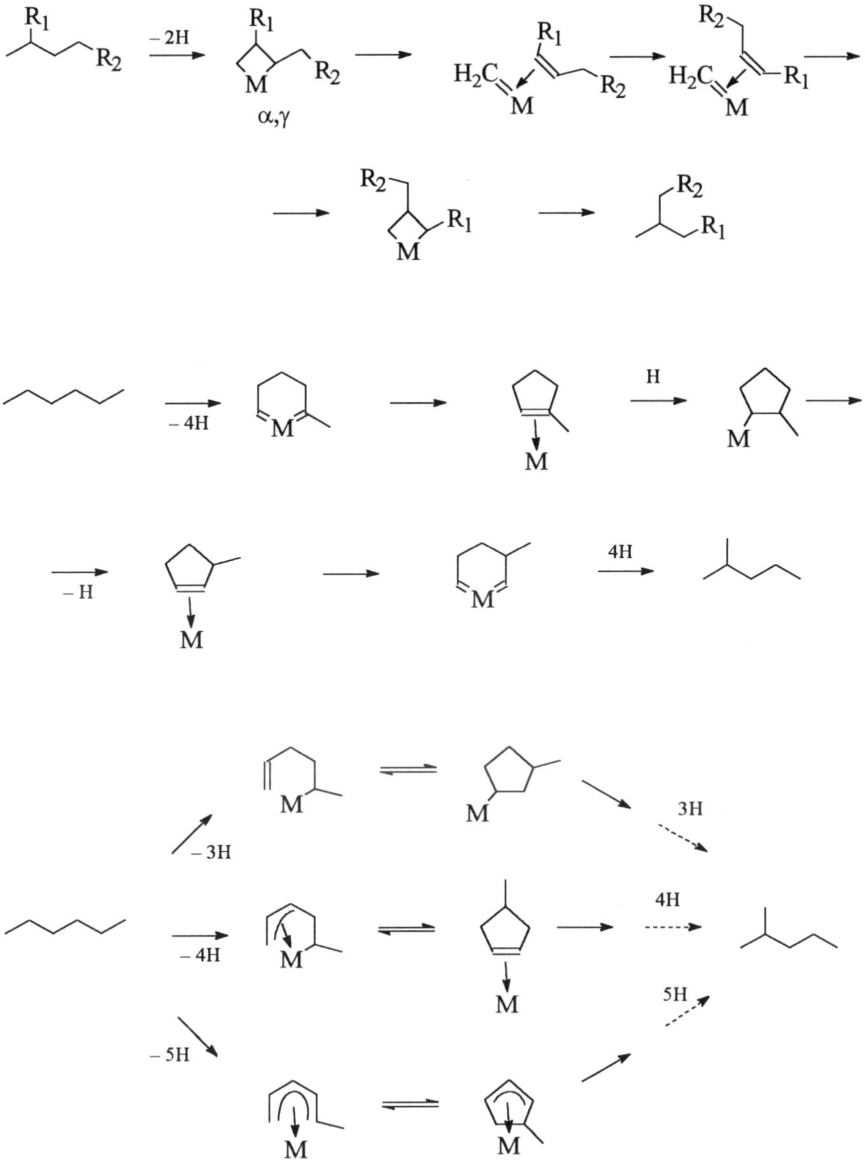

Scheme III.2. Mechanisms of isomerization on the metal, M, surface.

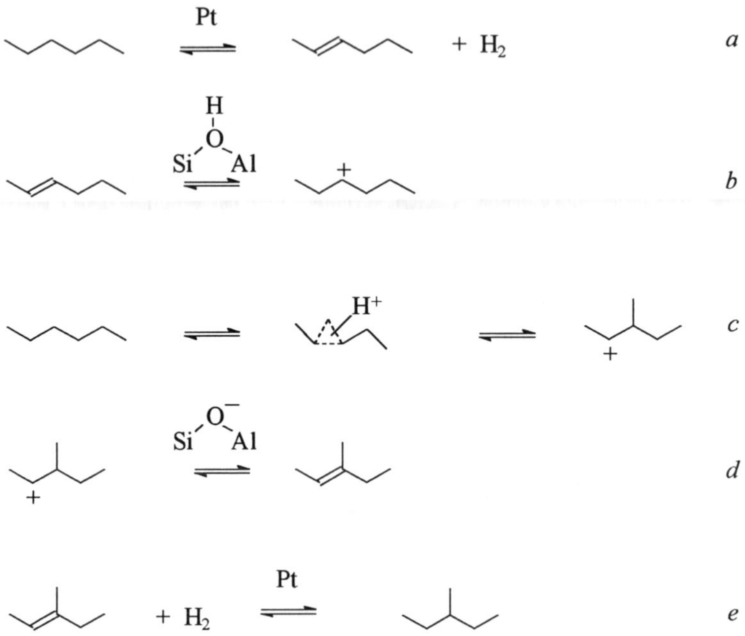

Scheme III.3. Steps of the bifunctional hydroisomerization of *n*-hexane.

III.4. DEHYDROGENATION AND DEHYDROCYCLIZATION

Catalytic (nonoxidative and oxidative) dehydrogenation is an important industrial method of hydrocarbon processing (for example, butene and butadiene production from butane [17a] and transformation of ethylbenzene to styrene [17b–f]), in view of which a great number of papers have been devoted to catalytic dehydrogenation (see, e.g., reviews and monographs [18], recent original publications [19] and patents [20]).

The mechanism of dehydrogenation and dehydrocyclyzation of alkanes on heterogeneous catalysts is connected with that of isomerization. Thus a study of the initial stages of propane activation over H-ZSM-5 zeolite by NMR, demonstrated that propane is protonated on the strong Brønsted sites of this zeolite to form ion type transition states.

Exocyclic alkyl shift

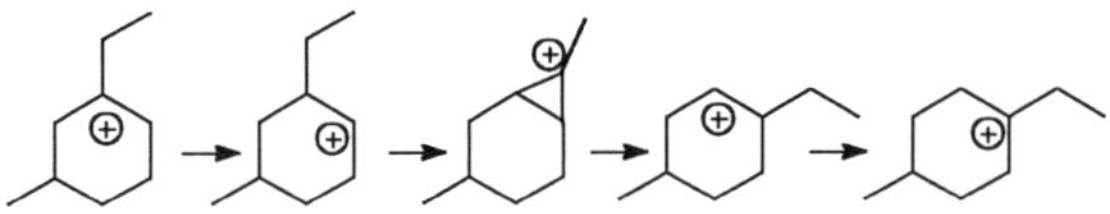

Endocyclic alkyl shift:

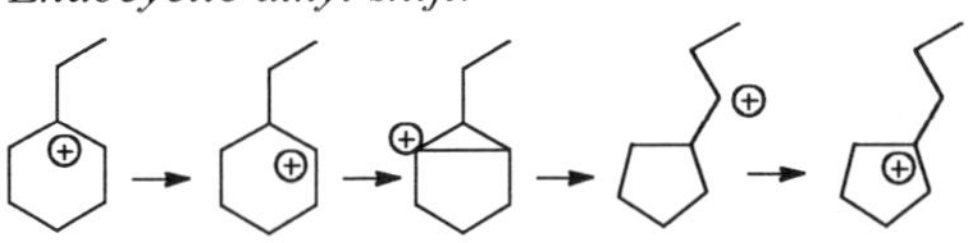

Ring branching:

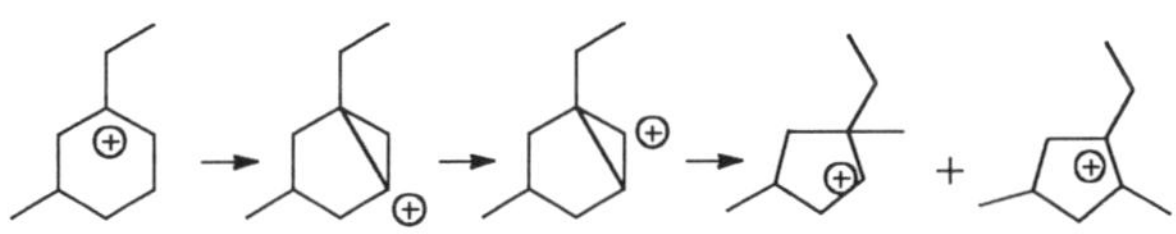

Side chain branching:

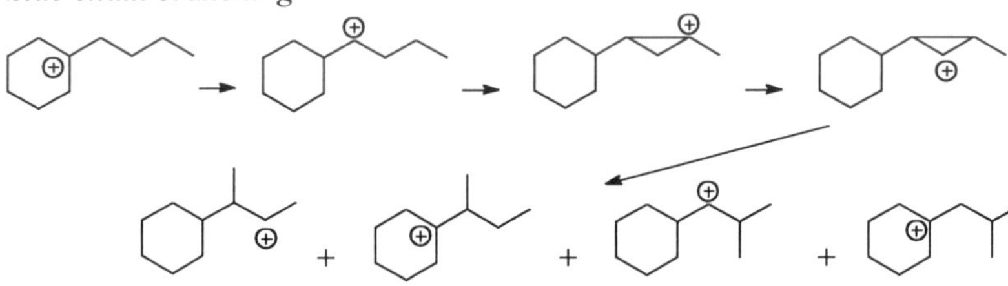

Alkyl shift in side chain:

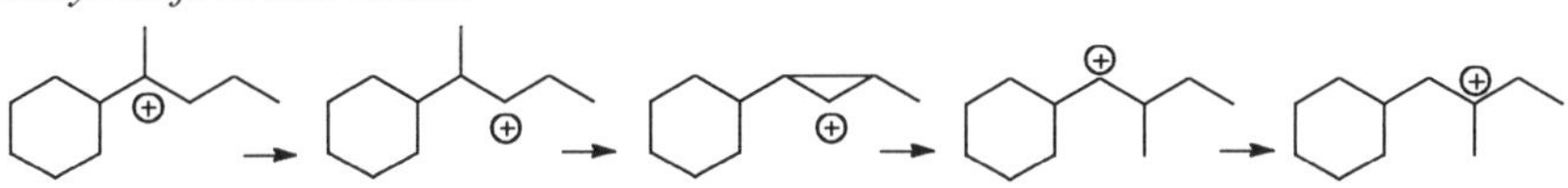

Scheme III.4. Mechanism of the bifunctional conversion of alkyl-cycloalkanes.

Further transformations include ^{13}C scrambling in propane, cracking, dehydrogenation and disproportionation [21a]. Iron- and manganese-promoted sulfated zirconia are assumed to be capable of protonating light alkanes to give transition states at temperatures as low as 200 °C. Although conventional carbenium ion reactions predominate at high conversions, proton-donation reactions are important at low alkane conversions [21b].

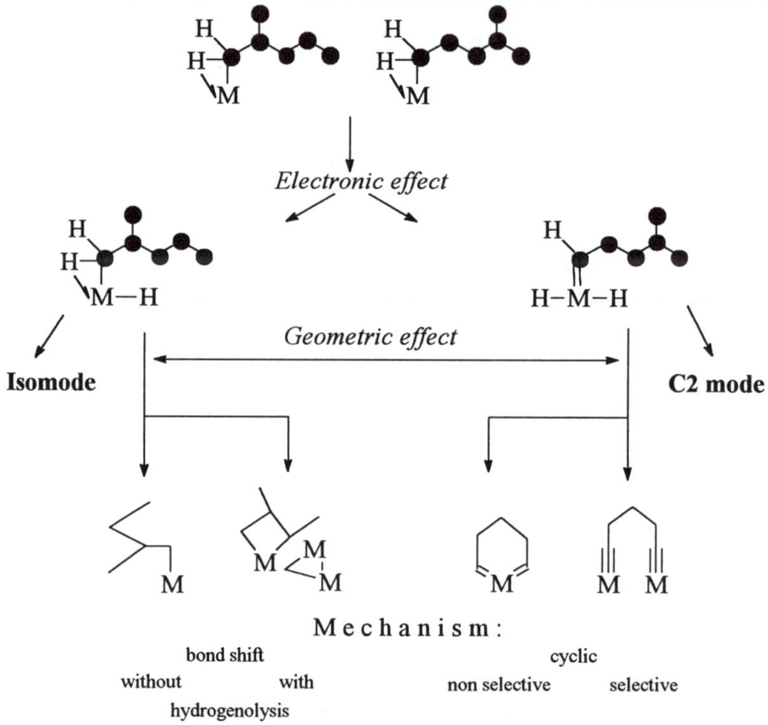

Scheme III.5. General mechanism for isomerization and hydrogenolysis of methylpentanes. Arrows " ⟶ " denote agostic bonds.

If an alkane contains a long chain of CH_2 groups, under the action of heterogeneous catalysts the transformation of this saturated hydrocarbon can be converted to cyclic or even aromatic hydrocarbon (dehydrocyclization) (books and reviews [22], certain papers [23]). Reactions of alkanes containing six and more hydrogen atoms occur via intermediate formation of olefins: alkane → olefin → diene → triene → alkylcyclohexadiene → aromatic hydrocarbon.

III.5. HYDROGENOLYSIS

Heating alkanes with molecular hydrogen in the presence of heterogeneous catalysts gives not only products of isomerization but also saturated hydrocarbons with lower molecular weights (hydrocracking or hydrogenolysis) (see recent books [24], papers [25] and patents [26]). The reactivity of 2-methylpentane on the surface of WC has been studied [27]. The various hydrogenolysis mechanisms have been interpreted by different reaction intermediates deduced from concepts of organometallic chemistry.

Two reaction mechanisms have been proposed to explain the single hydrogenolysis (the isomode and C2 mode according to Anderson) (Scheme III.5). The hydrogenolysis by isomode (Scheme III.6) is attributed to tungstenacylobutane intermediates and the C2 mode (Scheme III.7) of hydrogenolysis to the interconversion of dicarbene into dicarbyne species. The isomode mechanism is selectively poisoned by carbonaceous residues.

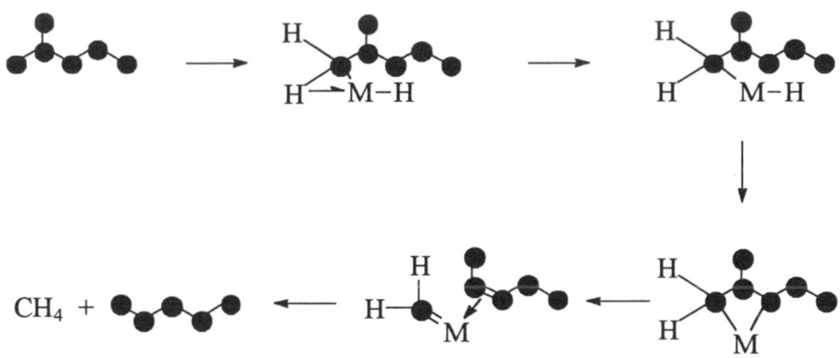

Scheme III.6. Single hydrogenolysis of 2-methylpentane leading via isomode to *n*-pentane.

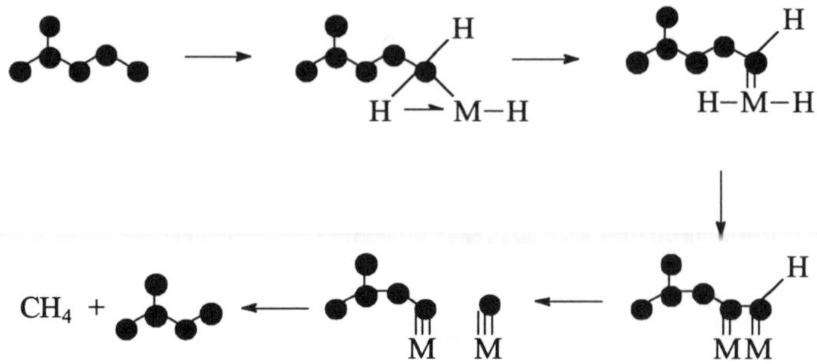

Scheme III.7. Single hydrogenolysis of 2-methylpentane leading via C2 mode to isopentane.

III.6. HETEROGENEOUS OXIDATION

Both metals and their oxides catalyze oxidation of hydrocarbons with molecular oxygen (see books [28], reviews [29], certain original papers [30] and recent patents [31]). Possible reactions of *n*-pentane in oxidative transformation catalyzed by vanadyl pyrophosphate are illustrated in Scheme III.8 [29d].

The catalytic oxidation of alkanes on heterogeneous catalysts is potentially a very important field, but even now it is still at the beginning of its development, particularly in comparison with the oxidation of olefins and aromatic hydrocarbons, which is already widely used in chemical industry (see Table III.1).

III.6.A. OXYGENATION WITH MOLECULAR OXYGEN

At present there are a few industrial alkane oxidation processes, for example, the synthesis of maleic anhydride by oxidation of butane [32], and the synthesis of butenes and 1,3-butadiene by oxidative dehydrogenation of butane [33]. The most difficult problem with alkanes is to stop the oxidation at the stage of a necessary product, since oxidation of gaseous hydrocarbon on a solid surface tends to proceed up to the stable final products (deep oxidation to produce carbon dioxide and water) [34]. In deep oxidation, the reactivity of alkanes, particularly those having a small number of carbon atoms, is usually lower than that of other compounds, for instance, olefins and acetylenes.

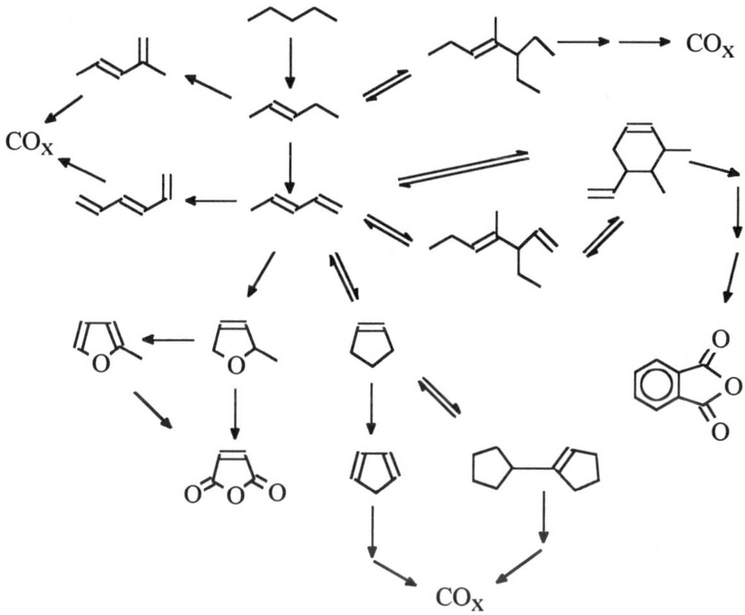

Scheme III.8. Transformations of *n*-pentane catalyzed by vanadyl pyrophosphate.

Heterogeneously catalyzed selective partial oxidation affords alcohols, ketones and carboxylic acids (see recent publications on oxidation of methane [35], ethane [36], propane [37] and higher alkanes [38]) as well as "synthesis gas", i.e., $CO + H_2$ [39]). Generally, the oxidation of methane can be depicted by the equation:

$$CH_4 + O_2 \longrightarrow CH_3OH + HCHO + HCOOH + CO + CO_2 + CH_3CH_3 + CH_2{=}CH_2$$

Table III.1. Industrial oxyfunctionalization of light alkanes in the petrochemical industry.

Process	Phase	Stage of development (1997)
Methane → Chloromethanes	Gas, heterog.	Industrial
Methane → Methanol	Gas, het./hom.	Pilot plant
Methane → Syngas	Gas, het./hom.	Research
Methane → Ethylene	Gas, het./hom.	Pilot plant
Ethane → Dichloroethane, Vinyl chloride	Gas, heterog.	Research
Ethane → Acetic acid	Gas, heterog.	Research
Ethane → Ethylene	Gas, het./hom.	Research
Propane → Acrylic acid	Gas or liquid	Research
Propane → Propyl alcohol	Liquid, het. or hom.	Research
Propane → Acrylonitrile	Gas, heterog.	Demonstrative plant
Propane → Propylene	Gas, heterog.	Research
n-Butane → Acetic acid	Liquid, homog.	Industrial
n-Butane → Maleic anhydride	Gas, heterog.	Industrial
n-Butane → Butadiene	Gas, heterog.	Industrial (abandoned)
Isobutane → Methacrylic acid	Gas, heterog.	Pilot plant
Isobutane → Isobutene	Gas, heterog.	Research
Isobutane → *t*-Butanol	Liquid, het. or hom.	Research
n-Pentane → Phthalic anhydride	Gas, heterog.	Research
Cyclohexane → Cyclohexanol, Cyclohexanone	Liquid, homog.	Industrial
Cyclohexane → Cyclohexanone	Liquid, het. or hom.	Research

The following equations are for deep oxidation:

$$CH_4 + 2\,O_2 = CO_2 + 2\,H_2O \qquad \Delta H^\circ_{298} = -\,191.5 \text{ kcal mol}^{-1}$$

and for the syngas production:

$$CH_4 + \tfrac{1}{2}\,O_2 = CO + 2\,H_2 \qquad \Delta H^\circ_{298} = -\,8.5 \text{ kcal mol}^{-1}$$

Products of methane partial oxidation on catalyst MoO_3–SiO_2 are presented in Table III.2 [35c].

Table III.2. Partial oxidation of methane on MoO_3–SiO_2.

Catalyst	Temperature, °C	Conversion CH_4, %	Selectivity, % HCHO	MeOH	CO	CO_2	Yield HCHO+ MeOH,%
Impregnated	500	0.58	32	2	25	41	0.20
	550	1.2	13	1	35	51	0.17
	600	4.0	12	1	33	54	0.52
	650	6.6	1	0	18	81	0.07
Alkoxide	500	1.0	73	13	0	14	0.86
	550	2.9	52	8	10	30	1.7
	600	8.2	35	11	17	37	3.8
	650	12	11	4	23	62	1.8

Alkylaromatic hydrocarbons give rise to the formation of corresponding aldehydes and carboxylic acids, for example, oxidation of toluene produces benzaldehyde[40], and phthalic anhydride can be obtained from xylene [41]. Oxidation of benzene produces phenol [42]. Deep oxidation of benzene gives rise to carbon dioxide and water as sole products [43].

Possible mechanisms of alkane oxidation in the presence of heterogeneous catalysts are discussed in publications [44]. The rate-determining step in the selective oxidation of methane is the rupture of the C–H bond. Methane activation can occur by a homolytic mechanism [44a,h]:

$$\text{Cat–[O}^-\text{]} + CH_4 \longrightarrow \text{Cat–OH} + CH_3\bullet$$

or heterolytically via interaction of methane with an acid–base pair [44b,h]:

$$M^{2+}O^{2-} + CH_4 \longrightarrow M^{2+}CH_3^- + H^+O^{2-}$$

In the last case a metal–methyl compound is formed, where the methyl is negatively charged. Further oxidation of these surface methyl anions would lead to methyl radicals:

$$M^{2+}CH_3^- \longrightarrow M^{2+} + CH_3\bullet + e^-$$

which can react with molecular oxygen or dimerize.

The oxidation of alkanes under light irradiation and in the presence of solid metal oxides (TiO_2, MoO_3, In_2O_3 etc.) is very interesting and many investigations have been devoted to this process [45]. Examples of photo-oxidations are given in Table III.3 [45f].

Table III.3. Photo-assisted catalytic oxidation of methane and ethane using silica-supported group V and VI metal oxide catalysts (493 K).

Catalyst	Conversion, %	Yield, μmol				Selectivity for aldehydes, %
		MeCHO	HCHO	CO	CO_2	
V_2O_5/SiO_2 [a,b]	0.48	–	60	12	tr	83
V_2O_5/SiO_2 [a,b]	0.60	–	62	16	12	69
V_2O_5/SiO_2 [a,b]	0.37	–	46	4	6	82
Nb_2O_5/SiO_2 [b]	0	–	tr	0	tr	0
MoO_3/SiO_2 [b]	0.24	–	26	6	4	72
V_2O_5/SiO_2 [a,c]	0.65	41	9	0	7	93
V_2O_5/SiO_2 [a,c]	0.89	62	7	1	2	98
MoO_3/SiO_2 [c]	0.77	42	22	2	7	92
WO_3/SiO_2 [c]	0.04	3	0	0	tr	~100

[a] Different catalyst preparation methods.
[b] Oxidation of methane, 2 h.
[c] Oxidation of ethane, 1 h.

Irradiation of molybdenum trioxide on titanium dioxide in an atmosphere of methane and molecular oxygen yields carbon monoxide and carbon dioxide [46]. It is assumed that the pair electron–hole is formed in the first stage of the reaction:

$$TiO_2 \xrightarrow{h\nu} TiO_2\,(e^-, h^+)$$

The interaction between the hole and the water molecule gives rise to the generation of hydroxyl radical:

$$h^+ + H_2O \longrightarrow HO\bullet + H^+$$

which reacts with the hydrocarbon:

$$RH + HO\bullet \longrightarrow R\bullet + H_2O$$

$$RH + HO\bullet \longrightarrow ROH + H\bullet$$

The electron produced in the TiO_2 photolysis is transferred to MoO_3 (which turns blue due to the formation of Mo^{5+}):

$$Mo^{VI}O_3 + e^- \longrightarrow Mo^{V}O_3^-$$

The last species reacts with the oxygen molecule to generate the radical HOO•:

$$Mo^{V}O_3^- + O_2 \longrightarrow (O_2^-)_{ads} \overset{H_2O}{\rightleftharpoons} HOO\bullet + HO^-$$

This radical can abstract the hydrogen from an alkane molecule, affording R•, in the case of methane – methyl radical:

$$HOO\bullet + CH_4 \longrightarrow H_2O_2 + CH_3\bullet$$

Further transformations of alkyl radicals give rise to the oxygenates:

$$CH_3\bullet + HOO\bullet \longrightarrow CH_3OOH \longrightarrow HCHO + H_2O$$

$$CH_3\bullet + O^- \longrightarrow CH_3O^- \xrightarrow{H_2O} CH_3OH + HO^-$$

$$HCHO + 2\,h^+ \longrightarrow 2\,H^+ + CO$$

$$CH_3OH + 4\,H^+ \longrightarrow 4\,H^+ + CO$$

$$CO + H_2O + 2\,h^+ \longrightarrow CO_2 + 2\,H^+$$

A new concept of selective photo-oxidation in zeolite cages has been recently proposed [47]. The hydrocarbon radical cation–O_2 charge-transfer pair is generated inside the cavities of alkali zeolites, in which the large electrostatic field stabilizes the highly polar charge transfer states of hydrocarbon–O_2 collisional pair and allows to control the pathways of further transformation.

III.6.B. OXYGENATION WITH OTHER OXIDANTS

Many oxygen-containing compounds are capable of transferring oxygen atoms to alkanes and other hydrocarbons. Oxidants can be generated *in situ* from molecular oxygen and a reducing reagent [48a]. Metal or metal oxides catalyze these reactions (or in some cases can oxidize hydrocarbons stoichiometrically [48b]). Some of them are of great practical importance, for example:

$$CH_4 + H_2O = CO + 3\,H_2 \qquad \Delta H^\circ_{298} = +49.3 \text{ kcal mol}^{-1}$$

$$CH_4 + CO_2 = 2\,CO + 2\,H_2 \qquad \Delta H^\circ_{298} = +59.1 \text{ kcal mol}^{-1}$$

Here we will very briefly survey reactions between alkanes and oxygen-containing compounds in the presence of heterogeneous catalysts. At high temperatures, alkanes can be oxidized not only by well-known oxidants but also by water and carbon dioxide.

Hydrogen Peroxide and Alkyl Peroxides

Hydrogen peroxide [49] and alkyl peroxides [50] oxidize alkanes and arenes at relatively low (< 100 °C) temperatures if heterogeneous catalysts, for example, titanium silicalites are used. A variety of oxygen based and protonated oxygen radicals, e.g., O^-, O^{2-}, HO•, are formed on the catalyst surface.

In the oxyfunctionalization of *n*-hexane by H_2O_2 (30 wt %) over titanium silicalites in methanol as a solvent [49a], the parallel-sequential reaction scheme was selected for transformation of the hydrocarbon into hexanol and hexanone:

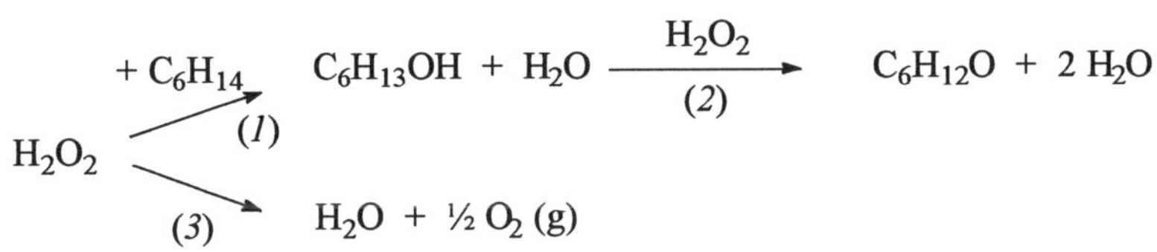

Designating the partial conversions of hydrogen peroxide through reactions *1*, *2*, and *3* as X_1, X_2, and X_3, respectively, the number of moles of reactants H_2O_2 (A) and C_6H_{14} (B) and products $C_6H_{13}OH$ (D) and $C_6H_{12}O$ (E) at any reaction time in the batch reactor will be:

$$N_A = N_{A0}(1 - X_1 - X_2 - X_3)$$
$$N_B = N_{B0} - N_{A0}X_1$$
$$N_D = N_{A0}(X_1 - X_2)$$
$$N_E = N_{A0}X_2$$

The partial conversions X_1, X_2, and X_3 are expressed in the equations:

$$X_1 = (N_D + N_E)/N_{A0}$$
$$X_2 = N_E/N_{A0}$$
$$X_3 = 2\,N_{O2}/N_{A0}$$

the selectivity of hydrogen peroxide conversion to oxygenate products is defined as:

$$S = \frac{N_D + N_E}{N_{A0} - N_A} = \frac{X_1}{(X_1 + X_2 + X_3)}$$

The following equations express the n-hexane consumption rate:

$$\frac{dX_1}{dt} = \frac{W_{cat}}{N_{A0}} \frac{k_1K_1[C_6H_{14}]_{aq}{}^{eq}[H_2O_2]^2}{1 + K_1[C_6H_{14}]_{aq}{}^{eq} + K_2[C_6H_{13}OH]_{org}{}^{n}}$$

the hexanone formation rate:

$$\frac{dX_2}{dt} = \frac{W_{cat}}{N_{A0}} \frac{k_2K_2[C_6H_{14}]_{aq}{}^{n}[H_2O_2]^2}{1 + K_1[C_6H_{14}]_{aq}{}^{eq} + K_2[C_6H_{13}OH]_{org}{}^{n}}$$

and the hydrogen peroxide catalytic decomposition rate:

$$\frac{dX_3}{dt} = 0.5 \frac{W_{cat}}{N_{A0}} k_3[H_2O_2]^2$$

where $[C_6H_{14}]_{aq}{}^{eq}$ is the solubility equilibrium concentration of n-hexane in the aqueous phase, $[H_2O_2]$ is the hydrogen peroxide concentration, $[C_6H_{13}OH]_{org}$ is the organic phase concentration of hexanol, and N_{A0} is the initial number of moles of hydrogen peroxide.

Water (Steam Reforming)

The reaction between alkanes (especially methane) and water is endothermic and produces syngas (see books, reviews [51], recent papers [52] and patents [53]). The photocatalytic oxidation of methane with water to produce methanol and molecular hydrogen has been patented [54].

Carbon Dioxide (Dry, CO_2 Reforming)

A mixture of carbon dioxide and methane can be converted to syngas (carbon monoxide and molecular hydrogen with a low H_2:CO ratio suitable for the synthesis of oxygenated chemicals) (see recent papers [55]). This reaction has attained wide research interest in the last years since it is a potential method of storage for solar energy [56].

The CO_2 reforming of methane over supported nickel catalysts proceeds according to the overall reaction stoichiometry

$$CH_4 + 2\,CO_2 = H_2 + H_2O + 3\,CO$$

and the following generalized reaction sequence has been proposed ($*$ represents a vacant site) [55f]:

$$CH_4 + * \rightleftharpoons CH_{x*} + [(4-x)/2]\,H_2$$

$$2\,[CO_2 + * \rightleftharpoons CO_{2*}]$$

$$H_2 + 2\,* \rightleftharpoons 2\,H_*$$

$$2\,[CO_{2*} + H_* \rightleftharpoons CO_* + OH_*]$$

$$OH_* + H_* \rightleftharpoons H_2O + 2\,*$$

$$CH_{x*} + OH_* \rightleftharpoons CH_xO_* + H_*$$

$$CH_xO_* \longrightarrow CO_* + (x/2)\,H_2$$

$$3\,[CO_* \rightleftharpoons CO + *]$$

Methane reduces carbon dioxide selectively to carbon monoxide under light irradiation in the presence of zirconium oxide [57].

Nitrous Oxide

Iron complexes stabilized in a ZSM-5 zeolite matrix produce a new form of surface oxygen (α-oxygen) upon decomposition of N_2O [58]. α-Oxygen selectively oxidizes methane and benzene:

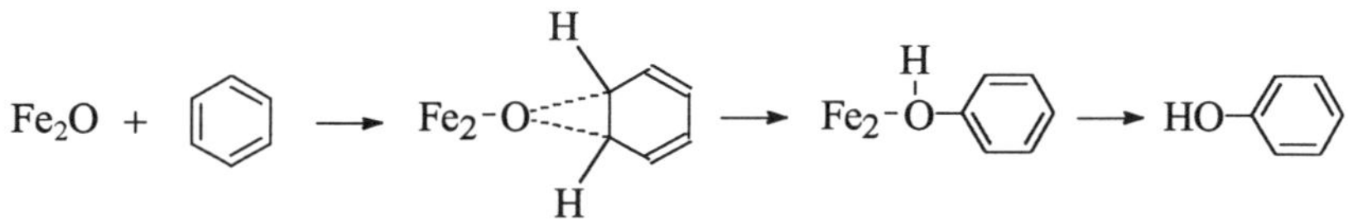

This direct conversion of benzene to phenol is of great practical importance [58c]. The surface oxygen radical anion, $O\bullet^-$, formed through a reaction of N_2O with electrons trapped on the surface of MgO, reacts with methane at 298 K [59]. The partial oxidation of ethane to ethanol and acetaldehyde over iron phosphate catalyst (573–773 K) by using nitrous oxide as an oxidant has been reported [60].

Nitrogen Oxides, NO_x

The catalytic reduction or decomposition of nitrogen oxides, particularly NO to N_2, is very important from the viewpoint of environmental air pollution. The selective reduction by hydrocarbons in the presence of air has been extensively studied as a potential process in NO_x emission control for Diesel and lean-burn engines (see certain recent publications [61]).

The reaction between methane and NO in the presence of O_2 over manganese-exchanged zeolite Mn-ZSM-5 [62a] begins with the adsorption of O_2 followed by NO to form a nitrito species:

$$Mn^{2+} + O_2 \rightleftharpoons Mn^{3+}\text{–}O_2^-$$

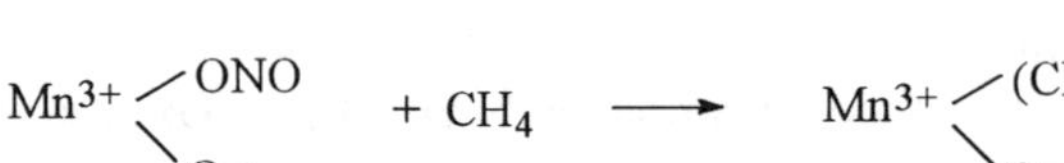

Absorbed nitrito species are then available to react with methane, forming adsorbed CH_3NO species with the release of hydroxyl radicals:

The following reactions are then possible:

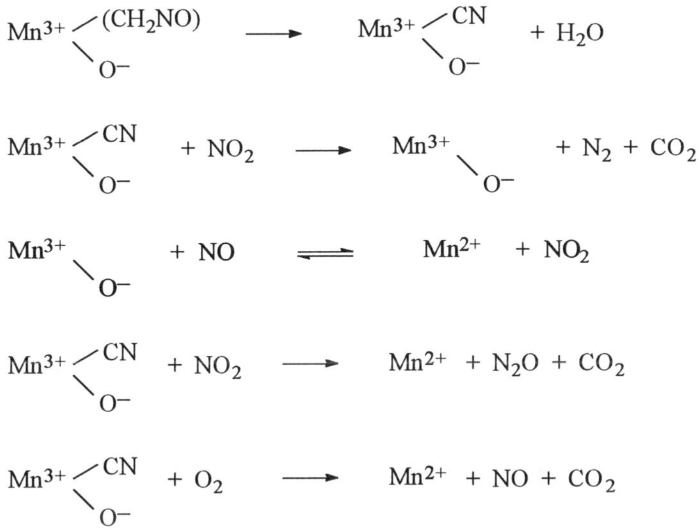

The proposed mechanism for the reduction of normal octane by NO in the presence of O_2 over Pt/Al_2O_3 is shown in Scheme III.9 [62b]. Reactions above the dotted line occur on the Pt surface, while reactions below occur on the Al_2O_3 support.

III.6.C. OXIDATIVE DEHYDROGENATION AND DEHYDROCYCLIZATION

Under the action of heterogeneous metal-containing catalysts, not only the insertion of the oxygen atom into a hydrocarbon molecule can be carried out, but also the oxidative dehydrogenation and dehydrocyclization can be achieved [63]. The general equation for these reactions is as follows:

$$C_nH_{2n+2} + \tfrac{1}{2} O_2 = C_nH_{2n} + H_2O$$

4-Vinylcyclohexene is readily converted to styrene when $PVMo_{10}O_{40}^{5-}$ supported on carbon is used as a catalyst, two scenarios being possible [64]:

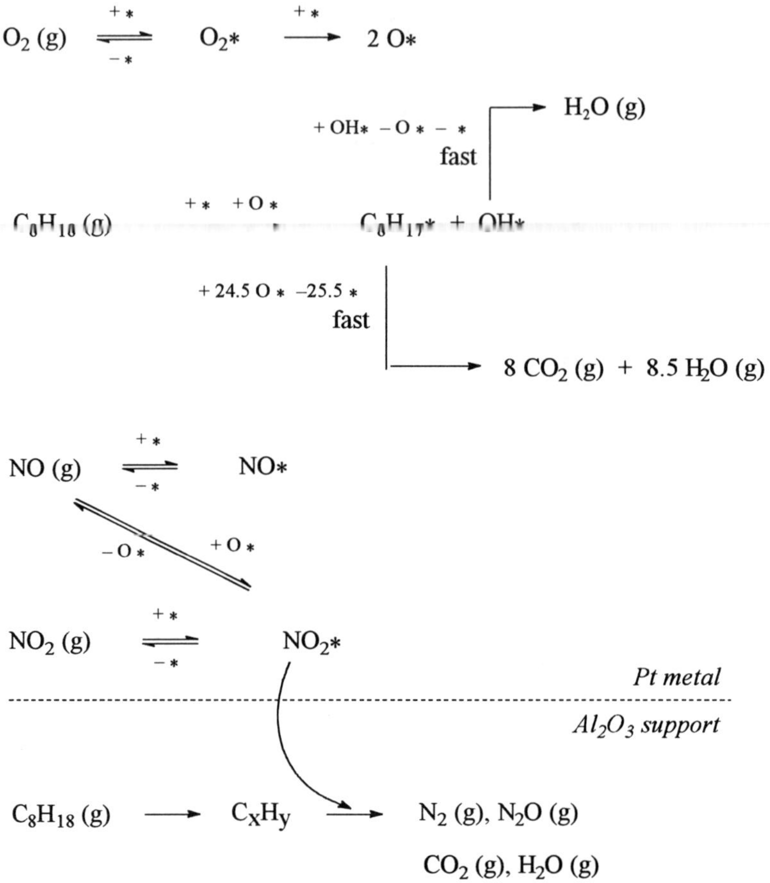

Scheme III.9. Mechanism proposed for the reaction of octane with NO and O_2 in the low [n-C_8H_{18}]/[O_2] concentration region.

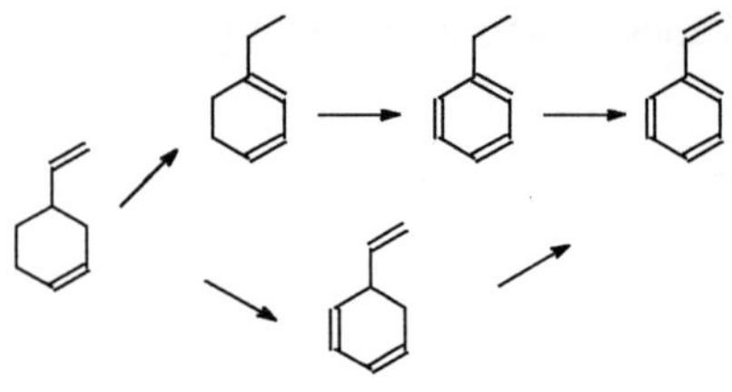

Oxidative dehydrogenation of ethane [65], propane [66], *n*-butane [33], and isobutane [67] has been described in recent papers (see also patents [68]). It is interesting that the oxidation of isobutane over Mo-containing Keggin-type heteropolycompounds and W-containing Keggin-type as well as Wells-Dawson-type heteropolycompounds occur via two different mechanisms [29d]. In both cases, the first stage of the reaction consists of the formation of an alkoxy species. The oxidation over $(NH_4)_3PW_{12}O_{40}$, in which the mechanism is essentially a radical one, gives rise to the predominant formation of isobutene (Scheme III.10). The reaction catalyzed by $(NH_4)_3PMo_{12}O_{40}$ proceeds via an ionic mechanism and affords methacrolein and methacrylic acid in expense of isobutene.

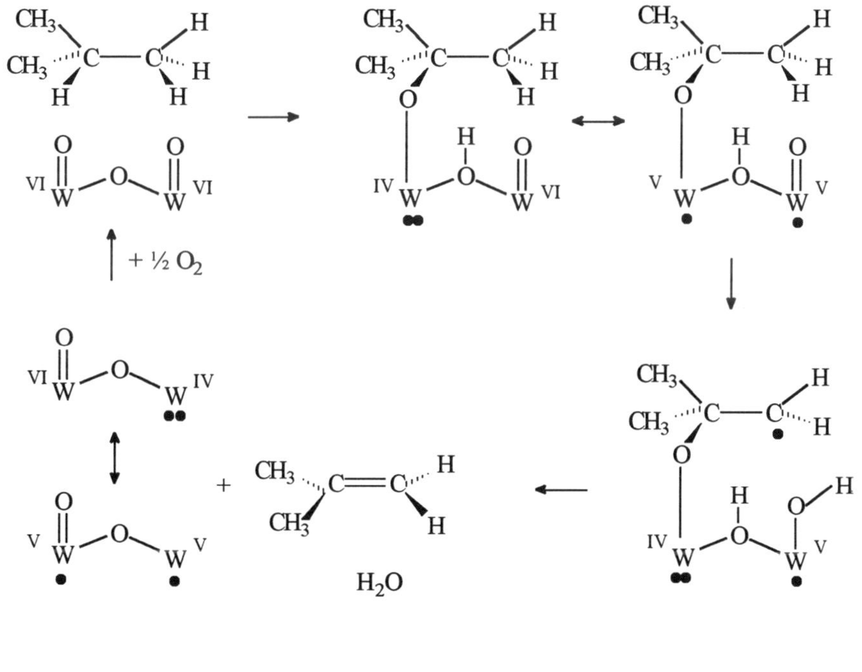

Scheme III.10. Proposed mechanism for isobutane oxidation on W-containing heteropolycompound.

Carbon dioxide can be used instead of molecular oxygen in transformation of ethylbenzene to styrene [69a] and ethane to ethene [69b]. The latter process occurs according to the equation:

$$C_2H_6 + CO_2 \longrightarrow CH_2{=}CH_2 + CO + H_2O$$

and gives also small amounts of methane and propane as shown in Table III.4.

Table III.4. Oxidative dehydrogenation of ethane under the action of CO_2.

Catalyst	Conversion of ethane, %	Selectivity, %			Yield of $CH_2{=}CH_2$ %
		$CH_2{=}CH_2$	CH_4	C_3H_8	
Ga_2O_3 (CO_2)	19.6	95.0	3.8	1.0	18.6
Ga_2O_3 (Ar)	9.6	94.0	5.0	0.7	9.0
Gr_2O_3 (CO_2)	12.1	93.8	5.8	0.4	11.4
Gr_2O_3 (Ar)	10.4	97.6	1.8	0.6	10.2
V_2O_5 (CO_2)	9.8	97.1	2.9	–	9.5
V_2O_5 (Ar)	12.5	91.7	7.4	0.9	11.5

It is clear that the yield of ethene with the Cr_2O_3 catalyst in the presence of CO_2 is slightly higher as compared to the run in an argon atmosphere. In the case of V_2O_5, the yield of ethene is even lower in the presence of CO_2, however, the selectivity of the process is higher.

III.6.D. OXIDATIVE DIMERIZATION OF METHANE

A great number of papers have been devoted to the oxidative condensation (coupling) of methane (see reviews [70], recent original reports [71] and patents [72]). The general equation of this reaction is as follows:

$$CH_4 + O_2 = CH_3CH_3 + CH_2{=}CH_2 + H_2O$$

Examples of oxidative coupling of methane on various metal oxides are presented in Table III.5 (conditions: $CH_4 : O_2 : He = 18 : 3 : 79$; 697 °C) [73]. Oxidative coupling of methane can be carried out at temperatures below 500 K if the process is stimulated by UV-irradiation [74].

Table III.5. Oxidative coupling of methane over various heterogeneous metal oxide catalysts.

	Conversion, %		Selectivity, %	
Catalyst	O_2	CH_4	$CH_2{=}CH_2$	C_2H_6
Y_2O_3	100.0	10.0	14.7	22.5
$LiTiO_3$	98.0	11.5	1.2	15.0
$LiFeO_2$	99.1	8.5	1.6	7.5
$LiNiO_2$	46.6	25.9	18.5	28.2
$NaFeO_4$	99.2	9.1	1.1	8.0
$AgCrO_4$	99.0	10.5	0	8.4

III.7. Oxidative and Nonoxidative Condensation of Alkanes

Under the action of heterogeneous catalysts, at relatively high temperatures and in the presence or absence of oxidants, alkanes can be introduced into various condensation reactions.

III.7.A. HOMOLOGATION

Methane and other alkanes can be condensed to produce higher alkanes and some other products [75]. In the absence of an oxidant, molecular hydrogen is formed along with higher alkanes, for example:

$$2\,CH_4 = C_2H_6 + H_2$$

Zeolites catalyze alkylation of alkanes with olefins [76]. The mechanism proposed for isobutane alkylation with butene is shown in Scheme III.11 (the alkoxy species shown schematically are bound to the catalyst surface) [76a].

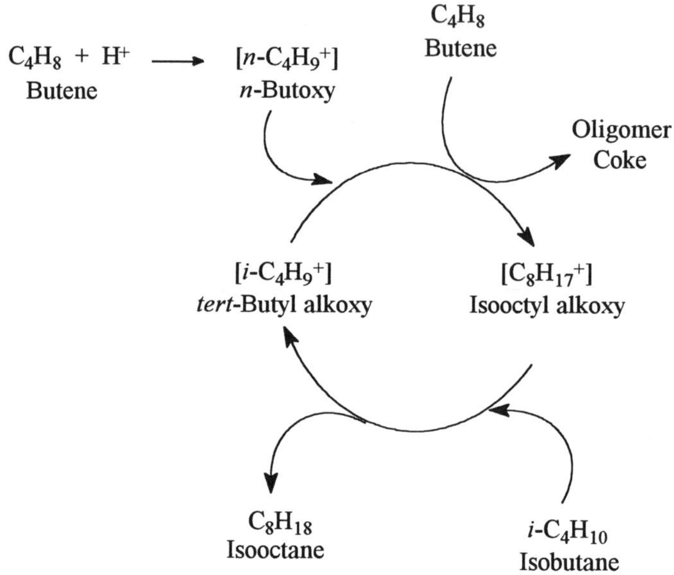

Scheme III.11. The classical mechanism of isobutene alkylation with butene.

III.7.B. AROMATIZATION OF LIGHT ALKANES

Transformations of methane [77] and other lower alkanes [78] into aromatic hydrocarbons, e.g., benzene, are very interesting. Table III.6 presents some catalytic systems for nonoxidative methane dehydrogenation over transition metal ion-loaded zeolites [77d].

Table III.6. Nonoxidative dehydrogenation over transition metal ion-loaded ZSM-5 zeolites.

Catalytic system	Active phase	Mechanism for catalytic activation of methane
Ru, X/H-ZSM-5 or Pt, X/H-ZSM-5 with X = Ga, Zn, or Cr	Metal/metal oxide	$MO + CH_4 \longrightarrow M\text{–}OH + CH_3\bullet$ $2\ CH_3\bullet \longrightarrow C_2H_6$
Mo/H-ZSM-5	MoO_3	$CH_4 + MO_3 \longrightarrow CH_3\text{–}H +$ $+ MoO_3 \longrightarrow H^{-}\text{–}CH_3^{+} + MoO_3$ $\longrightarrow Mo{=}CH_2$
Cu/H-ZSM-5	CuO	Carbenium ion mechanism
Zn/H-ZSM-5	ZnO	$CH_4 + Zn^{2+}\ (s) \longrightarrow CH^{+}\ (s) +$ $+ [Zn\text{–}H]^{+}$
Mo/H-ZSM-5	MoO_x	$CH_4 \longrightarrow CH_3\bullet + H\bullet$ $2\ CH_3\bullet \longrightarrow CH_2{=}CH_2 + H_2$
Mo/H-ZSM-5	Mo_2C/Mo_x–C–MoO_x	$CH_4 + O\ (s) \longrightarrow CH_3\bullet + OH\ (s)$ $2\ OH\ (s) \longrightarrow H_2O + O\ (s)$ $2\ CH_3\bullet \longrightarrow C_2H_6$
Mo/H-ZSM-5	Mo_2C	$CH_2{=}CH_2$ formed on Mo_2C and converted to aromatics at acid centers
Fe/H-ZSM-5	Fe_2O_3/Fe_3O_4	$CH_2{=}CH_2$ formed on transition metal ions and converted to aromatics at acid centers

The addition of carbon monoxide and carbon dioxide to the methane feed results in promotion of catalyst stability for direct benzene and naphthalene production from methane on Mo/H-ZSM-5 and Co modified Mo/H-ZSM-5 zeolites [79]. This effect is due to the efficient suppression of coke formation. A catalytic reaction between CH_4 and CO over supported Rh, Ru and Pd gives predominantly carbon dioxide as a result of CO disproportionation. However, addition of H_2 to the reaction mixture accelerates the benzene formation and increases the selectivity.

III.7.C. KHCHEYAN'S REACTION

Methane can also be co-condensed with some compounds in the presence of heterogeneous catalysts and molecular oxygen (Khcheyan's reaction [80]). Thus, oxidative coupling of methane and acetonitrile gives propionitrile and acrylonitrile:

$$CH_3CN + CH_4 + O_2 \longrightarrow CH_3CH_2CN + CH_2{=}CHCN + HCN + H_2O$$

Styrene may be obtained from methane and toluene in addition to some other compounds [81]:

$$C_6H_5CH_3 + CH_4 + O_2 \longrightarrow C_6H_5C_2H_5 + C_6H_5CH{=}CH_2 + C_6H_6 + C_6H_5OH$$

The proposed multistage mechanism includes formation of free radicals:

$$CH_4 + O_2 \longrightarrow CH_3\bullet + HOO\bullet$$

$$PhCH_3 + O_2 \longrightarrow PhCH_2\bullet + HOO\bullet$$

$$PhCH_2\bullet + CH_4 + O_2 \longrightarrow PhCH_2CH_3 + HOO\bullet$$

$$CH_3\bullet + PhCH_3 + O_2 \longrightarrow PhCH_2CH_3 + HOO\bullet$$

$$PhCH_2\bullet + CH_3\bullet \longrightarrow PhCH_2CH_3$$

Then catalyzed oxidative dehydrogenation of ethylbenzene to afford styrene occurs.

It is interesting that direct conversion of methane to styrene can be carried out over the mixed catalyst La_2O_3 + MoO_3/H-ZSM-5 [82]. Perspective homologation of olefins and acetylene with methane on transition metals has been demonstrated [83].

III.8. FUNCTIONALIZATION OF C–H COMPOUNDS

Catalyzed ammoxidation of ethane to acetonitrile [84]:

$$C_2H_6 + NH_3 + O_2 \longrightarrow CH_3\text{–}CN + H_2O,$$

propane to acrylonitrile [85]:

$$CH_3CH_2CH_3 + NH_3 + O_2 \longrightarrow CH_2\text{=}CH\text{–}CN + H_2O,$$

toluene to benzonitrile [86]:

$$C_6H_5CH_3 + NH_3 + O_2 \longrightarrow C_6H_5\text{–}CN + H_2O,$$

as well as some other organic compounds [87] are very important processes. The reaction network for ammoxidation of propane is shown in Scheme III.12 [85a].

Carbon monoxide carbonylates methane at 573–673 K in the presence of nitrous oxide on a rhodium-doped iron phosphate catalyst to produce methyl acetate [88a]. Vapor-phase carbonylation of toluene yields *p*-tolualdehyde, which can be easily oxidized to terephthalic acid [88b].

The catalytic process of oxidative chlorination of ethane to produce vinyl chloride occurs according to the overall equation

$$C_2H_6 + HCl + O_2 = CH_2\text{=}CHCl + 2\,H_2O$$

and involves the consecutive and parallel radical-chain and heterogeneously catalyzed stages: oxidation, chlorination, and dechlorination [89].

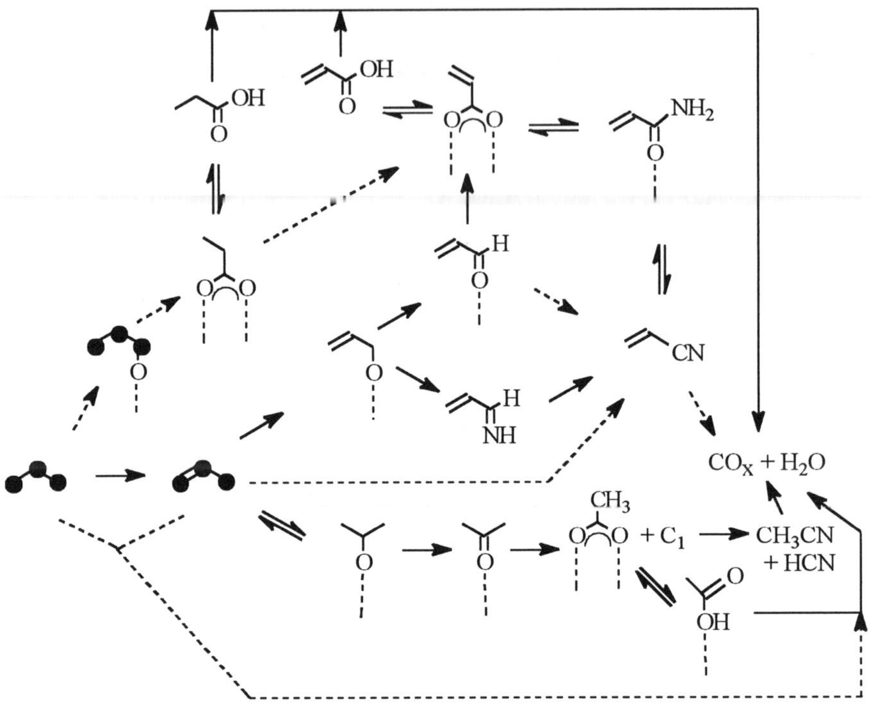

Scheme III.12. Surface reaction network for propane ammoxidation on V–Sb oxides (the dotted arrows are the less important pathways of transformation).

References for Chapter III

1. (a) Gonzalez, R. G. *Hart's Fuel Technol. Manage.* **1998**, *8*, 45. (b) Gesser, H. D.; Hunter, N. R. *Catalysis Today* **1998**, *42*, 183. (c) Gentry, J. C.; Kumar, C. S. *Hydrocarbon Processing* March **1998**, 69. (d) Wiltshire, J. *The Chemical Engineer* 12 March **1998**, 21. (e) Nakamura, I.; Fujimoto, K. *Petrotech.* **1998**, *21*, 34 (in Japanese). (f) Wilkinson, S. L. *C&EN*, February 23, **1998**, 55.

2. (a) Sen'kov, G. M.; Kozlov, N. S. *Industrial Reforming Catalysts*; Nauka i Tekhnika: Minsk, 1986 (in Russian). (b) *Methane Conversion*; Bibby, D. M. *et al.* Eds.; Elsevier: Amsterdam, 1988. (c) *Catalysis on Zeolites*; Kallo, D.; Minachev, Kh. M., Eds.; Akademia: Budapest, 1988. (d) Kung, H. H. *Transition Metal Oxides*; Elsevier: Amsterdam, 1989. (e) Minachev, Kh. M.; Dergachev, A. A.; Usachev, N. A. *Usp. Khim.* **1991**, *31*, 148 (in Russian). (f) *Catalysis*; Moulijn, J. A., van Leeuwen, P. W. N. M., van Santen, R. A., Eds.; Elsevier: Amsterdam, 1993. (g) Corma, *Chem. Rev.* **1995**, *95*, 559. (h) *11th International Congress on Catalysis – 40th Anniversary*; Hightower, J. W., Delgass, W. N., Iglesia, E., Bell, A. T., Eds.; Elsevier: Amsterdam, 1996. (i) Thomas, J. M.; Thomas, W. J. *Principles and Practice of Heterogeneous Catalysis*; VCH: Weinheim, Germany, 1997. (j) Masilamani, D. *3rd World Congress on Oxidation Catalysis*; Grasselli, R. K.; Oyama, S. T.; Gaffney, A. M.; Lyons, J. E., Eds.; Elsevier: Amsterdam, 1997, p. 1089. (k) Murugavel, R.; Roesky, H. W. *Angew. Chem., Int. Ed. Engl.* **1997**, *36*, 477. (l) Arends, I. W. C. E.; Sheldon, R. A.; Wallau, M.; Schuchardt, U. *Angew. Chem., Int. Ed. Engl.* **1997**, *36*, 1145. (m) Boudart, M. *J. Mol. Catal. A: Chem.* **1997**, *120*, 271. (n) Sheldon, R. A.; Wallau, M.; Arends, I. W. C. E.; Schuchardt, U. *Acc. Chem. Res.* **1998**, *31*, 485. (o) Cavani, F. *Catalysis Today* **1998**, *41* 73. (p) Jacobs, P. W.; Somorjai, G. A. *J. Mol. Catal. A: Chem.* **1998**, *131*, 5. (q) *Recent Advances in Basic and Applied Aspects of Industrial Catalysis*; Prasada Rao, T. S. R., Murali Dhar, G., Eds.; Elsevier: Amsterdam, 1998. (r) Vissokov, G. P.; Pirgov, P. S. *Appl. Catal. A: General* **1998**, *168*, 229. (s) Haber, J. *3rd World Congress on Oxidation Catalysis*; Grasselli, R. K.; Oyama, S. T.; Gaffney, A. M.; Lyons, J. E., Eds.; Elsevier: Amsterdam, 1997, p. 1. (t) Schmidt, L. D.; Goralski, C. T., Jr. *3rd World Congress on Oxidation Catalysis*; Grasselli, R. K.; Oyama, S. T.; Gaffney, A. M.; Lyons, J. E., Eds.; Elsevier: Amsterdam, 1997, p. 491.

3. (a) Wojciechowski, B. W.; Corma, A. *Catalytic Cracking: Catalysts, Chemistry and Kinetics*; Marcel Dekker: New York, 1986. (b) *Fluid Catalytic Cracking: Science and Technology*; Magee, J. S., Mitchell, M. M., Eds.; Elsevier: Amsterdam, 1993. (c) Wojciechowski, B. W. *Ind. Eng. Chem. Res.* **1997**, *36*,

3323. (d) *Hydrotreatment and Hydrocracking of Oil Fractions*; Froment, G. F., Delmon, B., Grange, P., Eds.; Elsevier: Amsterdam, 1997. (e) Watson, B. A.; Klein, M. T.; Harding, R. H. *Energy & Fuels* **1997**, *11*, 354. (f) Occelli, M. L.; O'Connor, P. *Fluid Cracking Catalysts*; Marcel Dekker: New York, 1997. (g) Cortright, R. D.; Dumesic, J. A.; Madon, R. J. *Topics Catal.* **1997**, *4*, 15. (h) Tynjälä, P.; Pakkanen, T. T., *J. Mol. Catal. A: Chem.* **1997**, *122*, 159. (i) Anikeev, V. I.; Gudkov, A. V.; Yermakova, A. *Fuel Proc. Technol.* **1997**, *53*, 115. (j) Koch, H.; Roos, K.; Stöcker, M.; Reschetilowski, W. *Chem. Eng. Technol.* **1998**, *21*, 401. (k) Lee, K.-h.; Ha, B.-H.; Lee, Y.-W. *Ind. Eng. Chem. Res.* **1998**, *37*, 1761. (l) Brait, A.; Koopmans, A.; Weinstabl, H.; Ecker, A.; Seshan, K.; Lercher, J. A. *Ind. Eng. Chem. Res.* **1998**, *37*, 873. (m) Zhang, T.; Amiridis, M. D. *Appl. Catal. A: General* **1998**, *167*, 161.

4. (a) Tanashev, Yu. Yu.; Fedoseev, V. I.; Aristov, Yu. I.; Pushkarev, V. V.; Avdeeva, L. B.; Zaikovskii, V. I.; Parmon, V. N. *Catalysis Today* **1998**, *42*, 333. (b) White, J. M. *J. Mol. Catal. A: Chem.* **1998**, *131*, 71.

5. van Santen, R. A. *J. Mol. Catal. A: Chem.* **1997**, *115*, 405.

6. (a) Muetterties, E. L. *Pure Appl. Chem.* **1982**, *54*, 83. (b) Baetzold, R. C. *J. Am. Chem. Soc.* **1983**, *105*, 4271. (c) Saillard, J.-Y.; Hoffmann, R. *J. Am. Chem. Soc.* **1984**, *106*, 2006. (d) Zheng, C.; Apeloig, Y.; Hoffmann, R. *J. Phys. Chem.* **1988**, *92*, 749. (e) Anderson, A. B.; Maloney, J. J. *J. Phys. Chem.* **1988**, *92*, 809. (f) Jarrold, M. F.; Bower, J. E. *J. Am. Chem. Soc.* **1988**, *110*, 6706. (g) Bobrov, A. V.; Kalushin, A. A.; Kimel'fel'd, Ya. M.; Plate, S. E. *Kinetika i Kataliz* **1988**, *29*, 1006 (in Russian). (h) Garin, F.; Maire, G. *Acc. Chem. Res.* **1989**, *22*, 100. (i) Somorjai, G. A. *Introduction to Surface Chemistry and Catalysis*; John Wiley & Sons: New York, 1994. (j) Zaera, F. *Chem. Rev.* **1995**, *95*, 2651. (k) Raut, J. S.; Fichthorn, K. A. *J. Chem. Phys.* **1998**, *108*, 1626. (l) Fuhrmann, D.; Wacker, D.; Weiss, K.; Hermann, K.; Witko, M.; Wöll, C. *J. Chem. Phys.* **1998**, *108*, 2651. (m) Carre, M.-N.; Jackson, B. *J. Chem. Phys.* **1998**, *108*, 3723.

7. (a) *Adsorption and Ion Exchange with Synthetic Zeolites*; Flank, W. H., Ed.; ACS Symposium Series No. 135, ACS: Washington, DC, 1987. (b) Anderson, A. B.; Maloney, J. J.; Yu. J. *J. Catal.* **1988**, *112*, 392. (c) Mastikhin, V. M.; Mudrakovskiy, I. L.; Pel'menschikov, A. G.; Zamaraev, K. I. *Khimich. Fizika* **1988**, *7*, 1096 (in Russian). (d) Golikov, A. P.; Avramenko, V. A.; Glushchenko, V. Yu. *Zhurn. Fiz. Kim.* **1988**, *62*, 2499 (in Russian). (e) Senchenya, I. N.; Kazanskiy, V. B. *Kinetika i Kataliz* **1989**, *30*, 339 (in Russian). (f) Jasra, R. V.; Bhat, S. G. T. *J. Catal.* **1989**, *115*, 605. (g) Tasi, G. Fejes, P. *J. Catal.* **1989**, *115*, 597. (h) Catlow, C. R. A.; Ackermann, L.; Bell, R. G.; Gay, D. H.; Holt, S.; Lewis, D. W.; Nygren, M. A.; Sastre, G.; Sayle, D.

C.; Sinclair, *J. Mol. Catal. A: Chem.* **1997**, *115*, 431. (i) Kissin, Yu. V. *J. Mol. Catal. A: Chem.* **1998**, *132*, 101. (j) Zscherpel, D.; Ranke, W.; Weiss, W.; Schlögl, R. *J. Chem. Phys.* **1998**, *108*, 9506. (k) Choso, T.; Tabata, K. *J. Mol. Catal. A: Chem.* **1998**, *129*, 225.

8. (a) Ozaki, A. *Isotopic Studies of Heterogeneous Catalysis*; Kadansha: Tokyo, 1976. (b) Cogen, J. M.; Maier, W. F. *J. Am. Chem. Soc.* **1986**, *108*, 7752. (c) Kemball, C. *Adv. Catalysis*; Academic Press: New York, 1959, p. 223.
9. Stepanov, A.; G.; Ernst, H.; Freude, D. *Catalysis Lett.* **1998**, *54*, 1.
10. (a) Rummel, S.; Ilatovskaya, M. A.; Mysov, E. I.; Lenenko, V. S.; Langguth, H.; Shur, V. B. *Angew. Chem., Int. Ed. Engl.* **1996**, *35*, 2489. (b) Rummel, S.; Yunusov, S. M.; Ilatovskaya, M. A.; Langguth, H.; Mysov, E. I.; Wahren, M.; Likholobov, V. A.; Shur, V. B. *J. Mol. Catal. A: Chem.* **1998**, *132*, 277. (c) Handa, H.; Baba, T.; Ono, Y. *J. Chem. Soc., Faraday Trans.* **1998**, *94*, 451.
11. (a) Anderson, J. R. *Adv. Catal.* **1973**, *23*, 1. (b) Clarke, J. K. A.; Rooney, J. J. *Adv. Catal.* **1976**, *25*, 125. (c) Gault, F. G. *Adv. Catal.* **1981**, *30*, 1. (d) Ponec, V. *Adv. Catal.* **1983**, *32*, 149. (e) Zhorov, Yu. M. *Hydrocarbon Isomerization: Chemistry and Technology*; Khimiya: Moscow, 1983 (in Russian). (f) Bragin, O.V.; Krasavin, S. A. *Usp. Khim.* **1983**, *52*, 1108 (in Russian). (g) Vysotskiy, A. V.; Sergeeva, O. V.; Shmidt, F. K. *Isomerization Transformations of Hydrocarbons on Zeolite Catalysts*; Irkutsk University: Irkustk, 1985 (in Russian). (h) Maire, G.; Garin, F. *J. Mol. Catal.* **1988**, *48*, 99. (i) Paal, Z.; Groeneweg, H.; Zimmer, H. *Catalysis Today* **1989**, *5*, 199. (j) Gubicza, L.; Uihidy, A.; Exner, H. *Catalysis Today* **1989**, *5*, 139.
12. Bond, G. C. *Ind. Eng. Chem. Res.* **1997**, *36*, 3173.
13. (a) Farcasiu, D.; Li J. Q.; Kogelbauer, A. *J. Mol. Catal. A: Chem.* **1997**, *124*, 67. (b) Bardin, B. B.; Davis, R. J. *Topics Catal.* **1998**, *6*, 77. (c) Gao, Z.; Xia, Y.; Hua, W.; Miao, C. *Topics Catal.* **1998**, *6*, 101. (d) Knözinger, H. *Topics Catal.* **1998**, *6*, 107. (e) Vera, C. R.; Yori, J. C.; Parera, J. M. *Appl. Catal. A: General* **1998**, *167*, 75. (f) Rigby, A. M.; Kramer, G. J.; van Santen, R. A. *J. Catal.* **1997**, *170*, 1. (g) Tran, M.-T.; Gnep, N. S.; Szabo, G.; Guisnet, M. *Appl. Catal. A: General* **1998**, *170*, 49. (h) Fogash, K. B.; Hong, Z.; Kobe, J. M.; Dumesic, J. A. *Appl. Catal. A: General* **1998**, *172*, 107. (i) Shimizu, K.; Kounami, N.; Wada, H.; Shishido, T.; Hattori, H. *Catalysis Lett.* **1998**, *54*, 3. (j) Fogash, K. B.; Hong, Z.; Dumesic, J. A. *J. Catal.* **1998**, *173*, 519. (k) Tran, M.-T.; Gnep, N. S.; Szabo, G.; Guisnet, M. *J. Catal.* **1998**, *174*, 185. (l) Adeeva, V.; Liu, H.-Y.; Xu, B.-Q.; Sachtler, M. H. *Topics Catal.* **1998**, *6*, 61. (m) Sun, Y.; Yue, Y. H.; Gao, Z. *React. Kinet. Catal. Lett.* **1998**, *63*, 349.
14. (a) Rezgui, S.; Gates, B. C. *Catalysis Lett.* **1996**, *37*, 5. (b) Matsuda, T. Niwayama, T.; Sakagami, H.; Takahashi, N. *Chemistry Lett.* **1997**, 373. (c)

Ryu, S. G.; Gates, B. C. *Ind. Eng. Chem. Res.* **1998**, *37*, 1786. (d) Hino, M.; Arata, K. *Appl. Catal. A: General* **1998**, *169*, 151. (e) Liu, Y.; Koyano, G.; Na, K.; Misono, M. *Appl. Catal. A: General* **1998**, *166*, L263. (f) Na, K.; Okuhara, T.; Misono, M. *J. Catal.* **1997**, *170*, 96. (g) Del Gallo, P.; Meunier, F.; Pham-Huu, C.; Crouzet, C.; Ledoux, M. J. (h) Kappenstein, C.; Guerin, M.; Lazar, K.; Matusek, K.; Paal, Z. *J. Chem. Soc., Faraday Trans.* **1998**, *94*, 2463. (i) Rezgui, S.; Jentoft, R. E.; Gates, B. C. *Catalysis Lett.* **1998**, *51*, 229. (j) Na, K.; Iizaki, T., Okuhara, T., Misono, M. *J. Mol. Catal. A: Chem.* **1997**, *115*, 449. (k) Souverijns, W.; Houvenaghel, A.; Feijen, E. J. P.; Martens, J. A.; Jacobs, P. A. *J. Catal.* **1998**, *174*, 201. (l) Canizares, P.; De Lucas, A.; Valverde, J. L.; Dorado, F. *Ind. Eng. Chem. Res.* **1997**, *36*, 4797. (m) Grillo, M. E.; Ramirez de Agudelo, M. M. *J. Mol. Catal. A: Chem.* **1997**, *119*, 105. (n) van de Runstraat, A.; Kamp, J. A.; Stobbelaar, P. J.; van Grondelle, J.; Krijnen, S.; van Santen, R. A. *J. Catal.* **1997**, *171*, 77. (o) Canizares, P.; De Lucas, A.; Valverde, J. L.; Dorado, F. *Ind. Eng. Chem. Res.* **1998**, *37*, 2592. (p) van de Runstraat, A.; van Grondelle, J.; Krijnen, S.; van Santen, R. A. *Ind. Eng. Chem. Res.* **1997**, *36*, 3116. (q) Sinha, A. K.; Sivasanker, S.; Ratnasamy, P. *Ind. Eng. Chem. Res.* **1998**, *37*, 2208. (r) Vaudry, F Di Renzo, F.; Fajula, F.; Schulz, P. *J. Chem. Soc., Faraday Trans.* **1998**, *94*, 617.

15. (a) Wu, A.-H.; Drake, C. A.; Melton, R. J. *Pat. US 5,707,918* (to Phillips Petroleum Company) [*J. Mol. Catal. A: Chem.* **1998**, *135*, 211. *Chem. Abstr.* **1998**, *128*, 130167g]. (b) Wu, A.-H.; Drake, C. A.; Melton, R. J. *Pat. US 5,707,921* (to Phillips Petroleum Company) [*J. Mol. Catal. A: Chem.* **1998**, *135*, 211. *Chem. Abstr.* **1998**, *128*, 130168h]. (c) Wu, A.-H. Pat. US 5,543,374 (to Phillips Petroleum Company) [*J. Mol. Catal. A: Chem.* **1997**, *120*, 287]. (d) Mizuno, K.; Wakamatsu, H.; Oi, M. *Pat. Jpn. Kokai Tokkyo Koho JP 10 15,399* (to Cosmo Oil Co.) [*Chem. Abstr.* **1998**, *128*, 104217d]. (e) Mizuno, K.; Wakamatsu, H.; Oi, M. *Pat. Jpn. Kokai Tokkyo Koho JP 10 17,872* (to Cosmo Oil Co.) [*Chem. Abstr.* **1998**, *128*, 104218e].

16. Röger, H. P.; Möller, K. P.; O'Connor, C. T. *J. Catal.* **1998**, *176*, 68.

17. (a) Sklyarov, A. V. *Kinetika i Kataliz* **1984**, *25*, 408 (in Russian). (b) Mihajlova, A.; Andreev, A.; Shopov, D.; Dimitrova, R. *Appl. Catal.* **1988**, *40*, 247. (c) Sato, S.; Ohhara, M.; Sodesawa, T.; Nozaki, F. *Appl. Catal.* **1988**, *37*, 207. (d) Luna, A. E. C.; Becerra, A. M. *React. Kinet. Catal. Lett.* **1998**, *63*, 335. (e) Yori, J. C.; Pieck, C. L.; Parera, J. M. *Catalysis Lett.* **1998**, *52*, 227. (f) Oganowski, W.; Hanuza, J.; Kepiski, L. *Appl. Catal. A: General* **1998**, *171*, 145.

18. (a) Skarchenko, V. K. *Usp. Khim.* **1971**, *40*, 2145 (in Russian). (b) Skarchenko, V. K. *Usp. Khim.* **1977**, *46*, 1411 (in Russian). (c) Skarchenko, V. K.

Hydrocarbon Dehydrogenation; Naukova Dumka: Kiev, 1981 (in Russian). (d) Sklyarov, A. V. *Usp. Khim.* **1986**, *55*, 450 (in Russian). (e) Bursian, N. P.; Kogan, S. B. *Usp. Khim.* **1989**, *58*, 451 (in Russian). (f) Vasil'ev, A. N.; Galich, P. N.; *Khim. Tekhnol. Topl. Masel* **1997**, No. 6, 45 (in Russian).

19. (a) Indovia, V. *Catalysis Today* **1998**, *41*, 95. (b) De Rossi, S.; Casaletto, M. P.; Ferraris, G.; Cimino, A.; Minelli, G. *Appl. Catal. A: General* **1998**, *167*, 257. (c) Hill, J. M.; Cortright, R. D.; Dumesic, J. A. *Appl. Catal. A: General* **1998**, *168*, 9. (d) Shimada, H.; Akazawa, T.; Ikenaga, N.; Suzuki, T. *Appl. Catal. A: General* **1998**, *168*, 243. (e) Brait, A.; Seshan, K.; Weinstabl, H.; Ecker, A.; Lercher, *Appl. Catal. A: General* **1998**, *169*, 315. (f) Price, G.L.; Kanazirev, V.; Dooley, K. M.; Hart, V. I. *J. Catal.* **1998**, *173*, 17. (g) Weckhuysen, B. M.; Bensalem, A.; Schoonheydt, R. A. *J. Chem. Soc., Faraday Trans.* **1998**, *94*, 2011. (h) Chandler, B. D.; Rubinstein, L. I.; Pignolet, L. H. *J. Mol. Catal. A: Chem.* **1998**, *133*, 267. (i) Li, R. X.; Wong, N. B.; Tin, K. C.; Chen, J. R.; Li, X. J. *Catalysis Lett.* **1998**, *50*, 219. (j) Arakawa, S. T.; Mulvaney, R. C.; Felch, D. E.; Petri, J. A.; Vandenbussche, K.; Dandekar, H. W. *Hydrocarbon Processing* March **1998**, 93. (k) Cortright, R. D.; Levin, P. E.; Dumesic, J. A. *Ind. Eng. Chem. Res.* **1998**, *37*, 1717.

20. (a) O'Young, C.-L.; Sawicki, R. A.; Yin, Y.-G.; Xu, W.-Q.; Suib, S. L. *Pat. US 5,597,944* [*J. Mol. Catal. A: Chem.* **1997**, *124*, 172]. (b) Minkkinen, A.; Burzynski, J.-P.; Saint, N. L. B. *Pat. US 5,711,919* (to Institut Francais du Petrole) [*J. Mol. Catal. A: Chem.* **1998**, *135*, 211]. (c) Dongara, R.; Barsur, A. C.; Gokak, D. T.; Rao, K. V.; Krishnamurthy, K. R.; Bhardwaj, I. S. Pat. US 5,677,260 (to Indian Petrochemicals) [*J. Mol. Catal. A: Chem.* **1998**, *132*, 118].

21. (a) Ivanova, I. I.; Pomakhina, E. B.; Rebrov, A. I.; Derouane, E. G. *Topics Catal.* **1998**, *6*, 49. (b) Cheung, T.-K.; Gates, B. C. *Topics Catal.* **1998**, *6*, 41.

22. (a) Somorjai, G. *Adv. Catal.* **1978**, *26*, 2. (b) Paal, Z. *Adv. Catal.* **1980**, *29*, 273. (c) Isagulyants, G. V.; Dubinskiy, Yu. G.; Rozengart, M. I. *Usp. Khim.* **1981**, *50*, 1541. (d) Isagulyants, G. V.; Rozengart, M. I.; Dubinskiy, Yu. G.; Rozengart, M. I. *Catalytic Aromatization of Aliphatic Hydrocarbons*; Nauka: Moscow, 1983 (in Russian). (e) Bragin, O. V.; Krasavin, S. A. *Usp. Khim.* **1983**, *52*, 1108 (in Russian).

23. Derouane, E. G.; Vanderveken, D. J. *Appl. Catal.* **1988**, *45*, L15. (b) Dass, D. V.; Odell, A. L. *J. Catal.* **1988**, *113*, 259. (c) Park, Y. K.; Kim, D. H.; Woo, S. I. *Korean J. Chem. Eng.* **1997**, 14, 249. (d) Kondratkova, N. I.; Ul'yanova, N. V. *Izv. Minist. Nauki – Akad. Nauk Resp. Kaz., Ser. Khim.* **1997**, No. 4, 8 (in Russian). (e) Choudhary, V. R.; Devadas, P. *Appl. Catal. A: General* **1998**, *168*, 187. (f) Viswanadham, N.; Dhar, G. M.; Rao, T. S. R. P. *J. Mol. Catal. A: Chem.* **1997**, *125*, 87. (g) Zheng, J.; Chun, Y.; Dong, J.; Xu, Q. *J. Mol. Catal.*

A: Chem. **1998**, *130*, 271. (h) Koel, B. E.; Blank, D. A.; Carter, E. A. *J. Mol. Catal. A: Chem.* **1998**, *131*, 39. (i) Takeguchi, T.; Kim, J.-B.; Kang, M.; Inui, T.; Cheuh, W.-T.; Haller, G. L. *J. Catal.* **1998**, *175*, 1. (j) Paal, Z.; Muhler, M.; Matusek, K. *J. Catal.* **1998**, *175*, 245.

24. (a) *Hydrotreatment and Hydrocracking of Oil Fractions*; Froment, G. F., Delmon, B., Grange, P., Eds.; Elsevier: Amsterdam, 1997. (b) Scherzer, J.; Gruia, A. L. *Hydrocracking Science and Technology*; Marcel Dekker: Monticello, 1997.

25. (a) Denayer, J. F.; Baron, G. V.; Souverijns, W.; Martens, J. A.; Jacobs, P. A. *Ind. Eng. Chem. Res.* **1997**, *36*, 3242. (b) Wu, H.; Leu, L.; Naccache, C.; Chao, K. *J. Mol. Catal. A: Chem.* **1997**, *127*, 143. (c) Lungstein, A.; Jentys, A.; Vinek, H. *Appl. Catal. A: General* **1998**, *166*, 29. (d) Egia, B.; Cambra, J. F.; Arias, P. L.; Güemez, M. B.; Legarreta, J. A.; Pawelec, B.; Fierro, J. L. G. *Appl. Catal. A: General* **1998**, *169*, 37. (e) Souverijns, W.; Martens, J. A.; Froment, G. F.; Jacobs, P. A. *J. Catal.* **1998**, *174*, 177. (f) Jackson, S. D.; Kelly, G. J.; Webb, G. *J. Catal.* **1998**, *176*, 225. (g) Campelo, J. M.; Lafont, F.; Marinas, J. M. *Appl. Catal. A: General* **1998**, *170*, 139. (h) Valyon, J.; Engelhardt, J. *React. Kinet. Catal. Lett.* **1998**, *63*, 27. (i) Siffert, S.; Garin, F. *J. Chim. Phys. Phys.-Chim. Biol.* **1998**, *95*, 536. (j) Furimsky, E. *Appl. Catal. A: General* **1998**, *171*, 177. (k) Aboul-Gheit, A. K.; Ghoneim, S. A.; Al-Owais, A. A. *Appl. Catal. A: General* **1998**, *170*, 277. (l) Yang, X.; Yuan, J. *Ranliao Huaxue Xuebao* **1998**, *26*, 61 (in Chinese).

26. (a) Poole, M.; Halbert, T. R.; Bearden, R.; Reynolds, S. D. *Pat. US 5,620,591* (to Exxon Research & Engineering Company) [*J. Mol. Catal. A: Chem.* **1998**, *129*, 113]. (b) Mulaskey, B. F.; Hise, R. L.; Trumbull, S. E.; Canella, W. J.; Innes, R. F. *Pat. US 5,620,937* (to Chevron Chemical Company) [*J. Mol. Catal. A: Chem.* **1998**, *129*, 113]. (c) Porter, M. K.; Clausen, G. A. *Pat. US 5,622,616* (to Texaco Development Corporation) [*J. Mol. Catal. A: Chem.* **1998**, *129*, 114]. (d) Ricardo, P.; Galiasso, R. T.; Romero, Y. Reyes, E.; Rodriguez, E. *Pat. US 5,558, 766* (to Intervep S.A.) [*J. Mol. Catal. A: Chem.* **1997**, *120*, 300].

27. Hemming, F.; Wehrer, P.; Katrib, A.; Maire, G. *J. Mol. Catal. A: Chem.* **1997**, *124*, 39.

28. (a) Margolis, L. Ya. *Oxidation of Hydrocarbons by Heterogeneous Catalysts*; Khimiya: Moscow, 1977 (in Russian). (b) Golodets, G. I. *Heterogeneous Catalytic Reactions Involving Molecular Oxygen*; Elsevier: Amsterdam, 1983. (c) Alkhazov, T. G.; Margolis, L. Ya. *Highly Selective Catalysts for Hydrocarbon Oxidation*; Khimiya: Moscow, 1988 (in Russian). (d) *New Developments in Selective Oxidation*; Centi, G., Trifiro, F., Eds.; Elsevier: Amsterdam, 1990. (e) *Methane and Alkane Conversion Chemistry*; Bhasin, M.

M., Slocum, D. W., Eds.; Plenum: New York, 1995. (f) *Heterogeneous Hydrocarbon Oxidation*; Warren, B. K., Oyama, S. T., Eds.; ACS Symp. Series 638: Washington, 1996. (g) Arutyunov, V. S.; Krylov, O. V. *Oxidative Conversion of Methane*; Nauka: Moscow, 1998 (in Russian).

29. (a) Sinev, M. Yu.; Korchak, V. N.; Krylov, O. V. *Usp. Khim.* **1989**, *58*, 38 (in Russian). (b) Sokolovskii, V. D. *Catal. Rev.– Sci. Eng.* **1990**, *32*, 1; (c) Eng, D.; Stoukides, M. *Catal. Rev.– Sci. Eng.* **1991**, *33*, 375. (d) Cavani, F.; Trifiro, F. *3rd Word Congress on Oxidation Catalysis*; Grasselli, R. K., Oyama, S. T., Gaffney, A. M., Lyons, J. E., Eds.; Elsevier: Amsterdam, 1997, p. 19. (e) Sinev, M. Yu., Margolis, L. Ya.; Bychkov, V. Yu.; Korchak, V. N. *3rd Word Congress on Oxidation Catalysis*; Grasselli, R. K., Oyama, S. T., Gaffney, A. M., Lyons, J. E., Eds.; Elsevier: Amsterdam, 1997, p. 327.

30. (a) Belyaev, V. D.; Sobyanin, V. A.; Demin, A. K.; Lipilin, A. S.; Zapesotskii, V. E. *Mendeleev Commun.* **1991**, 53. (b) Brown, M. J.; Parkyns, N. D. *Catalysis Today* **1991**, *8*, 305. (c) Wang, Y.; Otsuka, K. *J. Chem. Soc., Chem. Commun.* **1994**, 2209. (d) Suzuki, T.; Wada, K.; Shima, M.; Watanabe, Y. *J. Chem. Soc., Chem. Commun.* **1990**, 1059. (e) Anderson, L. C.; Xu, M.; Mooney, C. E.; Rosynek, M. P.; Lunsford, J. H. *J. Am. Chem. Soc.* **1993**, *115*, 6322. (f) Sojka, Z.; Herman, R. G.; Klier, K. *J. Chem. Soc., Chem. Commun.* **1991**, 185. (g) Wada, S.; Imai, H. *Catalysis Lett.* **1991**, *8*, 131. (h) Sun, H.; Blatter, F.; Frei, H. *J. Am. Chem. Soc* **1994**, *116*, 7951. (i) Jones, R. H.; Ashcroft, A. T.; Waller, D.; Cheetham, A. K.; Thomas, J. M. *Catalysis Lett.* **1991**, *8*, 169. (j) Otsuka, K.; Yamanaka, I.; Hosokawa, K. *Nature* **1990**, *345*, 697. (k) Yu, J.; Anderson, A. B. *J. Am. Chem. Soc.* **1990**, *112*, 7218. (l) Stiakaki, M.-A. D.; Tsipis, A. C.; Tsipis, C. A.; Xanthopoulos, C. E. *J. Mol. Catal.* **1993**, *82*, 425. (m) Shi, C.; Sun, Q.; Hu, H.; Herman, R.G.; Klier, K.; Wachs, I. E. *Chem. Commun.* **1996**, 663. (n) Zhang, Q.; Otsuka, K. *Chemistry Lett.* **1997**, 363. (o) Sokolovskii, V.; Arena, F.; Coluccia, S.; Parmaliana, A. *J. Catal.* **1998**, *173*, 238. (p) Centi, G.; Perathoner, S. *Catalysis Today* **1998**, *41*, 457. (q) Otsuka, K.; Yamanaka, I. *Catalysis Today* **1998**, *41*, 311. (r) Solomysi, F. *J. Mol. Catal. A: Chem.* **1998**, *131*, 121.

31. (a) Ushikubo, T. *Pat. Jpn. Kokai Tokkyo Koho JP 10 45,643* (to Mitsubishi Chemical Industies) [*Chem. Abstr.* **1998**, *128*, 204621z]. (b) De Jong, K. P.; Schoonebeek, R. J.; Vonkeman, K. A. *Pat. US 5,720,901* (to Shell Oil Company) [*J. Mol. Catal. A: Chem.* **1998**, *135*, 214]. (c) Kourrrtakis, K.; Sonnichsen, G. C. *Pat. US 5,543,532* (to E. I. Du Pont de Nemours and Company) [*J. Mol. Catal. A: Chem.* **1997**, *120*, 303].

32. (a) Centi, G.; Trifiro, F.; Ebner, J. R.; Franchetti, V. M. *Chem. Rev.* **1988**, *88*, 55. (b) Hutching, G. J.; Kiely C. J.; Sananes-Schulz, M. T.; Burrows, A.; Volta,

J. C. *Catalysis Today* **1998**, *40*, 135. (c) Zanthoff, H. W.; Sananes-Schultz, M.; Buchholz, S. A.; Rodemerck, U.; Kubias, B.; Baerns, M. *Appl. Catal. A: General* **1998**, *172*, 49. (d) Joly, J. P.; Mehier, C.; Bere, K. E.; Abon, M. *Appl. Catal. A: General* **1998**, *169*, 55.

33. (a) Sklyarov, A. V. *Kinetika i Kataliz* **1984**, *25*, 408 (in Russian). (b) Dejoz, A.; Nieto, J. M. L.; Melo, F.; Vasquez, I. *Ind. Eng. Chem. Res.* **1997**, *36*, 2588. (c) Lemonidou, A. A.; Stambouli, A. E. *Appl. Catal. A: General* **1998**, *171*, 325. (d) Nietom J. M. L.; Dejoz, A.; Vazquez, M. I.; O'Leary, W.; Cunningham, J. *Catalysis Today* **1998**, *40*, 73.

34. (a) Alkhazov, T. G.; Margolis, L. Ya. *Deep Catalytic Oxidation of Organic Compounds*; Khimiya: Moscow, 1985 (in Russian). (b) Methivier, C.; Beguin, B.; Brun, M.; Massardier, J.; Bertolini, J. C. *J. Catal.* **1998**, *173*, 374. (c) Carstens, J. N.; Su, S. C.; Bell, A. T. *J. Catal.* **1998**, *176*, 136. (d) Mehandjiev, D.; Zhecheva, E.; Ivanov, G.; Ioncheva, R. *Appl. Catal. A: General* **1998**, *167*, 277. (e) Merchetti, L.; Forni, L. *Appl. Catal. B: Environ.* **1998**, *15*, 179. (f) Ferri, D.; Forni, L. *Appl. Catal. B: Environ.* **1998**, *16*, 119. (g) Pecchi, G.; Reyes, P.; Gomez, R.; Lopez, T.; Fierro, J. L. G. *Appl. Catal. B: Environ.* **1998**, *17*, L7. (h) Baldi, M.; Escribano, V. S.; Amores, J. M. G.; Milella, F.; Busca, G. *Appl. Catal. B: Environ.* **1998**, *17*, L175. (i) Hua, W.; Gao, Z. *Appl. Catal. B: Environ.* **1998**, *17*, 37. (j) Widjaja, H.; Sekizawa, K.; Eguchi, K. *Chemistry Lett.* **1998**, 481. (k) Artizzu, P.; Guilhaume, N.; Garbowski, E.; Brull, Y.; Primet, M. *Catalysis Lett.* **1998**, *51*, 69. (l) Machida, M.; Taniguchi, H.; Kijima, T.; Nakatani, J. *J. Mater. Chem.* **1998**, *8*, 781.

35. (a) Takehira, K.; Namakawa, S.; Niyakawa, T.; Tsunoda, T.; Koizumi, M.; Sato, K.; Nakamura, I.; Ichijima, I. *Zh. Fiz. Khim.* **1997**, *71*, 1544 (in Russian). (b) Mizuno, N.; Ishige, H.; Seki, Y.; Misono, M.; Suh, D.-J.; Han, W.; Kudo, T. *Chem. Comm.* **1997**, 1295. (c) Ohmae, M.; Miyaji, K.; Azuma, N.; Takeishi, K.; Morioka, Y.; Ueno, A.; Ohfune, H.; Hayashi, H.; Udagawa, Y. *Chemistry Lett.* **1997**, 31. (d) Yu, L.; Yuan, S.; Wu, Z.; Wan, J.; Gong, M.; Pan, G.; Chen, Y. *Appl. Catal. A: General* **1998**, *171*, L172. (e) Sokolovskii, V. D.; Coville, N. J.; Parmaliana, A.; Eskendirov, I.; Makoa, M. *Catalysis Today* **1998**, *42*, 191. (f) de Lucas, A.; Valverde, J. L.; Canizares, P.; Rodrigez, *Appl. Catal. A: General* **1998**, *172*, 165. (g) Blick, K.; Mitrelias, T. D.; Hargreaves, J. S. J.; Hutchings, G. J. Joyner, R. W.; Kiely, K. J.; Wagner, F. E. *Catalysis Lett.* **1998**, *50*, 211. (h) Park, P. W.; Ledford, J. S. *Appl. Catal. B: Environ.* **1998**, *15*, 221. (i) Alemany, L. J.; Larrubia, M. A.; Blasco, J. M. *Appl. Catal. B: Environ.* **1998**, *16*, 139. (j) Basile, F.; Basini, L.; D'Amore, M. D.; Fornasari, G.; Guarinoni, A.; Matteuzzi, D.; Del Piero, G.; Trifiro, F.; Vaccari, A. *J. Catal.* **1998**, *173*, 247. (k) Hu, Y. H.; Ruckenstein, E. *Ind. Eng. Chem. Res.* **1998**, *37*,

2333. (l) Yang, C.; Xu, N.; Shi, J. *Ind. Eng. Chem. Res.* **1998**, *37*, 2601. (m) Yoshinaga, Y.; Okuhara, T. *J. Chem. Soc., Faraday Trans.* **1998**, *94*, 2235. (n) Tsiakaras, P.; Athanasiou, C.; Marnellos, G.; Stoukides, M.; ten Elshof, J. E.; Bouwmeester, H. J. M. *Appl. Catal. A: General* **1998**, *169*, 249. (o) Sexton, A. W.; Kartheuser, B.; Batiot, C.; Zanthoff, H. W.; Hodnett, B. K. *Catalysis Today* **1998**, *40*, 105. (p) Arutyunov, V. S.; Basevich, V. Ya.; Vedeneev, V. I.; Parfenov, Yu. V.; Sokolov, O. V.; Leonov, V. E. *Catalysis Today* **1998**, *42*, 241. (q) Taylor, S. H.; Hargreaves, J. S. J.; Hutchings, G. J.; Joyner, R. W.; Lembacher, C. W. *Catalysis Today* **1998**, *42*, 217.

36. (a) Ruth, K.; Kieffer, R.; Burch, R. *J. Catal.* **1998**, *175*, 16. (b) Sugiyama, S.; Abe, K.; Minami, T.; Hayashi, H.; Moffat, J. B. *Appl. Catal. A: General* **1998**, *169*, 77.

37. (a) Buyevskaya, O. V.; Baerens, M. *Catalysis Today* **1998**, *42*, 315. (b) Mmoru, A. *Catalysis Today* **1998**, *42*, 297. (c) Teng, Y.; Kobayashi, T. *Chemistry Lett.* **1998**, 327. (d) Dubien, C.; Schweich, D.; Mabilon, G.; Martin, B.; Prigent, M. *Chem. Eng. Sci.* **1998**, *53*, 471.

38. (a) Paul, S.; Le Courtois, V.; Vanhove, D. *Ind. Eng. Chem. Res.* **1997**, *36*, 3391. (b) Inumaru, K.; Ono, A.; Kubo, H.; Misono, M. *J. Chem. Soc., Faraday Trans.* **1998**, *94*, 1765. (c) Shigapov, A. N.; Hunter, N. R.; Gesser, H. D. *Catalysis Today* **1998**, *42*, 311. (d) Yao, C.-S.; Weng, H.-S. *Ind. Eng. Chem. Res.* **1998**, *37*, 2647. (e) Wildberger, M. D.; Mallat, T.; Göbel, U.; Baiker, A. *Appl. Catal. A: General* **1998**, *168*, 69. (f) Jacobs, L. L. G.; Lednor, P. W.; van Loon, P. J. M.; Oosterveld, M.; Vonkeman, K. A. *Pat. US 5,639,401* (to Shell Oil Company) [*J. Mol. Catal. A: Chem.* **1998**, *132*, 121].

39. (a) Nakagawa, K.; Ikenaga, N.; Suzuki, T.; Kobayashi, T.; Haruta, M. *Appl. Catal. A: General* **1998**, *169*, 281. (b) Fathi, M.; Hofstad, H. K.; Sperle, T.; Rokstad, O. A.; Holmen, A. *Catalysis Today* **1998**, *42*, 205. (c) Hegarty, M. E. S.; O'Connor, A. M.; Ross, J. R. H. *Catalysis Today* **1998**, *42*, 225. (d) Sobyanin, V. A.; Belyaev, V. D.; Gal'vita, V. V. *Catalysis Today* **1998**, *42*, 337. (e) Jiang, X. *Shiyou Huagong* **1998**, *27*, 4 (in Chinese). (f) Au, C.-T.; Liao, M.-S.; Ng, C.-F. *J. Phys. Chem. A* **1998**, *102*, 3959. (g) Yan, Q.-G.; Gao, L.-Z.; Chu, W.; Yu, Z.-L.; Yuan, S.-Y. *J. Nat. Gas Chem.* **1998**, *7*, 31. (h) Lu, Y.; Hou, R.-L.; Liu, Y.; Shen, S.-K. *J. Nat. Gas Chem.* **1998**, *7*, 38. (i) Wurzel, T.; Mleczko, L. *Chem. Eng. J.* 1998, 69, 127. (j) Otsuka, K.; Wang, Y.; Sunada, E.; Yamanaka, I. *J. Catal.* **1998**, *175*, 152. (k) Grunvald, V. R.; Piskunov, S. E.; Tolchinky, L. S.; Plate, N. A.; Kolbanovsky, Yu. A. *PCT Int. Appl. WO 98 06,663* [*Chem. Abstr.* **1998**, *128*, 182391v]. (l) Drago, R. S.; Jurczyk, K.; Kob, N.; Bhattacharyya, A.; Masin, J. *Catalysis Lett.* **1998**, *51*, 177. (m) Horstad, K. H.; Hoebink, J. H. B. J.; Holmen, A.; Marin, G. B. *Catalysis Today* **1998**, *40*,

15. (n) Ostrowski, T.; Giroir-Fendler, A.; Mirodatos, C.; Mleczko, L. *Catalysis Today* **1998**, *40*, 39, 49.

40. Fan, Y.; Kuang, W.; Chen, Y. *Chemistry Lett.* **1997**, 231.

41. Mongkhonsi, T.; Kershenbaum, L. *Appl. Catal. A: General* **1998**, *170*, 33.

42. Passoni, L. C.; Luna, F. J.; Wallau, M.; Buffon, R.; Schuchardt, U. *J. Mol. Catal. A: Chem.* **1998**, *134*, 229.

43. Becker, L.; Föster, H. *Appl. Catal. B: Environ.* **1998**, *17*, 43.

44. (a) Ito, T.; Lungford, H. H. *Nature* **1985**, *314*, 721. (b) Sokolovskii, V. D.; Aliev, G. M.; Buyevskaya, O. V.; Davydov, A. A. *Catalysis Today* **1989**, *4*, 293. (c) Zaera, F. *Acc. Chem. Res.* **1992**, *25*, 260. (d) Sokolovskii, V. D.; Mamedov, E. A. *Catalysis Today* **1992**, *14*, 331. (e) Bent, B. E. *Chem. Rev.* **1996**, *96*, 1361. (f) Wachs, I. E.; Jehng, J.-M.; Deo, G.; Weckhuysen, B. M.; Gulliants, V. V.; Benziger, J. B.; Sundaresan, S. *J. Catal.* **1997**, *170*, 75. (g) Ghoso, T.; Tabata, K. *J. Mol. Catal. A: Chem.* **1998**, *129*, 225. (h) Irigoyen, B.; Castellani, N.; Juan, A. *J. Mol. Catal. A: Chem.* **1998**, *129*, 297. (i) Solymosi, F. *J. Mol. Catal. A: Chem.* **1998**, *131*, 121.

45. (a) Augugliaro, V.; Coluccia, S.; Loddo, V.; Marchese, L.; Martra, G.; Palmisano, L.; Pantaleone, M.; Schiavello, M. *3rd Word Congress on Oxidation Catalysis*; Grasselli, R. K., Oyama, S. T., Gaffney, A. M., Lyons, J. E., Eds.; Elsevier: Amsterdam, 1997, p. 663. (b) Frei, H. *3rd Word Congress on Oxidation Catalysis*; Grasselli, R. K., Oyama, S. T., Gaffney, A. M., Lyons, J. E., Eds.; Elsevier: Amsterdam, 1997, p. 1041.(c) Wada, K.; Nakashita, M.; Yamamoto, A.; Wada, H.; Mitsudo, T. *Chemistry Lett.* **1997**, 1209. (d) Wada, K.; Nakashita, M.; Yamamoto, A.; Mitsudo, T. *Chem. Comm.* **1998**, 133. (e) Takenaka, S.; Tanaka, T.; Funabiki, T.; Yoshida, S. *J. Chem. Soc., Faraday Trans.* **1998**, *94*, 695. (f) Wada, K.; Yamada, H.; Watanabe, Y.; Mitsudo, T. *J. Chem. Soc., Faraday Trans.* **1998**, *94*, 1771.

46. Thampi, K. R.; Kiwi, J.; Grätzel, , M. *Catalysis Lett.* **1988**, *1*, 109.

47. Blatter, F.; Sun, H.; Vasenkov, S.; Frei, H. *Catalysis Today* **1998**, *41*, 297.

48. (a) Clerici, M. G.; Ingallina, P. *Catalysis Today* **1998**, *41*, 351. (b) Kuang, W.; Fan, Y.; Chen, K.; Chen, Y. *J. Chem. Res. (S)* **1997**, 366.

49. (a) Gallot, J. E.; On, D. T.; Kapoor, M. P.; Kaliaguin, S. *Ind. Eng. Chem. Res.* **1997**, *36*, 3458. (b) Vankelecom, I.; Vercruysse, K.; Moens, N.; Parton, R.; Reddy, J. S.; Jacobs, P. *Chem. Comm.* **1997**, 137. (c) Murphy, D. M.; Griffiths, E. W.; Rowlands, C. C.; Hancock, F. E.; Giamello, E. *Chem. Comm.* **1997**, 2177. (d) Tatsumi, T.; Koyano, K. A.; Igarashi, N. *Chem. Comm.* **1998**, 325. (e) Vayssilov, G. N.; van Santen, R. A. *J. Catal.* **1998**, *175*, 170.

50. Luna, F. J.; Ukawa, S. E.; Wallau, M.; Schuchardt, U. *J. Mol. Catal. A: Chem.* **1997**, *117*, 405.

51. (a) Temkin, M. I. *Adv. Catal.* **1979**, *28*, 173. (b) Veselov, V. V. *Kinetics and Catalysts of Hydrocarbon Conversion*; Naukova Dumka: Kiev, 1984 (in Russian).

52. (a) Barbieri, G.; Violante, V.; Di Maio, F. P.; Criscuoli, A.; Drioli, E. *Ind. Eng. Chem. Res.* **1997**, *36*, 3369. (b) Chan, K. Y. G.; Inal, F.; Senkan, S. *Ind. Eng. Chem. Res.* **1998**, *37*, 901. (c) Tjatjopoulos, G. J.; Vasalos, I. A.; *Ind. Eng. Chem. Res.* **1998**, *37*, 1410. (d) York, A. P. E.; Claridge, J. B.; Brungs, A. J.; Tsang, S. C.; Green, M. L. H. *Chem. Comm.* **1997**, 39. (e) Oklany, J. S.; Hou, K.; Hughes, R. *Appl. Catal. A: General* **1998**, *170*, 13. (f) Craciun, R.; Shereck, B.; Gorte, R. J. *Catalysis Lett.* **1998**, *51*, 149. (g) Bangala, D. N.; Abatzoglou, N.; Chornet, E. *AIChеE* **1998**, *44*, 927. (h) Wang, G.; Cui, Y.; Sun, Y.; Zhong, B. *Ranliao Huaxue Xuebao* **1998**, *26*, 56 (in Chinese).

53. (a) Maeno, H.; Matsumoto, H. *Pat. Jpn. Kokai Tokkyo Koho JP 10 52,639* (to Idemitsu Kosan Co.) [*Chem. Abstr.* **1998**, *128*, 232638d]. (b) Bangala, D. N.; Chornet, E. A. *Pat. US 5,679,614* (to University of Sherbrooke) [*J. Mol. Catal. A: Chem.* **1998**, *132*, 142].

54. Noceti, R. P.; Taylor, C. E.; D'Este, J. R. *Pat. US 5,720,858* [*CHEMTECH* May **1998**, 33].

55. (a) Olsbye, U.; Wurzel, T.; Mleczko, L. *Ind. Eng. Chem. Res.* **1997**, *36*, 5180. (b) Bitter, J. H.; Seshan, K.; Lercher, J. A. *J. Catal.* **1997**, *171*, 279. (c) Lu, Y.; Yu, C.; Xue, J.; Liu, Y.; Shen, S. *Chemistry Lett.* **1997**, 515. (d) Choudhary, V. R.; Uphade, B. S.; Mamman, A. S. *Appl. Catal. A: General* **1998**, *168*, 33. (e) Ruckenstein, E.; Hu, Y. H. *Ind. Eng. Chem. Res.* **1998**, *37*, 1744. (f) Bradford, M. C. J.; Vannice, M. A. *J. Catal.* **1998**, *173*, 157. (g) Wang, S.; Lu, G. Q. M. *Appl. Catal. B: Environ.* **1998**, *16*, 269. (h) Lu, Y.; Xue, J.-Z.; Liu, Y.; Shen, S.-K. *J. Nat. Gas Chem.* **1998**, *7*, 1. (i) Xu, L.-Y.; Wang, Q.-X.; Lin, L.-W.; Xu, Y.-D. *J. Nat. Gas Chem.* **1998**, *7*, 7. (j) Gamman, J. J.; Millar, G. J.; Rose, G.; Drennan, J. *J. Chem. Soc., Faraday Trans.* **1998**, *94*, 701. (k) Ferreira-Aparicio, P.; Guerrero-Ruiz, A.; Rodriguez-Ramos, I. *Appl. Catal. A: General* **1998**, *170*, 177. (l) Bitter, J. H.; Seshan, K.; Lercher, J. A. *J. Catal.* **1998**, *176*, 93. (m) Wu, Y.; Kawaguchi, O.; Matsuda, T. *Bull. Chem. Soc. Jpn.* **1998**, *71*, 563. (n) Wang, S.; Lu, G. Q. *Appl. Catal. A: General* **1998**, *169*, 271. (o) Chen, P.; Zhang, H.-B.; Lin, G.-D.; Tsai, K.-R. *Appl. Catal. A: General* **1998**, *166*, 343. (p) Osaki, T.; Horiuchi, T.; Sugiyama, T.; Suzuki, K.; Mori, T. *Catalysis Lett.* **1998**, *52*, 171. (q) Wang, S.; Lu, G. Q. *Energy & Fuels* **1998**, *12*, 248. (r) Hunter, N. R.; Prabawono, D.; Gesser, H. D. *Catalysis Today* **1998**, *42*, 347. (s)

Krylov, O. V.; Mamedov, A. Kh.; Mirzabekova, S. R. *Catalysis Today* **1998**, *42*, 211.

56. Wang, S.; Lu, G. Q.; Millar, G. J. *Energy & Fuels* **1996**, *10*, 896.

57. Kohno, Y.; Tanaka, T.; Funabiki, T.; Yoshida, S. *Chemistry Lett.* **1997**, 993.

58. (a) Dubkov, K. A.; Sobolev, V. I.; Talsi, E. P.; Rodkin, M. A.; Watkins, N. H.; Shteinman, A. A.; Panov, G. I. *J. Mol. Catal. A: Chem.* **1997**, *123*, 155. (b) Panov, G. I.; Uriarte, A. K.; Rodkin, M. A.; Sobolev, V. I. *Catalysis Today* **1998**, *41*, 365. (c) *Pat. US 5,672,777* (to Monsanto Company) [*CHEMTECH* January **1998**, 24].

59. Goto, A.; Aika, K. *Bull. Chem. Soc. Jpn.* **1998**, *71*, 95.

60. Wang, Y.; Otsuka, K. *J. Catal.* **1997**, *171*, 106.

61. (a) Novochinsky, I. I.; Chernavsky, P. A.; Ryabchenko, P. V.; Lunin, V. V. *Catalysis Lett.* **1998**, *51*, 191. (b) Delahay, G.; Ensuque, E.; Coq, B.; Figueras, F. *J. Catal.* **1998**, *175*, 7. (c) Burch, R.; Watling, T. C. *Appl. Catal. B: Environ.* **1998**, *17*, 131. (d) Huang, S.-J.; Walters, A. B.; Vannice, M. A. *Appl. Catal. B: Environ.* **1998**, *17*, 183. (e) Mitome, J.; Karakas, G.; Bryan, K. A.; Ozkan, U. S. *Catalysis Today* **1998**, *42*, 3. (f) Figueras, F.; Coq, B.; Ensuque, E.; Tachon, D.; Delahay, G. *Catalysis Today* **1998**, *42*, 117. (g) Efthimiadis, E. A.; Lionta, G. D.; Christoforou, S. C.; Vasalos, I. A. *Catalysis Today* **1998**, *40*, 15. (h) Ozkan, U. S.; Kumthekar, M. W.; Karakas, G. *Catalysis Today* **1998**, *40*, 3. (i) Stapf, D.; Ehrhardt, K.; Leuckel, W. *Chem. Eng. Technol.* **1998**, *21*, 412. (j) Lombardo, E. A.; Sill, G. A.; d'Itri, J. L.; Hall, W. K. *J. Catal.* **1998**, *173*, 440. (k) Shimizu, K.; Satsuma, A.; Hattori, T. *Appl. Catal. B: Environ.* **1998**, *16*, 319. (l) Praliaud, H.; Mikhailenko, S.; Chajar, Z.; Primet, M. *Appl. Catal. B: Environ.* **1998**, *16*, 359. (m) Maunula, T.; Kintaichi, Y.; Inaba, M.; Haneda, M.; Sato, K.; Hamada, H. *Appl. Catal. B: Environ.* **1998**, *15*, 291. (n) Burch, R.; Ramli, A. *Appl. Catal. B: Environ.* **199x**, *15*, 49. (o) Headon, K. A.; Zhang, D. *Ind. Eng. Chem. Res.* **1997**, *36*, 4595. (p) Yan, J.-Y.; Kung, H. H.; Sachtler, W. M. H.; Kung, M. C. *J. Catal.* **1998**, *175*, 294. (q) Ogura, M.; Hayashi, M.; Kikuchi, E. *Catalysis Today* **1998**, *42*, 159. (r) Abello, M. C.; Gomez, M. F.; Cadus, L. E. *Catalysis Lett.* **1998**, *53*, 185. (s) de Correa, C. M.; Villa de P, A. L. *Catalysis Lett.* **1998**, *53*, 205. (t) Halasz, I.; Brenner, A. *Catalysis Lett.* **1998**, *51*, 195. (u) Chajar, Z.; Le Chanu, V.; Primet, M.; Praliaud, H. *Catalysis Lett.* **1998**, *52*, 97.

62. (a) Aylor, A. W.; Lobree, L. J.; Reimer, J. A.; Bell, A. T. *J. Catal.* **1997**, *170*, 390. (b) Burch, R.; Fornasiero, P.; Watling, T. C. *J. Catal.* **1998**, *176*, 204.

63. Adesina, A. A.; Cant, N. W.; Saberi-Moghaddam, A.; Szeto, C. H. L.; Trimm, D. L. *J. Chem. Technol. Biotechnol.* **1998**, *72*, 19.

64. Neumann, R.; Dror, I. *Appl. Catal. A: General* **1998**, *172*, 67.

65. (a) Sugiyama, S.; Miyamoto, T.; Hayashi, H.; Tanaka, M.; Moffat, J. B. *J. Mol. Catal. A: Chem.* **1997**, *118*, 129. (b) Chen, Q.; Schweitzer, J. A.; Van Den Oooserkamp, P. F.; Berger, R. J.; De Smet, C. R. H.; Marin, G. B. *Ind. Eng. Chem. Res.* **1997**, *36*, 3248. (c) Sugiyama, S.; Abe, K.; Miyamoto, T.; Hayashi, H.; Moffat, J. B. *J. Mol. Catal. A: Chem.* **1998**, *130*, 297. (d) Kaddouri, A.; Anouchinsky, R.; Mazzocchia, C.; Madeira, L. M.; Portela, M. F. *Catalysis Today* **1998**, *40*, 59. (e) Brabec, L.; Jeschke, M.; Klik, R.; Novakova, J.; Kubelkova, L.; Freude, D.; Bosacek, V.; Meusinger, J. *Appl. Catal. A: General* **1998**, *167*, 309. (f) Sugiyama, S.; Miyamoto, T.; Hayashi, H.; Moffat, J. B. *J. Mol. Catal. A: Chem.* **1998**, *135*, 199. (g) Lin, S. W.; Kim, Y. C.; Ueda, W. *Bull. Chem. Soc. Jpn.* **1998**, *71*, 1089. (h) Dang, Z.; Gu, J.; Lin, J.; Yang, D. *Catalysis Lett.* **1998**, *54*, 129.

66. (a) Pantazidis, A.; Bucholz, S. A.; Zanthoff, H. W.; Schuurman, Y.; Mirodatos, C. *Catalysis Today* **1998**, *40*, 65. (b) Kaddouri, A.; Mazzacchia, C.; Tempesti, E. *Appl. Catal. A: General* **1998**, *169*, L3. (c) Del Rosso, R.; Kaddouri, A.; Anouchinsky, R.; Mazzacchia. C.; Gronchi, P.; Centola, P. *J. Mol. Catal. A: Chem.* **1998**, *135*, 181. (d) Rulkens, R.; Tilley, T. D. *J. Am. Chem. Soc.* **1998**, *120*, 9959.

67. (a) Tempesti, E.; Kaddouri, A.; Mazzocchia, C. *Appl. Catal. A: General* **1998**, *166*, L259. (b) Moriceau, P.; Grzybowska, B.; Barbaux, Y.; Wrobel, G.; Hecquet, G. *Appl. Catal. A: General* **1998**, *168*, 269. (c) Takita, Y.; Sano, K.; Muraya, T.; Nishiguchi, H.; Kawata, N.; Ito, M.; Akbay, T.; Ishihara, T. *Appl. Catal. A: General* **1998**, *170*, 23.

68. (a) Rajeshwer, D.; Gurudat, B. A.; Gokak, D. T.; Rao, K. V.; Krishmamurthy, K. R.; Bhadwaj, I. S. G. *Pat. US 5,677,260* (to Indian Petrochemicals Corp.) [*J. Mol. Catal. A: Chem.* **1998**, *132*, 118]. (b) Schmidt, R.; Jeanneret, J. J.; Raghuram, S.; McCulloch, B. Pat. *US 5,672,265* (to UOP) [*J. Mol. Catal. A: Chem.* **1998**, *132*, 138]. (c) Froment, G. F. A.; Dehertog, W. J.; Hippolyte, M. L. *Pat. US 5,672,796* (to Amoco Corporation) [*J. Mol. Catal. A: Chem.* **1998**, *132*, 138].

69. (a) Chang, J.-S.; Park, S.-E.; Park, M. S. *Chemistry Lett.* **1997**, 1123. (b) Nakagawa, K.; Okamura, M.; Ikenaga, N.; Suzuki, T.; Kobayashi, T. *Chem. Comm.* **1998**, 1025.

70. (a) Amenomiya, Y.; Birss, V. I.; Goledzinowski, M.; Galuszka, J.; Sanger, A. R. *Catal. Rev. – Sci. Eng.* **1990**, *32*, 163. (b) Krylov, O. V. *Usp. Khim.* **1992**, *61*, 1550 (in Russian). (c) Minachev, Kh. M.; Dergachev, A. A. *Izvest. Akad. Nauk, Ser. Khim.* **1998**, 1071 (in Russian).

71. (a) Wendt, G.; Sommer, M.; Wolf, D. *React. Kinet. Catal. Lett.* **1991**, *44*, 457. (b) Woldman, L. S.; Sokolovskii, V. D. *Catalysis Lett.* **1991**, *8*, 61. (c) Sugiyama, S.; Shimodan, K.; Ookubo, A.; Shigemoto, A.; Hayashi, H.; Moffat, J. B. *Bull. Chem. Soc. Jpn.* **1993**, *66*, 2391. (d) Andrianova, Z. S.; Ivanova, A. N.; Matkovskiy, P. E.; Startseva, G. P. *Kinetika i. Kataliz* **1996**, *37*, 264 (in Russian). (e) Balint, I.; Dobrescu, G.; Vass, M. *Rev. Roum. Chim.* **1997**, *42*, 1009. (f) Srivasan, C.; Viswanathan, B.; Mulla, S. A. R.; Choudhary, V. R. *Indian J. Chem., Sect. A: Inorg., Bio-inorg., Phys., Theor. Anal. Chem.* **1997**, *36A*, 926. (g) Stansch, Z.; Mleczko, L.; Baerns, M. *Ind. Eng. Chem. Res.* **1997**, *36*, 2568. (h) Onal, I.; Senkam, S. *Ind. Eng. Chem. Res.* **1997**, *36*, 4028. (i) Taniewski, M.; Czechowicz, D. *Ind. Eng. Chem. Res.* **1997**, *36*, 4193. (j) Liu, Y.; Liu, X.-X.; Hou, R.-L.; Xue, J.-Z.; Yu, C.-C.; Shen, S.-K. *J. Nat. Gas Chem.* **1998**, *7*, 43. (k) Kurosaka, T.; Matsuhashi, H.; Arata, K. *Chemistry Lett.* **1998**, 265. (l) Choudhary, V. R.; Mulla, S. A. R.; Rane, V. H. *J Chem. Technol. Biotechnol.* **1998**, *71*, 167. (m) Kou, Y.; Zhang, B.; Niu, J.; Li, S.; Wang, H.; Tanaka, T.; Yoshida, S. *J. Catal.* **1998**, *173*, 399. (n) Burrows, A.; Kiely, C. J.; Hargeaves, J. S. J.; Joyner, R. W.; Hutchings, G. J.; Sinev, M. Yu.; Tulenin, Yu. P. *J. Catal.* **1998**, *173*, 383. (o) Au, C. T.; Chen, K. D.; Ng, C. F. *Appl. Catal. A: General* **1998**, *170*, 81. (p) Wan, H. L.; Weng, W. Z.; Iwasawa, Y. *Hyomen* **1998**, *36*, 53 (in Japanese). (q) Vereschagin, S. N.; Gupalov, V. K.; Anisimov, L. N.; Terekhin, N. A.; Kovrigin, L. A.; Kirik, N. P.; Kondratenko, E. V.; Anshits, A. G. *Catalysis Today* **1998**, *42*, 361. (r) Weng, W.; Chen, M.; Wan, H.; Liao, Y. *Catalysis Lett.* **1998**, *53*, 43. (s) Au, C. T.; Chen, K. D.; Ng, C. F. *Appl. Catal. A: General* **1998**, *171*, 283. (t) Anshits, A. G.; Voskresenskaya, E. N.; Kondratenko, E. V.; Fomenko, E. V.; Sokol, E. V. *Catalysis Today* **1998**, *42*, 197. (u) Pyatnitskiy, Yu. I.; Ilchenko, N. I.; Pavlenko, M. V. *Catalysis Today* **1998**, *42*, 233. (v) Maksimov, N. G.; Selytin, G. E.; Anshits, A. G.; Kondratenko, E. V.; Roguleva, V. G. *Catalysis Today* **1998**, *42*, 279. (w) Liu, Y.; Liu, X.; Xue, J.; Hou, R.; Zhang, B.; Yu, C.; Shen, S. *Appl. Catal. A: General* **1998**, *168*, 139. (x) Traykova, M.; Davidova, N.; Tsaih, J.-S.; Weiss, A. H. *Appl. Catal. A: General* **1998**, *169*, 237. (y) Pak, S.; Lunsford, J. H. *Appl. Catal. A: General* **199x**, *168*, 131. (z) Sugiyama, S.; Iguchi, Y.; Nishioka, H.; Minami, T.; Moriga, T.; Hayashi, H.; Moffat, J. B. *J. Catal.* **1998**, *176*, 25.

72. Fornasari, G.; Bellussi, G. C. *Pat. US 5,670,442* (to Eniricherche S. p A.) [*J. Mol. Catal. A: Chem.* **1998**, *132*, 116].

73. Ungar, R. K.; Zhang, X.; Lambert, R. M. *Appl. Catal.* **1988**, *42*, L1.

74. Okabe, K.; Sayama, K.; Kusama, H.; Arakawa, H. *Chemistry Lett.* **1997**, 457.

75. (a) Kodama, T.; Shimizu, T.; Aoki, A.; Kitayama, Y. *Energy & Fuels* **1997**, *11*, 1257. (b) Rezgui, S.; Liang, A.; Cheung, T.-K.; Gates, B. C. *Catalysis Lett.* **1998**, *53*, 1. (c) Zadeh, J. S. M.; Smith, K. J. *J. Catal.* **1998**, *176*, 115. (d) Vereschagin, S. N.; Shishkina, N. N.; Anshits, A. G. *Catalysis Today* **1998**, *42*, 303. (e) Sawa, H.; Abe, H. *Eur. Pat. Appl. EP 814,071* (to Furukawa Denchi Kabushiki Kaisha) [*Chem. Abstr.* **1998**, *128*, 103600m]. (f) Pareja, P.; Molina, S.; Amariglio, A.; Amariglio, H. *Appl. Catal. A: General* **1998**, *168*, 289. (g) Didenko, L. P.; Linde, V. R.; Savchenko, V. I. *Catalysis Today* **1998**, *42*, 367.

76. (a) Nivarthy, G. S.; He, Y.; Seshan, K.; Lercher, J. A. *J. Catal.* **1998**, *176*, 192. (b) Sun, M.; Sun, J.; Li, Q. *Chemistry Lett.* **1998**, 519.

77. (a) Wang, D.; Lunsford, J. H.; Rosynek, M. P. *Topics Catal.* **1996**, *3*, 289. (b) Weckhuysen, B. M.; Wang, D.; Rosynek, M. P.; Lunsford, J. H. *Angew. Chem., Int. Ed. Engl.* **1997**, *36*, 2374. (c) Weckhuysen, B. M.; Rosynek, M. P.; Lunsford, J. H. *Catalysis Lett.* **1998**, *52*, 31. (d) Weckhuysen, B. M.; Wang, D.; Rosynek, M. P.; Lunsford, J. H. *J. Catal.* **1998**, *175*, 338. (e) Guczi, L.; Borko, L,; Koppany, Zs.; Mizukami, F. *Catalysis Lett.* **1998**, *54*, 33. (f) Liu, W.; Xu, Y.; Wong, S.-T.; Wang, L.; Qiu, J.; Yang, N. *J. Mol. Catal. A: Chem.* **1997**, *120*, 257.

78. (a) Minachev, Kh. M.; Bragin, O. V.; Bondarenko, T. N. *Neftekhimiya* **1989**, *29*, 480 (in Russian). (b) Solymosi, F.; Szoke, A. *Appl. Catal. A: General* **1998**, *166*, 225. (c) Pierella, L. B.; Eimer, G. A.; Anunziata, O. A. *React. Kinet. Catal. Lett.* **1998**, *63*, 271.(d) Haizmann, R. S.; Park, J. Y. G.; Russ, M. B. *Pat US 5,683,573* (to UOP) [*J. Mol. Catal. A: Chem.* **1998**, *132*, 142].

79. (a) Liu, S.; Dong, Q.; Ohnishi, R.; Ichikawa, M. *Chem. Comm.* **1998**, 1217. (b) Naito, S.; Karaki, T.; Iritani, T. *Chemistry Lett.* **1997**, 877.

80. Revenko, O. M.; Khcheyan, Kh. E.; Tikhonova, M. P.; Borisoglebskaya, A. V. *Neftekhimiya* **1985**, *25*, 475 (in Russian).

81. (a) Osada, Y.; Ogasawara, S.; Fukushima, T.; Shikada, T.; Ikariya, T. *J. Chem. Soc., Chem. Commun.* **1990**, 1434. (b) Osada, Y.; Okino, N.; Ogasawara, S.; Fukushima, T.; Shikada, T.; Ikariya, T. *Chem. Lett.* **1990**, 281. (c) Khan, A. Z.; Ruckenstein, E. *J. Chem. Soc., Chem. Commun.* **1993**, 587. (d) Arishtirova, K.; Kovacheva, P.; Davidova, N. *Appl. Catal. A: General* **1998**, *167*, 271. (e) Takahashi, Y. *Pat. Jpn. Kokai Tokkyo Koho JP 10 25, 257* (to Toyota Motor Corp.) [*Chem. Abstr.* **1998**, *128*, 127744n].

82. Liu, Y.; Lin, J.; Tan, K. L. *Catalysis Lett.* **1998**, *50*, 165.

83. Koerts, T.; Leclercq, P. A.; van Santen, R. A. *J. Am. Chem. Soc.* **1992**, *114*, 7272.

84. (a) Li, Y.; Armor, J. N. *Chem. Comm.* **1997**, 2013. (b) Li, Y.; Armor, J. N. *J. Catal.* **1998**, *173*, 511.

85. (a) Centi, G.; Perathoner, S. *CHEMTECH* February **1998**, 13. (b) Albonetti, S., Blanchard, G.; Burratin, P.; Masetti, S.; Tagliani, A.; Trifiro, F. *Catalysis Lett.* **1998**, *50*, 17. (b) Nilsson, R.; Andersson, A. *Ind. Eng. Chem. Res.* **1997**, *36*, 5207. (c) Albonetti, S.; Blanchard, G.; Burratin, P.; Cavani, F.; Masetti, S.; Trifiro, F. *Catalysis Today* **1998**, *42*, 283. (d) Sanati, M.; Akbari, R.; Masetti, S.; Trifiro, F. *Catalysis Today* **1998**, *42*, 325. (e) Bradzil, J. F., Jr.; Cavalcanti, F. A. P. *Pat. US 5,576,469* (to Standard Oil Co.)[*CHEMTECH* March **1997**, 4]. (f) Bradzil, J. F., Jr.; Cavalcanti, F. A. P.; Padolewski, J. P. *Pat. US 5,693,587* (to Standard Oil Co.) [*Chem. Abstr.* **1998**, *128*, 49799t].

86. (a) Martin, A.; Berndt, H.; Lucke, B.; Meisel, M. *Topics Catal.* **1996**, *3*, 377. (b) Hannour, F. K.; Martin, A.; Bruckner, A.; Wolf, G.; Lucke, B. *React. Kinet. Catal. Lett.* **1998**, *63*, 225. (c) Martin, A.; Hannour, F. K.; Bruckner, A.; Lucke, B. *React. Kinet. Catal. Lett.* **1998**, *63*, 245.

87. Manohar, B.; Reddy, B. M. *J. Chem. Technol. Biotechnol.* **1998**, *71*, 141.

88. (a) Wang, Y.; Katagiri, M.; Otsuka, K. *Chem. Comm.* **1997**, 1187. (b) Bruce, D. A. *et al. Pat US 5,679,867* (to Hoechst Celanese Corp.) [*CHEMTECH* January **1998**, 25].

89. Flid, M. R.; Kurlyandskaya, I. I.; Treger, Yu. A.; Guzhnovskaya, T. D. *3rd Word Congress on Oxidation Catalysis*; Grasselli, R. K., Oyama, S. T., Gaffney, A. M., Lyons, J. E., Eds.; Elsevier: Amsterdam, 1997, p. 305.

CHAPTER IV

ACTIVATION OF C–H BONDS BY LOW-VALENT METAL COMPLEXES ("THE ORGANOMETALLIC CHEMISTRY")

In the end of the 1960s the leading specialist in homogeneous catalysis Jack Halpern wrote [1]: "to develop a successful approach for activation of C–H bonds, particularly saturated hydrocarbons, this problem being at present one of the most important and challenging in the entire field of homogeneous catalysis". By that time, a large quantity of data had accumulated, indicating the possibility of activating alkanes by transition metal complexes, where the alkane molecule reacts directly with the metal center by entering into the coordination sphere of metal complexes to form intermediately alkyl derivatives.

Indeed, it had been known, that

(i) a large number of complexes react with various compounds, containing carbon–hydrogen bonds, particularly with aromatic hydrocarbons, with rupture of the C–H bond and the formation of a metal-carbon fragment;

(ii) an aliphatic carbon-hydrogen bond can be activated if it is present in a ligand attached to a metal center in the complex;

(iii) reactions of alkanes on the surface of metals or their compounds, which undoubtedly involve the C–H bond cleavage in an alkane molecule, also indicate the possibility of such reactions with metal complexes in homogeneous solutions, at least for polynuclear cluster complexes containing several metal atoms;

(iv) the reactions of alkanes in homogeneous solutions with electrophilic oxidizing agents containing metal in a high oxidation state sometimes suggest the possibility of direct participation of the metal ion in the reaction with the alkane to form a metal-carbon bond.

Starting from analogies with the activation of other compounds that do not contain any π-bonds, in particular dihydrogen:

$$H–H + M \longrightarrow H–M–H$$

it was possible to assume that the initial alkane reaction with compounds of transition metals in a low oxidation state would be an oxidative addition of the C–H bond to a metal center, M, according to the general equation:

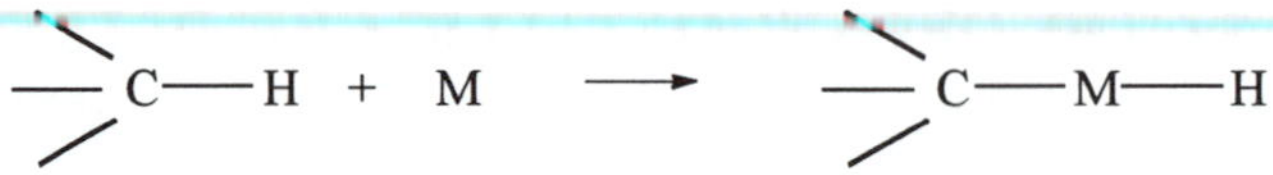

Numerous processes of such type affording organyl hydride metal complexes became known in the end of 1970s and in 1980s. Many processes discovered at that time occur via intermediate formation of organyl hydrides. In σ-organyl hydride complexes of formula σ-R–M–H, the organyl group R can be alkyl, aryl, acyl, vinyl etc. Oxidative addition of simple alkanes gives rise to the formation of alkyl hydride complexes (R = alkyl).

The reactions between C–H compounds and metal complexes in a low oxidation state, which produce organometallic derivatives (i.e., compounds containing direct carbon–metal bonds) will be considered in this chapter. In some cases the intermediate σ-organyl and organyl hydride complexes can not be isolated, although the evidence for their formation is obtained. Reactions of such type (often-catalytic processes) will be considered in this chapter.

IV.1. Formation of σ-Organyl Hydride Complexes

In reactions between a C–H compound, RH, and a metal complex, M, to produce R–M–H, the oxidation state of a metal ion is changed and it is two units higher in the organyl hydride than it was in the initial metal compound. Alkanes, arenes, alkenes and monosubstituted acetylenes enter into such processes. The reactions frequently occur in solution at room temperature, but sometimes heating is required. Certain reactions are stimulated by irradiation. An increase in temperature or light is essential for the abstraction of several ligands from the initial complex and the formation of a coordinatively unsaturated species capable of effecting the oxidative addition of the C–H compound to itself.

IV.1.A. CYCLOMETALATION

Due to the chelate effect, the *intra*molecular cleavage of the C–H bond occurs much more readily than *inter*molecular activation and gives rise to a more stable σ-organyl hydride complex. The general schematic equation of this cyclometalation is depicted as follows:

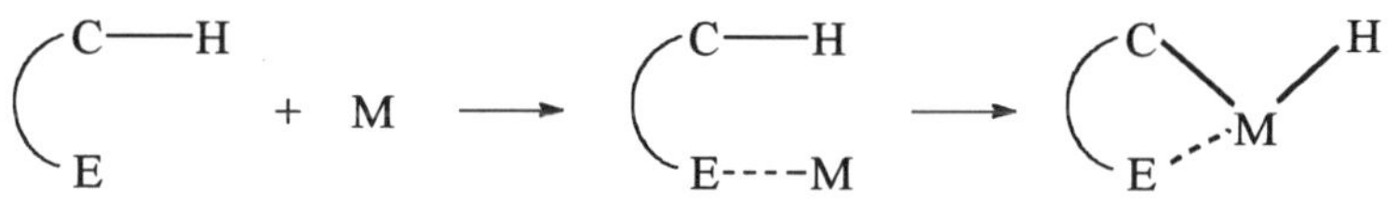

Here E is a donor atom, e.g., N, P, As etc.

A book [2] and reviews [3] have been devoted to the cyclometalation reaction with participation of metal complexes in a low oxidation state. Many examples of cyclometalation into *ortho*-position of monosubstituted aromatic ring are known, some of them are presented in Scheme IV.1 [4]. Sometimes the cyclometalation can occur with splitting either at an aryl or methyl group in the same molecule of a ligand. Thus the reaction of 6-R-2,2′-bipyridine (R = CH_2Ph, CHMePh, CMe_2Ph, CH_2Me, CMe_3, CH_2CMe_3) with $HAuCl_4$ gives the cyclometalated species which arise from direct activation of a C–H bond either of a phenyl or a methyl substituent [5]. Cyclometalation reactions of heterocyclic aromatic ligands are shown in Scheme IV.2 [6]. Not all of the cyclometalation reactions shown in these schemes proceed via oxidative addition mechanism.

The most interesting (from the point of view of alkane activation) cyclometalation reactions of sp^3-C–H bonds are demonstrated in Scheme IV.3 [7] (in addition to references [7] depicted in this scheme, see also [8]). Scheme IV.4 gives some examples of the cyclometalation of vinylic C–H bonds, occurring with the splitting of bonds between carbon and some elements, and other special cases [9]. For cyclometalation of the C–H bonds at the C=C bonds, see also [10a,b].

The activation of the aldehyde C–H bond gives an acyl complex [10c]:

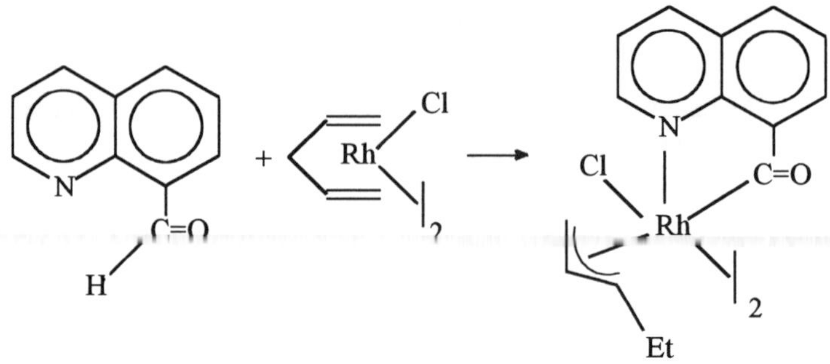

In some cases, the functionalization of C–H compounds involves the cyclometalation stage. For example, $CoH_3(PPh_3)_3$ catalyzes the regiospecific double and monoinsertion of tolan into the *ortho*-C–H bonds of 1,2-diarylazenes (Scheme IV.5) [11a] and L_3RhCl catalyzes the alkylation of 2-vinylpyridines with alkenes (Scheme IV.6) [11b]. Carefully investigated cyclopalladation (orthopalladation) reactions which proceed most likely as electrophilic substitution, as well as cyclometalation with complexes of metals in a high oxidation state will be considered in Chapter VIII.

IV.1.B. INTERMOLECULAR OXIDATIVE ADDITION

The σ-organyl complexes formed in oxidative addition [12] of alkanes, arenes as well as alkenes and monosubstituted acetylenes can be fairly stable and in many cases have been isolated. Thus, upon heating or photolysis, the complexes Cp_2WH_2, Cp_2WCO, and Cp_2WHCH_3 give rise to a coordinatively unsaturated tungstocene species Cp_2W, which readily combines with aromatic or alkylaromatic hydrocarbons [13].

In the reaction with toluene, the products are $Cp_2W(H)C_6H_4CH_3$ and $Cp_2W(CH_2Ph)C_6H_4CH_3$, i.e., tungsten is inserted both in a C–H bond in the aromatic ring and in a C–H bond at sp^3-hybridized carbon atom. However, cyclohexane and neopentane do not give rise to insertion products in this instance.

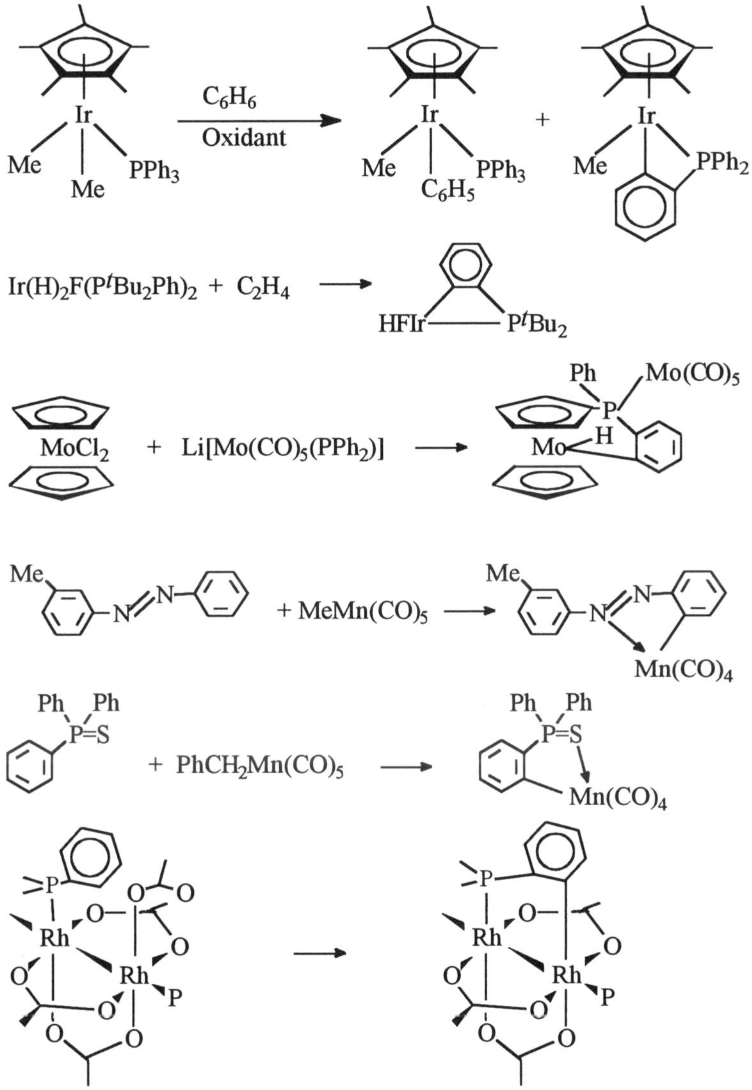

[4a] [4b] [4c] [4d] [4e] [4f]

Scheme IV.1 (part 1). Examples of the cyclometalation of aromatic compounds by metal complexes in a low oxidation state. Reference numbers are in brackets.

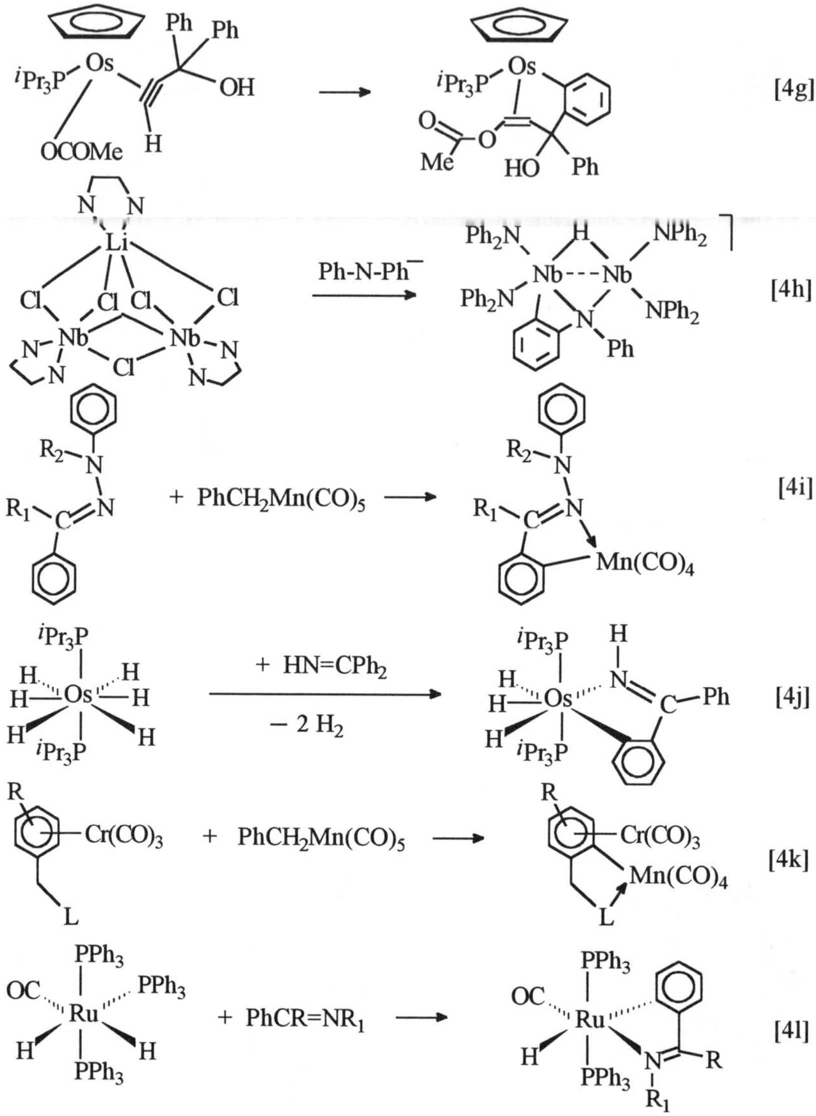

Scheme IV.1 (part 2). Examples of the cyclometalation of aromatic compounds by metal complexes in a low oxidation state. Reference numbers are in brackets.

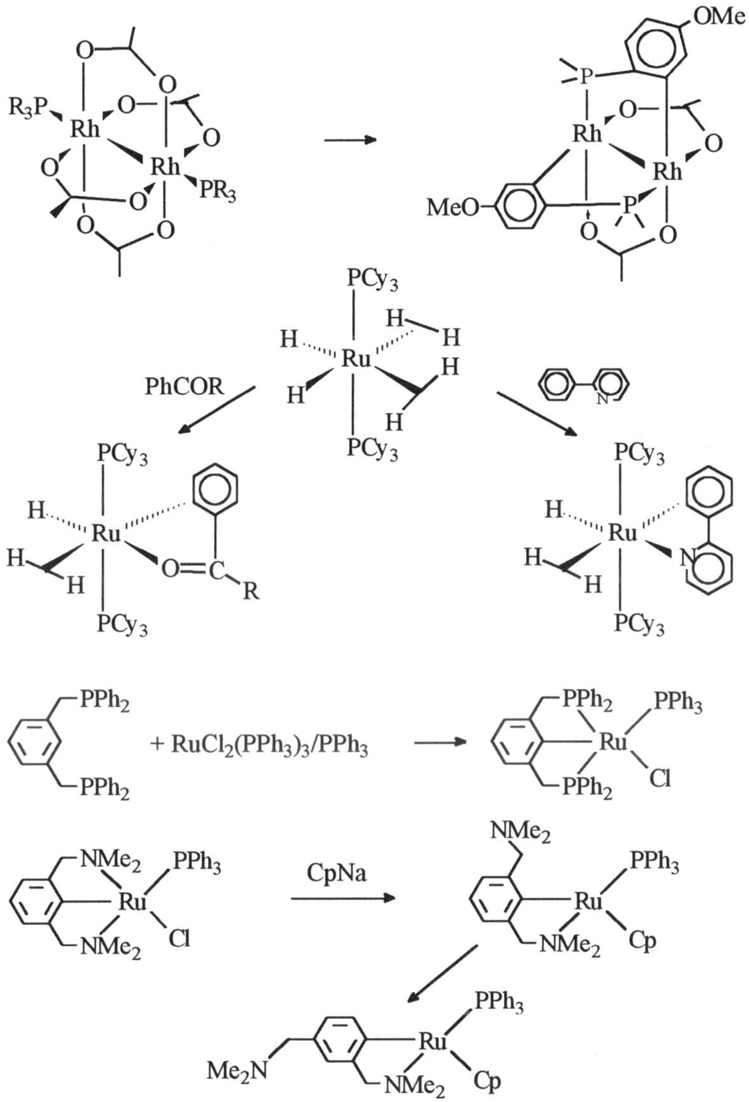

[4m]

[4n]

[4o,p]

[4q]

Scheme IV.1 (part 3). Examples of the cyclometalation of aromatic compounds by metal complexes in a low oxidation state. Reference numbers are in brackets.

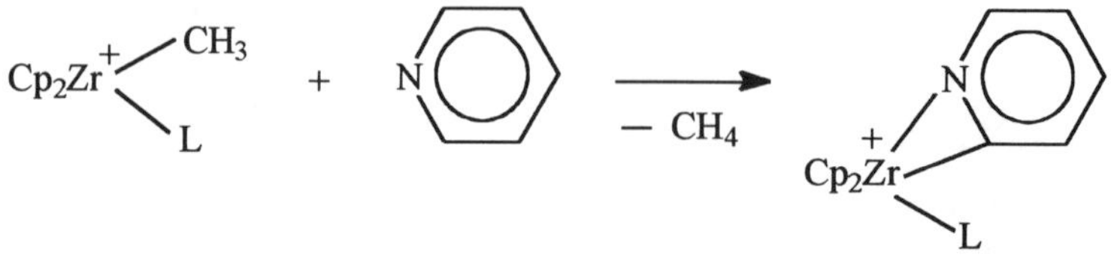

[6a]

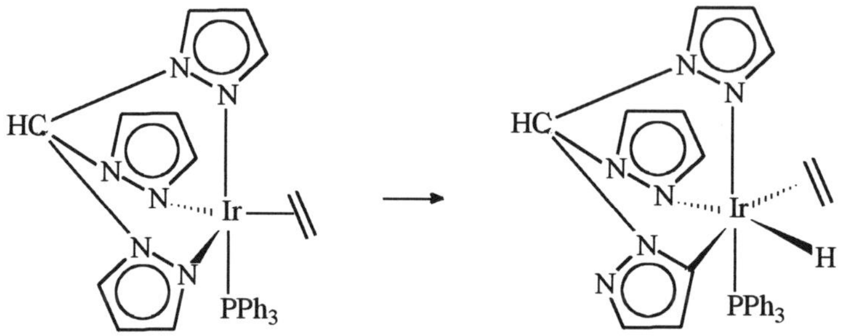

[6b]

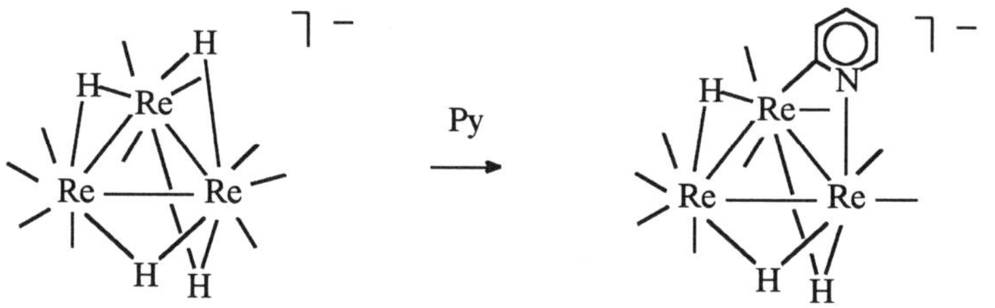

[6c]

Scheme IV.2. Examples of the cyclometalation of aromatic heterocycles by metal complexes in a low oxidation state. Reference numbers are in brackets.

The oxidative addition of alkanes with formation of alkyl hydride complexes was first demonstrated directly in studies using iridium complexes [14]. Thus the iridium dihydride derivative $Cp^*Ir(H)_2PMe_3$ (Cp^* = pentamethylcyclopentadienyl) after irradiation in solution in cyclohexane or neopentane, produces the complexes $Cp^*(PMe_3)Ir(H)(C_6H_{11})$ and $Cp^*(PMe_3)Ir(H)CH_2CMe_3$ in a satisfactory yield. Other saturated hydrocarbons and benzene also readily add to an iridium complex.

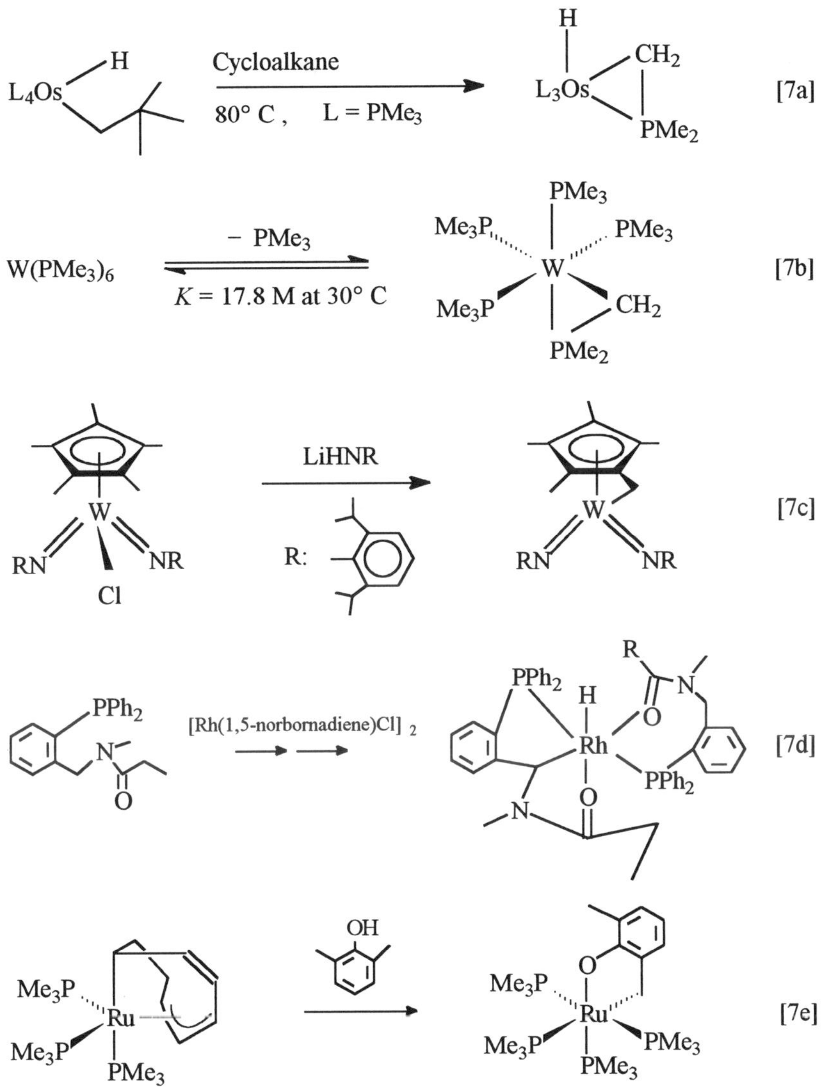

Scheme IV.3 (part 1). Examples of the cyclometalation reactions between metal complexes in a low oxidation state and aliphatic C–H bonds. Reference numbers are in brackets.

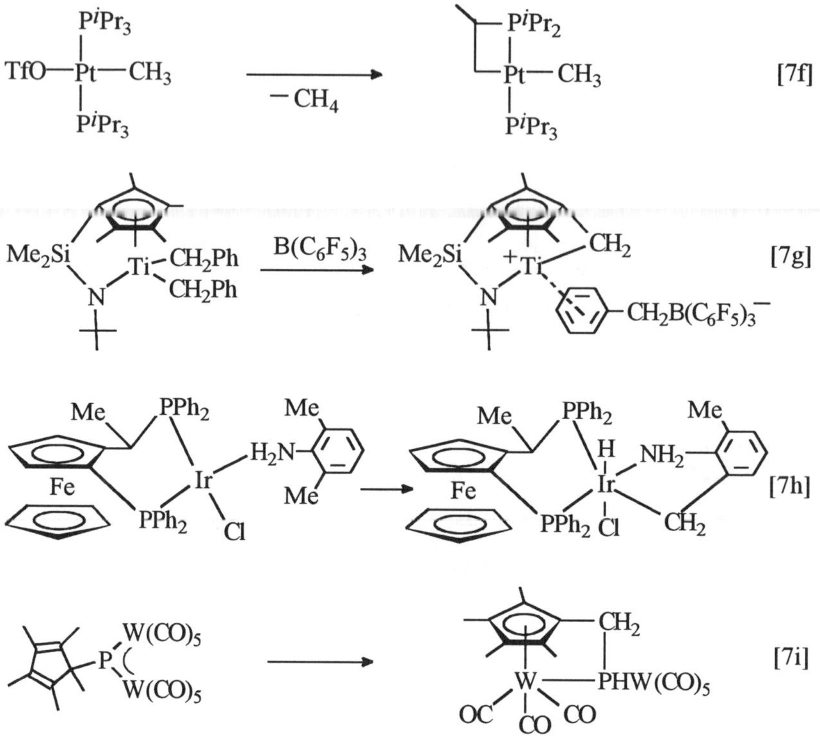

Scheme IV.3 (part 2). Examples of the cyclometalation reactions between metal complexes in a low oxidation state and aliphatic C–H bonds. Reference numbers are in brackets.

By treatment with $CHBr_3$ at –60 °C, alkyl hydride complexes were converted into the more stable derivatives $Cp^*(PMe_3)Ir(Br)R$. Irradiation of $Cp^*Ir(H)_2PMe_3$ in CMe_4+C_6D_{12} mixture led to the formation of $Cp^*(PMe_3)$-$Ir(H)CH_2CMe_3$ and $Cp^*(PMe_3)Ir(D)C_6D_{11}$ with very small admixtures of cross-addition products.

The reaction of alkanes with the complex $Cp^*Rh(H)_2PMe_3$ takes place analogously under irradiation at a temperature below –30 °C [14b]. The selectivity (both substrate and positional selectivity) proved to be much higher for the rhodium complex than for the iridium one.

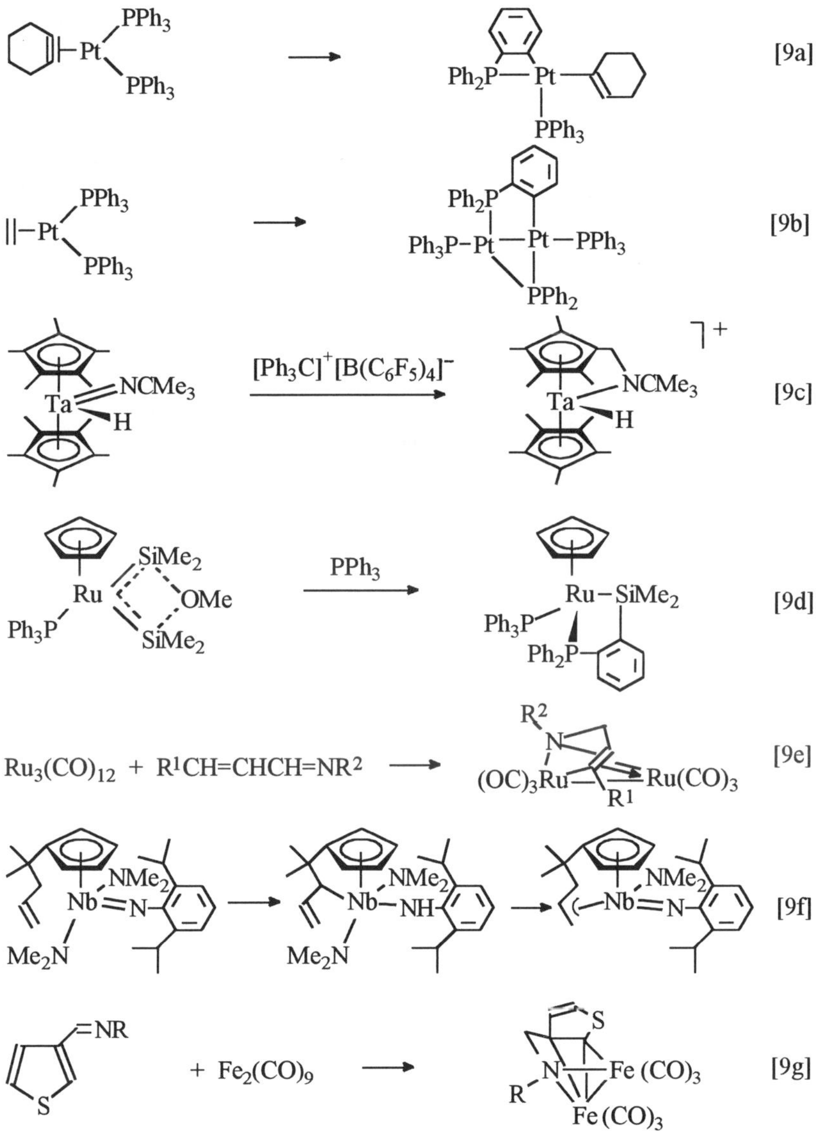

Scheme IV.4. Some other examples of the cyclometalation reactions. Reference numbers are in brackets.

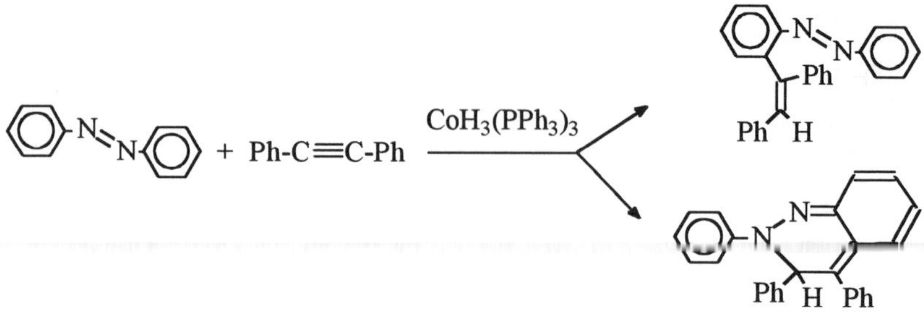

Scheme IV.5. The insertion of tolan into the *ortho*-C–H bonds of 1,2-diarylazenes.

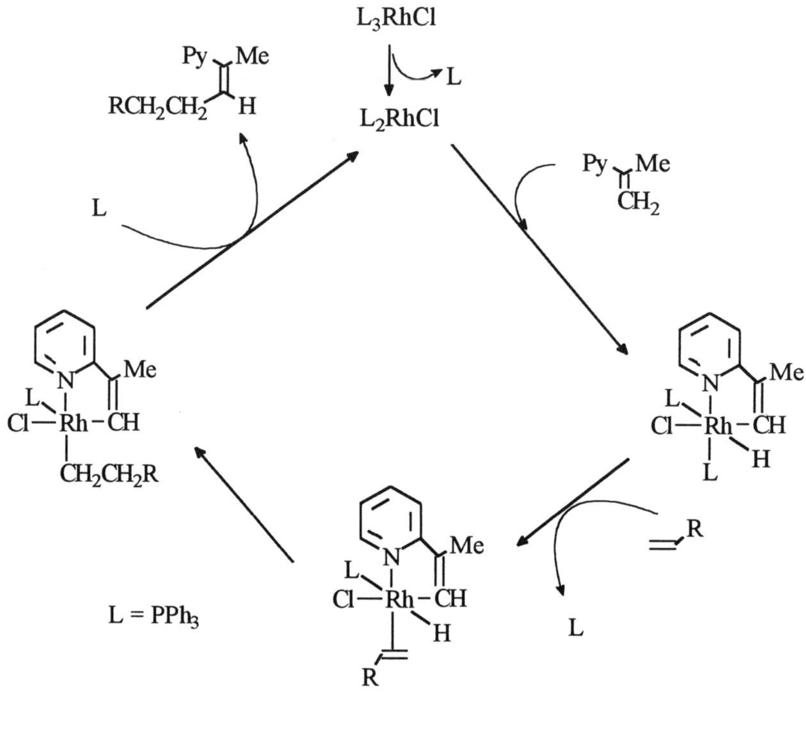

Scheme IV.6. The alkylation of 2-vinylpyridines with alkenes.

When the cyclohexyliridium hydride complex **IV-1** or the *n*-pentyliridium hydride derivative **IV-2** was heated for 50 h at 140 °C in a mixture of 91.5 % of cyclohexane and 8.5 % of *n*-pentane, the following equilibrium was established:

$$Cp^*(PMe_3)Ir(H)C_6H_{11} + CH_3(CH_2)_3CH_3 \overset{K}{\rightleftharpoons}$$

IV-1

$$\rightleftharpoons \quad Cp^*(PMe_3)Ir(H)(CH_2)_4CH_3 + C_6H_{12} \qquad (IV.1)$$

IV-2

The ratio **IV-2** : **IV-1** was 1.0 ± 0.1. Hence the constant

$$K = [\mathbf{IV\text{-}2}][C_6H_{12}]/[\mathbf{IV\text{-}1}][n\text{-}C_5H_{12}] = 10.6,$$

which corresponds to $\Delta G° = -2.0$ kcal/mol. It has been suggested that the entropy changes in the reaction are small and, assuming that the CH_2 and CH_3 bond energies in cyclohexane and *n*-pentane are 94.5 and 98 kcal mol^{-1} respectively, we can find that the energy of the M–C bond in the complex **IV-2** is higher by 5.5 kcal mol^{-1} than in the complex **IV-1**. It follows from the estimates that the methyliridium hydride complex $Cp^*(PMe_3)Ir(H)CH_3$ should also be very thermodinamically stable. It was obtained in 58% yield on heating a solution of the complex **IV-1** in cyclooctane in the presence of methane [14b].

A large number of reactions are known in which, as in reactions just discussed, a coordinatively unsaturated species is formed by elimination of molecular hydrogen or a molecule RH (R = alkyl, aryl etc.). Examples of reactions that occur according to equation (IV.2) [15, 16] are demonstrated in Scheme IV.7.

$$R\text{–}M\text{–}H \xrightarrow{-RH} M \xrightarrow{+R'H} R'\text{–}M\text{–}H \qquad (IV.2)$$

Here both R and R' are H, alkyl, aryl etc.

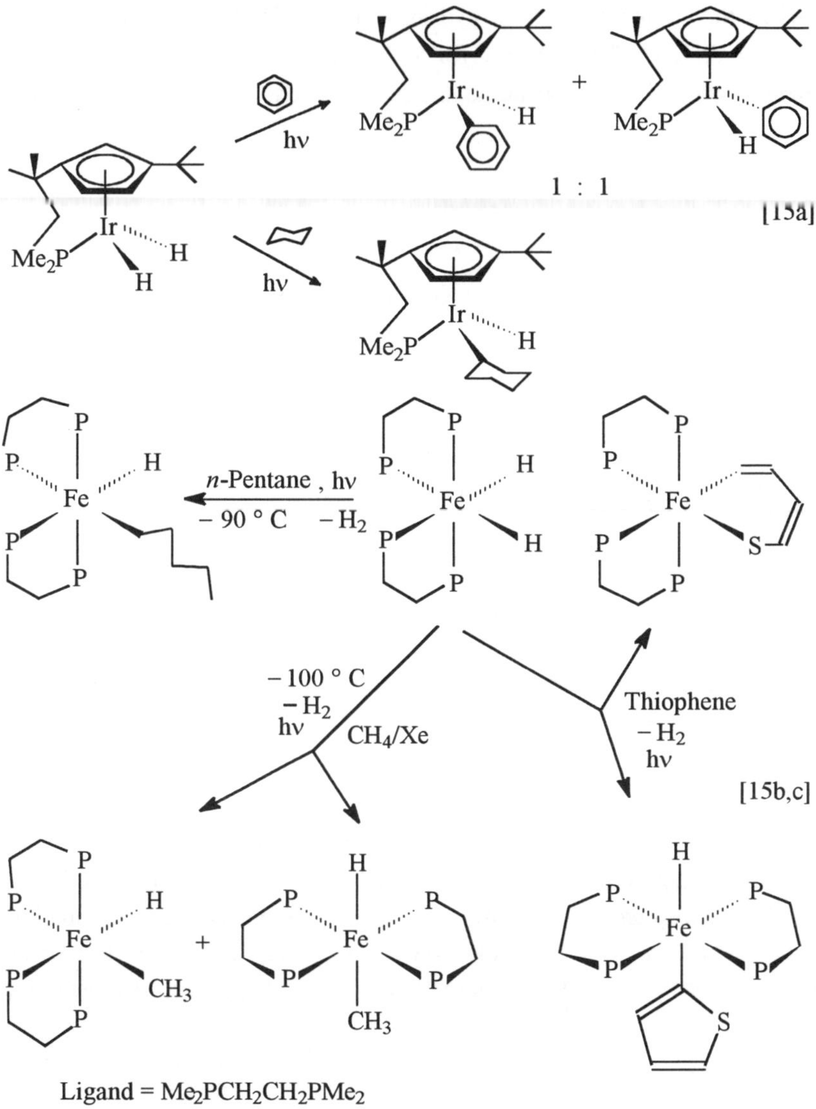

Scheme IV.7 (part 1). Reductive elimination of H–H or R–H from metal complexes followed by oxidative addition of C–H compounds. Reference numbers are in brackets.

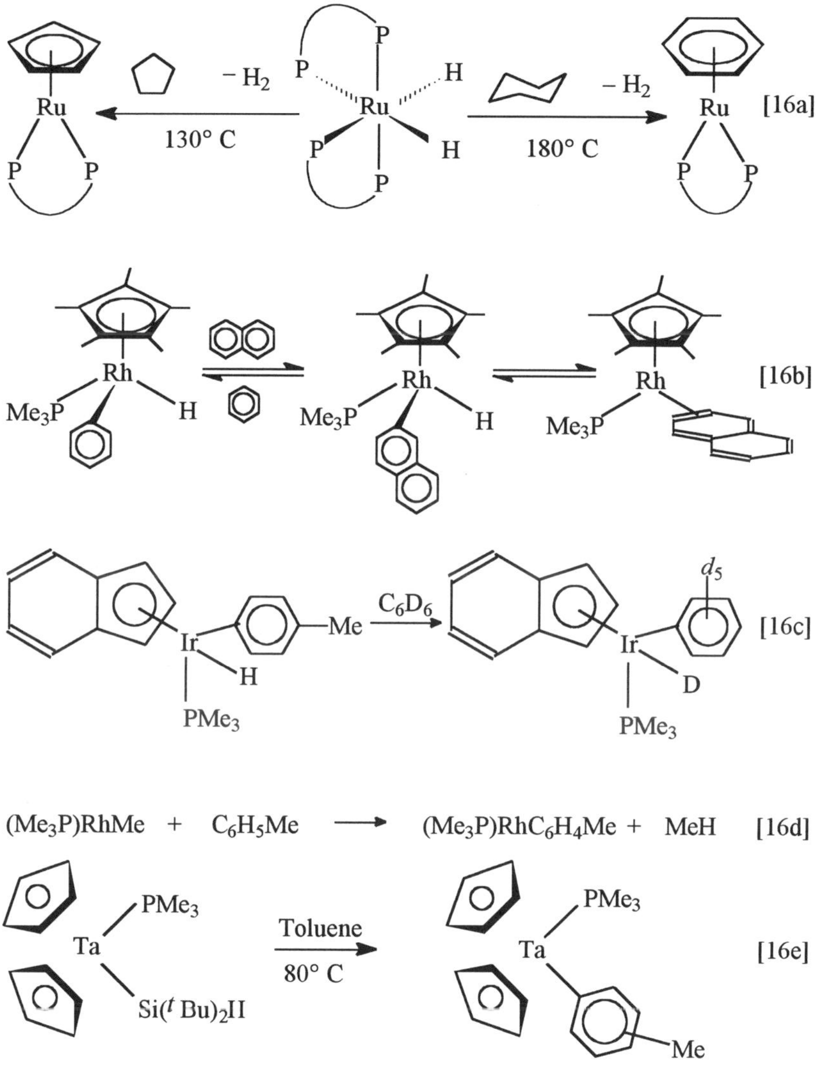

Scheme IV.7 (part 2). Reductive elimination of H–H or R–H from metal complexes followed by oxidative addition of C–H compounds. Reference numbers are in brackets.

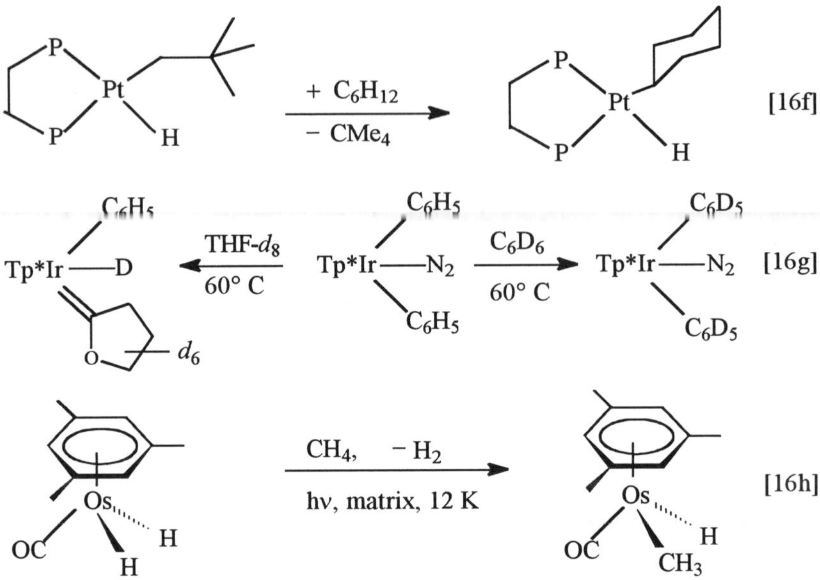

Scheme IV.7 (part 3). Reductive elimination of H–H or R–H from metal complexes followed by oxidative addition of C–H compounds. Reference numbers are in brackets.

A coordinatively unsaturated species capable of adding RH can also be generated by extrusion from such neutral ligands as phosphine (Scheme IV.8) [17], carbon monoxide (Scheme IV.9) [18], olefin and, as well as, some other ligands (Scheme IV.10) [19].

IV.1.C. FORMATION OF SOME OTHER PRODUCTS

Examples of oxidative addition coupled with some other transformations were mentioned in previous sections. Here we discuss other reactions of oxidative addition, which are rather complicated and do not lead to the formation of organyl hydrides.

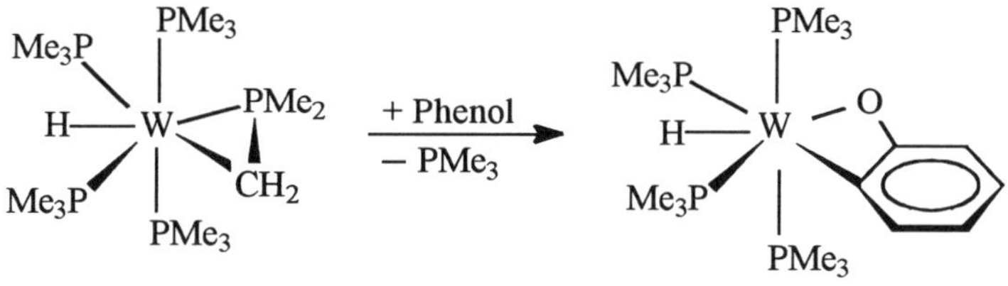

[17a]

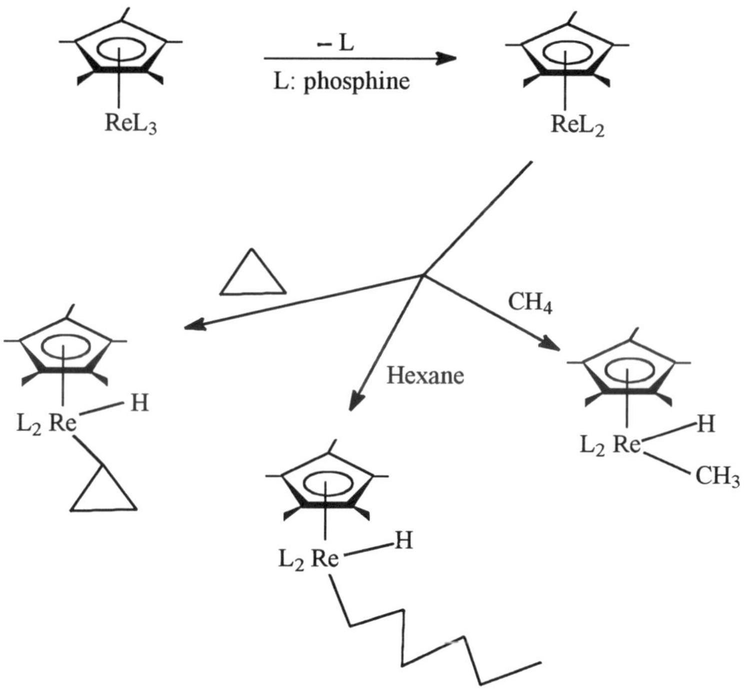

[17b]

Scheme IV.8. Oxidative addition of C–H compounds to coordinatively unsaturated species generated by extrusion of phosphines. Reference numbers are in brackets.

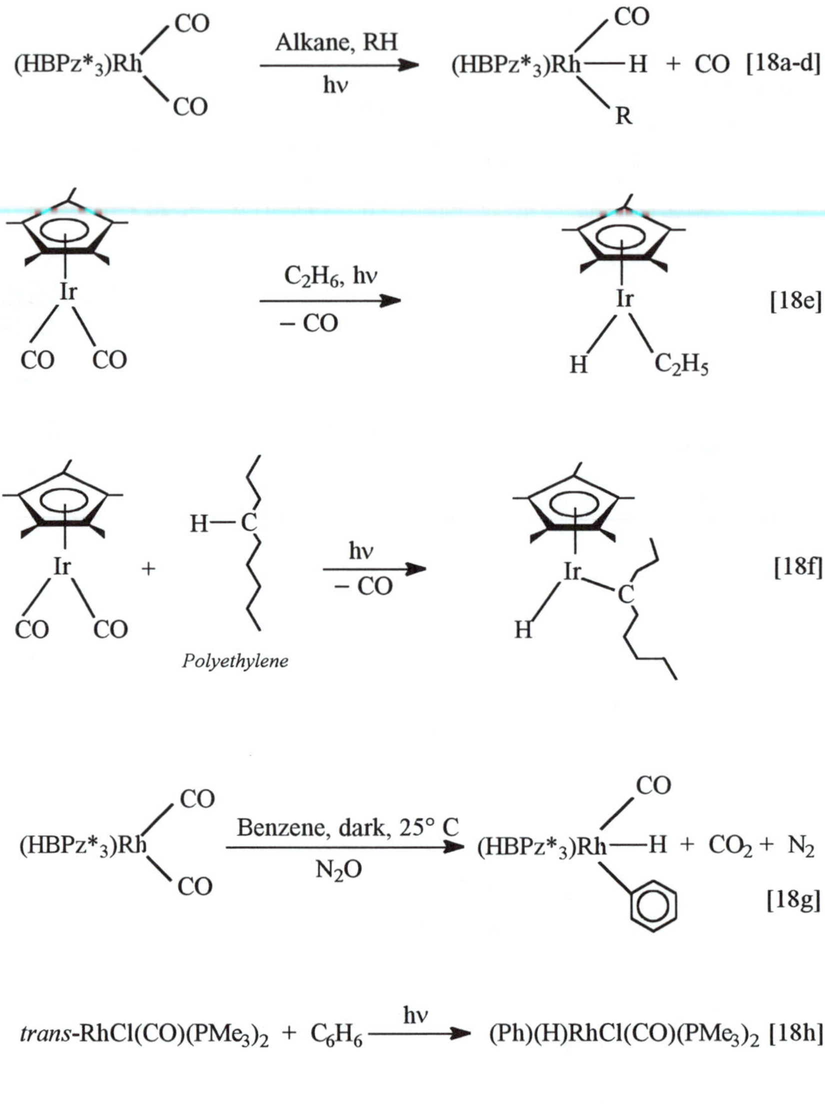

Scheme IV.9. Oxidative addition of C–H compounds to coordinatively unsaturated species generated by extrusion of CO. Reference numbers are in brackets.

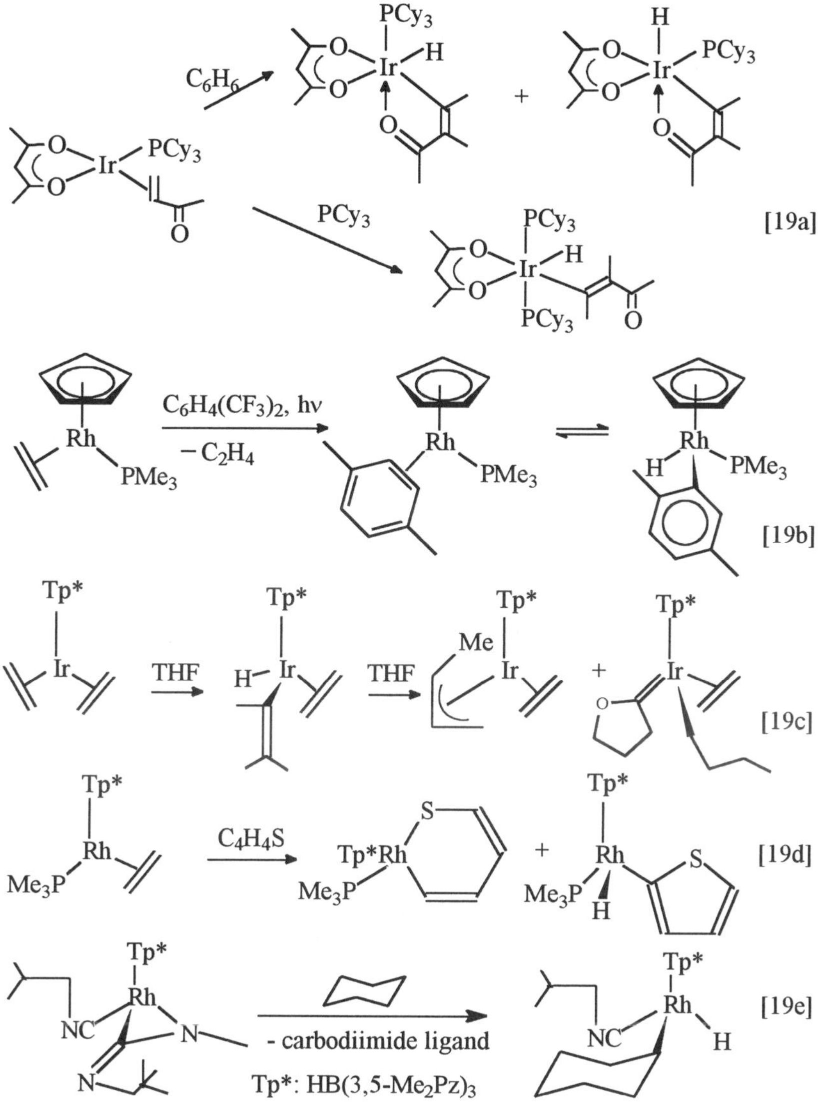

Scheme IV.10 (part 1). Oxidative addition of C–H compounds to coordinatively unsaturated species generated by extrusion of olefins and some other ligands. Reference numbers are in brackets.

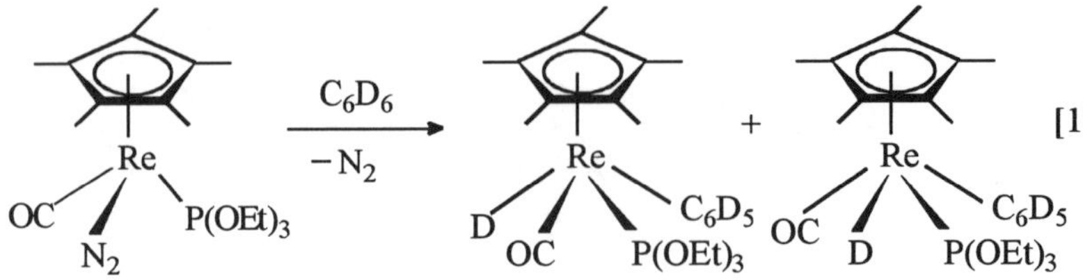

[19f]

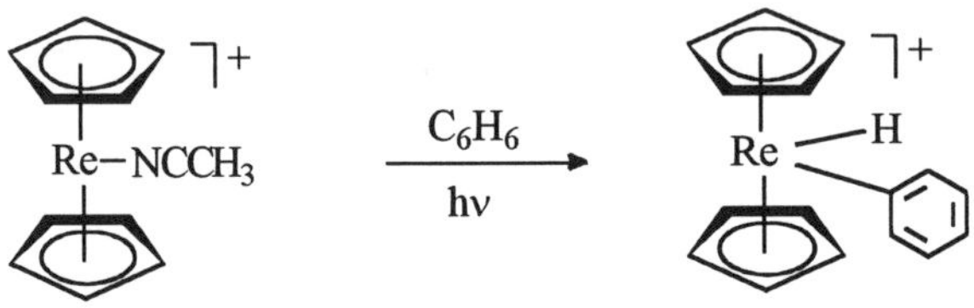

[19g]

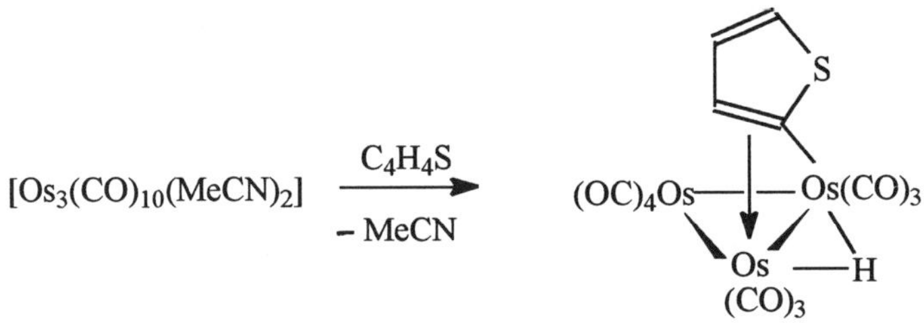

[19h]

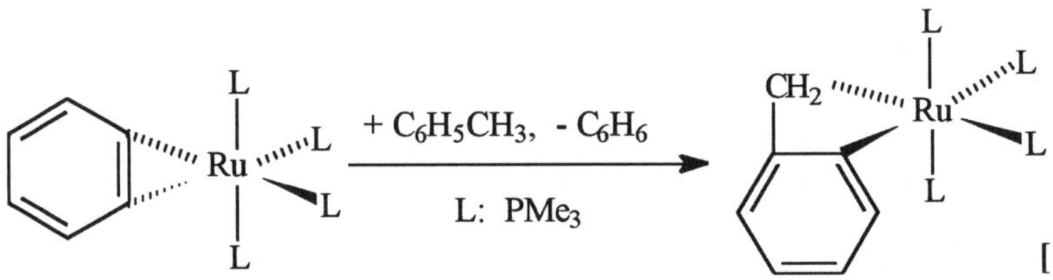

[19i]

Scheme IV.10 (part 2). Oxidative addition of C–H compounds to coordinatively unsaturated species generated by extrusion of olefins and some other ligands. Reference numbers are in brackets.

Sometimes thermal and photochemical activation can result in the formation of different products. For example, heating the solution of the ethylrhodium carbonyl complex $RhHBPz^*_3(CO)(C_2H_4)$ (Pz^* = 3,5-dimethylpyrazole) in benzene entails the elimination of the ethylene π-ligand and the formation of phenylrhodium hydride (complex **IV-3** in Scheme IV.11) [20a]. On irradiation of this solution, the hydride ligand adds not to the metal atom but to the ethylene molecule, which results in the appearance of an σ-ethyl group in the complex **IV-4** [20b].

In some cases a coordinatively unsaturated species and then a product of oxidative addition can be formed via a rearrangement of the complex molecule without elimination of any particles (e.g., olefin complex **IV-5** may be converted into the π-allyl hydride derivative **IV-6** in Scheme IV.11 [20c]). Analogously, photolysis of butadiene iridium complex **IV-7** gives rise to the formation of a hydrido-allyl isomer **IV-8** [20d].

Transition metal complexes readily cleave the C–H bonds not only in arenes and alkanes, as described above, but can also activate vinylic C–H bonds [21] (Scheme IV.12) and the C–H bonds in acetylenes [22] (Scheme IV.13).

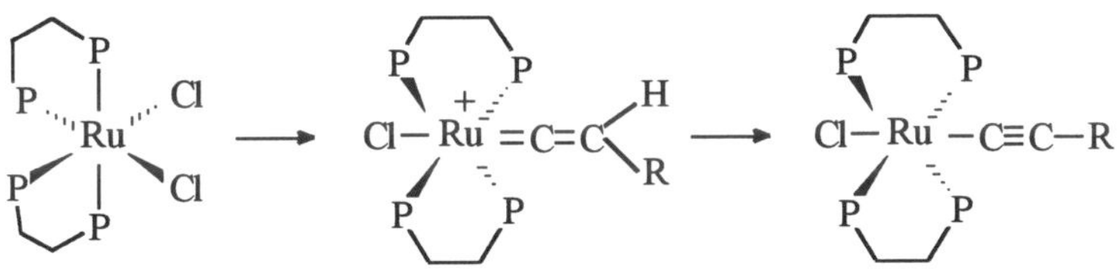

IV-9 **IV-10** **IV-11**

It is important to note that some reactions leading formally to σ-vinyl- and σ-alkynyl complexes of transition metals can be produced via different mechanisms. For example, the reaction of the ruthenium complex **IV-9** with terminal alkynes gives rise to the formation of vinylidene-metal complexes **IV-10,** which can be converted to the alkynyl-ruthenium derivatives **IV-11** by the action of triethylamine [23].

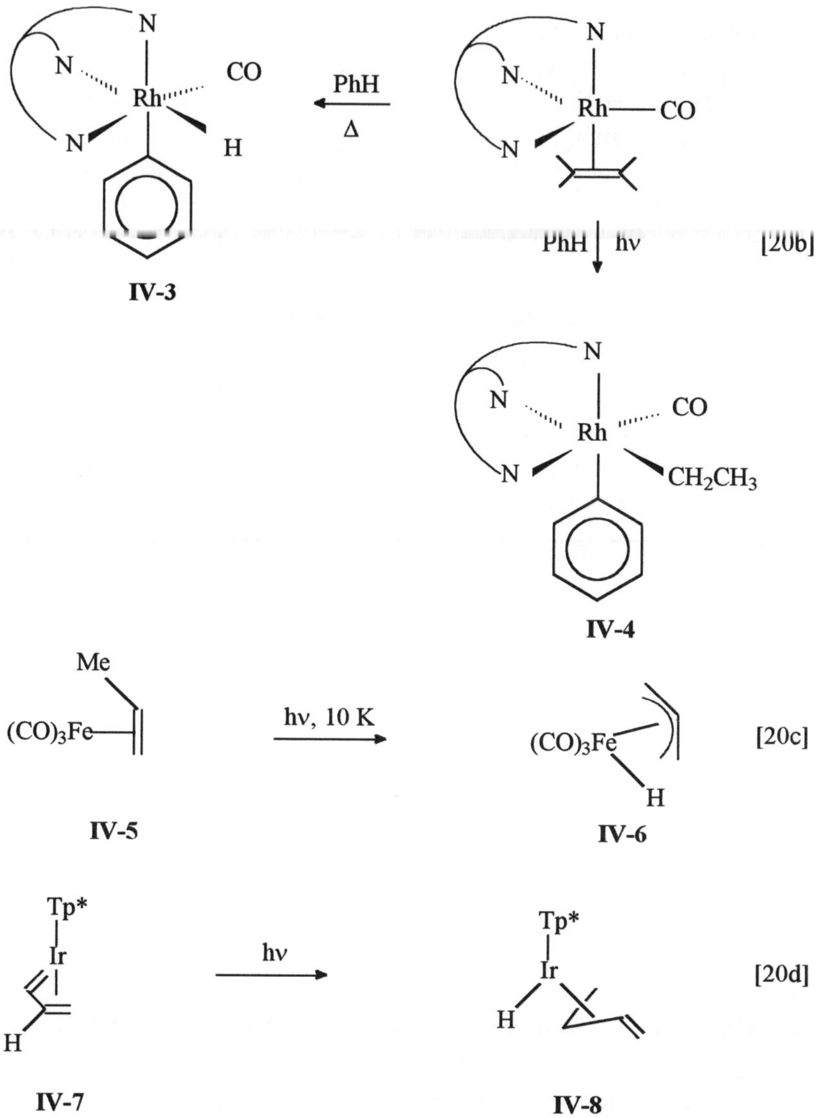

Scheme IV.11. Oxidative addition reactions followed by addition of hydride to olefin and by the formation of allyl complexes. Reference numbers are in brackets.

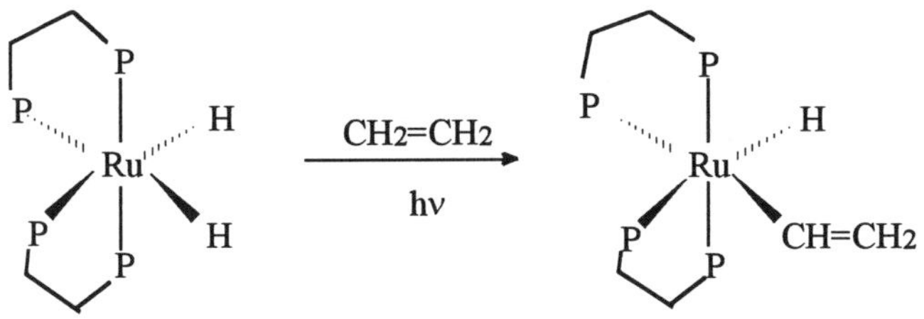

[21a]

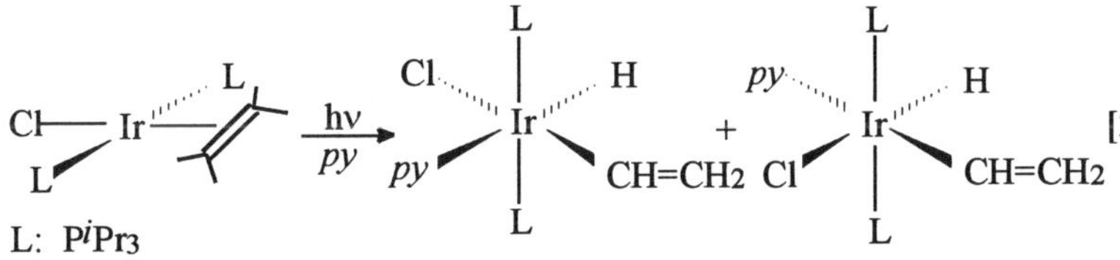

[21b]

L: P^iPr_3
py: pyridine

[21c]

$[Ir]^* = Tp^{Me_2}Ir$ $Tp = HB(pz)_3$

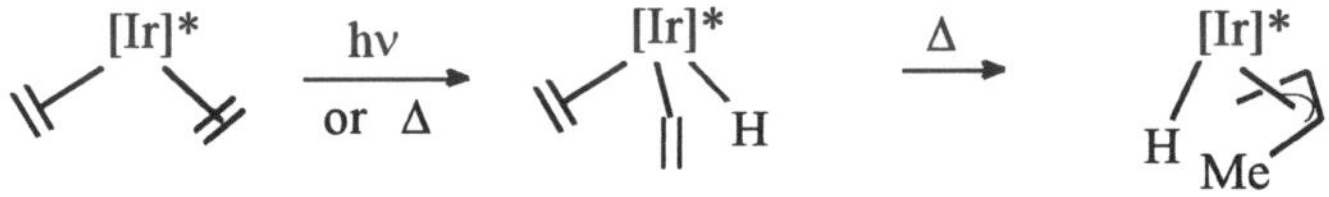

[21d]

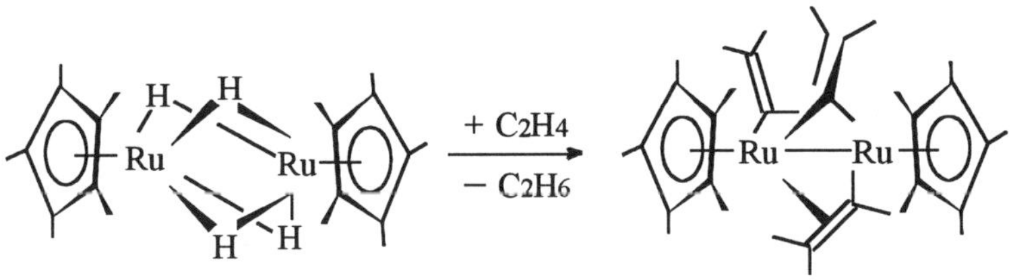

[21e]

Scheme IV.12 (part 1). Activation of vinylic C–H bonds by metal complexes in a low oxidation state via an oxidative addition mechanism. Reference numbers are in brackets.

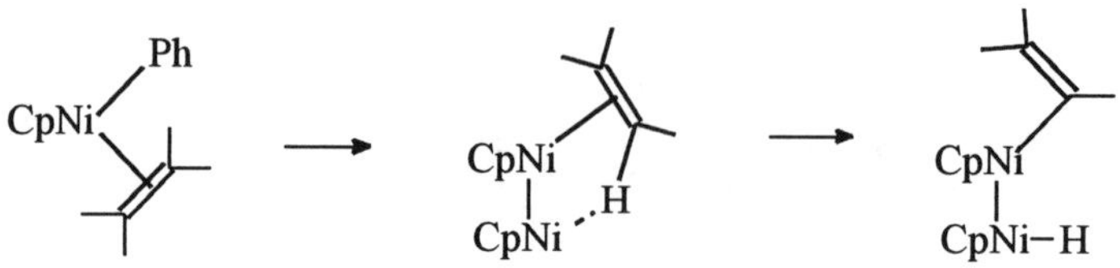

[21f]

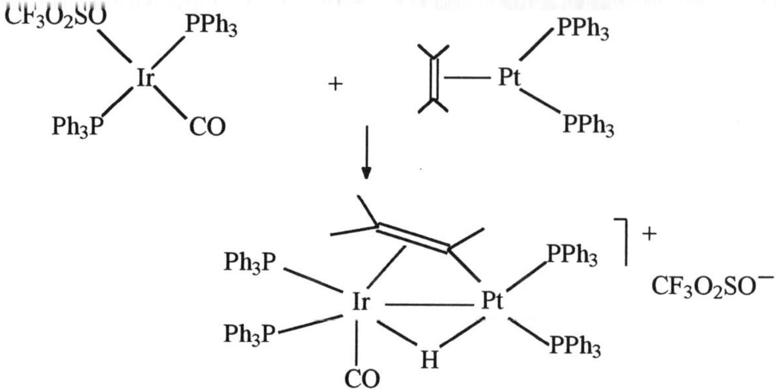

[21g]

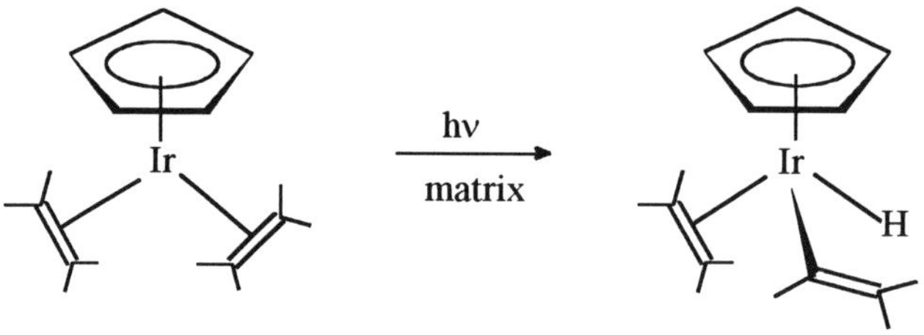

[21h]

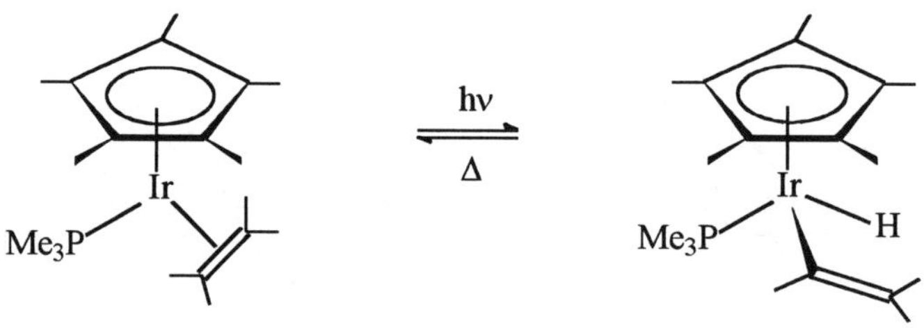

[21i]

Scheme IV.12 (part 2). Activation of vinylic C–H bonds by metal complexes in a low oxidation state via an oxidative addition mechanism. Reference numbers are in brackets.

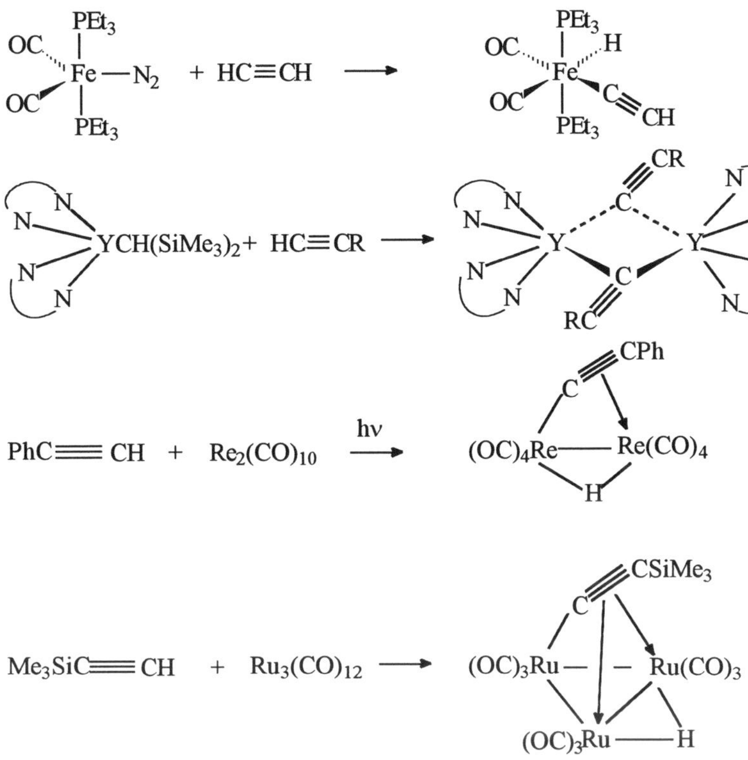

[22a] [22b] [22c] [22d]

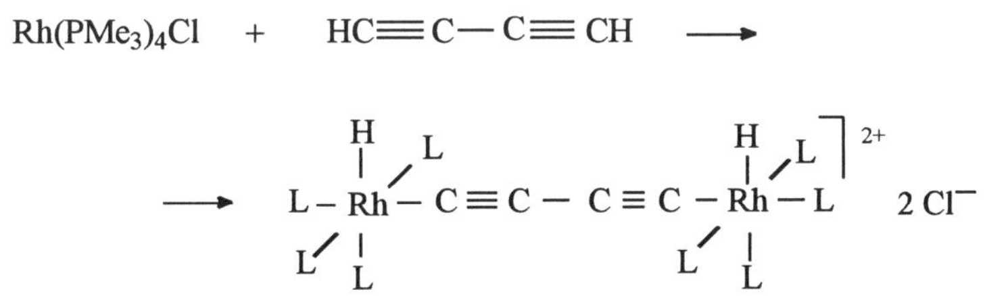

[22e]

$$(triphos)Ir(H)_2C_2H_5 + PhC{\equiv}CH \longrightarrow (triphos)\ Ir(H)_2C \equiv CPh \qquad [22f]$$

Scheme IV.13 (part 1). Activation of C–H bonds in acetylenes by metal complexes in a low oxidation state via an oxidative addition mechanism. Reference numbers are in brackets.

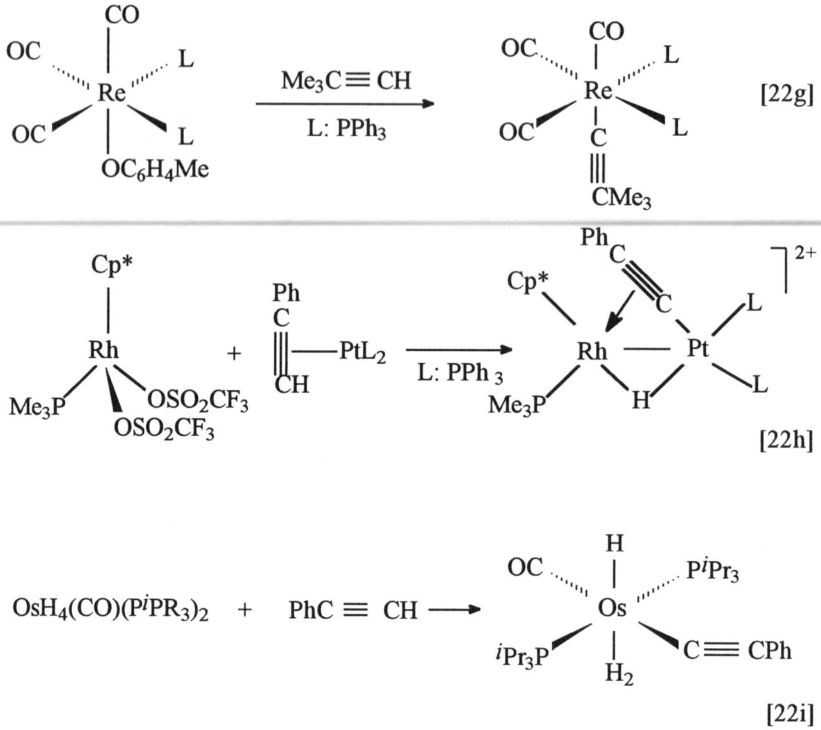

Scheme IV.13 (part 2). Activation of C–H bonds in acetylenes by metal complexes in a low oxidation state via an oxidative addition mechanism. Reference numbers are in brackets.

Iridium π-allylhydride complex reacts with various C–H compounds (Scheme 14) [24], oxidative addition of benzene and cyclopropane occurring in the presence of phosphine and giving rise to the formation of products of the hydrogen atom migration from the substrate to allyl ligand. The proposed mechanism includes addition of hydride atom to allyl ligand as the first stage (to produce coordinatively unsaturated species **IV-12** in Scheme IV-14) followed by usual oxidative addition of RH. After this π–σ-rearrangement of olefin-hydride complex **IV-13** occurs to give σ-alkyl derivative **IV-14**. The last stage is addition of phosphine yielding coordinatively saturated stable complex **IV-15**.

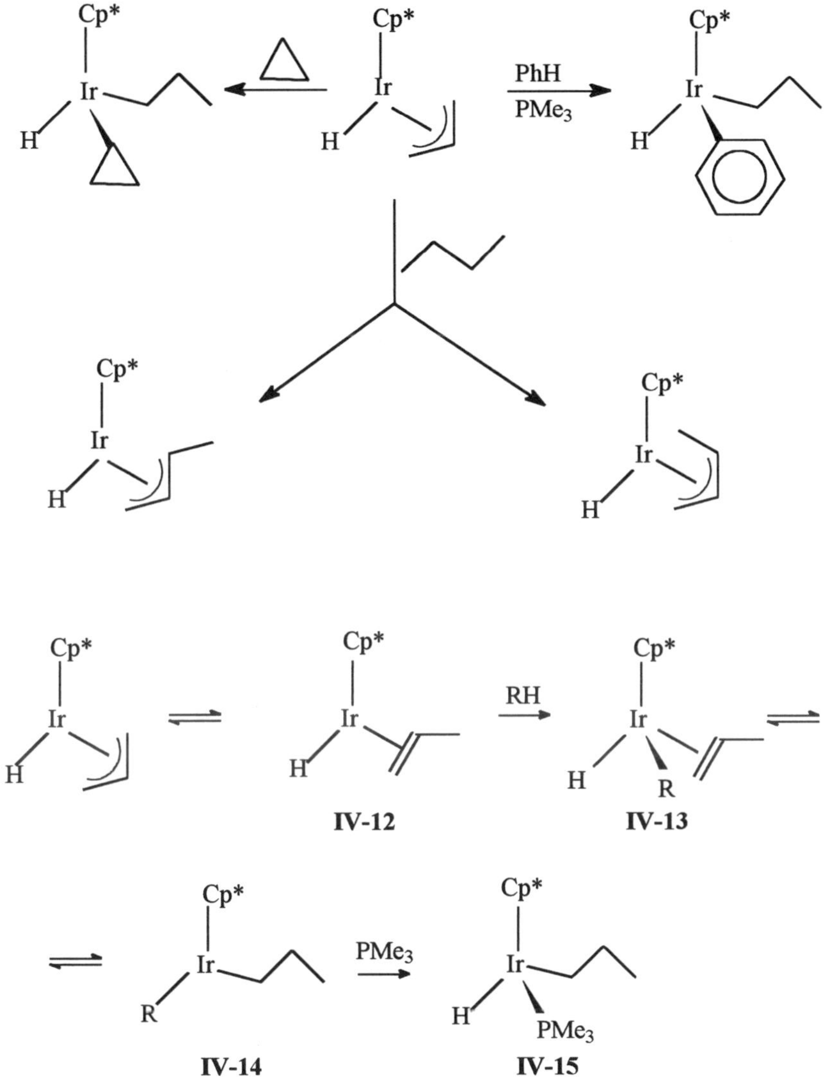

Scheme IV.14. Reactions of the iridium π-allylhydride complex with C–H compounds.

When a cyclohexane solution of thorium complex **IV-16** (Scheme IV-15) containing neopentyl ligands is heated one of the ligands can be eliminated and cyclometalation of a remaining alkyl groups proceeds [25a]. Metallacycle **IV-17** thus formed metalates intermolecularly tetramethylsilane. The reaction does not stop when compound **IV-18** is formed, and cyclometalated complex **IV-19** can be obtained. The thorium complex also readily activates methane. It should be noted that formally thorium is not a low-valent ion in these reactions.

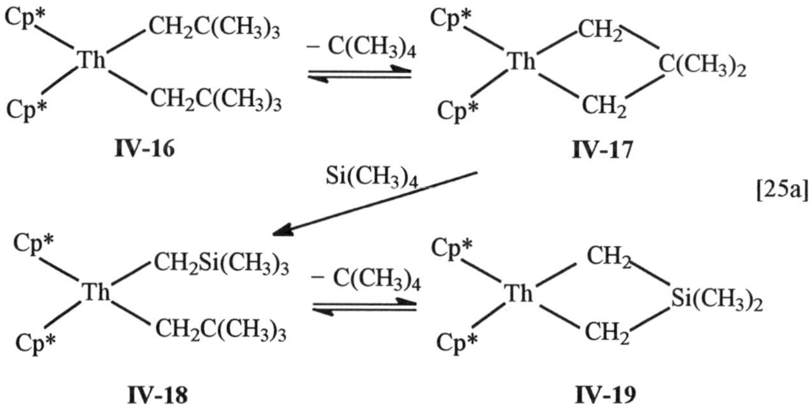

Scheme IV.15. Reactions of the thorium complex.

The hydrogen atom from alkane cannot add to the metal ion but can add to a ligand instead. For example, heating a solution of complex **IV-20** and methane in deuterated cyclohexane leads to the replacement of the CD_3 group by methyl from methane and gives also CD_3H [25b]. Species **IV-21** is assumed to be an intermediate in this process.

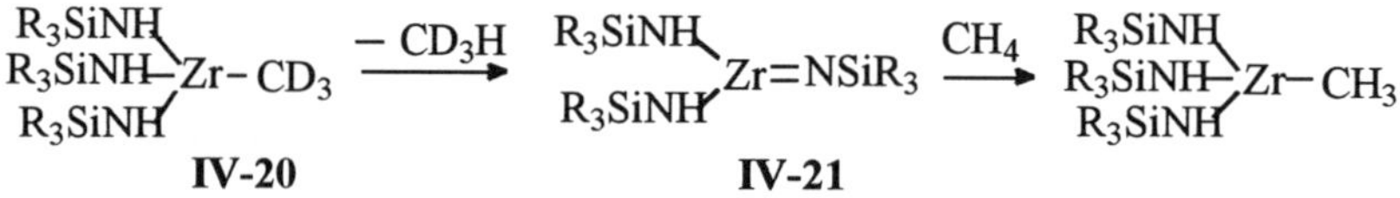

The reaction of the tantalum hexamethyl complex **IV-22** with deuterated 3,3-dimethyl-1-butyne proceeds as an intramolecular oxidative addition and gives rise to the σ-organyl complex **IV-23** [25c].

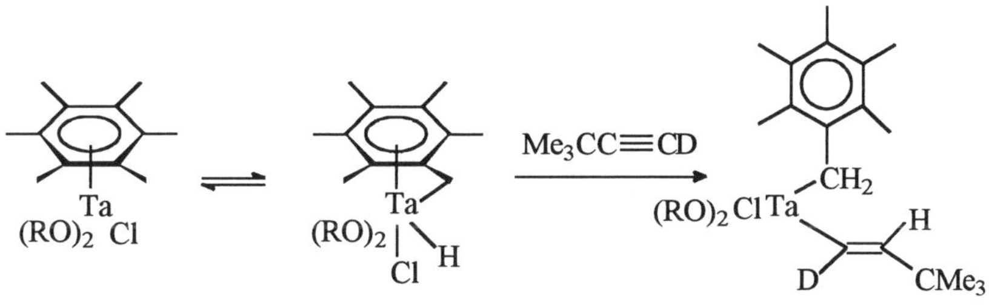

IV-22 **IV-23**

A $Cp^*Ir(PMe_3)(CH_3)^+$ complex reacts at low temperatures with an alkane, RH, to yield the σ-alkyl derivative $Cp^*Ir(PMe_3)(R)^+$ and methane [26a,b]. Two different mechanisms were proposed for this metathesis:

(i) formation of a four-center adduct between the methyl complex and RH, and

(ii) oxidative addition followed by the reductive elimination of a methane molecule.

Theoretical calculations support a low-energy oxidative addition mechanism [26c]. Reaction of the unsolvated cationic complex $Cp^*Ir(PMe_3)(CH_3)^+$ with pentane, cyclohexane or benzene in the gas phase also gives $Cp^*Ir(PMe_3)(R)^+$ as the product. However, a mechanistic investigation of this process by electrospray tandem spectrometry has demonstrated that neither the oxidative addition–elimination mechanism nor the concerted σ-bond metathesis mechanism is operative. Instead, the authors proposed a dissociative elimination–addition mechanism which proceeds through a series of 16-electron Ir(III) intermediates [26d].

Various other examples of the activation of C–H bonds by conventional, low-valent metal complexes have been described [27]. Not all of these reactions proceed via an unambiguous, simple oxidative addition mechanism.

IV.1.D. SPLITTING THE C–H BOND ACTIVATED BY POLAR SUBSTITUENTS

When an electron-withdrawing substituent is introduced into an alkane molecule, reactions of such substituted alkanes with electrophilic oxidizing reagents are more difficult in comparison with unsubstituted substrates. For example, permanganate ion in a solution of aqueous trifluoroacetic acid, which reacts with alkanes, does not oxidize nitroethane. This property allows carrying out reactions of alkanes with strong oxidants in organic solvents and stopping the reaction at the initial step when a polar functional group is incorporated into alkane molecule. On the contrary, in the reactions with low-valence electron-donor metal complexes, electron-withdrawing substituents facilitate the oxidative addition of the C–H bond components to the metal complex and stabilize the derivative in the higher oxidation state formed.

Oxidative addition of acetonitrile to a complex of iridium(I) gives a product which can be detected by NMR (dmpe: dimethyldiphosphinoethane):

$$Ir(dmpe)_2^+ + CH_3CN \rightleftharpoons \underset{\displaystyle CH_2CN}{\overset{\displaystyle H}{Ir^+(dmpe)_2}}$$

The equilibrium can be shifted to the right if CO_2 is present in the reaction mixture. In this case, the cyanoacetate complex formed can be easily isolated:

$$\underset{\displaystyle CH_2CN}{\overset{\displaystyle H}{Ir^+(dmpe)_2}} + CO_2 \longrightarrow \underset{\displaystyle O{-}CO{-}CH_2CN}{\overset{\displaystyle H}{Ir^+(dmpe)_2}}$$

The interaction between acetonitrile and iridium(0) complex generated electrochemically produces iridium(I) hydride and cyanomethyl radical [28a]. The palladium(I) dimer containing dimethylphosphinomethane as ligands reacts with nitromethane to yield (nitromethyl)palladium(II) dimer [28b]:

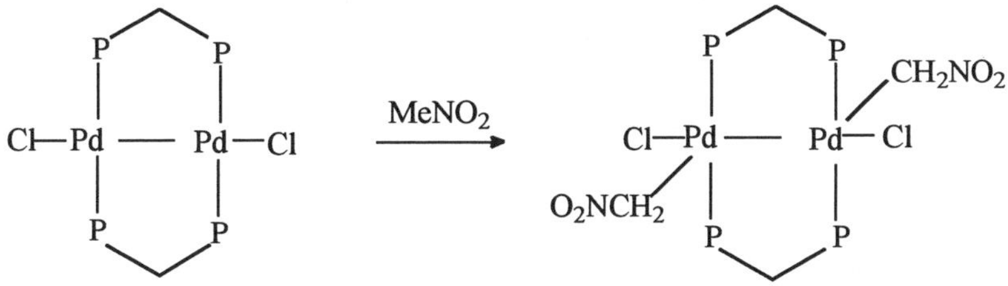

Phenylazophenylgold(III) complex activates the C–H bond in acetone [29a]:

$$[(C_6H_5N{=}NC_6H_4\text{-}2)Au(bipy)Cl] + MeCOMe \xrightarrow{AgClO_4} [(C_6H_5N{=}NC_6H_4\text{-}2)Au(bipy)CH_2COMe]^+ ClO_4^-$$

Sometimes a functional group present in the C–H compound reacts also with the metal center. For example, the reaction of platinum(0) complex with nitromethane has been proposed to proceed via oxidative addition according to the scheme [29b]:

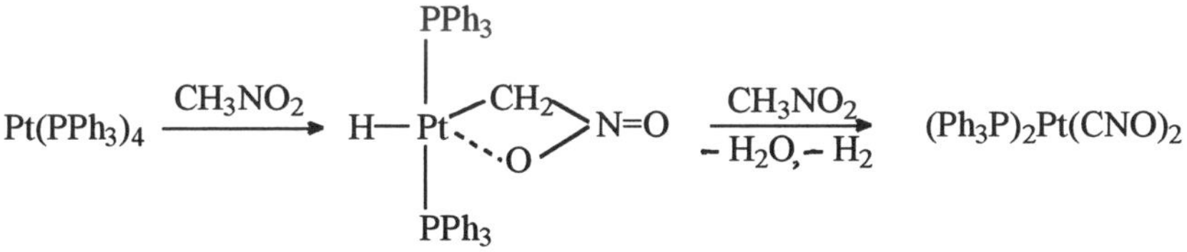

IV.2. REPLACING HYDROGEN ATOMS BY VARIOUS GROUPS

Metal complexes induce a whole series of reactions involving the substitution of hydrogen atoms in hydrocarbons by the atoms of other elements. The complex very often catalyzes such reactions. Considerable proportions of the processes proceed via the oxidative addition mechanism. After this first stage the alkyl hydride complex formed is involved in transformations which follow. A resulting reaction may be H–D exchange, dehydrogenation of alkane or alkyl group, and the introduction of a functional group into a C–H compound.

The epimerization of secondary alcohols catalyzed by rhenium alkoxide complexes proceeds via the C–H bond activation: the first step of the reaction is the hydrogen atom migration from alcohol to the metal ion (Scheme IV-16) [30].

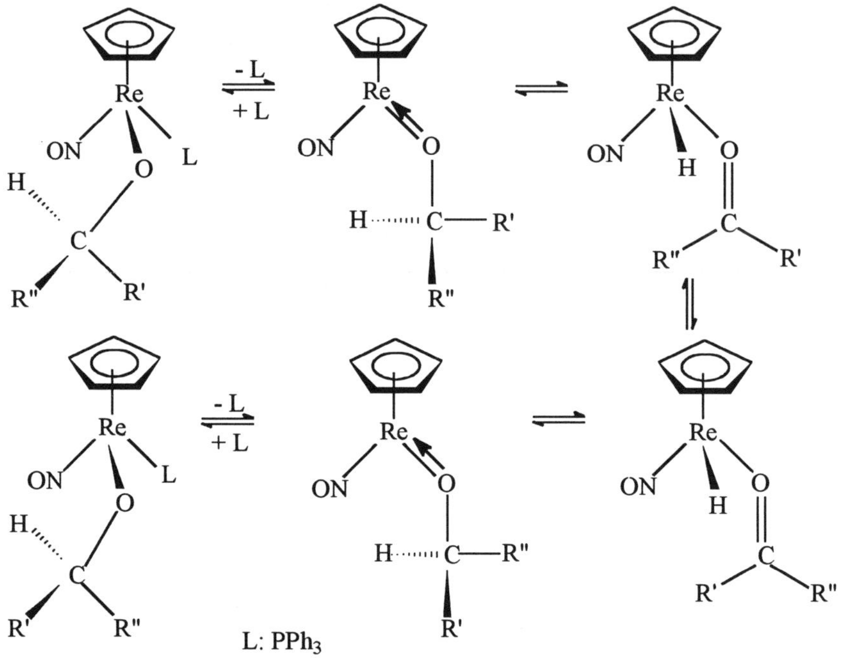

Scheme IV.16. Rhenium-catalyzed epimerization of *sec*-alcohols.

IV.2.A. ISOTOPE EXCHANGE

As previously mentioned (see Section I.3), in the first paper described the "true" activation of alkane by metal complex, formation of a small amount of monodeuteromethane was observed when methane and deuterium were in prolonged contact with a benzene solution of tris-triphenylphosphine cobalt trihydride at room temperature [31]. Many works devoted to the isotope exchange appeared in recent decades. This exchange may occur between alkane or arene and molecular deuterium, between deuterated and undeuterated hydrocarbons, between hydrocarbon and another deuterated inorganic (e.g., water) or organic substance. Two different groups within the same molecule can exchange their hydrogen and deuterium atoms. Some examples of all such reactions (see a review [32] and recent papers [33]) are shown in Scheme IV-17.

Dicyclopentadienylvanadium(II) catalyzes H–D exchange between methane and molecular deuterium (or tetradeuteroethylene) in a benzene solution at 70 °C. Benzene also exchanges its hydrogens. There is multiple H–D exchange, i.e., each hydrocarbon molecule entering into the reaction can exchange several hydrogen atoms at a time without leaving the complex. The crucial step of the reaction is the reversible oxidative addition of methane to vanadium atom and the ethylene ligand [34]:

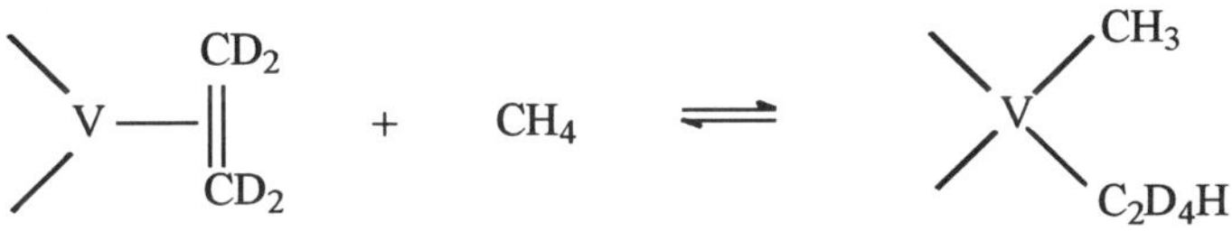

Then the hydrogen transfer from methyl group occurs to produce ethane and a carbene complex of vanadium.

The hydrogen exchanges among the terminal hydride, the agostic hydrogen, and a non-coordinating methylene hydrogen of 1,4-bis(diphenylphosphino)butane ligand (dppb) in formally five-coordinate ruthenium complex $[RuH(dppb)_2]PF_6$ were proven by NMR method [35a]. When the partially deuterated ligand $Ph_2P(CD_2)_4PPh_2$ was used and the reaction was carried out at a high temperature, the hydrogen scrambling between *ortho* hydrogens on the phenyl groups, all the methylene hydrogens, and the terminal hydride in this complex was proved. Deuterium incorporation takes place at *ortho* and all methylene positions of diphosphine ligand if the complex is in the contact with D_2 in solution. The deuterium content of each site is in the order of:

β-CH_2 > α-CH_2 > *o*-CH.

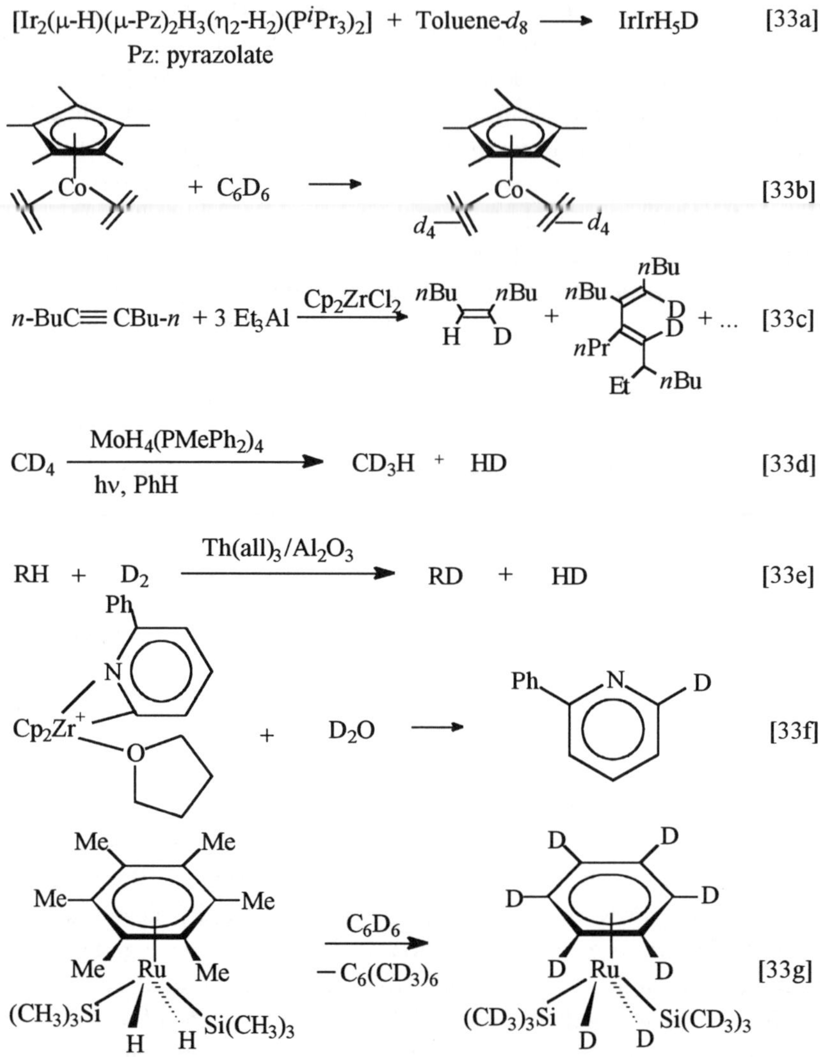

Scheme IV.17. Examples of H–D exchange in hydrocarbons and organyl groups induced by low-valent metal complexes. Reference numbers are in brackets.

The proposed mechanism of the deuterium incorporation and the hydrogen scrambling is depicted in Scheme IV-18.

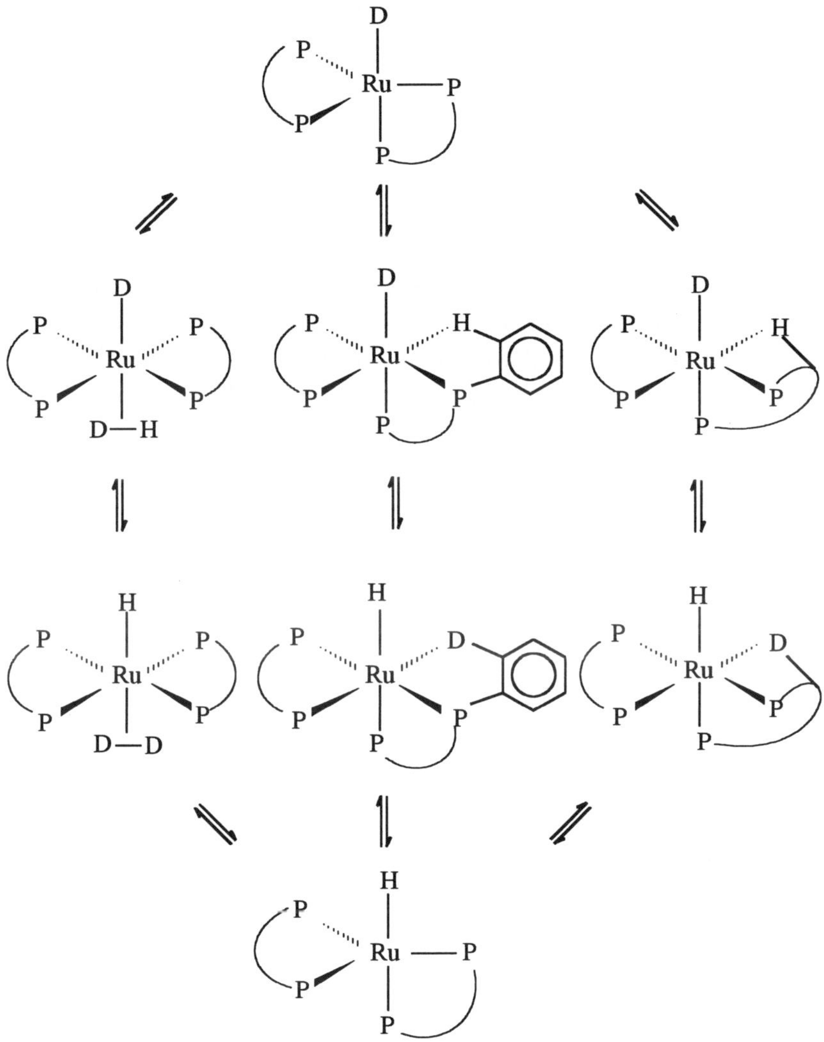

Scheme IV.18. The proposed mechanism of the of H–D exchange.

The complex $MnH_3(dmpe)_2$ [dmpe = 1,2-bis(dimethylphosphino)ethane] activates aromatic and benzylic C–H bonds under both thermal and photochemical conditions as determined by catalytic H–D exchange with C_6D_6 and D_2 [35b]. Electron-deficient ruthenium(II) porphyrins also catalyze H–D exchange into methane and benzene, and into both the benzyl and aryl positions of toluene [36].

IV.2.B. DEHYDROGENATION

The intramolecular dehydrogenation of alkyl groups in the σ-alkyl metal complex molecule is well known to produce chelated π-olefin complexes [37] (examples see in Scheme IV-19).

Various transition metal complexes readily abstract two hydrogen atoms from C–H compounds yielding π-olefin derivatives. If the π-complexes of unsaturated hydrocarbons formed in the dehydrogenation process are relatively unstable, the π-ligand easily dissociates and the reaction becomes catalytic with respect to metal complex:

$$R'CH_2CH_2R'' \overset{[M]}{\rightleftharpoons} R'CH{=}CHR'' + H_2 \qquad (IV.3)$$

Thermally induced reactions of this type [38, 39] are shown in Schemes IV-20 and IV-21 (in cases of cyclopentane and cyclohexane π-cyclopentadienyl and π-benzene complexes can be obtained). Continuous removal of molecular hydrogen evolved in the reflux process displaces the equilibrium (IV.3) toward olefin. Hydrogen evolving in the course of the dehydrogenation can be efficiently removed if the reaction is carried out by refluxing in perfluorodecane or by bubbling argon [38c].

The equilibrium may be also shifted to the right by adding an acceptor of molecular hydrogen, for example, olefin or carbonyl compound:

$$R'CH_2CH_2R'' + R^*CH{=}CHR^{**} \overset{[M]}{\rightleftharpoons} R'CH{=}CHR'' + R^*CH_2CH_2R^{**} \qquad (IV.4)$$

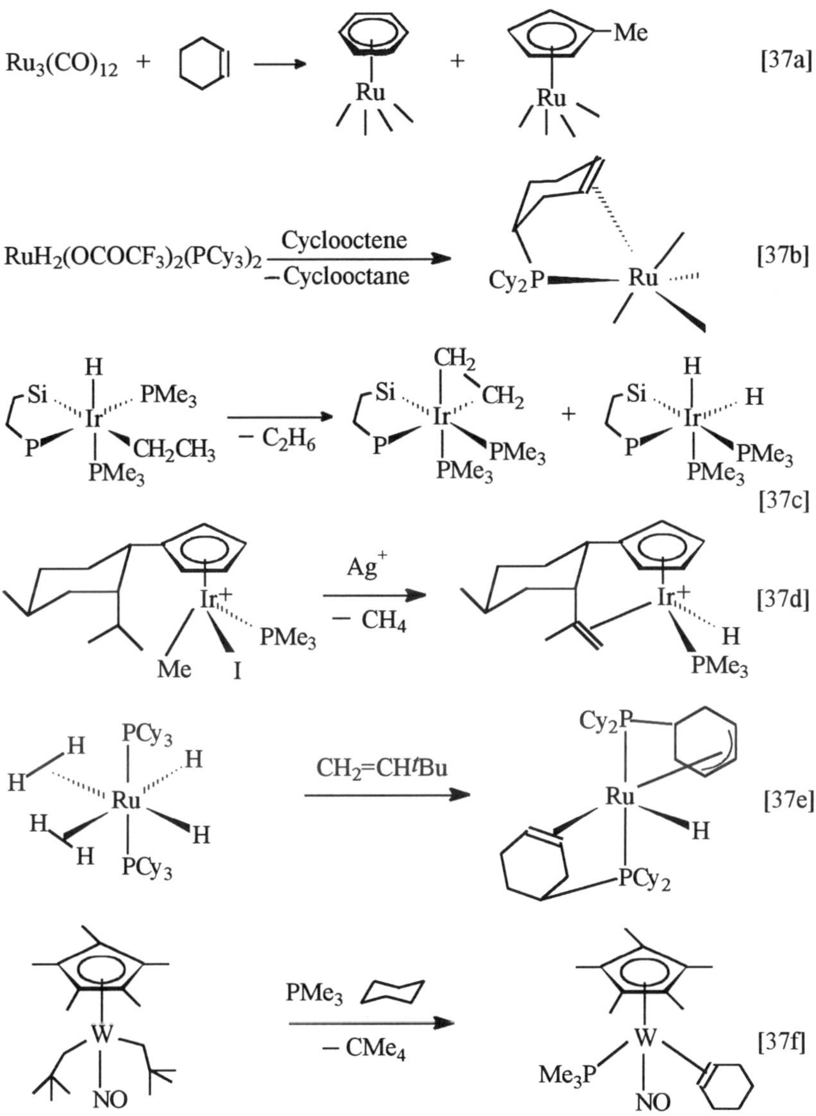

Scheme IV.19. Examples of dehydrogenation of alkanes and alkyl groups to produce π-olefin complexes. Reference numbers are in brackets.

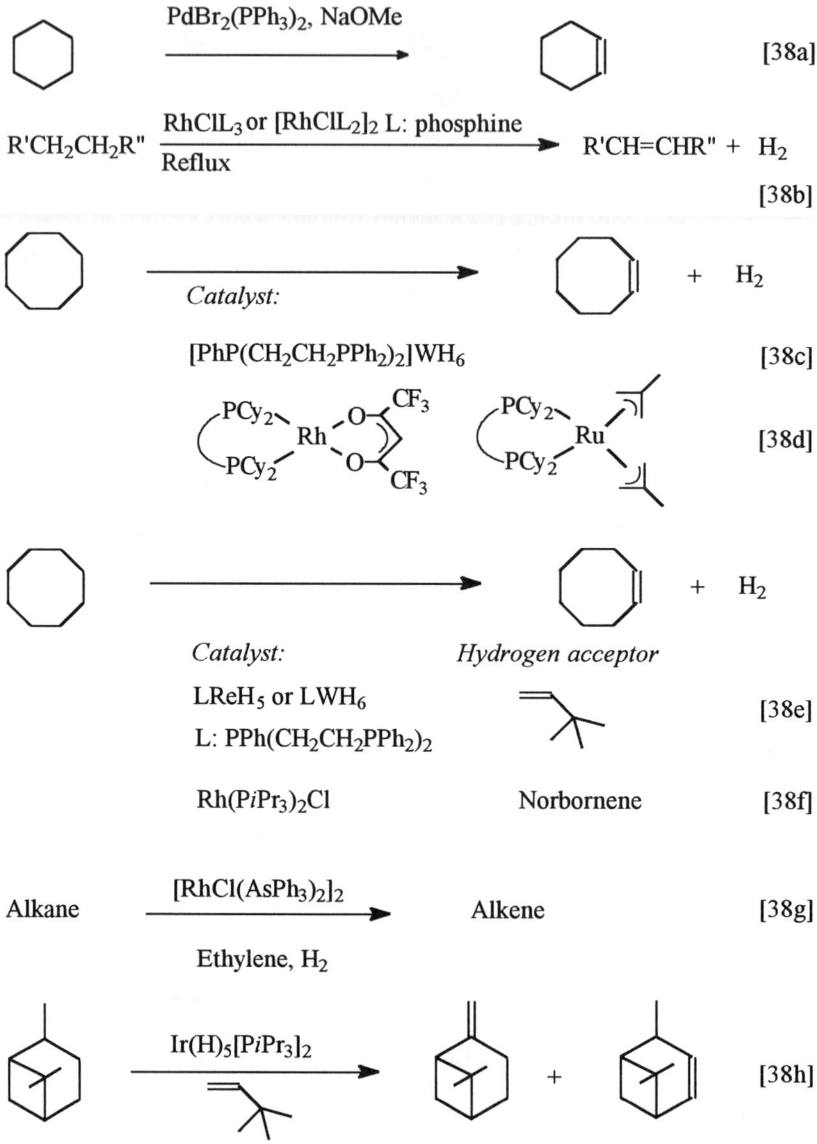

Scheme IV.20. Examples of alkane dehydrogenation to produce olefins. Reference numbers are in brackets.

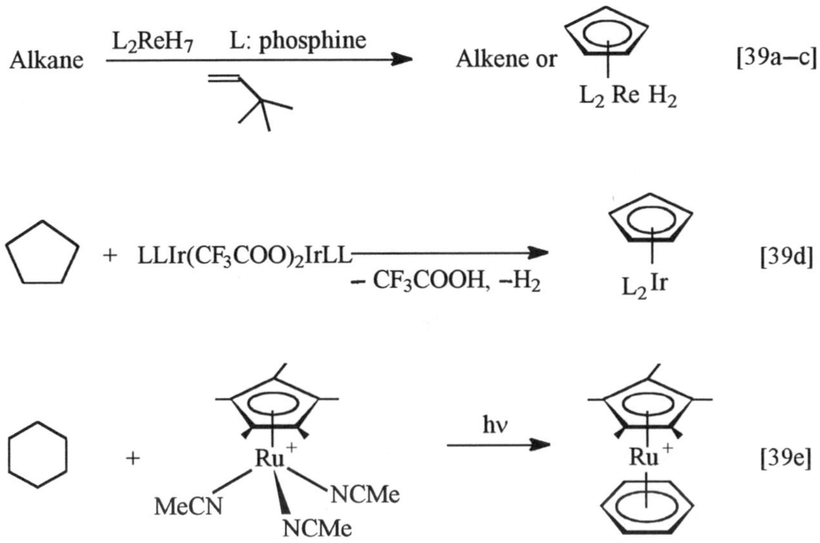

Scheme IV.21. The cycloalkane dehydrogenation to produce cyclopentadienyl and arene π-complexes. Reference numbers are in brackets.

If for example 3,3-dimethylbutene is present in the reaction mixture, the turnover number is usually noticeably increased. Thermal reactions of type (IV.4), in which molecular hydrogen is not evolved but adds to an acceptor, are shown in Schemes IV-20 and IV-21.

Efficient low-temperature dehydrogenation of alkanes catalyzed by complex $Rh(PMe_3)_2Cl(CO)$ proceeds under a high pressure dihydrogen atmosphere [40a]. Norbornene has been used as a hydrogen acceptor. Under 1000 psi of dihydrogen at 60 °C for 25 h, a cyclooctane solution of the complex and norbornene yielded 560 turnovers of cyclooctene and norbornane. It is noteworthy that at 100 °C a similar solution afforded 950 turnovers of products in under 15 min. Even greater activity, with the use of much lower hydrogen pressures, is obtained using other complexes containing the RhL_2Cl fragment, e.g., $RhL_2Cl(P^iPr_3)$ and $[RhL_2Cl]_2$ (L: PMe_3) [40b]. The thermal system $RhL_2Cl(P^iPr_3)$–H_2 affords a higher ratio of 4- to 3-isopropylcyclohexene, which has been attributed to participation of an intermediate such as $RhLCl(P^iPr_3)$. This species could be more sensitive to steric factors than RhL_2Cl. The authors proposed that the role of dihydrogen is to add to the four-coordinate rhodium centers, giving octahedral dihydride complexes which then lose L', or dissociate in the case of $[RhL_2Cl]_2$. The resulting intermediate, H_2RhL_2Cl, then hydrogenates a sacrificial olefin to generate the active species RhL_2Cl.

Olefin complex $RhCl(CH_2=CH_2)(PMe_3)_2$ catalyzes hydrogen transfer from cycloalkanes to ethylene under ethylene pressure at a temperature over 170 °C. Ethylene insertion into the C–H bond of cycloalkanes occurs at 230 °C to produce ethylcycloalkanes [40c]. It is interesting that uranium hydrogen sponge may be used as a hydrogen acceptor in combination with catalysts for dehydrogenation of cycloalkanes [40d]. Neutral and cationic rhodium complexes containing keto phosphine ligands have been reported to catalyze the transfer dehydrogenation of cyclooctane [40e].

Complex $IrClH_2(PPr_3)_2$ catalyzes the dehydrogenation of alkanes to alkenes at 150 °C [41a]. The P–C–P pincer complex $(PCP)IrH_2$, where PCP = η^3-$C_6H_3(P^tBu_2)_2$-1,3, catalyzes the very efficient dehydrogenation of alkanes and cycloalkanes to give corresponding alkenes and arenes, respectively [41b–f]. The turnover number attains several hundred moles of product per mole of catalyst.

Numerous publications devoted to photochemical, metal-induced dehydrogenations of C–H compounds appeared during the last decade. For example, irradiation (3 h) of a solution of 2-cyclohexen-1-one and the complex Cp*Ru-$(MeCN)_3PF_6$ in methylene chloride with full light of a high-pressure mercury lamp, leads to the formation of the derivative Cp*Ru(π-C_6H_6)PF_6 with a yield of 40% based on Ru [39e]. Under analogous conditions, acetonitrile ruthenium complex reacts with cyclohexene and even cyclohexane to produce the same π-benzene derivative (yield ca. 10%).

Complexes of rhodium and iridium are known to be the most effective photocatalysts for the alkane dehydrogenation and have been investigated thoroughly. Irradiation of a solution of complex $RhCl(CO)(PMe_3)_2$ in cyclohexane at room temperature, with full light from a high-pressure mercury lamp, induces the formation of cyclohexene (138 mols per 1 mol of the catalyst after 16.5 h) and molecular hydrogen [42a,b]. In addition, a small amount of benzene (3 mols) is detected. Analogously, dehydrogenation of *n*-hexane affords a mixture of hexenes (155 moles per 1 mol of the catalyst after 27 h). The ratio of isomers -1 : -2 : -3 is 1 : 79 : 20; for 2-hexene the *cis/trans* ratio is 1 : 3. Isomerization of olefins formed occurs apparently in the course of the reaction, with the double bond migrating along the hydrocarbon chain. As a result, the ratio of hexene-1/hexene-2 gradually decreases. Under conditions described, this reaction is reversible; however, if dinitrogen is bubbled through the reaction solution during irradiation, the dihydrogen formed is removed and the rate of the photodehydrogenation is increased.

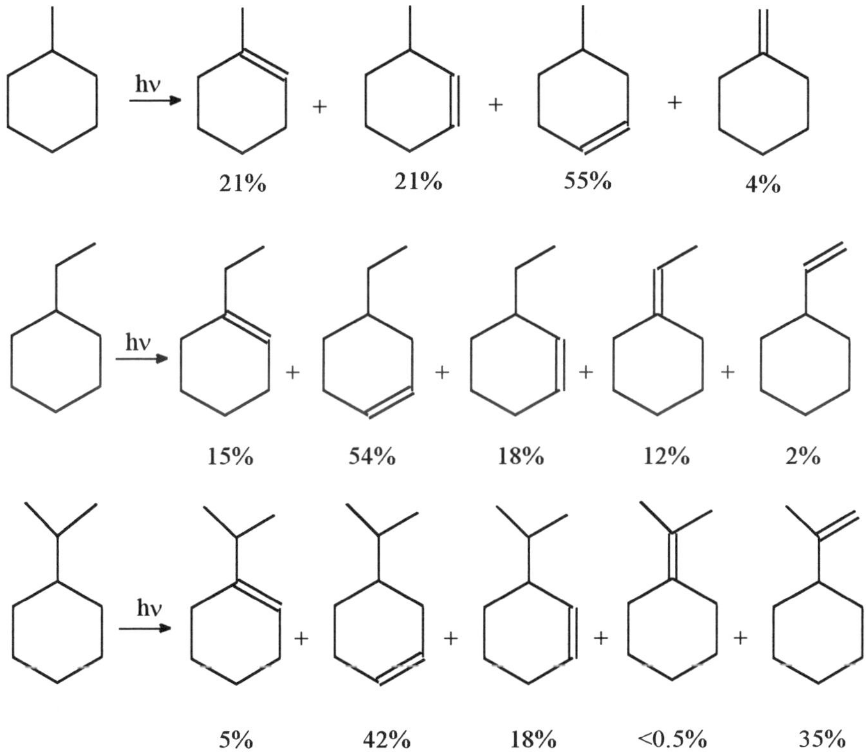

Scheme IV.22. The cycloalkane photo-dehydrogenation catalyzed by $RhCl(CO)(PMe_3)_2$.

Dehydrogenation of alkylcyclohexanes photocatalyzed by complex RhCl-(CO)(PMe_3)$_2$ [43] affords mixtures of products (Scheme IV-22). The proposed mechanism of this reaction includes oxidative addition of alkane to the species RhCl(PMe_3)$_2$ and then elimination of hydrogen from the β-position of alkyl chain of the alkylhydride derivative [43c,d].

Photodehydrogenation of alkanes is usually carried out using the alkane as a solvent; however, bulky hydrocarbons like 2,2,3,3-tetramethylhexane and 1,3,5-tri-*tert*-butylbenzene may be employed as inert solvents [44a]. Naturally, the rate of photodehydrogenation depends dramatically on the wavelength of light used for stimulation of the reaction [44b]. Almost in all cases, UV light has been used, but complexes (CH_2=CH_2)RhCl(PMe_3)$_2$ [44c] and $Ir_2(CO)_2(Ph_2PCH_2$-$PPh_2)_2$ [44d,e] do catalyze dehydrogenation of alkanes under irradiation with visible light.

Carbonyl compounds can be used as hydrogen acceptors in the alkane dehydrogenation induced by light. Thus, complex RhCl(CO)(PMe_3)$_2$ catalyzes the reduction of aldehydes to their corresponding alcohols at room temperature, in the presence of cyclooctane [45a]. Like alkanes, some organic compounds containing alkyl groups can be dehydrogenated photocatalytically. For example, irradiation of methylpropionate in the presence of RhCl(CO)(PMe_3)$_2$ gives rise to the formation of dimeric products [45b]:

$$CH_3CH_2COOCH_3 \longrightarrow CH_3CH_2COO(CH_2)_3COOCH_3 + H_3COCO(CH_2)_4COOCH_3$$

Yields of the first and second dimers are 2035% and 116%, respectively, based on Rh.

IV.2.C. INTRODUCTION OF CARBONYL GROUPS INTO HYDROCARBON MOLECULES

The complex RhCl(CO)(PMe_3)$_2$ successfully used for the dehydrogenation of alkanes, turned out to be an efficient catalyst for photochemical introducing a CO group into alkanes and arenes. Scheme IV-23 shows products of carbonylation of *n*-pentane, *n*-decane, 2-methylpentane, as well as benzene (yields are given based on Rh) [46a].

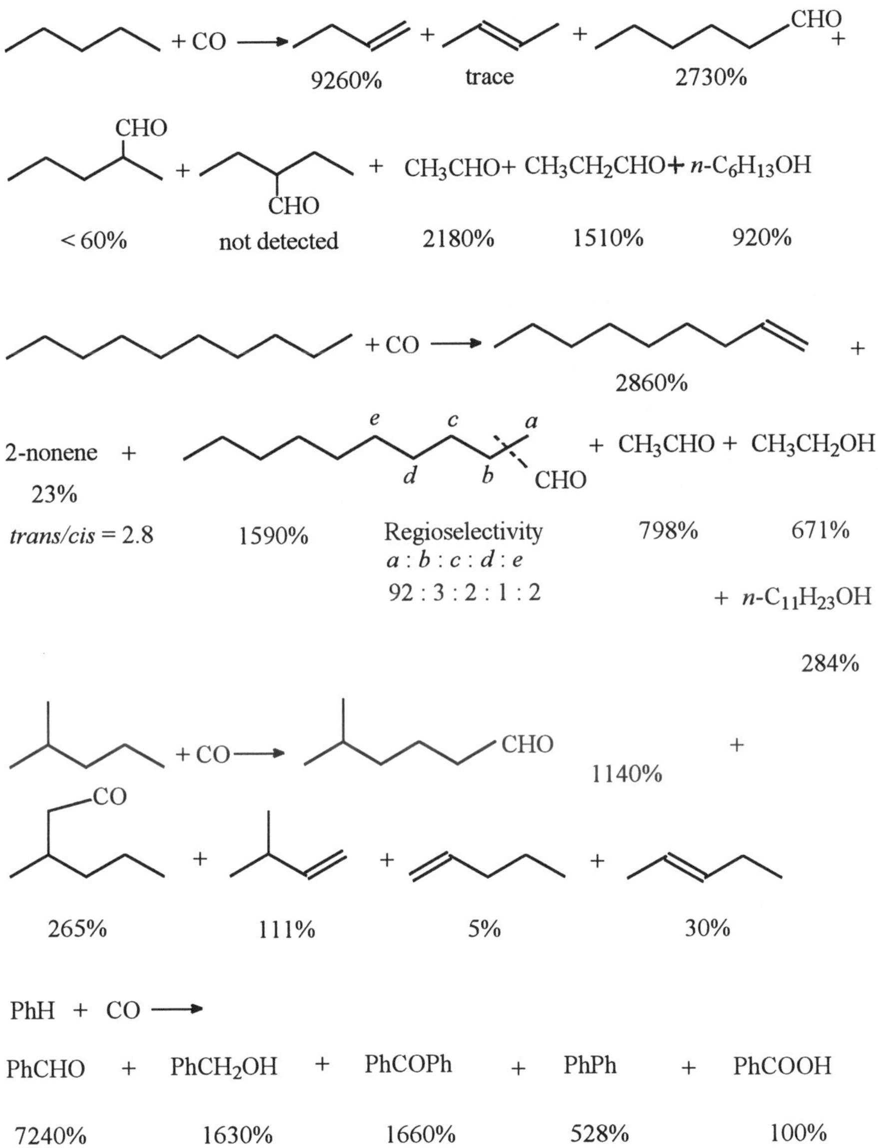

Scheme IV.23. Some alkane photo-carbonylation reactions catalyzed by $RhCl(CO)(PMe_3)_2$.

The isomeric ratio of tolualdehydes formed in carboxylation of toluene was *ortho* : *meta* *para* = 0 : 2 : 1. The formation of phenylacetaldehyde was only about 1% that of tolualdehyde. Competitive reactions revealed the following reactivity order:

benzene > cyclohexane > pentane ≈ decane > cyclooctane.

The mechanism proposed for the photocarbonylation of benzene is presented in Scheme IV-24 [46b] (also see [46c–e]).

Photochemical cyclohexane carbonylation was found to be co-catalyzed by d^8 transition metal carbonyls, and aromatic ketones and aldehydes [47]. The following mechanism has been suggested for this reaction.

$$\mathrm{CyH + ArR'C{=}O \xrightarrow{h\nu} Cy\bullet + ArR'C(OH)\bullet}$$

$$\mathrm{Cy\bullet + M(CO) \longrightarrow \bullet M[C(O)Cy]}$$

$$\mathrm{\bullet M[C(O)Cy] + ArR'C(OH)\bullet \longrightarrow HM[C(O)Cy] + ArR'C{=}O}$$

$$\mathrm{HM[C(O)Cy] + CO \longrightarrow M(CO) + HC(O)Cy}$$

It is interesting that upon illumination, the complex $RhCl(CO)(PMe_3)_2$ catalyzes carbonylation of liquified propane at room temperature to yield butanal with high regioselectivity [48]. The reaction of some C–H compounds with carbon monoxide and ethylene or other olefins in the presence of metal carbonyls results in carbonylation at a C–H bond [49] (Scheme IV-25). The carbonyl $Ru_3(CO)_{12}$ catalyzes the cyclocarbonylation of yne-aldehydes to bicyclic α,β-unsaturated γ-butyrolactones [49d].

IV.2.D. OTHER FUNCTIONALIZATIONS

Complexes of transition metals can posses the possibility to insert various functional groups catalytically into C–H bonds. For example, complexes $RhX[P(i\text{-}Pr)_3]_2(CNR)$ (X = Cl, I; R = neopentyl, 2,6-xylyl) catalyze isocyanide insertion into aromatic C–H bonds [50]:

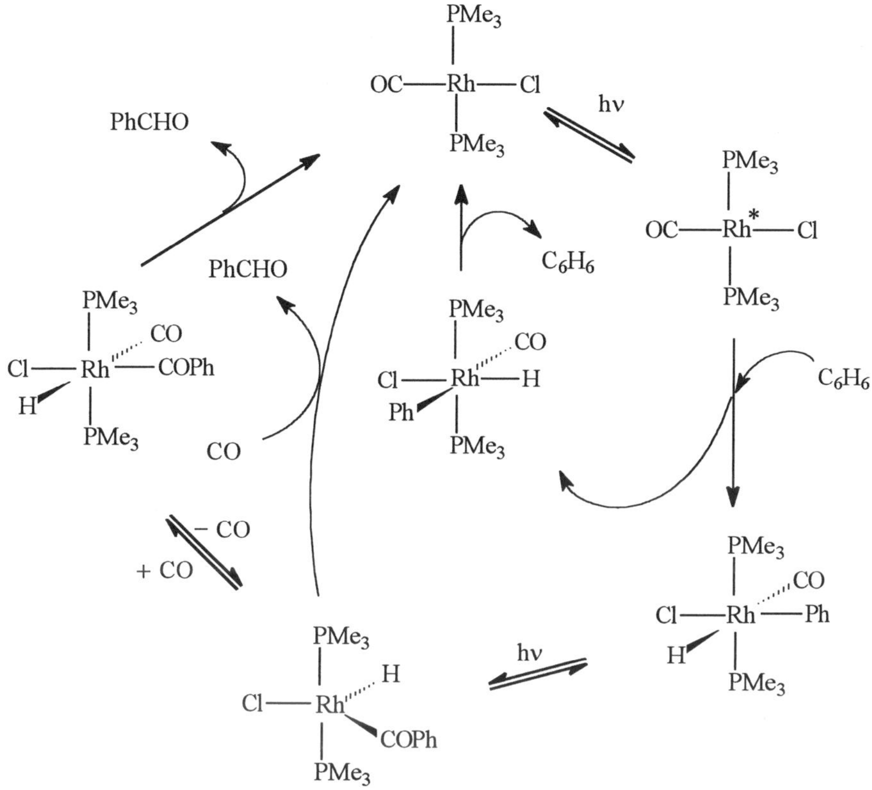

Scheme IV.24. The mechanism of benzene photocarbonylation catalyzed by $RhCl(CO)(PMe_3)_2$.

$$C_6H_6 + CNR \xrightarrow{h\nu,\ [M]} C_6H_5CH{=}NR$$

The reaction of aromatic and aliphatic C–H bonds with terminal alkynes gives 1,1-disubstituted ethenes when photocatalyzed by $RhCl(CO)(PMe_3)_2$ [51].

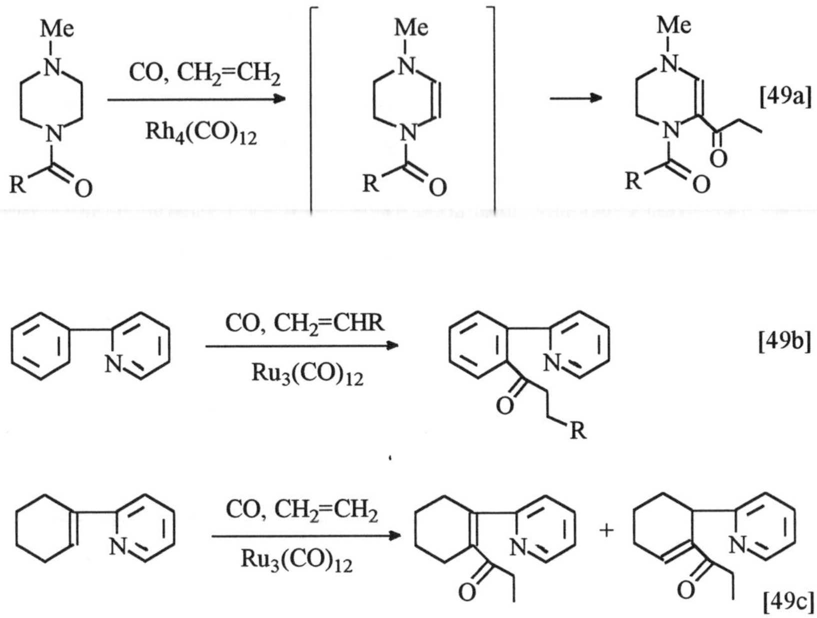

Scheme IV.25. The carbonylation of C–H compounds catalyzed by metal carbonyls. Reference numbers are in brackets.

Thus, the reaction of hexane with phenylacetylene affords the following products (yields are based on Rh):

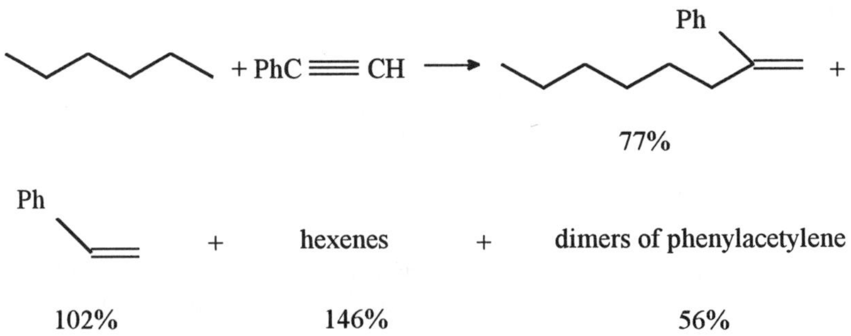

Complexes $Cp^*M(CO)_nBR_2$ (M= Fe, Ru, $n = 2$; M = W, $n = 3$) react with alkanes and release products resulting from selective functionalization of an alkane at the terminal position [52a,b] (Scheme IV-26). Aromatic hydrocarbons can be functionalized by the complex $(CO)_5MnBcat$ (cat = $O_2C_6H_4$) [52b,c].

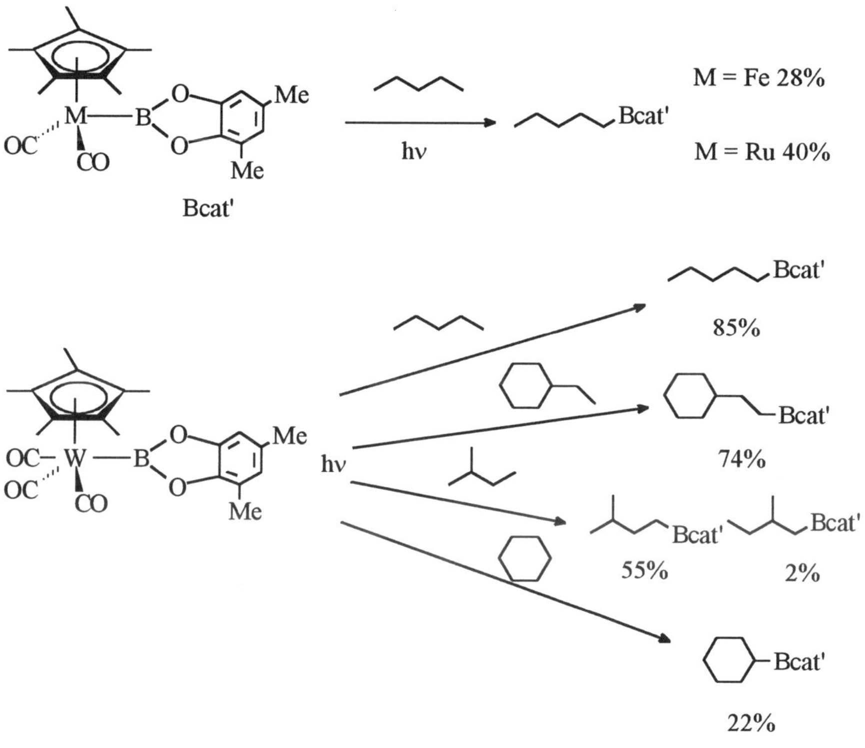

Scheme IV.26. Selective functionalization of alkanes by transition-metal boryl complexes.

Intramolecular hydroacylation of unsaturated aldehydes is catalyzed by complex $Co_2(N_2)(PPh_3)_6$ and some other derivatives of transition metals. The proposed mechanism of the reaction includes, as a crucial step, the cleavage of C–H bond of aldehyde group (Scheme IV-27) [53].

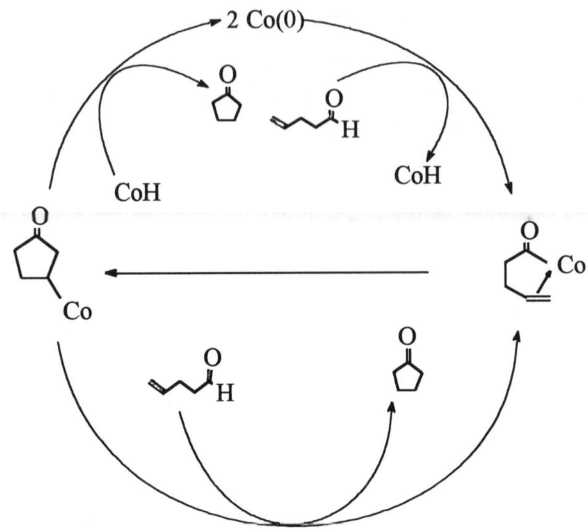

Scheme IV.27. Intramolecular hydroacylation catalyzed by $Co_2(N_2)(PPh_3)_6$.

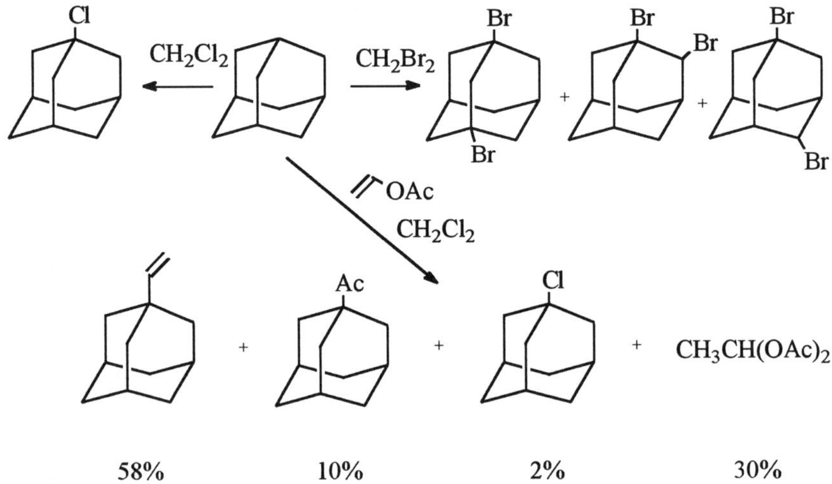

Scheme IV.28. Adamantane functionalization in the presence of complex $Rh(PPh_3)_3Cl$.

Rhodium complexes, particularly $Rh(PPh_3)_3Cl$, induce functionalization of adamantane as shown in Scheme IV-28 [54]. Activation of an α-C–H bond in the coordinated Et_2O of $[Cp^*W(CO)_3(OEt_2)]^+$ by a phosphine, PR_3, to yield $[CH_3(EtO)CH(PR_3)]^+$ has been described [55]. Other recently reported examples of replacing C–H bonds by various groups [56] are depicted in Scheme IV-29.

IV.3. FUNCTIONALIZATION OF C–H BONDS WITH INTERMEDIATE FORMATION OF RADICALS AND CARBENES

IV.3.A. RADICALS IN C–H BOND FUNCTIONALIZATION

Rhodium(II) porphyrin complexes were observed to react with C–H bonds of alkylbenzenes and even methane [57]:

$$[PorphRh]_2 + R\text{–}H \rightleftharpoons PorphRh\text{–}R + PorphRh\text{–}H$$

In this case, components of the C–H bond add simultaneously to two species of activating complex. The authors proposed that the crucial step of the reaction is the attack of metal-centered radicals on the C–H compound. Formation of a transition state that contains two PorhRh• units and methane

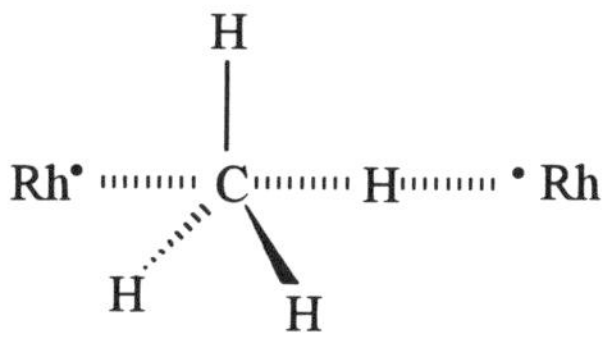

could occur by a single step (Mechanism *A*):

$$[PorphPh]_2 \rightleftharpoons 2\ PorphRh\bullet$$

$$2\ PorphRh\bullet + CH_4 \rightleftharpoons \{PorphRh{\cdot}CH_4{\cdot}RhPorph\}$$

$$\{PorphRh{\cdot}CH_4{\cdot}RhPorph\} \rightleftharpoons PorphRh\text{–}H + PorphRh\text{–}CH_3$$

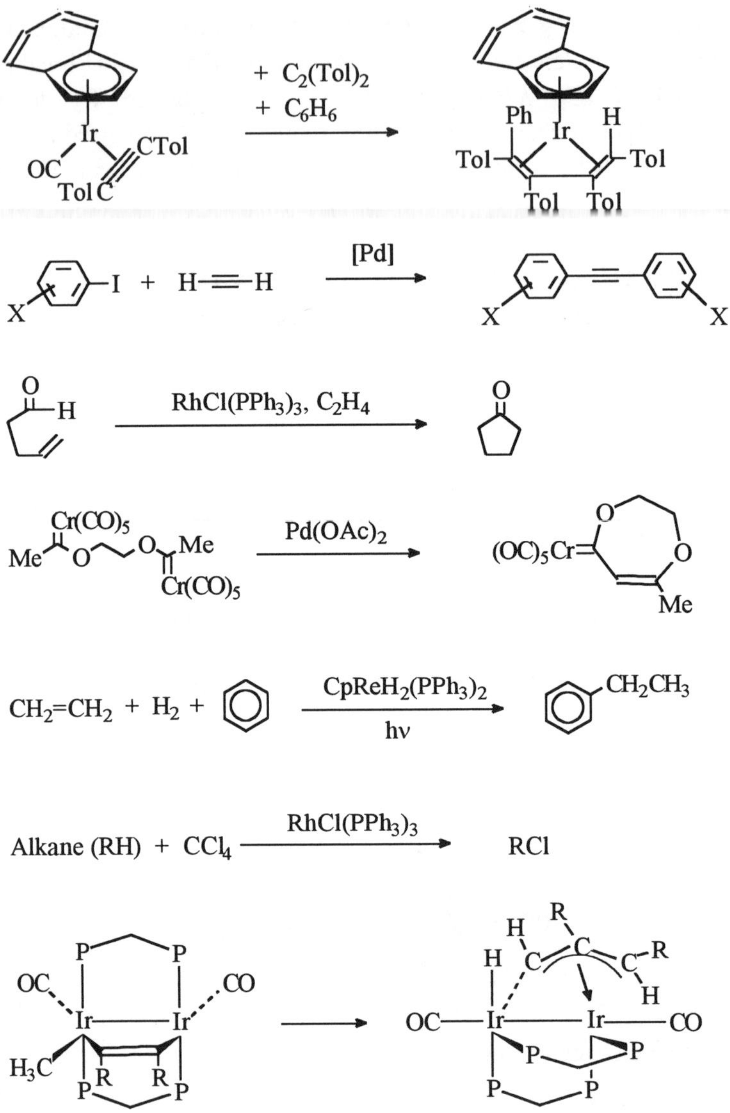

[56a]

[56b]

[56c]

[56d]

[56e]

[56f]

[56g]

Scheme IV.29. Various examples of C–H functionalizations under the action of metal complexes. Reference numbers are in brackets.

or a series of bimolecular steps involving an intermediate (Mechanism *B*):

$$[PorphRh]_2 \rightleftharpoons 2\ PorphRh\bullet$$

$$PorphRh\bullet + CH_4 \rightleftharpoons \{PorphRh{\cdot}CH_4\}$$

$$\{PorphRh{\cdot}CH_4\} + PorphRh\bullet \rightleftharpoons \{PorphRh{\cdot}CH_4{\cdot}RhPorph\}$$

$$2\ PorphRh\bullet \rightleftharpoons \{PorphRh\bullet\ \bullet RhPorph\}$$

$$\{PorphRh\bullet\ \bullet RhPorph\} + CH_4 \rightleftharpoons \{PorphRh{\cdot}CH_4{\cdot}RhPorph\}$$

$$\{PorphRh{\cdot}CH_4{\cdot}RhPorph\} \rightleftharpoons PorphRh\text{–}H + PorphRh\text{–}CH_3$$

It is interesting that the selective activation of a *meta* C–H bond of benzonitrile via the reaction of $RhCl_3$ with porphyrins in refluxing benzonitrile gives rise to the formation of (*meta*-cyanophenyl) rhodium porphyrins [58].

IV.3.B. INSERTION OF CARBENES INTO C-H BONDS

The intermolecular and especially intramolecular carbenoid C–H insertion is a very useful method for the synthesis of cyclic organic compounds as well as other products [59–63]. For example, the remote functionalization of C–H bonds by α-diazoketones, which is efficiently catalyzed by rhodium(II) complexes, gives cyclopentanones, lactams and lactones according to the general equation:

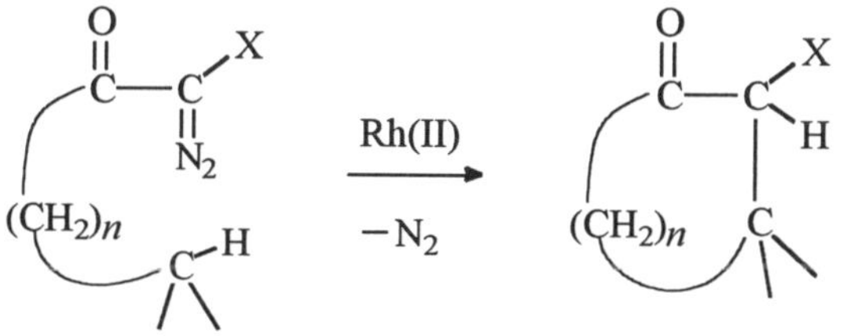

Examples of such processes are presented in Scheme IV-30. Valuable cyclic compounds including natural products can be prepared by this method (Scheme IV-31).

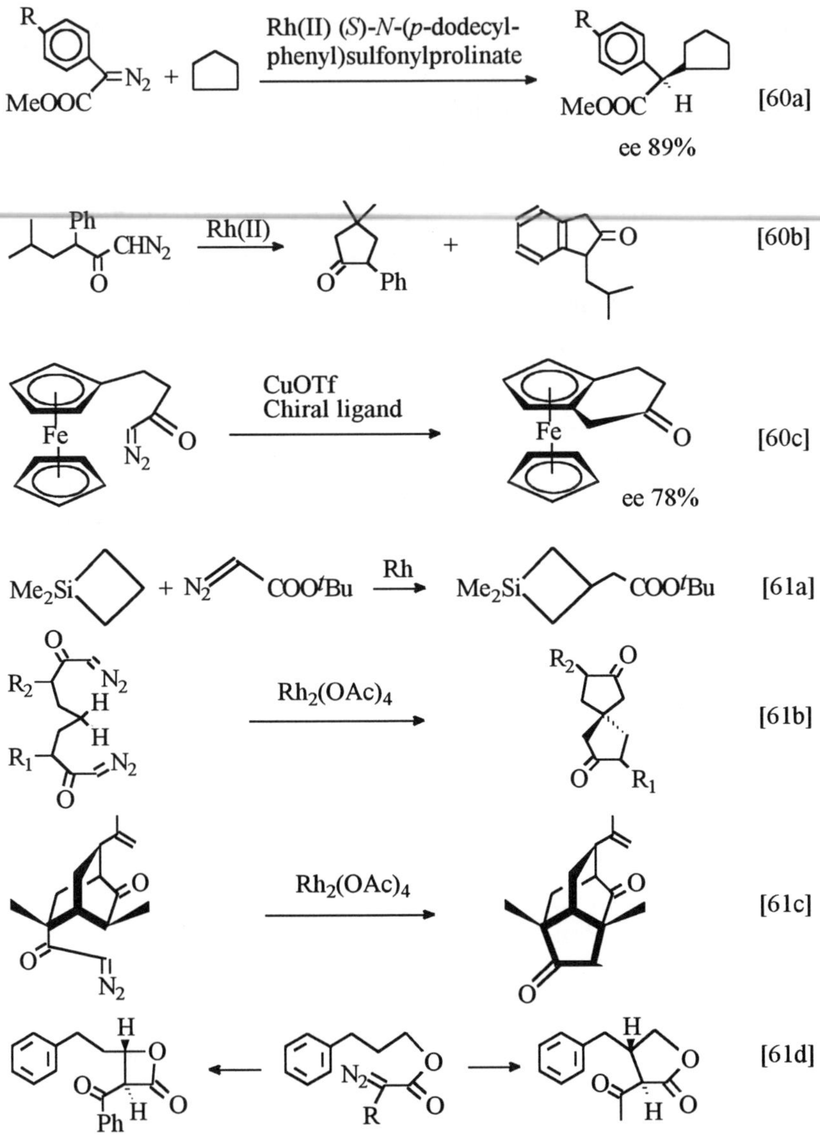

Scheme IV.30. Some examples of carbenoid C–H insertions catalyzed by metal complexes. Reference numbers are in brackets.

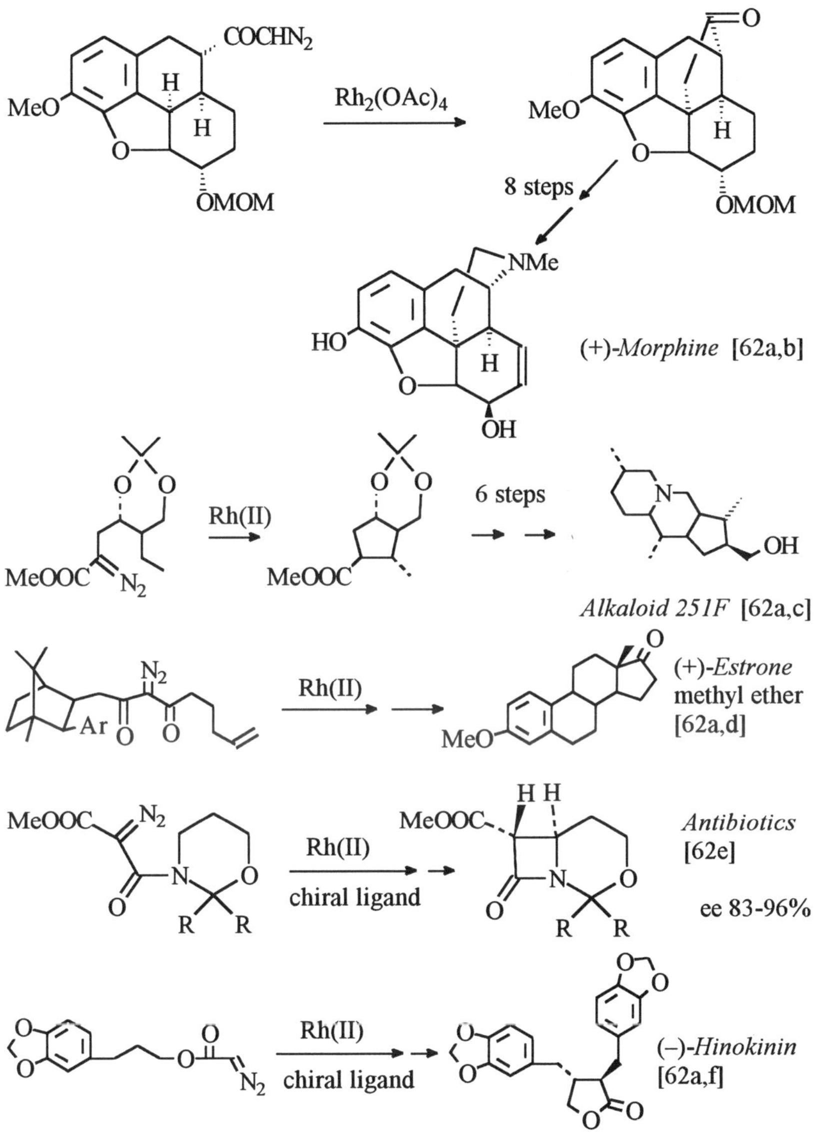

Scheme IV.31. The synthesis of natural products via the insertion of rhodium carbenoids into the C–H bonds. Reference numbers are in brackets.

The mechanism proposed for the rhodium-catalyzed intramolecular insertion of carbenoids into C–H bonds adjacent to ethers [63] is more complex than previously recognized for such reactions. According to this mechanism (Scheme IV-32), the acetal product arises by oxygen-assisted hydride transfer to the electrophilic carbon of the carbenoid.

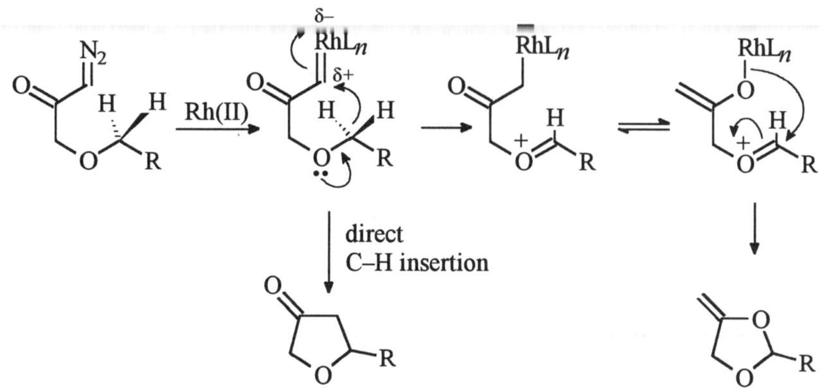

Scheme IV.32. The mechanism proposed for the insertion of rhodium carbenoids into the α-C–H bonds of ethers.

Transition-metal carbene complexes undergo insertion reactions with C–H compounds. For example, under the action of PF_3 or HCl, the rhodium cyclopentadienyl carbene complex gives the products of an migratory insertion of the CPh_2 unit into an sp^2-C–H bond of the cyclopentadienyl ring [64]:

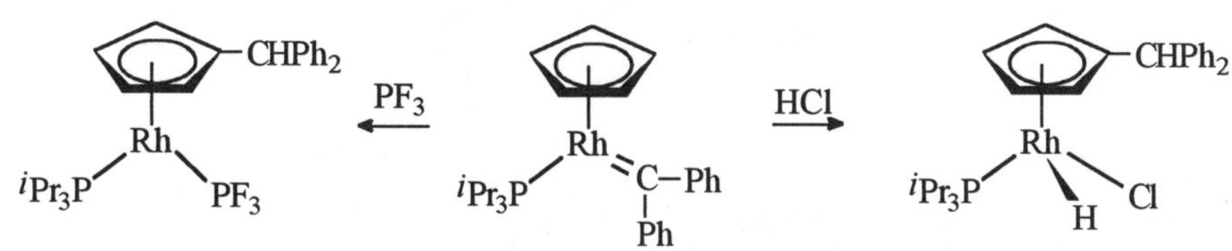

The authors assume that the initial addition step of HCl to the Rh=CPh_2 bond takes place and the formed intermediate then reacts by migration of the $CHPh_2$ fragment to the cyclopentadienyl ligand to afford the substituted cyclopentadienerhodium(I) species.

IV.4. Cleavage of Some Other Bonds

Compounds containing bonds between some other elements can react with transition metal complexes via oxidative addition mechanism to produce various derivatives. In some aspects, these reactions are related to the oxidative addition reaction of C–H bonds to low-valent metal complexes and consequently it would be interesting to consider them briefly in this section of the book. Among these processes, the activation of C–C bonds is the most important for us.

IV.4.A. ACTIVATION OF C–C BONDS

A strong bond between two carbon atoms can be also oxidatively added to soluble metal complexes; although, the number of examples of C–C bond splitting by metal complexes is noticeably lower in comparison with reaction of C–H compounds.

The rupture of sp^3-C–C bonds by metal complexes is known mainly for strained hydrocarbons, cyclopropane and cyclobutane and their derivatives [65]. In some cases, reactions occur according with the simplest equation:

$$\gt C{-}C\lt \; + \; M \longrightarrow \; \gt C{-}M{-}C\lt \qquad \text{(IV.5)}$$

The examples of such type processes [66] are summarized in Scheme IV.33 (it can be seen from this scheme that examples of splitting C–C bonds between aromatic rings or between an aromatic ring and an alkyl group are investigated in more detail). However, usually reactions with metal complexes give different types of compounds [67] (Scheme IV.34) rather than simple dialkyl (diorganyl) derivatives, R–M–R.

Complexes of metals in low oxidation state induce rearrangements and isomerizations of strained and unsaturated hydrocarbons which occur with C–C bond cleavage [68] (Scheme IV.35). 1,4-Pentadiene can be isomerized (with splitting C–C bonds) under the action of hydride cyclopentadienyl complexes of scandium [69a]. Hydrogenolysis of C–C bonds in saturated hydrocarbons has been described in the presence of the system "$H_2 - Re_2(CO)_{10} - Al_2O_3$" [69b],

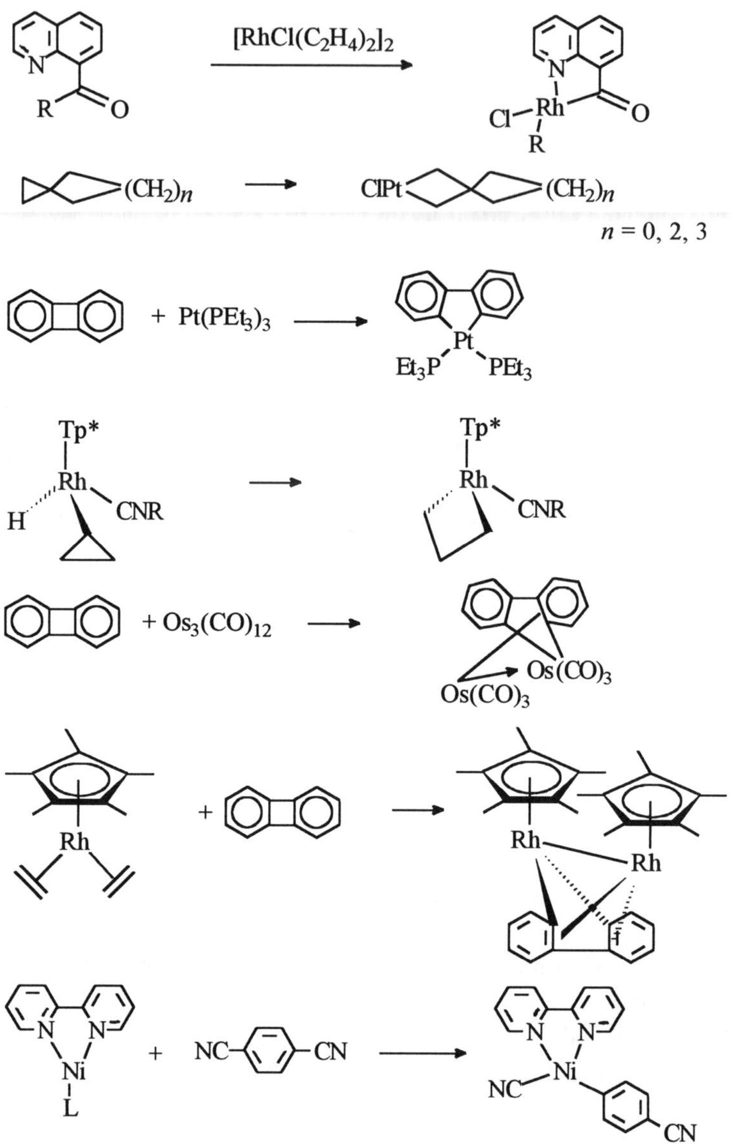

[66a]

[66b]

[66c]

[66d]

[66e]

[66f]

[66g]

Scheme IV.33 (part 1). Activation of C–C bonds by metal complexes in accordance with equation (IV.5). Reference numbers are in brackets.

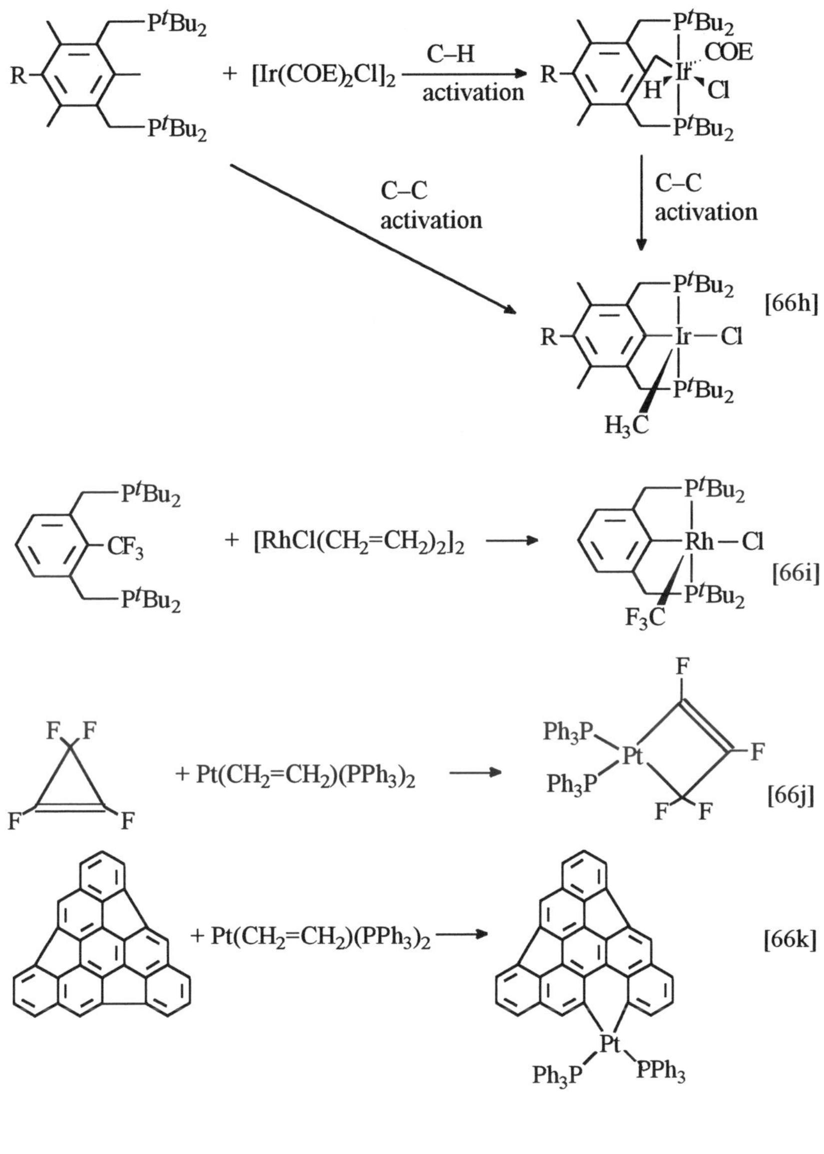

Scheme IV.33 (part 2). Activation of C–C bonds by metal complexes in accordance with equation (IV.5). Reference numbers are in brackets.

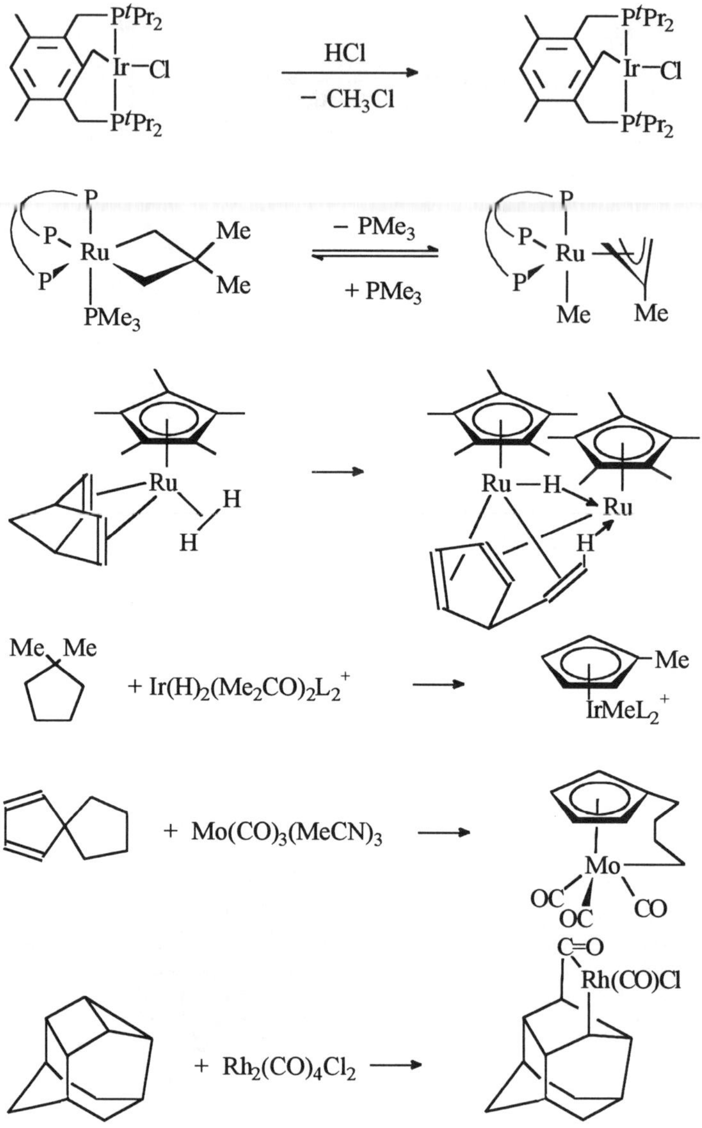

[67a,b]

[67c]

[67d]

[67e]

[67f]

[67g]

Scheme IV.34. Splitting C–C bonds by metal complexes in low valent state. Reference numbers are in brackets.

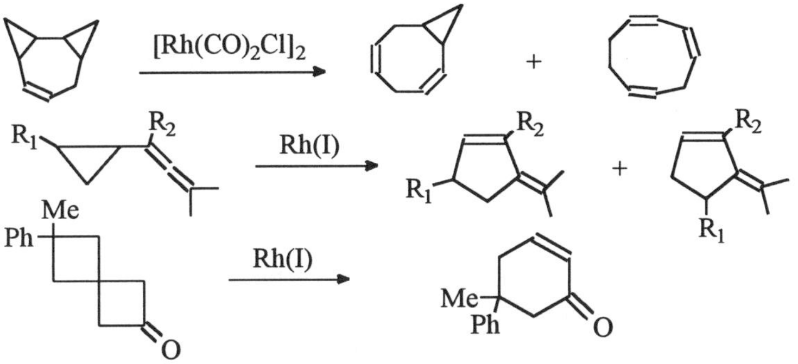

Scheme IV.35. Rearrangements and isomerizations of strained and unsaturated hydrocarbons under the action of metal complexes. Reference numbers are in brackets.

which includes a heterogeneous catalyst: either particles of rhenium metal or solid cluster species of the type $Re_x(CO)_y$. Catalytic hydrogenolysis and isomerization of light alkanes were induced by the silica-supported titanium hydride complex [69c]. Low conversions of neopentane give methane, isobutane, and *n*-butane as primary products. The presence and importance of -butane were a clear indication of carbon skeletal rearrangement prior to the formation of products. The reaction of isobutane with hydrogen at 150 °C gave ethane as a primary product. This indicates that the initially formed surface isobutyl fragment has rearranged to an *n*-butyl fragment. An initial step of the reaction is the C–H bonds activation. Then the surface alkyl fragment undergoes β-methyl elimination. The intermediate metal–alkyl–olefin complex persists long enough to undergo reinsertion of the rotated olefin into the metal–carbon bond.

IV.4.B. ACTIVATION OF Si–H BONDS

Since silicon is an analog of carbon in the Periodic System, it is important to survey reactions of Si–H compounds with complexes of metals in low oxidation states. Similar to C–H compounds, derivatives containing Si–H bonds easily react with low-valent metal complexes via oxidative addition mechanism to afford hydridosilyl or silyl compounds. Typical examples of such reactions [70] are shown below.

$RhClL_2 + HSiAr_3 \longrightarrow Ar_3SiRh(H)ClL_2$ [70a]

$(Me_3P)_4RhMe + H_2SiPh_2 \longrightarrow (Me_3P)_4SiHPh_2 + CH_4$ [70b]

$RhClL_2 + HSiR_3 \longrightarrow [RhL(H)_2\{\mu\text{-}SiR_2\}Rh(H)_2L]$ [70c]

IV.4.C. ACTIVATION OF C–F BONDS

Oxidative addition of C–F bonds to low-valent metal complexes constitute one of the routes in activation of fluoroalkanes and fluoroarenes [71, 72]. The chemical activation of carbon–fluorine bonds has attracted much attention since functionalization of C–F compounds gives various fluoroorganic derivatives, which have many technological applications. Typical examples of oxidative addition of C–F bonds [73] are shown in Scheme IV.36.

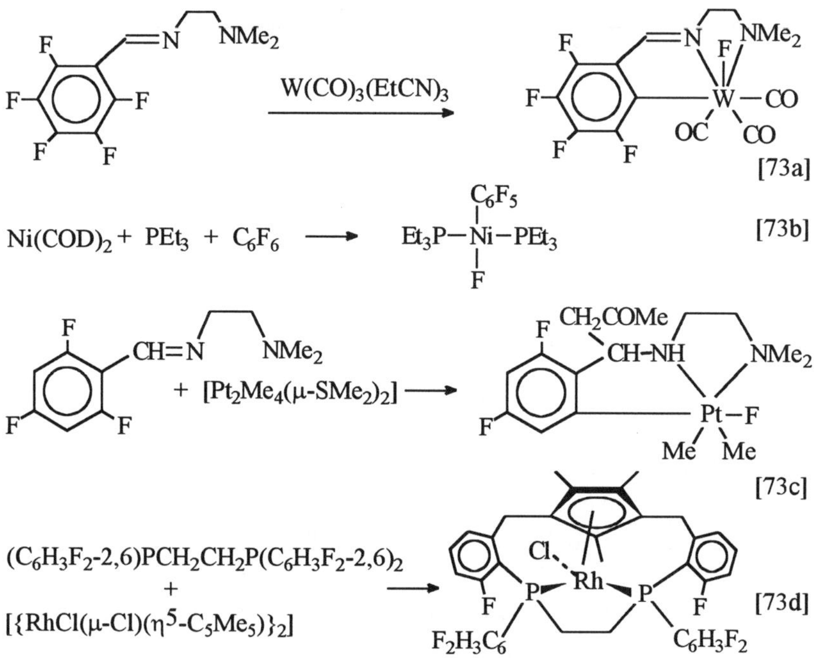

Scheme IV.36. Oxidative addition of C–F compounds to metal complexes.

IV.4.D. ACTIVATION OF CARBON–ELEMENT, ELEMENT–ELEMENT, AND ELEMENT–HYDROGEN BONDS

Metal complexes can add components of bonds between carbon and some other elements via oxidative addition mechanism to produce derivatives, which contain both metal–carbon and metal–element bonds. Here the element is Si, S, P, etc. [74]. Analogously, metal complexes split bonds between two atoms of an element [75] and between element and hydrogen [76]. Examples of such reactions published in recent years [77] are depicted in Scheme IV.37.

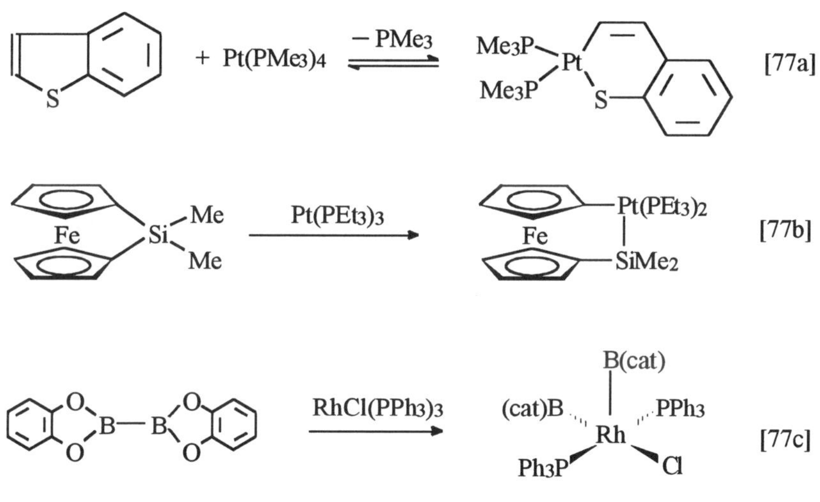

Scheme IV.37.. Oxidative addition of compounds containing C–Element and Element–Element bonds to metal complexes. Reference numbers are in brackets.

References for Chapter IV

1. Halpern, J. *Disc. Faraday Soc.* **1968**, *46*, 7.

2. Omae, I. *Organometallic Intramolecular-coordination Compounds*; Elsevier: Amsterdam, 1986.

3. (a) Dehand, J.; Pfeffer, M. *Coord. Chem. Rev.* **1976**, *18*, 327. (b) Abicht, H.-P.; Issleib, K. *Ztschr. Chem.* **1977**, *17*, 1. (c) Newkome, G. R.; Puckett, W. E.; Gupta, V. K.; Kiefer, G. E. *Chem. Rev.* **1986**, *86*, 451. (d) Evans, D. W.; Baker, G. R.; Newkome, G. R. *Coord. Chem. Rev.* **1989**, *93*, 155.

4. (a) Diversi, P.; Iacoponi, S.; Ingrosso, G.; Laschi, F.; Lucherini, A.; Zanello, P. *J. Chem. Soc., Dalton Trans.* **1993**, 35. (b) Cooper, A. C.; Folting, K.; Huffman, J. C.; Caulton, K. G. *Organometallics* **1997**, *16*, 505. (c) Rigny, S.; Leblanc, J.-C.; Nuber, B.; Moise, C. *J. Chem. Soc., Dalton Trans.* **1997**, 1187. (d) Morton, C.; Duncaff, D. J.; Rourke, J. P. *J. Organometal. Chem.* **1997**, *530*, 19. (e) Depree, G. J.; Childerhouse, N. D.; Nicholson, B. K. *J. Organometal. Chem.* **1997**, *533*, 143. (f) Alarcon, C. J.; Estevan, F.; Lahuerta, P.; Ubeda, M. A.; Gonzalez, G.; Martinez, M.; Stirba, S. E. *Inorg. Chim. Acta* **1998**, *278*, 61. (g) Crochet, P.; Esteruelas, M. A.; Gutirrez-Puebla, E. *Organometallics* **1998**, *17*, 3141. (h) Tayebani, M.; Gambarotta, S.; Yap, G. *Organometallics* **1998**, *17*, 3639. (i) Dinger, M. B.; Main, L.; Nicholson, B. K. *J. Organometal. Chem.* **1998**, *565*, 125. (j) Barea, G.; Esteruelas, M. A.; lledos, A.; Lopez, A. M.; Onate, E.; Tolosa, J. I. *Organometallics* **1998**, *17*, 4065. (k) Djukic, J.-P.; Maisse, A.; Pfeffer, M. *J. Organometal. Chem.* **1998**, *567*, 65. (l) Hiraki, K.; Koizumi, M.; Kira, S.; Kawano, H. *Chemistry Lett.* **1998**, 47. (m) Pruchnik, F. P.; Starosta, R.; Lis, T.; Lahuerta, P. *J. Organometal. Chem.* **1998**, *568*, 177. (n) Guari, Y.; Sabo-Etienne, S.; Chaudret, B. *J. Am. Chem. Soc.* **1998**, *120*, 4228. (o) Karlen, T.; Dani, P.; Grove, D. M.; Steenwinkel, P.; van Koten, G. *Organometallics* **1996**, *15*, 5687. (p) Dani, P.; Karlen, T.; Gossage, R. A.; Smeets, W. J. J.; Spek, A. L.; van Koten, G. *J. Am. Chem. Soc.* **1997**, *119*, 11317. (q) Steenwinkel, P.; James, S. L.; Gossage, R. A.; Grove, D. M.; Kooijman, H.; Smeets, W. J. J.; Spek, A. L.; van Koten, G. *J. Organometallics* **1998**, *17*, 4680.

5. Cinellu, M. A.; Zucca, A.; Stoccoro, S.; Minghetti, G.; Manassero, M.; Sansoni, M. *J. Chem. Soc., Dalton Trans.* **1996**, 4217.

6. (a) Jordan, R. F.; Guram, A. S. *Organometallics* **1990**, *9*, 2116. (b) Heinekey, D. M.; Oldham, W. J., Jr.; Wiley, J. S. *J. Am. Chem. Soc.* **1996**, *118*, 12842. (c)

Beringhelli, T.; D'Alfonso, G.; Panigati, M. *J. Organometal. Chem.* **1997**, *527*, 215.

7. (a) Shinimoto, R.; Desrosiers, P. J.; Harper, T. G. P.; Flood, T. C. *J. Am. Chem. Soc.* **1990**, *112*, 704. (b) Rabinovich, D.; Parkin, G. *J. Am. Chem. Soc.* **1990**, *112*, 5381. (c) Huber, S. R.; Baldwin, T. C.; Wigley, D. E. *Organometallics* **1993**, *12*, 91. (d) Sjövall, S, Kloo, L.; Nikitidis, A.; Andersson, C. *Organometallics* **1998**, *17*, 579. (e) Hirano, M.; Kurata, N.; Marumo, T.; Komiya, S. *Organometallics* **1998**, *17*, 501. (f) Thorn, D. L. *Organometallics* **1998**, *17*, 348. (g) Chen, Y.-X.; Marks, T. J. *Organometallics* **1997**, *16*, 3649. (h) Dorta, R.; Togni, A. *Organometallics* **1998**, *17*, 3423. (i) Scheer, M.; Leiner, E.; Kramkowski, P.; Schiffer, M.; Baum, G. *Chem. Eur. J.* **1998**, *4*, 1917.

8. (a) Horacek, M.; Hiller, J.; Thewalt, U.; Polasek, M.; Mach, K. *Organometallics* **1997**, *16*, 4185. (b) Mauthner, K.; Slugovc, C.; Mereiter, K.; Schmid, R.; Kirchner, K. *Organometallics* **1997**, *16*, 1956.

9. (a) Bennett, M. A.; Dirnberger, T.; Hockless, D. C. R.; Wenger, E.; Willis, A. C. *J. Chem. Soc., Dalton Trans.* **1998**, 271. (b) Bennett, M. A.; Berry, D. E.; Dirnberger, T.; Hockless, D. C. R.; Wenger, E.; *J. Chem. Soc., Dalton Trans.* **1998**, 2367. (c) Blake, R. E., Jr.; Antonelli, D. A.; Henling, L. M.; Schaefer, W. P.; Hardcastle, K. I.; Bercaw, J. E. *Organometallics* **1998**, *17*, 718. (d) Wada, H.; Tobita, H.; Ogino, H. *Organometallics* **1997**, *16*, 3870. (e) Mul, W. P.; Elsevier, C. J.; Vuurman, M. A.; Smeets, W. J. J.; Spek, A. L.; de Boer, J. L. *J. Organometal. Chem.* **1997**, *532*, 89. (f) Djakovitch, L.; Herrmann, W. A. *J. Organometal. Chem.* **1997**, *545-546*, 399. (g) Imhof, W.; *J. Organometal. Chem.* **1997**, *533*, 31.

10. (a) Voskoboynikov, A. Z.; Osina, M. A.; Shestakova, A. K.; Kazankova, M. A.; Trostyanskaya, I. G.; Beletskaya, I. P.; Dolgushin, F. M.; Yanovsky, A. I.; Struchkov, Yu. T. *J. Organometal. Chem.* **1997**, *545-546*, 71. (b) Giordano, R.; Sappa, E.; Predieri, G.; Tiripicchio, A. *J. Organometal. Chem.* **1997**, *547*, 49. (c) Jun, C.-H. *J. Organomet. Chem.* **1990**, *390*, 361.

11. (a) Dürr, U.; Heinemann, F. W.; Kisch, H. *J. Organometal. Chem.* **1997**, *541*, 307. (b) Lim, Y.-G. Kang, J.-B.; Kim, Y. H. *J. Chem. Soc., Perkin Trans. 1* **1998**, 699.

12. (a) Bergman, R. G. *J. Organomet. Chem.* **1990**, *400*, 273. (b) Bergman, R. G. Presented at the *217th Am. Chem. Soc. Natl. Meeting*; Anaheim, March **1999**; paper INOR No. 431.

13. Green, M. L. H. *J. Chem. Soc., Dalton Trans.* **1986**, 2469.

14. (a) Periana, R. A.; Bergman, R. G. *Organometallics* **1984**, *3*, 508. (b) Wax, M. J.; Stryker, J. M.; Buchanan, J. M.; Kovac, C.A.; Bergman, R. G. *J. Am. Chem. Soc.* **1984**, *106*, 1121.

15. (a) Mobley, T. A.; Bergman, R. G. *J. Am. Chem. Soc.* **1998**, *120*, 3253. (b) Whittlesey, M. K.; Mawby, R. J.; Osman, R.; Perutz, R. N.; Field, L. D.; Wilkinson M. P.; George, M. W. *J. Am. Chem. Soc.* **1993**, *115*, 8627. (c) Buys, I. E.; Field, L. D.; Hambley T. W.; McQueen, A. E. D. *J. Chem. Soc., Chem. Commun.* **1994**, 557.

16. (a) Koola, J. D.; Roddick, D. M. *J. Am. Chem. Soc.* **1991**, *113*, 1450. (b) Chin, R. M.; Dong, L.; Duckett, S. B.; Partridge, M. G.; Jones, W. D.; Perutz, R. N. *J. Am. Chem. Soc.* **1993**, *115*, 7685. (c) Foo, T.; Bergman, R. G.; *Organometallics* **1992**, *11*, 1801. (d) Price, R. T.; Andersen, R. A.; Muetterties, E. L. *J. Organomet. Chem.* **1989**, *376*, 407. (e) Jiang, Q.; Pestana, D. C.; Carroll, P. J.; Berry, D. H. *Organometallics* **1994**, *13*, 3679. (f) Squires, M. E.; Sardella, D. J.; Kool, L. B. *Organometallics* **1994**, *13*, 2970. (g) Gutiérrez, E.; Monge, A.; Nicasio, M. C.; Poveda, M. L.; Carmona, E. *J. Am. Chem. Soc.* **1994**, *116*, 791. (h) Perutz, R. N.; Belt, S. T.; McCamley, A.; Whittlesey, M. K. *Pure Appl. Chem.* **1990**, *63*, 1539. (i) Peters, R. G.; White, S.; Roddick, D. M. *Organometallics* **1998**, *17*, 4493.

17. (a) Rabinovich, D.; Zelman, R.; Parkin, G. *J. Am. Chem. Soc.* **1990**, *112*, 9632. (b) Bergman, R. G.; Seidler, P. F.; Wenzel, T. T. *J. Am. Chem. Soc.* **1985**, *107*, 4358.

18. (a) Purwoko, A. A.; Lees, A. J. *Inorg. Chem.* **1995**, *34*, 424. (b) Purwoko, A. A.; Tibensky, S. D.; Lees, A. J. *Inorg. Chem.* **1996**, *35*, 7049. (c) Dunwoody, N.; Lees, A. J. *Organometallics* **1997**, *16*, 5770. (d) Lees, A. J. *J. Organometal. Chem.* **1998**, *554*, 1. (e) Jobling, M.; Howdle, S. M.; Healy, M. A.; Poliakoff, M. *J. Chem. Soc., Chem. Commun.* **1990**, 1287. (f) Jobling, M.; Howdle, S. M.; Poliakoff, M. *J. Chem. Soc., Chem. Commun.* **1990**, 1762. (g) Barrientos, C.; Ghosh, C. K.; Graham, W. A. G. *J. Organomet. Chem.* **1990**, *394*, C31. (h) Bridgewater, J. S.; Lee, B.; Bernhard, S.; Schoonover, J. R.; Ford, P. C. *Organometallics* **1997**, *16*, 5592.

19. (a) Castarlenas, R.; Esteruelas, M. A.; Martin, M.; Oro, L. A. *J. Organometal. Chem.* **1998**, *564*, 241. (b) Belt, S. T.; Dong, L.; Duckett, S. B.; Jones, W. D.; Partridge, M. G.; Perutz, R. N. *J. Chem. Soc., Chem. Commun.* **1991**, 266. (c) Boutry, O.; Gutierrez, E.; Monge, A.; Nicasio, M. C.; Perez, P. J.; Carmona, E.

J. Am. Chem. Soc. **1992**, 114, 7288. (d) Paneque, M.; Taboada, S.; Carmona, E. *Organometallics* **1996**, *15*, 2678. (e) Jones, W. D.; Hessell, E. T. *J. Am. Chem. Soc.* **1993**, *115*, 554. (f) Leiva, C.; Sutton, D. *Organometallics* **1998**, *17*, 1700. (g) Tobita, H.; Hashidzume, K.; Endo, K.; Ogino, H. *Organometallics* **1998**, *17*, 3405. (h) Arce, A. J.; Deeming, A. J.; De Sanctis, Y.; Machado, R.; Manzur, J.; Rivas, C. *J. Chem. Soc., Chem. Commun.* **1990**, 1568. (i) Hartwig, J. F.; Bergman, R. G.; Andersen, R. *J. Am. Chem. Soc.* **1991**, *113*, 3404.

20. (a) Ghosh, C. K.; Rodgers, D. P. S.; Graham, W. A. G. *J. Chem. Soc., Chem. Commun.* **1988**, 1511. (b) Ghosh, C. K.; Graham, W. A. G. *J. Am. Chem. Soc.* **1989**, *111*, 375. (c) Barnhart, T. M.; McMahon, R. J. *J. Am. Chem. Soc.* **1992**, *114*, 5434. (d) Boutry, O.; Poveda, M. L.; Carmona, E. *J. Organometal. Chem.* **1997**, *528*, 143.

21. (a) Hall C.; Jones, W. D.; Mawby, R. J.; Osman, R.; Perutz, R. N.; Whittlesey, M. K. *J. Am. Chem. Soc.* **1992**, *114*, 7425. (b) Schulz, M.; Werner, H. *Organometallics* **1992**, *11*, 2790. (c) Gutierrez-Puebla, E.; Monge, A.; Nicasio, M. C.; Perez, P. J.; Poveda, M. L.; Rey, L.; Ruiz, C.; Carmona, E. *Inorg. Chem.* **1998**, *37*, 4538. (d) Alvarado, Y.; Boutry, O.; Gutierrez, E.; Monge, A.; Nicasio, M. C.; Poveda, M. L.; Perez, P. J.; Ruiz, C.; Bianchini, C.; Carmona, E. *Chem. Eur. J.* **1997**, *3*, 845. (e) Suzuki, H.; Omori, H.; Lee, D.H.; Yoshida, Y.; Fukushima, M.; Tanaka, M., Moro-oka, Y. *Organometallics* **1994**, *13*, 1129. (f) Pasynkiewicz, S.; Buchowicz, W.; Pietrzykowski, A.; Glowiak, T. *J. Organometal. Chem.* **1997**, *536-537*, 249. (g) Huang, Y.-H.; Stang, P. J.; Arif, A. M. *J. Am. Chem. Soc.* **1990**, *112*, 5648. (h) Bell, T. W.; Haddleton, D. M.; McCamley, A.; Partridge, M. G.; Perutz, R. N.; Willner, H. *J. Am. Chem. Soc.* **1990**, *112*, 9212. (i) Bell, T. W.; Brough, S.-A.; Partridge, M. G.; Perutz, R. N.; Rooney A. D. *Organometallics* **1993**, *12*, 2933. (j) Wadepohl, H.; Borchert, T.; Pritzkow, H. *Chem. Ber./Recueil*, **1997**, *130*, 593.

22. (a) Gauss, C.; Veghini, D.; Berke, H. *Chem. Ber./Recueil*, **1997**, *130*, 183. (b) Duchateau, R.; van Wee, C. T.; Teuben, J. H. *Organometallics* **1996**, *15*, 2291. (c) Shaposhnikova,A. D.; Stadnichenko,R. A.; Bel'skiy, V. K.; Pasynskiy, A. A. *Metalloorg. Khim.* **1988**, *1*, 945 (in Russian). (d) Koridze, A. A.; Astakhova, N. M.; Yanovskiy, A. I.; Struchkov, Yu. T. *Metalloorg. Khim.* **1992**, *5*, 886 (in Russian). (e) Fyfe, H. B.; Mlekuz, M.; Zargarian, D.; Taylor, N. J.; Marder, T. B. *J. Chem. Soc., Chem. Commun.* **1991**, 188. (f) Bianchini, C.; Barbaro, P.; Meli, A.; Peruzzini, M.; Vacca, A.; Vizza, F. *Organometallics* **1993**, *12*, 2505. (g) Simpson, R. D.; Bergman, R. G. *Angew. Chem., Int. Ed. Engl.* **1992**, *31*, 220. (h) Stang, P. J.; Cao, D. *Organometallics* **1993**, *12*, 996. (i) Espuelas, J.;

Esteruelas, M. A.; Lahoz, F. J.; Oro, L. A.; Valero, C. *Organometallics* **1993**, *12*, 663. (j) Burkey, D. J.; Hanusa, T. P. *Organometallics* **1996**, *15*, 4971. (k) Whittall, I. R.; Humphrey, M. G.; Houbrechts, S.; Maes, J.; Persoons, A.; Schmid, S.; Hockless, D. C. R. *J. Organometal. Chem.* **1997**, *544*, 277. (l) Mihan, S.; Weidmann, T.; Weinrich, V.; Fenske, D.; Beck, W. *J. Organometal. Chem.* **1997**, *541*, 423. (m) Irwin, M. J.; Jia, G.; Vittal, J. J.; Puddephatt, R. J. *Organometallics* **1996**, *15*, 5321. (n) Vicente, J.; Chicote, M.-T.; Abrisqueta, M.-D.; Jones, P. G. *Organometallics* **1997**, *16*, 5628.

23. (a) Touchard, D.; Haquette, P.; Guesmi, S.; Le Pichon, L.; Daridor, A.; Toupet, L.; Dixneuf, P. H. *Organometallics* **1997**, *16*, 3640. (b) Touchard, D.; Haquette, P.; Daridor, A.; Romero, A.; Dixneuf, P. H. *Organometallics* **1998**, *17*, 3844.

24. McGhee, W. D.; Bergman, R. G. *J. Am. Chem. Soc.* **1986**, *108*, 8097.

25. (a) Fendrick, C. M.; Marks, T. J. *J. Am. Chem. Soc.* **1986**, *108*, 425. (b) Cummins, C. C.; Baxter, S. M., Wolczanski, P. T. *J. Am. Chem. Soc.* **1988**, *110*, 8731. (c) Ballard, K. R.; Gardiner, I. M.; Wigley, D. E. *J. Am. Chem. Soc.* **1989**, *111*, 2159.

26. (a) Arndtsen, B. A.; Bergman, R. G. *Science* **1995**, *270*, 1970. (b) Luecke, H. F.; Arndtsen, B. A.; Burger, P.; Bergman, R. G. *J. Am. Chem. Soc.* **1996**, *118*, 2517. (c) Strout, D. L.; Zaric, S.; Niu, S.; Hall, M. B. *J. Am. Chem. Soc.* **1996**, *118*, 6068. (d) Hinderling, C.; Plattner, D. A.; Chen, P. *Angew. Chem., Int. Ed. Engl.* **1997**, *36*, 243.

27. (a) Brough, S.-A.; Hall, C.; McCamley, A.; Perutz, R. N.; Stahl, S.; Wecker, U.; Werner, H. *J. Organomet. Chem.* **1995**, *504*, 33. (b) Esteruelas, M. A.; Lahoz, F. J.; Oñate, E.; Oro, L.; Sola, E. *J. Am. Chem. Soc.* **1996**, *118*, 89. (c) Diversi, P.; Iacoponi, S.; Ingrosso, G.; Laschi, F.; Lucherini, A.; Pinzino, C.; Ucello-Barretta, G.; Zanello, P. *Organometallics* **1995**, *14*, 3275. (d) Banister, J. A.; Cooper, A. I.; Howdle, S. M.; Jobling, M.; Poliakoff, M. *Organometallics* **1996**, *15*, 1804. (e) Ferrari, A.; Polo, E.; Rüegger, H.; Sostero, S.; Venanzi, L. M. *Inorg. Chem.* **1996**, *35*, 1602. (f) Diversi, P.; Ferrarini, A.; Ingrosso, G.; Lucherini, A.; Uccello-Barretta, G.; Pinzino, C.; de Biani Fabrizi, F.; Laschi, F.; Zanello, P. *Gazz. Chim. Ital.* **1996**, *126*, 391. (g) Lian, T.; Bromberg, S. E.; Yang, H.; Proulz, G.; Bergman, R. G.; Harris, C. B. *J. Am. Chem. Soc.* **1996**, *118*, 3769. (h) Debad, J. D.; Legzdins, P.; Lumb, S. A.; Batchelor, R J.; Einstein, F. W. B. *J. Am. Chem. Soc.* **1995**, *117*, 3288. (i) van der Heijden, H.; Hessen, B. *J. Chem. Soc., Chem. Commun.* **1995**, 145. (j) Schaller, C. P.; Cummins, C. C.; Wolczanski, P. T. *J. Am. Chem. Soc.* **1996**, *116*, 591. (k)

Vidal, V.; Théolier, A.; Thivolle-Cazat, J.; Basset, J.-M. *J. Chem. Soc., Chem. Commun.* **1995**, 991. (l) Chisholm, M. H.; Huang, J.-H.; Huffman, J. C. *J. Organometal. Chem.* **1997**, *528*, 221. (m) Bianchini, C.; Casares, J. A.; Osman, R.; Pattison, D. I.; Peruzzini, M.; Perutz, R. N.; Zanobini, F. *Organometallics* **1997**, *16*, 4611. (n) Legzdins, P.; Lumb, S. A. *Organometallics* **1997**, *16*, 1825.

28. (a) Teo, B. K.; Ginsberg, A. P.; Calabrese, J. T. *J. Am. Chem. Soc.* **1976**, *98*, 3027. (b) Davies, J. A.; Dutremez, S.; Pinkerton, A. A.; Vilmer, M. *Organometallics* **1991**, *10*, 2956.

29. (a) Vicente, J.; Bermúdez, M.-D.; Chicote, M.-T.; Sánchez-Santano, M.-J. *J. Chem. Soc., Dalton Trans.* **1990**, 1945. (b) Bech, W.; Schropp, K.; Kern, F. *Angew. Chem., Int. Ed. Engl.* **1971**, *10*, 66.

30. Saura-Llamas, I.; Gladysz, J. A. *J. Am. Chem. Soc.* **1992**, *114*, 2136.

31. Gol'dshleger, N. F.; Tyabin, M. B.; Shilov, A. E.; Shteinman, A. A. *Zh. Fiz. Khim.* **1969**, *43*, 2174 (in Russian).

32. Junk, T.; Catallo, W. J. *Chem. Soc. Rev.* **1997**, *26*, 401.

33. (a) Sola, E.; Bakhhhmutov, V. I.; Torres, F.; Eldugue, A.; Lopez, J. A.; Lahoz, F. J.; Werner, H.; Oro, L. A. *Organometallics* **1998**, *17*, 683. (b) Lenges, C. P.; Brookhart, M.; Grant, B. E. *J. Organometal. Chem.* **1997**, *528*, 199. (c) Negishi, E.; Kondakov, D. Y. *Chem. Soc. Rev.* **1996**, *25*, 417. (d) Pivovarov, A. P.; Ioffe, L. M.; Gak, Yu. V.; Makhaev, V. D.; Borisov, A. P.; Borod'ko, Yu. G. *Izv. Akad. Nauk SSSR, Ser. Khim.* **1987**, 1008 (in Russian). (e) Eisen, M. S.; Marks, T. J. *Organometallics* **1992**, *11*, 3939. (f) Jordan, R. F.; Guram, A. S. *Organometallics* **1990**, *9*, 2116. (g) Djurovich, P. I.; Carroll, P. J.; Berry, D. H. *Organometallics* **1994**, *13*, 2551.

34. (a) Grigoryan, E. A.; Dyachkovskiy, F. S.; Zhuk, S. Ya.; Vyshinskaya, L. I. *Kinet. Katal.* **1978**, *19*, 1063 (in Russian). (b) Shestakov, A. F.; Zhuk, S. Ya.; Papoyan, A. T.; Grigoryan, E. A. *Kinet. Katal.* **1982**, *23*, 597 (in Russian).

35. (a) Ogasawara, M.; Saburi, M. *Organometallics* **1994**, *14*, 1911. (b) Perthuisot, C.; Fan, M.; Jones, W. D. *Organometallics* **1992**, *11*, 3622.

36. Collman, J. P.; Fish, H. T.; Wagenknecht, P. S.; Tyvoll, D. A.; Chng, L.-L.; Eberspracher, T. A.; Brauman, J. I.; Bacon, J. W.; Pignolet, L. H. *Inorg. Chem.* **1996**, *35*, 6746.

37. (a) Brown, D. B.; Dyson, P. J.; Johnson, B. F. G.; Martin, C. M.; Parker, D. G.; Parsons, S. *J. Chem. Soc., Dalton Trans.* **1997**, 1909. (b) Arliguie, T.;

Chaudret, B.; Chung, G.; Dahan, F. *Organometallics* **1991**, *10*, 2973. (c) Okazaki, M.; Tobita, H.; Ogino, H. *Chemistry Lett.* **1998**, 69. (d) Ma, Y.; Bergman, R. G. *Organometallics* **1994**, *13*, 2548. (e) Borowski, A. F.; Sabo-Etienne, S.; Christ, M. L.; Donnadieu, B.; Chaudret, B. *Organometallics* **1996**, *15*, 1427. (f) Tran, E.; Legzdins, P. *J. Am. Chem. Soc.* **1997**, *119*, 5071.

38. (a) Vedernikov, A. N.; Sayakhov, M. D.; Solomonov, B. N. *Mendeleev Commun.* **1997**, 205. (b) Fujii, T.; Higashino, Y.; Saito, Y. *J. Chem. Soc., Dalton Trans.* **1993**, 517. (c) Aoki, T.; Crabtree, R. H. *Organometallics* **1993**, *12*, 294. (d) Leitner, W.; Six, C. *Chem. Ber./Recueil* **1997**, *130*, 555. (e) Michos, D.; Luo, X.-L.; Faller, J. W.; Crabtree, R. H. *Inorg. Chem.* **1993**, *32*, 1370. (f) Shih, K.-C.; Goldman, A. S. *Organometallics* **1993**, *12*, 3390. (g) Miller, J. A.; Knox, L. K. *J. Chem. Soc., Chem. Commun.* **1994**, 1449. (h) Lu, X. Y.; Lin, Y. G., Ma, D. W. *Pure Appl. Chem.* **1988**, *60*, 1299.

39. (a) Baudry, D.; Ephritikhine, M.; Felkin, H.; Zakrzewski, J. *J. Chem. Soc., Chem. Commun.* **1982,** 1235. (b) Baudry, D.; Ephritikhine, M.; Felkin, H.; Holmes-Smith, R. *J. Chem. Soc., Chem. Commun.* **1983**, 788. (c) Baudry, D.; Ephritikhine, M.; Felkin, H.; Zakrzewski, J. *Tetrahedron Lett.* **1984**, *25*, 1283. (d) Schnabel, R. C.; Roddick, D. M. *Organometallics* **1996**, *15*, 3550. (e) Nizova, G. V.; Chaudret, B.; He.; X.-D.; Shul'pin, G. B. *Bull. Russ. Acad. Sci., Div. Chem. Sci.* **1992,** *41*, 1138.

40. (a) Maguire, J. A.; Goldman, A. S. *J. Am. Chem. Soc.* **1991**, *113*, 6706. (b) Maguire, J. A.; Petrillo, A.; Goldman, A. S. *J. Am. Chem. Soc.* **1992**, *114*, 9492. (c) Sakakura, T.; Abe, F.; Tanaka, M. *Chem. Lett.* **1991**, 359. (d) Mimoun, H.; Brazi, E.; Cameron, C. J.; Benazzi, E.; Meunier, P. *New J. Chem.* **1989**, *13*, 713. (e) Braunstein, P.; Chauvin, Y.; Nähring, J.; DeCian, A.; Fischer, J.; Tiripicchio, A.; Ugozzoli, F. *Organometallics* **1996**, *15*, 5551.

41. (a) Belli, J.; Jensen, C. M. *Organometallics* **1996**, *15*, 1532. (b) Gupta, M.; Hagen, C.; Flesher, R. J.; Kaska, W. C.; Jensen, C. M. *Chem. Comm.* **1996**, 2083. (c) Xu, W.; Rosini, G. P.; Gupta, M.; Jensen, C. M.; Kaska, W. C.; Krogh-Jespersen, K.; Goldman, A. S. *Chem. Comm.* **1997**, 2273. (d) Gupta, M.; Hagen, C.; Kaska, W. C.; Cramer, R. E.; Jensen, C. M. *J. Am. Chem. Soc.* **1997**, *119*, 840. (e) Rouhi, M. *C&EN* December **1997**, 27. (f) Goldman, A. S.; Liu, F.; Rosini, G.; Singh, B.; Xu, W.-W. Presented at the *217th Am. Chem. Soc. Natl. Meeting*; Anaheim, March **1999**; paper INOR No. 352.

42. (a) Sakakura, T.; Sodeyama, T.; Tokunaga, Y.; Tanaka, M. *Chem. Lett.* **1988**, 263. (b) Sakakura, T.; Sodeyama, T.; Tanaka, M. *New J. Chem.* **1989**, *13*, 737.

43. (a) Maguire, J.A.; Boese, W.T.; Goldman, A.S. *J. Am. Chem. Soc.* **1989**, *111*, 7088. (b) Maguire, J.A.; Boese, W.T.; Goldman, M.E.; Goldman, A.S. *Coord. Chem. Rev.* **1990**, *97*, 179. (c) Rosini, G. P.; Soubra, S.; Vixamar, M.; Wang, S.; Goldman, A. S. *J. Organometal. Chem.* **1998**, *554*, 41. (d) Spillett, C. T., Ford, P. C. *J. Am. Chem. Soc.* **1989**, *111*, 7088.

44. (a) Sakakura, T.; Ishida, K.; Tanaka, M. *Chem. Lett.* **1990**, 585. (b) Iwamoto, A.; Itagaki, H.; Saito, Y. *J. Chem. Soc., Dalton Trans.* **1991**, 1093. (c) Sakakura, T.; Abe, F.; Tanaka, M. *Chem. Lett.* **1991**, 297. (d) Itagaki, H.; Einaga, H.; Saito, Y. *Chem. Lett.* **1993**, 2097. (e) Itagaki, H.; Einaga, H.; Saito, Y. *J. Chem. Soc., Dalton Trans.* **1993**, 1689.

45. (a) Sakakura, T.; Abe, F.; Tanaka, M. *Chem. Lett.* **1990**, 583. (b) Sakakura, T.; Sodeyama, T.; Tanaka, M. *Chem. Lett.* **1988**, 683.

46. (a) Sakakura, T.; Sodeyama, T.; Sasaki, K.; Wada, K.; Tanaka, M. *J. Am. Chem. Soc.* **1990**, *112*, 7221. (b) Rosini, G. P.; Boese, W. T.; Goldman, A. S. *J. Am. Chem. Soc.* **1994**, *116*, 9498. (c) Boyd, S. E.; Field, L. D.; Partridge, M. G. *J. Am. Chem. Soc.* **1994**, *116*, 9492. (d) Rosini, G. P.; Zhu, K.; Goldman, A. S. *J. Organomet. Chem.* **1995**, *504*, 115. (e) Margl, P.; Ziegler, T.; Blöchl, P. E. *J. Am. Chem. Soc.* **1996**, *118*, 5412.

47. Boese, W. T.; Goldman, A. S. *J. Am. Chem. Soc.* **1992**, *114*, 350.

48. Sakakura, T.; Ishiguro, K.; Okano, M.; Sako, T. *Chem. Lett.* **1997**, 1089.

49. (a) Ishii, Y.; Chatani, N.; Kakiuchi, F.; Murai, S. *Tetrahedron Lett.* **1997**, *38*, 7565. (b) Chatani, N.; Ie, Y.; Kakiuchi, F.; Murai, S. *J. Org. Chem.* **1997**, *62*, 2604. (c) Chatani, N.; Ishii, Y.; Ie, Y.; Kakiuchi, F.; Murai, S. *J. Org. Chem.* **1998**, *632*, 5129. (d) Chatani, N.; Morimoto, T.; Fukumoto, Y.; Murai, S. *J. Am. Chem. Soc.* **1998**, *120*, 5335. (e) Fujii, N.; Kakiuchi, F.; Yamada, A.; Chatani, N.; Murai, S. *Bull. Chem. Soc. Jpn.* **1998**, *71*, 285. (f) Ishii, Y.; Chatani, N.; Kakiuchi, F.; Murai, S. *Organometallics* **1997**, *16*, 3615.

50. (a) Jones, W. D.; Hessell, E. T. *New J. Chem.* **1990**, *14*, 481. (b) Jones, W. D.; Hessell, E. T. *Organometallics* **1990**, *9*, 718.

51. (a) Tokunaga, Y.; Sakakura, T.; Tanaka, M. *J. Mol. Catal.* **1989**, *56*, 305. (b) Boese, W. T.; Goldman, A. S. *Organometallics* **1991**, *10*, 782.

52. (a) Waltz, K. M.; Hartwig, J. F. *Science* **1997**, *277*, 211. (b) Hartwig, J. F.; Waltz, K. M.; Chen, H.; Anistasi, N. Presented at the *217th Am. Chem. Soc. Natl. Meeting*; Anaheim, March **1999**; paper INOR No. 350. (c) Waltz, K. M.; He, X.; Muhoro, C.; Hartwig, J. F. *J. Am. Chem. Soc.* **1995**, *117*, 11357.

53. (a) Vinogradov, M. G.; Tuzikov, A. B.; Nikishin, G. I. *Izv. Akad. Nauk SSSR, Ser. Khim.* **1985**, 356 (in Russian). (b) Vinogradov, M. G.; Tuzikov, A. B.; Nikishin, G. I. *Izv. Akad. Nauk SSSR, Ser. Khim.* **1985**, 2557

54. Khusnutdinov, R. I.; Shchadneva, N. A.; Dzhemilev, U. M. *Izv. Akad. Nauk SSSR, Ser. Khim.* **1991**, 2897 (in Russian).

55. Yi,; C. S.; Wódka, D.; Rheingold, A. L.; Yap, G. P. A. *Organometallics* **1996**, *15*, 2.

56. (a) Comstock, M. C.; Shapley, J. R. *Organometallics* **1997**, *16*, 4816. (b) Li, C.-J.; Chen, D.-L.; Costello, C. W. *Org. Proc. Res. Develop.* **1997**, *1*, 325. (c) Trost, B. M.; Krische, M. J. *Synlett* **1998**, 1. (d) Sierra, M. A.; Mancheño, M. J.; Saez, E.; del Amo, J. C. *J. Am. Chem. Soc.* **1998**, *120*, 6812. (e) Jones, W. D.; Maguire, J. A.; Rosini, G. P. *Inorg. Chim. Acta* **1998**, *270*, 77. (f) Vedernikov, A. I.; Sharifullin, R. R.; Solomonov, B. N. *Mendeleev Commun.* **1996**, 54. (g) Torkelson, J. R.; McDonald, R.; Cowie, M. *J. Am. Chem. Soc.* **1998**, *120*, 4047. (h) Scheffknecht, C.; Peringer, P. *J. Organometal. Chem.* **1997**, *535*, 77. (i) Slugovc, C.; Mauthner, K.; Kacetl, M.; Mereiter, K.; Schmid, R.; Kirchner, K. *Chem. Eur. J.* **1998**, *4*, 2043. (j) Slugovc, C.; Wiede, P.; Mereiter, K.; Schmid, R.; Kirchner, K. *Organometallics* **1997**, *16*, 2768. (k) Yi, C. S.; Liu, N. *Organometallics* **1998**, *17*, 3158. (l) Cenini, S.; Ragaini, F.; Tollari, S.; Paone, D. *J. Am. Chem. Soc.* **1996**, *118*, 11964. (m) Fantacci, S.; Re, N.; Rosi, M.; Sgamellotti, A.; Guest, M. F.; Sherwood, P.; Floriani, C. *J. Chem. Soc., Dalton Trans.* **1997**, 3845. (n) Yoshida, M.; Jordan, R. F. *Organometallics* **1997**, *16x*, 4508.

57. (a) Sherry, A. E.; Wayland, B. B. *J. Am. Chem. Soc.* **1990**, *112*, 1259. (b) Wayland, B. B.; Ba, S.; Sherry, A. E. *J. Am. Chem. Soc.* **1991**, *113*, 5305. (c) Zhang, X.-X.; Wayland, B. B. *J. Am. Chem. Soc.* **1994**, *116*, 7897.

58. Zhou, X.; Li, Q.; Mak, C. W.; Chan, K. S. *Inorg. Chim. Acta* **1998**, *270*, 551.

59. (a) Herrmann, W. A. In *Applied Homogeneous Catalysis with Organometallic Compounds*; Cornils, B.; Herrmann, W. A., Eds.; VCH: Weinheim, 1996, p. 1149. (b) Doyle, M. P.; McKervey, M. A.; Ye, T. *Reaction Synthesis with α-Diazocarbonyl Compounds*; Wiley: New York, 1997. (c) Doyle, M. P.; McKervey, M. A.; Ye, T. *Modern Catalytic Methods for Organic Synthesis with Diazo Compounds: From Cyclopropanes to Ylides*; Wiley: New York, 1998. (d) Doyle, M. P. In *Catalysis by Di- and Polynuclear Metal Cluster Complexes*; Adams, R. D.; Cotton, F. A., Eds.; Wiley-VCH, 1998, p. 249. (e) Ye, T.; McKervey, M. A. *Chem. Rev.* **1994**, *94*, 1091. (f) Doyle, M. P.; Forbes, D. C. *Chem. Rev.* **1998**, *98*, 911.

60. (a) Davies, H. M. L.; Hansen, T. *J. Am. Chem. Soc.* **1997**, *119*, 9075. (b) Estevan, F.; Lahuerta, P.; Perez-Prieto, J.; Sanau, M.; Stiriba, S.-E.; Ubeda, M.

A. *Organometallics* **1997**, *16*, 880. (c) Siegel, S.; Schmalz, H.-G. *Angew. Chem., Int. Ed. Engl.* **1997**, *36*, 2456. (d) Yakura, T.; Yamada, S.; Kunimune, Y.; Ueki, A.; Ikeda, M. *J. Chem. Soc., Perkin Trans. 1* **1997**, 3643.

61. (a) Hatanaka, Y.; Watanabe, M.; Onozawa, S.; Tanaka, M.; Sakurai, H. *J. Org. Chem.* **1998**, *63*, 422. (b) Aburel, P. S.; Undheim, K. *Tetrahedron Lett.* **1998**, *39*, 3813. (c) Srikrishna, A.; Gharpure, S. J. *Chem. Comm.* **1998**, 1589. (d) Balaji, B. S.; Chanda, B. M. *Tetrahedron Lett.* **1998**, *39*, 6381. (e) Taber, D. F.; Malcolm, S. C. *J. Org. Chem.* **1998**, *63*, 3717. (f) Davies, H. M. L.; Hodges, L. M.; Matasi, J. J.; Hansen, T.; Stafford, D. G. *Tetrahedron Lett.* **1998**, *39*, 4417. (g) Wang, J.; Chen, B.; Bao, J. *J. Org. Chem.* **1998**, *63*, 1853. (h) Wee, A. G. H.; Liu, B.; McLeod, D. D. *J. Org. Chem.* **1998**, *63*, 4218. (i) Estevan, F.; Lahuerta, Perez-Prieto, J.; Pereira, I.; Stiriba, S.-E. *Organometallics* **1998**, *17*, 3442.

62. (a) Taber, D. F.; Stiriba, S.-E. *Chem. Eur. J.* **1998**, *4*, 990. (b) White, J. D.; Hrnciar, P.; Stappenbeck, F. *J. Org. Chem.* **1997**, *62*, 5250. (c) Taber, D. F.; You, K. K. *J. Am. Chem. Soc.* **1995**, *117*, 5757. (d) Taber, D. F.; Raman, K.; Gaul, M. D. *J. Org. Chem.* **1987**, *52*, 28. (e) Anada, M.; Watanabe, N.; Hashimoto, S. *Chem. Comm.* **1998**, 1517. (f) Bode, J. W.; Doyle, M. P.; Protopopova, M. N.; Zhou, Q.-L. *J. Org. Chem.* **1996**, *61*, 9146.

63. Clark, J. S.; Dossetter, A. G.; Russell, C. A.; Whittingham, W. G. *J. Org. Chem.* **1997**, *62*, 4910.

64. Herber, U.; Bleuel, E.; Gevert, O.; Laubender, M.; Werner, H. *Organometallics* **1998**, *17*, 10.

65. (a) Liebman, J. F.; Greenberg, A. *Chem. Rev.* **1976**, *76*, 311. (b) Minkin, V. I.; Glukhovtsev, M. N.; Simkin, B. Ya. *J. Mol. Struct. (Theochem.)* **1988**, *181*, 93.

66. (a) Suggs, J. W.; Jan, C.-H. *J. Am. Chem. Soc.* **1984**, *106*, 3054. (b) Brown, D. B.; Viens, V. A. *J. Organometal. Chem.* **1977**, *142*, 117. (c) Edelbach, B. L.; Lachocotte, R. J.; Jones, W. D. *J. Am. Chem. Soc.* **1998**, *120*, 2843. (d) Wick, D. D.; Northcutt, T. O.; Lachicotte, R. J.; Jones, W. D. *Organometallics* **1998**, *17*, 4484. (e) Yeh, W.-Y.; Hsu, S. C. N.; Peng, S.-M.; Lee, G.-H. *Organometallics* **1998**, *17*, 2477. (f) Perthuisot, C.; Edelbach, B. L.; Zubris, D. L.; Jones, W. D. *Organometallics* **1997**, *16*, 2016. (g) Abla, M.; Yamamoto, T. *J. Organometal. Chem.* **1997**, *532*, 267. (h) Rybtchinski, B.; Vigalok, A.; Ben-David, Y.; Milstein, D. *J. Am. Chem. Soc.* **1996**, *118*, 12406. (i) van der Boom, M. E.; Ben-David, Y.; Milstein, D. *Chem. Comm.* **1998**, 917. (j) Hemond, R. C.; Hughes, R. P.; Robinson, D. J.; Rheingold, A. L. *Organometallics* **1988**, *7*,

2239. (k) Shaltout, R. M.; Sygula, R.; Sygula, A.; Fronczek, F. R.; Stanley, G. G.; Rabideau, P. W. *J. Am. Chem. Soc.* **199x**, *120*, 835.

67. (a) van der Boom, M. E.; Kraatz, H.-B.; Ben-David, Y.; Milstein, *J. Chem. Soc., Chem. Commun.* **1996**, 2167. (b) Gandelman, M.; Vigalok, A.; Shimon, L. J. W.; Milstein, D. *Organometallics* **1997**, *16*, 3981. (c) McNeill, K.; Andersen, R. A.; Bergamn, R. G. *J. Am. Chem. Soc.* **1997**, *119*, 11244. (d) Jia, G.; Ng, W. S.; Lau, C. P. *Organometallics* **1998**, *17*, 4538. (e) Crabtree, R. H.; Dion, R. P.; Gibbini, D. J. *et al. J. Am. Chem. Soc.* **1986**, *108*, 7222. (f) Eilbracht, P. *Chem. Ber.* **1980**, *113*, 2211. (g) Yamaguchi, R.; Kawanisi, M. *J. Org. Chem.* **1984**, *49*, 4460. (h) Edwards, A. J.; Esteruelas, M. A.; Lahoz, F. J.; Lopez, A. M.; Oñate, E.; Oro, L. A.; Tolosa, J. I. *Organometallics* **1997**, *16*, 1316.

68. (a) Bishop, K. C. *Chem. Rev.* **1976**, *76*, 461. (b) Paquette, L. A. *Synthesis* **1975**, 347. (c) Paquette, L. A.; Detty, M. R. *Tetrahedron Lett.* **1978**, 713. (d) Hayashi, M.; Ohmatsu, T.; Meng, Y.-P.; Saigo, K. *Angew. Chem., Int. Ed. Engl.* **1998**, *37*, 837. (e) Murakami, M.; Takahashi, K.; Amii, H.; Ito, Y. *J. Am. Chem. Soc.* **1997**, *119*, 9307.

69. (a) Bunel, E.; Burger, B. J.; Bercaw, J. E. *J. Am. Chem. Soc.* **1988**, *110*, 976. (b) Akhrem, I. S.; Grushin, V. V.; Reznichenko, S. V.; Vol'pin, M. E. *Doklady Akad. Nauk SSSR* **1987**, *294*, 363 (in Russian). (c) Rosier, C.; Niccolai, G. P.; Basset, J.-M. *J. Am. Chem. Soc.* **1997**, *119*, 12408.

70. (a) Osakada, K,; Koizumi, T.; Yamamoto, T. *Organometallics* **1997**, *16*, 2063. (b) Mitchell, G. P.; Tilley, T. D. *Organometallics* **1998**, *17*, 2912. (c) Osakada, K.; Koizumi, T.; Yamamoto, T. *Bull. Chem. Soc. Jpn.* **1997**, *70*, 189. (d) Yang, H.; Kotz, K. T.; Asplund, M. C.; Harris, C. B. *J. Am. Chem. Soc.* **1997**, *119*, 9564.

71. (a) Saunders, G. C. *Angew. Chem., Int. Ed. Engl.* **1996**, *35*, 2615. (b) Plenio, H. *Chem. Rev.* **1997**, *97*, 3363. (c) Burdeniuc, J.; Jedlicka, B.; Crabtree, R. H. *Chem. Ber./Recueil* **1997**, *130*, 145.

72. (a) Patel, B. P.; Crabtree, R. H. *J. Am. Chem. Soc.* **1996**, *118*, 13105. (b) Rumin, R.; Guennou, K.; Petillon, F. Y.; Muir, K. W. *J. Chem. Soc., Dalton Trans.* **1997**, 1381. (c) Hughes, R. P.; Lindner, D. C.; Rheingold, A. L.; Liable-Sands, L. M. *J. Am. Chem. Soc.* **1997**, *119*, 11544. (d) Rumin, R.; Guennou, K.; Pichon, R.; Petillon, F. Y.; Muir, K. W.; Yufit, D. S. *J. Organometal. Chem.* **1997**, *533*, 177. (e) Su, M.-D.; Chu, S.-Y. *J. Am. Chem. Soc.* **1997**, *119*, 10178. (f) Crespo, M.; Martinez, M.; de Pablo, E. *J. Chem. Soc., Dalton Trans.* **1997**, 1231. (g) Chernega, A. N.; Graham, A. J.; Green, M. L. H.; Haggitt, J.; Lloyd, J.; Mehnert, C. P.; Metzler, N.; Souter, J. *J. Chem. Soc., Dalton Trans.* **1997**,

2293. (h) Klahn, A. H.; Oelckers, B.; Godoy, F.; Garland, M. T.; Vega, A.; Perutz, R. N.; Higgitt, C. L. *J. Chem. Soc., Dalton Trans.* **1998**, 3079. (i) Arroyo, M.; Bernes, S.; Brianzo, J. L.; Mayoral, E.; Richards, R. L.; Rius, J.; Torrens, H. *Inorg. Chem. Commun.* **1998**, *1*, 273. (j) Atherton, M. J.; Fawcett, J.; Holloway, J. H.; Hope, E. G.; Martin, S. M.; Russell, D. R.; Saunders, G. C. *J. Organometal. Chem.* **1998**, *555*, 67. (k) Beck, C. M.; Park, Y.-J.; Crabtree, R. H. *Chem. Comm.* **1998**, 693. (l) Ishii, Y.; Chatani, N.; Yorimitsu, S.; Murai, S. *Chem. Lett.* **1998**, 157.

73. (a) Kiplinger, J. L.; King, M. A.; Fechtenkötter, A.; Arif, A. M.; Richmond, T. G. *Organometallics* **1996**, *15*, 5292. (b) Cronin, L.; Higgitt, C. L.; Karch, R.; Perutz, R. N. *Organometallics* **1997**, *16*, 4920. (c) Lopez, O.; Crespo, M.; Font-Bardia, M.; Solans, X. *Organometallics* **1997**, *16*, 1233. (d) Fawcett, J.; Friedrichs, S.; Holloway, J. H.; Hope, E. G.; McKee, V.; Nieuwenhuyzen, M.; Russell, D. R.; Saunders, G. C. *J. Chem. Soc., Dalton Trans.* **1998**, 1477.

74. (a) Bianchini, C.; Casares, J. A.; Masi, D.; Meli, A.; Pohl, W.; Vizza, F. *J. Organometal. Chem.* **1997**, *541*, 143. (b) Kourkine, I. V.; Sargent, M. D.; Glueck, D. S. *Organometallics* **1998**, *17*, 125. (c) Dullaghan, C. A.; Zhang, X.; Greene, D. L.; Carpenter, G. B.; Sweigart, D. A.; Camiletti, C.; Rajaseelan, E. *Organometallics* **1998**, *17*, 3316. (d) Firth, A. V.; Stephan, D. W. *Organometallics* **1997**, *16*, 2183. (e) Steenwinkel, P.; James, S. L.; Grove, D. M.; Kooijman, H.; Spek, A. L.; van Koten, G. *Organometallics* **1997**, *16*, 513. (f) Tanaka, Y.; Yamashita, H.; Shimada, S.; Tanaka, M. *Organometallics* **1997**, *16*, 3246. (g) Huang, D.; Heyn, R. H.; Bollinger, J. C.; Caulton, K. G. *Organometallics* **1997**, *16*, 292.

75. (a) Sharma, H. K.; Pannell, K. H. *Chem. Rev.* **1995**, *95*, 1351. (b) Suginome, M.; Ito, Y. *J. Chem. Soc., Dalton Trans.* **1998**, 1925.

76. Huang, J.; Li, C.; Nolan, S. P.; Peterson, J. L. *Organometallics* **1998**, *17*, 3516.

77. (a) Iretskii, A.; Adams, H.; Garcia, J. J.; Picazo, G.; Maitlis, P. M. *Chem. Comm.* **1998**, 61. (b) Temple, K.; Lough, A. J.; Sheridan, J. B.; Manners, I. *J. Chem. Soc., Dalton Trans.* **1998**, 2799. (c) Clegg, W.; Lawlor, F. J.; Marder, T. B.; Nguyen, P.; Norman, N. C.; Orpen, A. G.; Quayle, M. J.; Rice, C. R.; Robins, E. G.; Scott, A. J.; Souza, F. E. S.; Stringer, G.; Whittell, *J. Chem. Soc., Dalton Trans.* **1998**, 301.

CHAPTER V

HYDROCARBON ACTIVATION BY METAL IONS, ATOMS, AND COMPLEXES IN THE GAS PHASE AND IN A MATRIX

Taking into account the experimental details of the reactions between metals and hydrocarbons in the gas phase and at low temperatures as well as their reactions in solutions, one will notice that they are very different. However it is reasonable to discuss such reactions of naked metal atoms in a special chapter of this book because the mechanisms of interaction between complexes in solutions and atoms or ions may have much in common. Again, from the experimental point of view, the chemistry of naked atoms at low temperature and the chemistry of gas phase reactions between ions and hydrocarbons are different, but the mechanisms are quite similar. In both cases the crucial step of such processes is oxidative addition of the C–H bond components to a metal atom or low-valent ion.

V.1. REACTIONS WITH METAL IONS, ATOMS, AND COMPLEXES IN THE GAS PHASE

Metal ions are known to react with hydrocarbons, particularly alkanes, in the gas phase (see reviews [1]). A brief survey of these reactions will be given in the present section.

V.1.A. THERMAL REACTIONS WITH NAKED IONS AND ATOMS

Alkanes

Usually oxidative addition of C–H and C–C bonds occurs at the first step of the reaction between metal ions and alkanes. The alkyl hydride or dialkyl derivative thus formed provides an entry to further transformations.

"Naked" metal ions (for example, Fe^+, Cr^+, Co^+, Ti^+, V^+) react with alkanes, alkanes bearing functional substituents, and silanes [2]. Some reactions of this type are shown in Scheme V.1.

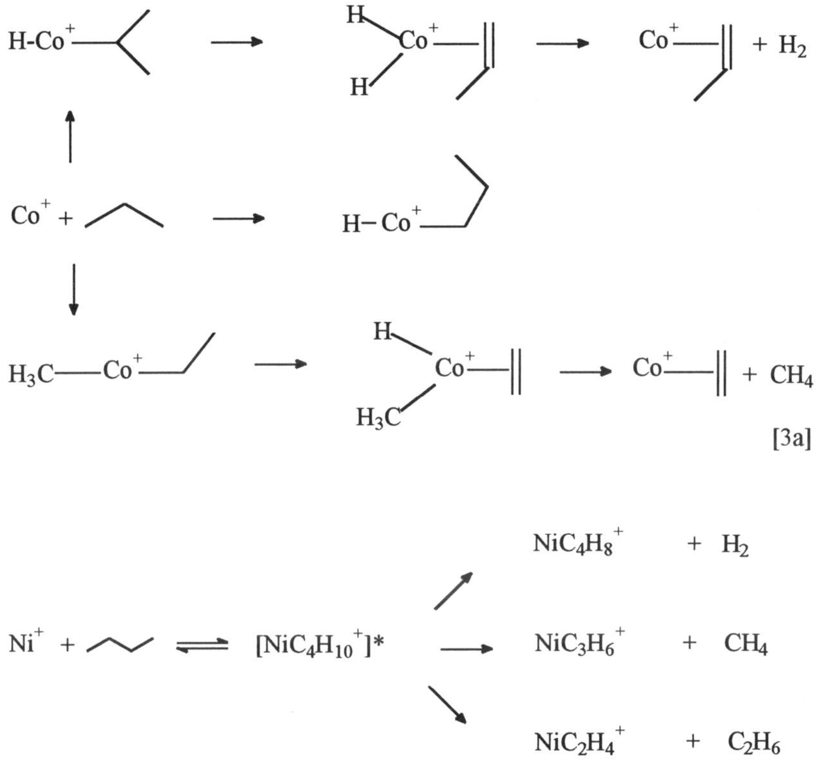

Scheme V.1. Examples of the gas-phase reactions between metal ions and simple alkanes. Reference numbers are in brackets.

Bare Pt^+ cations activate methane in the gas phase [4a,b] according to the equation:

$$Pt^+ + CH_4 \longrightarrow H\text{–}Pt\text{–}CH_3^+ \longrightarrow PtCH_2^+ + H_2$$

A cationic hydrido-methyl complex, $HPtCH_3^+$, is formed upon the almost barrierless insertion of a Pt^+ ion into a C–H bond of methane and then 1,2-migration of a hydrogen atom and subsequent elimination of H_2 from the transition-metal center gives the platinum carbene cation $PtCH_2^+$ with an apparent rate constant of $k = 8.2 \times 10^{-10}$ cm^3 molecule^{-1} s^{-1}. The reaction of ground-state platinum atoms produced by the photodissociation of [PtMe$_3$-(C_5H_4Me)] follows a termolecular mechanism [4c].

$$Pt + CH_4 \ [Ar] \longrightarrow H\text{–}Pt\text{–}CH_3$$

The room-temperature limiting low-pressure third-order rate constant in argon buffer is $(2.1 \pm 0.9) \times 10^{-28}$ molecule^{-2} sm^6 s^{-1} and the limiting high-pres-sure second-order rate constant is $(6.3 \pm 0.3) \times 10^{-12}$ molecule^{-1} sm^3 s^{-1}.

The analogous reaction of metastable Mo($^5S_2s^1d^5$) with methane, which has been studied using the method of crossed molecular beams [5], proceeds via insertion of Mo into CH_4 with potential energy barrier less than or equal to 7.2 kcal mol^{-1}:

$$Mo + CH_4 \longrightarrow H\text{–}Mo\text{–}CH_3 \longrightarrow MoCH_2 + H_2$$

Reactions of methane with metal ions, M^+, can lead to oligomerization products as follows [6]:

$$M^+ + n\,CH_4 \longrightarrow \longrightarrow MC_nH_{2n}^+ + n\,H_2$$

For example, the cation $PtCH_2^+$ reacts with an excess of methane to produce species $Pt(C_2H_4)^+$ and higher ions $Pt(C_nH_m)^+$.

The interaction of higher alkanes with metal ions also induces dehydrogenation and affords various π-complexes (Schemes V.2 and V.3) [7]. The activation of various alkanes (as well as some unsaturated and aromatic hydrocarbons) by ions W^+ is shown in Table V.1 [8].

It is noteworthy that while ground state rhodium atoms insert into methane molecules [9a], there is no evidence of chemical reaction for the ground state of rhenium with CH_4 up to a temperature of 548 K [9b]. Reactions of metal cluster ions, for example, $MgFe^+$ [10a], Pt_n^+ [10b], Nb_n^-, and Rh_n^- [10c] with saturated and aromatic hydrocarbons in the gas phase have been recently described.

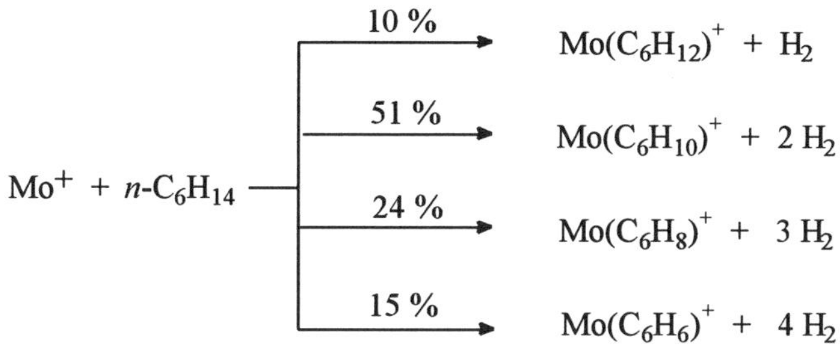

Scheme V.2. Dehydrogenation of *n*-hexane under the action of Mo^+.

Table V.1. Reaction of W^+ with hydrocarbons.

Hydrocarbon	Products	Comments
CH_4	$WCH_2^+ + H_2$	
C_2H_2	No reaction	
C_2H_4	$WC_2H_2^+ + H_2$	
C_2H_6	$WC_2H_4^+ + H_2$	
	$WC_2H_2^+ + 2\ H_2$	
C_3H_6	$WC_3H_2^+ + 2\ H_2$	
c-C_3H_6	$WC_3H_2^+ + 2\ H_2$	major
	$WC_2H_2^+ + CH_4$	minor
C_3H_8	$WC_3H_4^+ + 2\ H_2$	major
	$WC_2H_2^+ + CH_4 + H_2$	minor
n-C_4H_{10}	$WC_4H_6^+ + 2\ H_2$	
	$WC_4H_4^+ + 3\ H_2$	
	$WC_3H_4^+ + CH_4 + H_2$	
	$WC_3H_2^+ + CH_4 + 2\ H_2$	
i-C_4H_{10}	$WC_4H_6^+ + 2\ H_2$	major
	$WC_3H_2^+ + CH_4 + 2\ H_2$	minor
	$WC_3H_4^+ + CH_4 + H_2$	minor
C_6H_6	$WC_6H_4^+ + H_2$	
c-C_6H_{12}	$WC_6H_6^+ + 3\ H_2$	
n-C_6H_{14}	$WC_6H_8^+ + 3\ H_2$	major
	$WC_6H_6^+ + 4\ H_2$	
	$WC_5H_6^+ + CH_4 + 2\ H_2$	minor

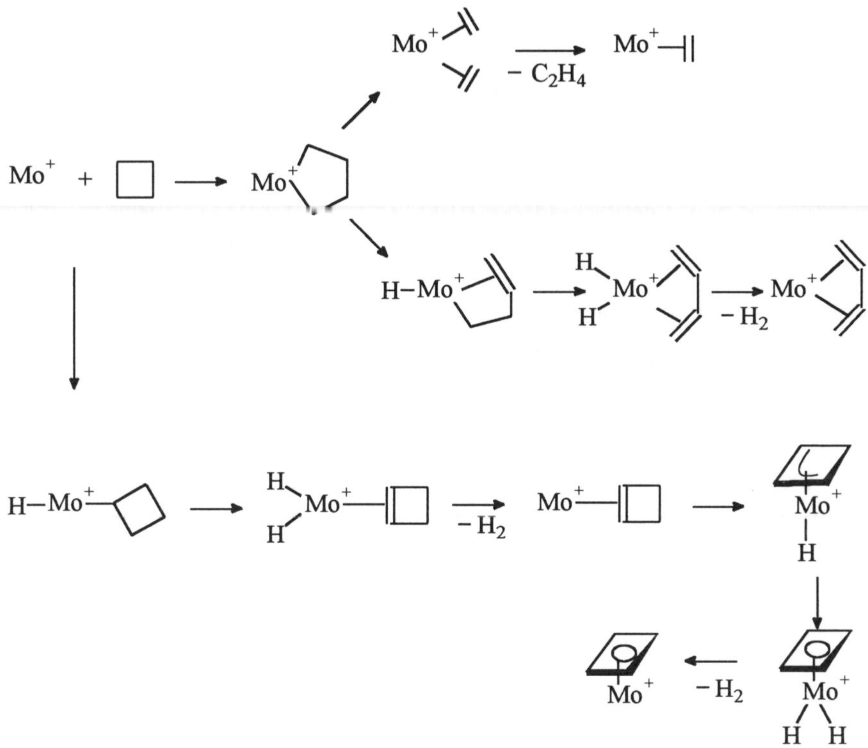

Scheme V.3. Reaction of Mo^+ with cyclobutane.

Very interesting alkane functionalizations in the presence of metal ions in the gas phase have also been reported. Schwarz *et al.* have described recently [11a] a gas-phase model for the platinum-catalyzed coupling of methane and ammonia. Endothermic by 61 kcal mol^{-1} reaction

$$CH_4 + NH_3 = HCN + 3\,H_2$$

can be an economically attractive basis for industrial synthesis of hydrogen cyanide as well as molecular hydrogen. In the Degussa process, the reaction is performed at ca. 1500 K while in the Andrussow synthesis, added molecular oxygen transformed hydrogen atoms from methane into water. Both processes use platinum catalysts [12]. In the model study, the reaction of atomic platinum cations with methane and ammonia has been examined using Fourier transform ion cyclotron resonance (FTICR) mass spectrometry. Mass-selected ion $PtCH_2^+$, formed in the reaction between Pt^+ and methane, reacts rapidly with ammonia ($k = 6.2 \times 10^{-10}$ sm^3 $molecule^{-1}$ s^{-1}) to generate three different products in a ratio of 70 : 25 : 5 as shown below:

$$PtCH_2^+ + NH_3 \longrightarrow CH_2NH_2^+ + [PtH]$$
$$PtCH_2^+ + NH_3 \longrightarrow [Pt,C,N,H_3]^+ + H_2$$
$$PtCH_2^+ + NH_3 \longrightarrow NH_4+ + [PtCH]$$

The sequence of all steps leading to the final product, hydrogen cyanide, includes the dehydrogenation of CH_2NH:

$$Pt^+ + CH_4 \longrightarrow PtCH_2^+ + H_2$$
$$PtCH_2^+ + NH_3 \longrightarrow CH_2NH_2^+ + [PtH]$$
$$CH_2NH_2^+ + NH_3 \longrightarrow NH_4^+ + CH_2NH$$
$$CH_2NH \longrightarrow HCN + H_2$$
$$[PtH] + NH_4^+ \longrightarrow Pt^+ + NH_3 + H_2$$

Taking into account the results on gas-phase reaction between methane and ammonia, a working model for the heterogeneous catalysis has been proposed (Figure V.1) [11a].

The oxidation of methane with molecular oxygen is catalyzed by the atomic platinum cation [11b]. A key step in the catalytic cycle is the reaction of $PtCH_2^+$ with molecular oxygen to mainly (70%) regenerate Pt^+ via liberation of neutral species $[C,H_2,O_2]$ which either represents vibrationally excited formic acid and/or its decomposition products ($CO+H_2O$) and (CO_2+H_2). Final oxygenates are methanol, formaldehyde as well as higher oxidation products (Scheme V.4). Experimentally determined reaction energies for the elemental steps are summarized in Table V.2 [11b]. The coupling of carbon dioxide and aromatic C–H bonds mediated by ion SiF_3^+ has also been observed [11c].

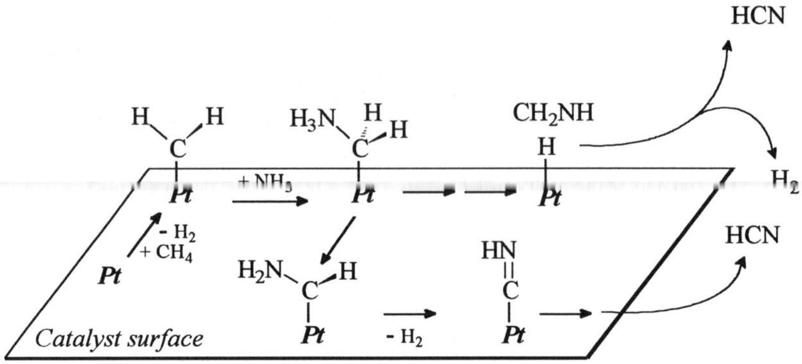

Figure V.1. Proposed mechanism for the methane–ammonia coupling catalyzed by a platinum surface.

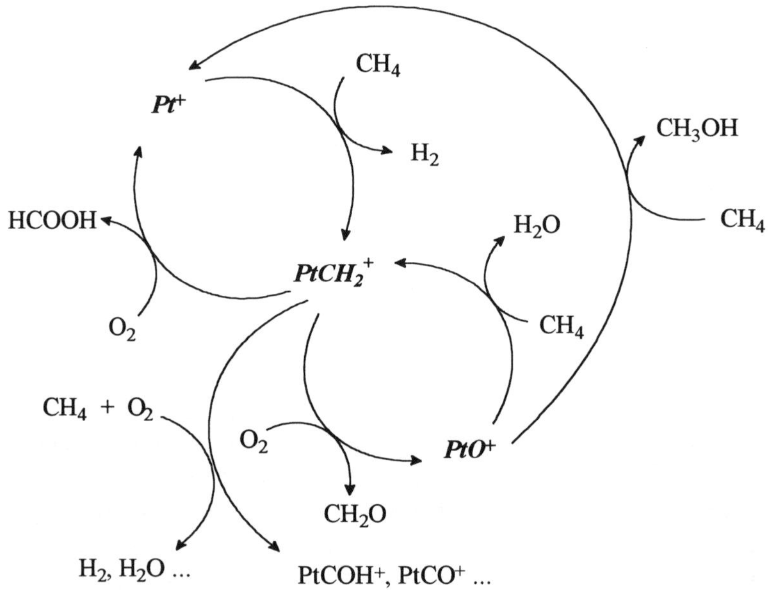

Scheme V.4. The catalytic cycle proposed for the Pt^+-catalyzed oxidation of methane in the gas phase.

Table V.2. Energies of steps of the Pt^+-catalyzed methane oxidation.

No	Reaction	ΔE, kcal mol^{-1}
1	$Pt^+ + CH_4 \longrightarrow Pt(CH_2)^+ + H_2$	– 2.9
2	$Pt(CH_2)^+ + O_2 \longrightarrow PtO^+ + CH_2O$	– 24.1
3	$Pt(CH_2)^+ + O_2 \longrightarrow Pt^+ + HCOOH$	– 71.1
4	$Pt(CH_2)^+ + O_2 \longrightarrow Pt^+ + CO + H_2O$	– 65.1
5	$Pt(CH_2)^+ + O_2 \longrightarrow Pt^+ + CO_2 + H_2$	– 75.1
6	$PtO^+ + CH_2O \longrightarrow Pt^+ + HCOOH$	– 47
7	$PtO^+ + CH_2O \longrightarrow Pt^+ + CO + H_2O$	– 41
8	$PtO^+ + CH_2O \longrightarrow Pt^+ + CO_2 + H_2$	– 51
9	$PtO^+ + CH_4 \longrightarrow Pt(CH_2)^+ + H_2O$	– 46
10	$Pt^+ + O \longrightarrow PtO^+$	– 77
11	$Pt^+ + CH_2 \longrightarrow Pt(CH_2)^+$	–112.5
12	$CH_4 \longrightarrow CH_2 + H_2$	– 109.6
13	$O_2 + 2\,CH_4 \longrightarrow H_2 + CH_2O + CH_3OH$	– 38.3
14	$O_2 + CH_4 \longrightarrow 2\,H_2 + CO_2$	–78.0
15	$O_2 + CH_4 \longrightarrow H_2 + H_2O + CO$	– 68.0
16	$O_2 + CH_4 \longrightarrow H_2 + HCOOH$	– 74.4

Remote Functionalization of Substituted Alkanes

In remote functionalization of alkanes bearing various groups, a bare transition-metal cation is first complexed to the functional group and then directed toward a certain region of the aliphatic backbone of the substrate [13]. For example, in the reaction of Fe^+ and Co^+ with nonanitrile, decanitrile, and undecanitrile [13d], both metal ions exhibit an overall similar reactivity pattern Molecular hydrogen, methane, and small olefins are formed as major neutral products. According to the theoretical study, the most favorable pathway of bond activation proceeds via initial insertion into the C–H bond at position C(8), followed by exocyclic activation of a C–H bond and reductive elimination of H_2 via a multicentered transition structure (Scheme V.5).

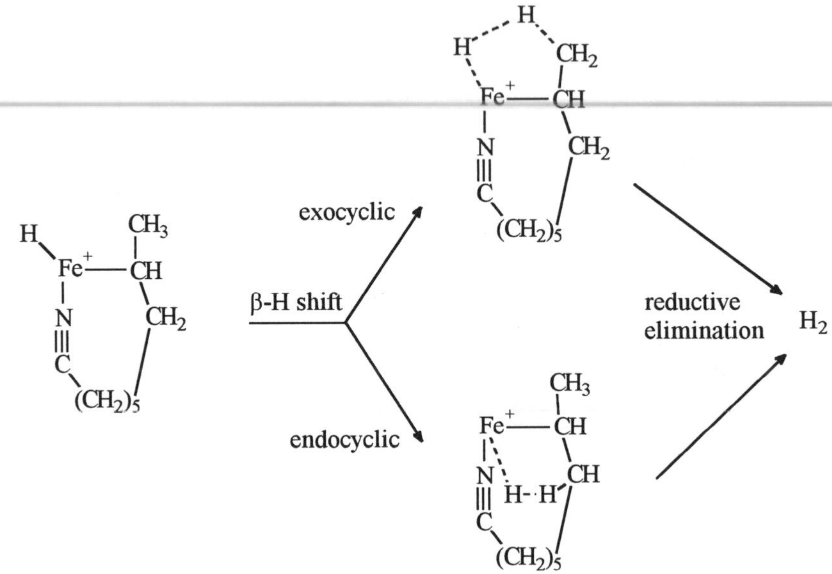

Scheme V.5. Remote functionalization of nonanitrile by Fe^+.

Aromatic and Other Hydrocarbons

Thorium and uranium metal and oxide ions react with several arenes (benzene, naphthalene, toluene, mesitylene, hexamethylbenzene) in the gas phase [14]. For M^+ ions, C–H and/or C–C bond activation was observed in the primary reactions for all the arenes.

The gas-phase reactions of bare lanthanide monocations Ln with fluorobenzene occur with C–F bond activation to produce LnF^+ and phenyl radical [15]. The reactions of various fluorocarbons with calcium monocation proceed according to the following equations [16]:

$$Ca^+ + CH_3F \longrightarrow Ca^+\text{–}F + CH_3\bullet$$

$Ca^+ + CH_3CHF_2 \longrightarrow Ca^+{-}F + CH_3CHF\bullet$

$Ca^+ + C_6H_5F \longrightarrow Ca^+{-}F + C_6H_5\bullet$

$Ca^+ + CHF_3 \longrightarrow Ca^+{-}F + CHF_2\bullet$

$Ca^+ + CF_4 \longrightarrow Ca^+{-}F + CHF_3\bullet$

V.1.B. THERMAL REACTIONS WITH LIGATED METAL IONS

Reactions of oxometal cations (e.g., CoO^+, ScO^+, FeO^+, ThO^+, UO^+, CeO_2 [17]) and positively or even negatively charged complex ions, for example, $Co_4(CO)_n^+$ [18] and $(OC)_2Fe^-$ [19], with various hydrocarbons in the gas phase have been described. The reaction between MoO^+ and methane is presented in Scheme V.6 [17j].

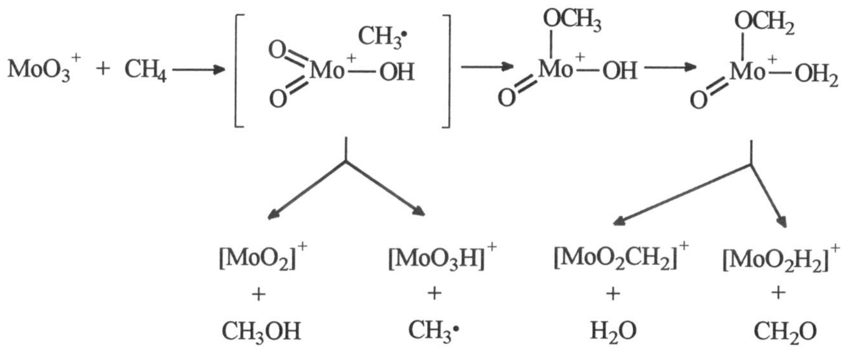

Scheme V.6. Reaction between MoO^+ and methane in the gas phase.

The following gas-phase reactions of metal derivatives with alkanes have also been described:

$Sc(CH_3)_2^+ + c\text{-}C_6H_{12} \longrightarrow CH_3{-}Sc^+{-}c\text{-}C_6H_{11} + CH_4$ [20a]

$Sc(CH_3)_2^+ + c\text{-}C_5H_{10} \longrightarrow CH_3{-}Sc^+{-}c\text{-}C_5H_9 + CH_4$ [20a]

$$FeCF_3^+ + n\text{-}C_4H_{10} \longrightarrow F\text{–}Fe^+\text{–}(C_4H_{10}) + CF_2 \quad (92\%)$$
$$\longrightarrow C_4H_9^+ + FeCF_3H \quad (8\%) \quad [20b]$$

$$CoCF_3^+ + n\text{-}C_4H_{10} \longrightarrow CoCF_3(C_3H_6)^+ + CH_4 \quad (75\%)$$
$$\longrightarrow CoF(C_4H_{10})^+ + CF_2 \quad (18\%)$$
$$\longrightarrow CoCF_3(C_4H_{10})^+ \quad (7\%) \quad [20b]$$

The gas-phase reactions of ligated iron ions $Fe(CH_2X)+$ with alkanes give different products depending on X [20c]. Thus like bare Fe, C–C insertion, particularly terminal C–C insertion, is predominant for the reactions of $Fe(CH_2O)^+$, while C–H insertion is preferred for $Fe(CH_2S)^+$.

V.1.C. REACTIONS WITH PHOTOEXCITED METAL IONS

Vapors of mercury, cadmium and zinc sensitize photochemical alkane transformations [21]. Thus, irradiation of propane with light of $\lambda = 307.6$ nm at 633 K and pressure 67–40000 Pa in the presence of zinc vapor gives rise to the formation of hydrogen, methane, ethylene and dimethylbutane [21a]. The first step in the reaction is a hydrogen atom transfer from the alkane to the excited zinc atom:

$$Zn^* + RH \longrightarrow Zn\cdots H\cdots R \longrightarrow ZnH + R\bullet$$

Alkane functionalization on a preparative scale by mercury-photosensitized C–H bond activation has been recently developed by Crabtree [22]. Mercury absorbs 254-nm light to generate a $^3P_1$ excited state which homolyzes a C–H bond of the substrate with a 3° > 2° > 1° selectivity. Radical disproportionation gives an alkene, but this intermediate is recycled back into the radical pool via H-atom attack, which is beneficial in terms of yield and selectivity. The reaction gives alkane dimers and products of cross-dehydrodimerization of alkanes with various C–H compounds:

$$R'\text{–}H \xrightarrow{Hg^*} R'\bullet + H\bullet$$

$$R''\text{–}H \xrightarrow{Hg^*} R''\bullet + H\bullet$$

$$R'\bullet + R''\text{–}H \longrightarrow R''\bullet + R'\text{–}H$$

$$2\,R'\bullet \longrightarrow R'\text{–}R'$$

$$2\,R''\bullet \longrightarrow R''\text{–}R''$$

$$R'\bullet + R''\bullet \longrightarrow R'\text{–}R''$$

For example, the interaction between cyclohexane and methanol proceeds according to the equation:

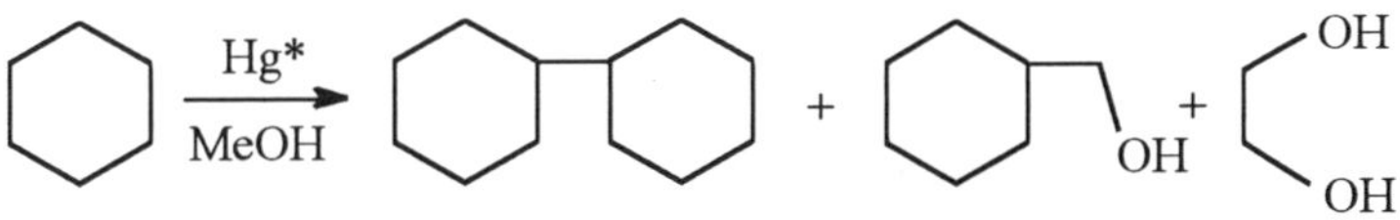

When a mixture of alkane and ammonia containing a trace of mercury vapor was exposed to UV light, an oligomeric high-boiling liquid containing C, H, and N was formed. The authors proposed that imines are produced in the reaction:

$$RCH_3 + NH_3 \xrightarrow{h\nu} RHC{=}NH + 2\,H_2$$

Here R = H and CH_3. The analogous dehydrodimerization of aromatic substrates proceeded with the formation of an organometallic exciplex, $^3[Hg(\eta^2\text{-}C_6H_6)]$.

V.2. Reactions with Metal Atoms in a Matrix

At high temperature metals react with alkanes. For example, at temperatures > 1600 °C, tungsten interacts with methane [23] according to the equation:

$$W + CH_4 \longrightarrow W_2C + WC + H_2$$

It is interesting that in recent decades the numerous reactions of metal atoms with hydrocarbons occurring at very low temperatures have been discovered and thoroughly studied [24]. A method of cocondensing metal atoms with various compounds has been used for instance in order to investigate the reaction of zirconium atoms with isobutane and neopentane at 77 K [25]. Metal atoms insert into both the C–H and C–C bonds:

$$Zr + C(CH_3)_4 \longrightarrow HZrCH_2C(CH_3)_3 + H_3CZrC(CH_3)_3$$

The reaction of Fe_2 dimers with methane in a matrix at 77 K gives rise to the formation of species H–FeFe–CH_3 or FeFe–CH_3 [26a]. Analogously, the C–H bond in alkanes is cleaved by nickel clusters [26b]. Methane activation has also been detected during the co-condensation of methane with aluminum atoms at 10 K [26c]. It is noteworthy that under the same conditions, atoms of Mg, Ti, Cr, Fe, Ga, Pd and some other metals do not react with methane.

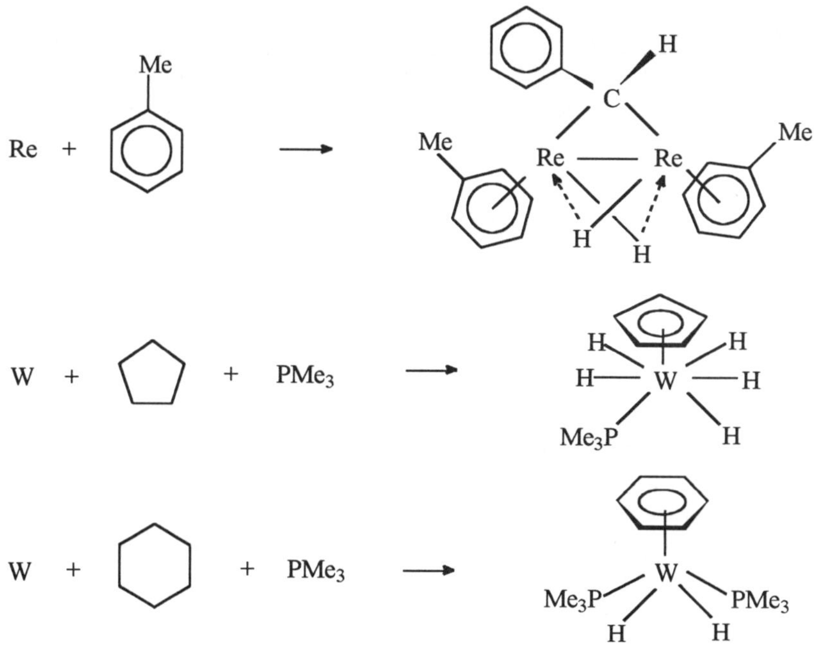

Scheme V.7. Synthesis of organometallic compounds by the co-condensation of metals with hydrocarbons.

Various organometallic compounds can be prepared by the co-condensation of transition metals with some hydrocarbons (for examples, see Scheme V.7) [27].

In the cases when metal atoms in their ground state do not react with alkane at low temperature, an active species may be generated by photoexcitation of metal atoms. The excited atoms which are formed are capable of inserting into the C–H bond of alkane [28]. For example, irradiated ($\lambda < 360$ nm) iron atoms react with methane to produce the species CH_3–Fe–H. Analogously, atoms of Mn, Co, Cu, Zn, Ag and Au insert into the C–H bond of methane. However, atoms of Ca, Ti, Cr and Ni are inactive in this reaction [29]. Species containing M–C bonds can be detected by IR spectroscopy (Table V.3) [29].

Table V.3. IR spectra of species formed in the photolysis of metal atoms in a methane matrix at 15 K.

	ν, cm^{-1}				
Atom	C–H	M–H	CH_3		M–C
Mn	2932.9	1582.6	1142.3	550.3	500.7
Fe	2933.5	1653.1	1153.4	548.8	521.1
	2908.6	–	–	545.9	–
Co	–	1699.5	–	585.4	527.5
	–	–	–	576.7	–
Cu	–	1855.7	1200.1	613.8	433.9
	–	–	–	–	417.4
Zn	–	1845.8	1069.5	689.1	447.1
Ag	2907.6	1725.8	1232.4	6144.7	–
	2900.1	–	–	–	–
Au	–	2195.8	1202.8	610.9	–

Photoexcited particles Cu and Cu_2 can also activate methane [30]. An investigation of the reactions of excited gallium atoms with methane in Ar, Kr, and neat CH_4 matrices has shown that H–Ga–CH_3 was the only photoreaction product [31]. Activation of methane in a matrix by atoms of Zn, Cd, Hg, Mg, Ca, and Be has been reported [32]. Co-condensation of beryllium atoms produced by laser ablation with methane–argon mixtures onto a substrate at 10 K gave the following organoberrylium products: CH_3BeH, CH_3BeCH_3, CH_3Be, H_2CBeH, HCBeH. It is interesting that in the analogous reaction of magnesium or calcium atoms, only CH_3MH was identified [32d]. Theoretical investigations of reactions between photoexcited metal atoms and alkanes have been carried out [33].

References to Chapter V

1. (a) Beauchamp, J. L. In *High-Energy Processes in Organometallic Chemistry*; Suslick, K. S., Ed.; ACS: Washington, 1987, p. 11. (b) Dolgoplosk, B. A.; Oreshkin, I. A.; Soboleva, T. V. *Usp. Khim.* **1987**, *56*, 50 (in Russian). (c) Czekay, G.; Drewello, T.; Eller, K.; Lebrilla, C. B.; Prüsse, T.; Schulze, C.; Steinrück, N.; Sülzle, D.; Weiske, T.; Schwarz, H. In *Organometallics in Organic Synthesis*; Werner, H., Erker, G., Eds.; Springer: Heidelberg, 1989; Vol. 2, p. 203.(d) Schwarz, H. *Acc. Chem. Res.* **1989**, *22*, 282. (e) Armentrout, P. B.; Beauchamp, J. L. *Acc. Chem. Res.* **1989**, *22*, 315. (f) Eller, K.; Schwarz, H. *Chimia* **1989**, *43*, 371. (g) Armentrout, P. B. In *Gas Phase Inorganic Chemistry*; Russell, D. H., Ed.; Plenum: New York, 1989, p. 1. (h) Simões, J. A. M.; Beauchamp, J. L. *Chem. Rev.* **1990**, *90*, 629. (i) Eller, K.; Schwarz, H. *Chem. Rev.* **1991**, *91*, 1121. (j) *Gas-Phase Metal Reactions*; Fontijn, A., Ed.; Elsevier: Amsterdam, 1992. (k) Eller, K. *Coord. Chem. Rev.* **1993**, *126*, 93. (l) Freiser, B. S. *Acc. Chem. Res.* **1994**, *27*, 353. (m) Schröder, D.; Schwarz, H. *Angew. Chem., Int. Ed. Engl.* **1995**, *34*, 1973. (n) Schröder, D.; Heinemann, C.; Koch, W.; Schwarz, H. *Pure Appl. Chem.* **1997**, *69*, 273.
2. (a) Kickel, B. L.; Armentrout, P. B.. *J. Am. Chem. Soc.* **1995**, *117*, 764. (b) Schalley, C. A.; Schröder, D.; Schwarz, H. *J. Am. Chem. Soc.* **1994**, *116*, 11089. (c) van Koppen, P. A. M.; Kemper, P. R.; Bowers, M. T. *J. Am. Chem. Soc.* **1992**, *114*, 1083. (d) Fisher, E. R.; Armentrout, P. B. *J. Am. Chem. Soc.* **1992**, *114*, 2039. (e) Kickel, B. L.; Armentrout, P. B. *J. Am. Chem. Soc.* **1994**, *116*, 10742. (f) Carroll, J. J.; Haug, K. L.; Weisshaar, J. C. *J. Am. Chem. Soc.* **1993**, *115*, 6962. (g) Noll, R. J.; Yi, S. S.; Weishaar, J. C. *J. Phys. Chem. A* **1998**, *102*, 386. (h) Armentrout, P. B.; Tjelta, B. L. *Organometallics* **1997**, *16*, 5372. (i) Schroeter, K.; Schalley, C. A.; Schröder, D.; Schwarz, H. *Helv. Chim. Acta.* **1997**, *80*, 1205. (j) Hertwig, R. H.; Seemeyer, K.; Schwarz, H.; Koch, W. *Chem. Eur. J.* **1997**, *3*, 1315.
3. (a) van Koppen, P. A. M.; Brodbelt-Lustig, J.; Bowers, M. T.; Dearden, D. V.; Beauchamp, J. L.; Fisher, E. R.; Armentrout, P. B. *J. Am. Chem. Soc.* **1991**, *113*, 2359. (b) Noll, R. J.; Weisshaar, J. C. *J. Am. Chem. Soc.* **1994**, *116*, 10288.
4. (a) Irikura, K. K.; Beauchamp, J. L. *J. Phys. Chem.* **1991**, *95*, 8344. (b) Heinemann, C.; Wesendrup, R.; Schwarz, H. *Chem. Phys. Lett.* **1995**, *239*, 75. (c) Campbell, M. L. *J. Chem. Soc., Faraday Trans.* **1998**, *94*, 353.

5. Willis, P. A.; Stauffer, H. U.; Hindrichs, R. Z.; Davis, H. F. *J. Chem. Phys.* **1998**, *108*, 2665.

6. Irikura, K. K.; Beauchamp, J. L. *J. Am. Chem. Soc.* **1991**, *113*, 2769.

7. Schilling, J. B.; Beauchamp, J. L. *Organometallics* **1988**, *7*, 194.

8. Mourgues, P.; Ferhati, A.; McMahon, T. B.; Ohanessian, G. *Organometallics* **1997**, *16*, 210.

9. (a) Campbell, M. L. *J. Am. Chem. Soc.* **1997**, *119*, 5984. (b) Campbell, M. L. *J. Phys. Chem. A* **1998**, *102*, 892.

10. (a) Roth, L. M.; Freiser, B. S.; Bauschlicher, C. W., Jr.; Partridge, H.; Langhoff, S. R. *J. Am. Chem. Soc.* **1991**, *113*, 3274. (b) Jackson, G. S.; White, F. M.; Hammill, C. L.; Clark, R. J.; Marshall, A. G. *J. Am. Chem. Soc.* **1997**, *119*, 7567. (c) Berg, C.; Beyer, M.; Achatz, U.; Joos, S.; Niedner-Schattenburg, G.; Bondybey, V. E. *J. Chem. Phys.* **1998**, *108*, 5398.

11. (a) Aschi, M.; Brönstrup, M.; Diefenbach, M.; Harvey, J. N.; Schröder, D.; Schwarz, H. *Angew. Chem., Int. Ed. Engl.* **1998**, *37*, 829. (b) Pavlov, M.; Blomberg, M. R. A.; Siegbahn, P. E. M.; Wesendrup, R.; Heinemann, C.; Schwarz, H. *J. Phys. Chem. A* **1997**, *101*, 1567. (c) Cecchi, P.; Crestoni, M. E.; Grandinetti, F.; Vinciguerra, V. *Angew. Chem., Int. Ed. Engl.* **1996**, *35*, 2522.

12. (a) Waletzko, N.; Schmidt, L. D. *AIChE J.* 1988, 34, 1146. (b) Bockholt, A.; Harding, I. S.; Nix, R. M. *J. Chem. Soc., Faraday Trans.* **1997**, 93, 3869.

13. (a) Stepnowski, R. M.; Allison, J. *Organometallics* **1988**, *7*, 2097. (b) Prüsse, T.; Drewello, T.; Lebrilla, C. B.; Schwarz, H. *J. Am. Chem. Soc.* **1989**, *111*, 2857. (c) Hankinson, D. J.; Miller, C. B.; Allison, J. *J. Phys. Chem.* **1989**, *93*, 3624. (d) Holthausen, M. C.; Hornung, G.; Schröder, D.; Sen, S.; Koch, W.; Schwarz, *Organometallics* **1997**, *16*, 3135.

14. Marçalo, J.; Leal, J. P.; de Matos, A. P.; Marshall, A. G. *Organometallics* **1997**, *16*, 4581.

15. Cornehl, H. H.; Hornung, G.; Schwarz, H. *J. Am. Chem. Soc.* **1996**, *118*, 9960.

16. Harvey, J. N.; Schröder, D.; Koch, W.; Danovich, D.; Shaik, S.; Schwarz, H. *Chem. Phys. Lett.* **1997**, *278*, 391.

17. (a) Schröder, D.; Eller, K.; Schwarz, H. *Helv. Chim. Acta.* **1990**, *73*, 380. (b) Ryan, M. F.; Fiedler, A.; Schröder, D.; Schwarz, H. *Organometallics* **1994**, *13*, 4072. (c) Schröder, D.; Schwarz, H. *Angew. Chem., Int. Ed. Engl.* **1993**, *32*, 1420. (d) Schröder, D.; Fiedler, A.; Hrušák, J.; Schwarz, H. *J. Am. Chem. Soc.* **1992**, *114*, 1215. (e) Ryan, M.F.; Stöckigt, D.; Schwarz, H. *J. Am. Chem. Soc.*

1994, *116*, 9565. (f) Gibson, J. K. *Organometallics* **1997**, *16*, 4214. (g) Cornehl, H. H.; Wesendrup, R.; Diefenbach, M.; Schwarz, H. *Chem. Eur. J.* **1997**, *3*, 1083. (h) Heinemann, C.; Cornehl, H. H.; Dolg, M.; *Inorg. Chem.* **1996**, *35*, 2463. (i) Fiedler, A.; Kretzschmar, I.; Schröder, D.; Schwarz, H. *J. Am. Chem. Soc.* **1996**, *118*, 9941. (j) Kretzschmar, I.; Fiedler, A.; Harvey, J. N. Schröder, D.; Schwarz, H. *J. Phys. Chem. A* **1997**, *101*, 6252.

18. Pan, Y. H.; Sohlberg, K.; Ridge, D. P. *J. Am. Chem. Soc.* **1991**, *113*, 2406.
19. McDonald, R. N.; Jones, M. T.; Chowdhury, A. K. *J. Am. Chem. Soc.* **1991**, *113*, 476.
20. (a) Crellin, K. C.; Beauchamp, J. L.; Geribaldi, S.; Decouzon, M. *Organometallics* **1996**, *15*, 5368. (b) Chen, Q.; Freiser, B. S. *J. Phys. Chem. A* **1998**, *102*, 3343. (c) Chen, Q.; Chen, H.; Kais, S.; Freiser, B. S. *J. Am. Chem. Soc.* **1997**, *119*, 12879.
21. (a) Yamamoto, S.; Kozasa, M.; Sueishi, Y.; Nishimura, N. *Bull. Chem. Soc. Jpn.* **1988**, *61*, 3439. (b) Yamamoto, S.; Hokamura, H. *Chem. Lett.* **1989**, 1429.
22. (a) Brown, S. H.; Crabtree, R. H. *J. Am. Chem. Soc.* **1989**, *111*, 2935. (b) Brown, S. H.; Crabtree, R. H. *J. Am. Chem. Soc.* **1989**, *111*, 2946. (c) Muedas, C. A.; Ferguson, R. R.; Brown, S. H.; Crabtree, R. H. *J. Am. Chem. Soc.* **1991**, *113*, 2233. (d) Lee, J. C., Jr.; Boojamra, C. G.; Crabtree, R. H. *J. Org. Chem.* **1993**, *58*, 3895. (e) Ferguson, R. R.; Crabtree, R. H. *New J. Chem.* **1989**, *13*, 647. (f) Michos, D.; Sassano, C. A.; Krajnik, P.; Crabtree, R. H. *Angew. Chem., Int. Ed. Engl.* **1993**, *32*, 1491. (g) Fowley, L. A.; Lee, J. C., Jr.; Crabtree, R. H.; Siegbahn, P. E. M. *J. Organomet. Chem.* **1995**, *504*, 57. (h) Burdeniuc, J.; Chupka, W.; Crabtree, R. H. *J. Am. Chem. Soc.* **1995**, *117*, 10119.
23. Kharatyan, S. L.; Chatilyan, A. A.; Merzhanov, A. G. *Khim. Fiz.* **1989**, *8*, 521 (in Russian).
24. (a) Power, W. J.; Ozin, G. A. *Adv. Inorg. Chem. Radiochem.* **1980**, *23*, 80. (b) Klabunde, K. J. *Chemistry of Free Atoms and Particles*; New York: Acad. Press. 1980.
25. Remick, R. J.; Asunta, T. A.; Skell, P. S. *J. Am. Chem. Soc.* **1979**, *101*, 1320.
26. (a) Barrett, P. H.; Pasternak, M.; Pearson, R. G. *J. Am. Chem. Soc.* **1979**, *101*, 222. (b) Davies, S. C.; Severson, S. J.; Klabunde, K. J. *J. Am. Chem. Soc.* **1981**, *103*, 3024. (c) Klabunde, K. J.; Tanaka, Y. *J. Am. Chem. Soc.* **1983**, *105*, 3544.
27. (a) Green, M. L. H. *Pure Appl. Chem.* **1984**, *56*, 47. (b) Green, M. L. H.; O'Hare, D. In: *High-Energy Processes in Organometallic Chemistry*; Suslick, K. S., Ed.; ACS: Washington, 1987, p. 260.

28. (a) Ozin, G. A.; McCaffrey, J. G.; McIntosh, D. F. *Pure Appl. Chem.* **1984**, *56*, 111. (b) Perutz, R. N. *Chem. Rev.* **1985**, *85*, 77. (c) Ozin, G. A.; McCaffrey, J. G.; Parnis, J. M. *Angew. Chem.* **1986**, *98*, 1076.

29. Billups, W. E.; Konarskii, M. M.; Hauge, R. H.; Margrave, J. L. *J. Am. Chem. Soc.* **1980**, *102*, 7393.

30. Ozin, G. A.; Mitchell, S. A.; Prieto, J. G. *Angew. Chem., Suppl.* **1982**, *369*, 798.

31. Xiao, Z. L.; Hauge, R. H.; Margrave, J. L. *Inorg. Chem.* **1993**, *32*, 642.

32. (a) McCaffrey, J. G.; Parnis, J. M.; Ozin, G. A.; Breckenridge, W. H. *J. Phys. Chem.* **1985**, *89*, 4945. (b) Greene, T. M.; Andrews, L.; Downs, A. J. *J. Am. Chem. Soc.* **1995**, *117*, 8180. (c) Breckenridge, W. H. *J. Phys. Chem. A* **1996**, *100*, 14840. (d) Greene, T. M.; Lanzisera, D. V.; Andrews, L.; Downs, A. J. *J. Am. Chem. Soc.* **1998**, *120*, 60097.

33. (a) Sevin, A.; Chaquin, P. *Nouv. J. Chim.* **1983**, *7*, 353. (b) Lefcourt, M. A.; Merritt, C. L.; Peterson, M. R.; Csizmadia, I. G. *J. Mol. Struct.* **1988**, *181*, 315. (c) Kolbanovskiy, Yu. A.; Gagarin, S. G. *Kinet. Katal.* **1988**, *29*, 88 (in Russian).

CHAPTER VI

MECHANISMS OF C–H BOND SPLITTING BY LOW-VALENT METAL COMPLEXES

Reactions of metal complexes with saturated hydrocarbons are very important not only from the standpoint of applications but are also extremely interesting for theoretical chemistry. So it is not surprising that many papers devoted to the theoretical aspects of C–H bond activation and especially oxidative addition of C–H compounds to metal complexes (as well as ions and atoms), have been published in recent decades. Some of their results are summarized in books [1] and reviews [2].

VI.1. WEAK COORDINATION OF METAL IONS WITH H–H AND C–H BONDS

Reactions between metal complexes and organic substrates in the condensed phase usually begin with the coordination of these reactants. Due to π- and n-electrons present in the molecules of substances frequently employed in catalytic processes (such as olefins, acetylenes, carbon monoxide) they are capable of forming rather stable complexes with transition metals. For example, the complex $TpRe(CO)_2(THF)$ (Tp = hydridotris(pyrazolyl)borate; THF = tetrahydrofuran) reacts with a variety of aromatic molecules to form stable binuclear complexes of the form $\{TpRe(CO)_2\}_2(\mu\text{-}\eta^2\text{-}\eta^{2'}\text{-L})$, where L = furan, N-methylpyrrole, or naphthalene [3a]:

The pentaammineosmium(II) fragment $[(NH_3)_5Os]^{2+}$ is also known to coordinate easily to double bonds in arenes thereby activating them to various transformations[3b].

From a classical viewpoint, the formation of any complexes of saturated hydrocarbons is still difficult to imagine due to the lack of π- or n-electrons in their molecules. However, compounds containing hydrogen and other three-center bonds, for example B⋯H⋯B or Al⋯CH_3⋯Al, are known (see discussions of hydrogen and other "nonclassical" bonds [4]). Fairly stable complexes of molecular hydrogen have been prepared and thoroughly investigated recently (see, for example, [5]). If these data are taken into account, one can foresee the possibility of the existence of complexes of alkanes.

On the other hand, bearing in mind that saturated hydrocarbons are extremely weak electron donors and poor electron acceptors, one can postulate that alkane adducts with metal-containing species should be extremely unstable. Thus, a quantum-chemical calculation for one of the most simple metal-free systems (CH_4–O_2) showed that if the C–H σ-bond is located along the axis of the π orbital of oxygen in such a way that the O⋯H distance is 2.05 Å, a minimum corresponding to a bond energy between the components of ~ 0.5 kcal mol^{-1} appears on the potential curve [6]. According to the calculation for another model system (CH_3H⋯ONH_2) by the INDO method, the stabilization energy of the adduct for a O⋯H distance of ~2.0 Å does not exceed 1 kcal mol^{-1}.

VI.1. A. FORMATION OF "AGOSTIC" BONDS

Complexes in which there is an intermolecular bond between the metal atom and one of the C–H groups of the ligand, according to X-ray diffraction data and confirmed by IR and NMR spectra, have been discovered in recent decades [7]. The bonds formed by saturated hydrocarbon fragments, especially methyl groups, are of special interest. It has been suggested that such a bond be referred to as "agostic" and to designate it by a half arrow: C–H ⇀ M. Thus the term "agostic" bond refers to the case where the hydrogen atom is simultaneously bound covalently by a three-center two-electron bond to carbon and the transition metal atom. When the agostic bond is formed, the C–H bond usually lengthens by 5–10%, but in the case where it is formed by the C(sp^3)–H fragment virtually no increase in length is observed. The M–H bond is also somewhat longer (by 15–20%) than in the usual hydride. The M–C bond length is always appreciably smaller than the sum of the van der Waals radii of M and C. The appearance of the agostic bond is reflected in the NMR spectra (upfield shift of the ^{1}H and ^{13}C signals) and IR spectra (decrease of the stretching vibration frequency of the C–H bond to 2700-2350 cm^{-1}). A few examples of complexes with agostic bonds are presented in Scheme VI.1.

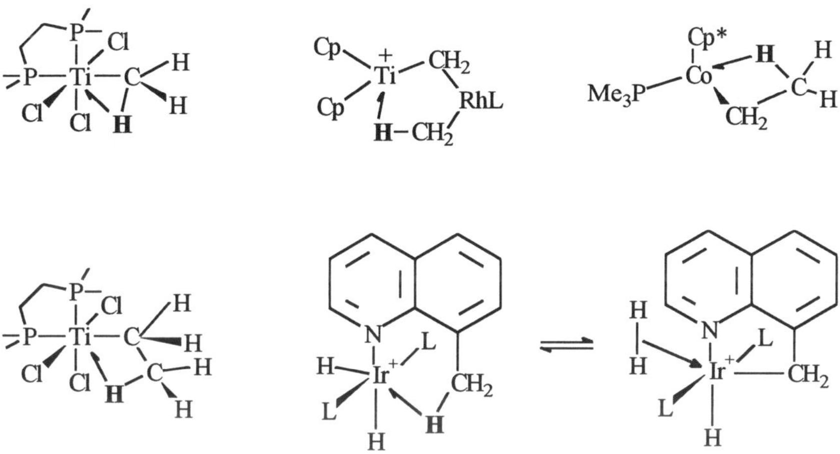

Scheme VI.1. Examples of complexes containing agostic C–H → M bonds.

Thus, a system with a C–H → M agostic bond is, as it were, on the reaction pathway between the C–H + M system and the alkyl hydride derivative C–M–H. In recent years, compounds containing agostic bonds have been proposed as intermediates in reactions that involve C–H bond activation. For example, in the thermolytic rearrangement of *cis*-bis(silylmethyl)platinum(II) complexes, a mechanism is proposed which involves preliminary dissociation of one Pt–P bond compensated by an agostic interaction between the coordinatively unsaturated metal and a phosphine substituent (VI.2) [8].

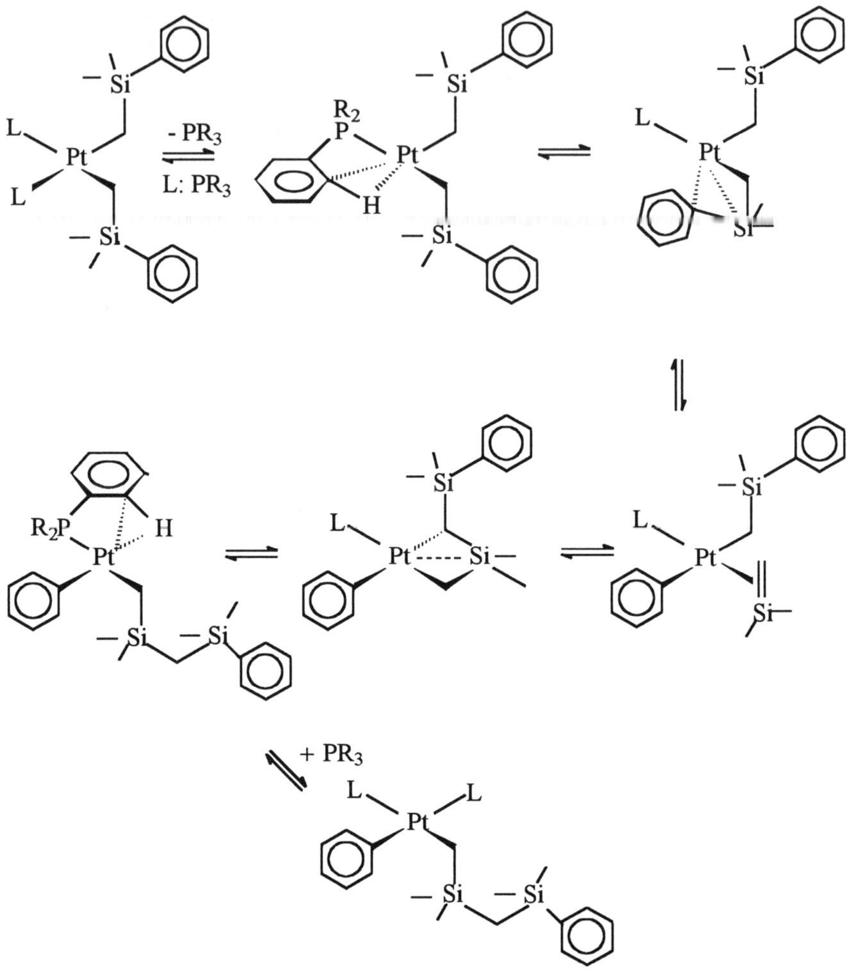

Scheme VI.2. The mechanism of the rearrangement of *cis*-bis(silylmethyl)platinum(II).

A dynamic motion of the metalated C–H bond and/or exchange of the agostic hydrogen atom in the iridium complex is reflected in the ^{13}C NMR equivalence of the tertiary butyl group quaternary carbon atoms. In the proposed [9] slow exchange that H_1 undergoes with iridium atom in the equilibrium **VI-1**

$\rightleftharpoons$ **VI-2** the agostic atom H_1 moves to the iridium while one of the iridium hydrogen atoms moves to C_1.

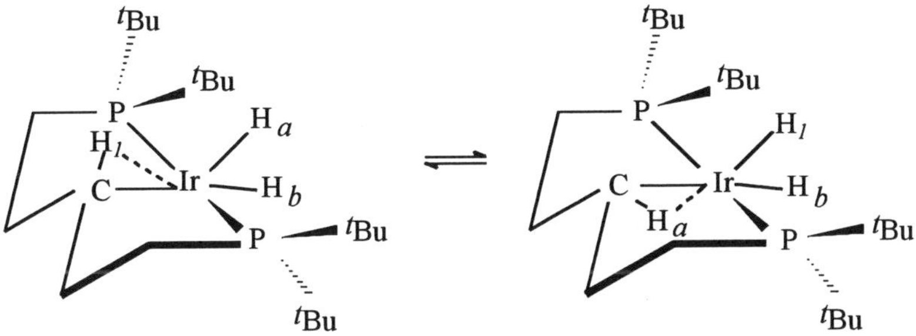

VI-1 **VI-2**

Reversible on/off switching of C–H → Rh interactions in rhodathiaboranes has been decribed [10]. Theoretical investigations of complexes containing agostic bonds have been carried out [11]. The conclusion that formula **VI-3** (in comparison with **VI-4**) describes more precisely the "true" structure of the complex $[Fe_2(CH_3)(CO)(Ph_2PCH_2PPh_2)Cp_2]^+$ has been made on the basis of calculations by the INDO method [12].

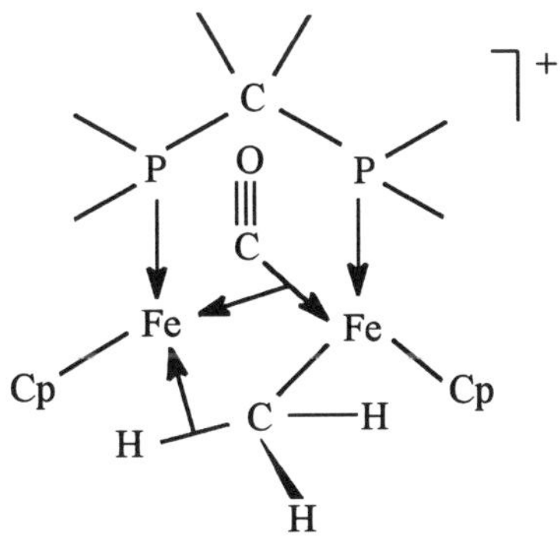

VI-3

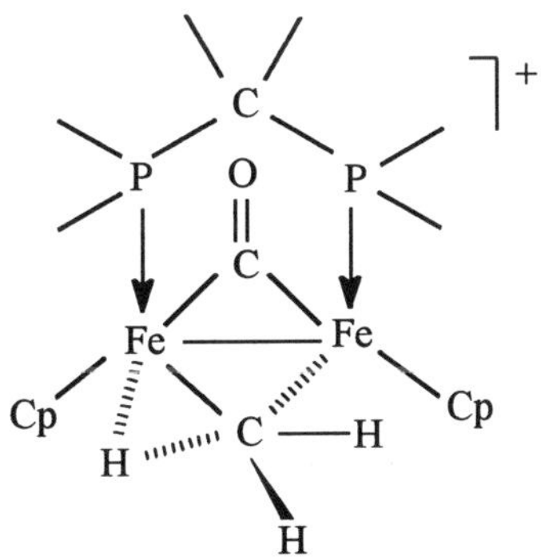

VI-4

Ab initio calculations on the complex $IrH_2[P(Et)H_2]_2^+$ showed that the methyl group of the phosphine ligand does not form any agostic bond with the strongly electron-deficient metal, while the analogous complex $Ir(H_2)(PtBu_2Ph)_2^+$, containing bulky substituents, has two agostic *tert*-butyl groups [13].

VI.1.B. UNSTABLE ADDUCTS BETWEEN ALKANES AND METAL COMPLEXES

Weak adducts of alkanes and metal derivatives (the alkane molecules play the role of "token" ligands in these complexes) have been detected and even isolated using a number of methods [14]. These complexes are unstable at room temperature. Matrix isolation is one of the best established methods for the stabilization and characterization of intermediates. Complexes of alkanes with metal atoms and ions have been detected in the gas phase. All these adducts belong to the larger class of σ-complexes, which has been defined as complexes where the donor is a σ-bond [14c]. Dihydrogen and silane complexes are also from this class.

The following nomenclature has been proposed in review [14c] and publication [15]. If the alkane is bound in an end-on fashion through one, two, or three hydrogen atoms, the complexes are termed η^1-*H*, η^2-*H,H*, and η^3-*H,H,H*, respectively (see Scheme VI.3). The η^2-*H,H* and η^3-*H,H,H* structures involve close contact between the metal and carbon. An alkane may also bind side-on through a single C–H σ-bond: η^2-*C,H* mode. This mode is distinguishable from an end-on η^1-*H* interaction.

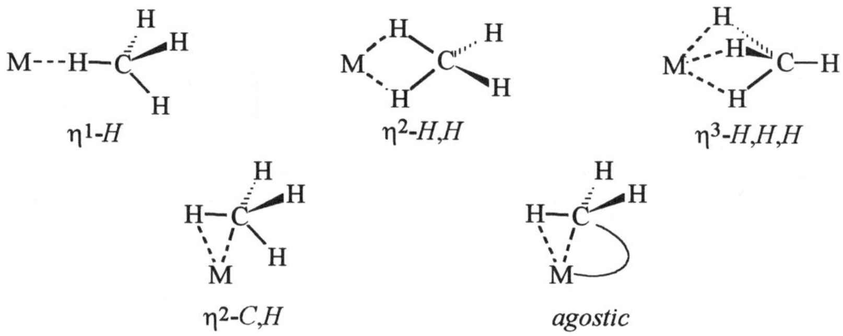

Scheme VI.3. Coordination modes of methane to metal atoms.

Weak adducts of alkanes with metal complexes have been detected by ^{13}C NMR. Dissolution of chromium *tris*-acetylacetonate in 1-chlorobutane entails a paramagnetic shift of the ^{13}C signals of the solvent in the NMR spectrum [16a]. It is of interest that the shift ($\Delta\delta$) along the chain initially diminishes and then again increases for the terminal CH_3 group; the ratio of $\Delta\delta$ is C(1) : C(2): C(3) : C(4) = 1.00 : 0.30 : 0.14 : 0.62. These results indicate the transfer of electron spin density from the complex to the methyl group of the chlorobutane coordinated to it and, since the complex is coordinatively saturated, the interaction apparently takes place via the acetonylacetonate ligand of the complex (outer-sphere coordination). Acetylacetonate complexes of chromi-um(III) and iron(III) also induce changes of chemical shifts in ^{13}C NMR spectra of various alkanes [16b]. The application of the NMR method led to the detection of the short-lived adducts of *bis*[hydro*tris*(pyrazolyl)borato]cobalt with alkanes in solutions [16c,d].

Complexation of a free alkane has been stabilized by creating a binding cavity in the vicinity of a vacant coordination site. This has been seen in the interaction of *n*-heptane with iron(II) in a crystalline complex with the double A-frame porphyrin at room temperature, as confirmed by X-ray analysis [16e].

It has been shown that solid samples of the complex $IrXH_2(H_2)(P^iPr_3)_2$, which are dissolved in hydrocarbons, liberate hydrogen [17a]. The following equilibrium has been proposed for this reaction:

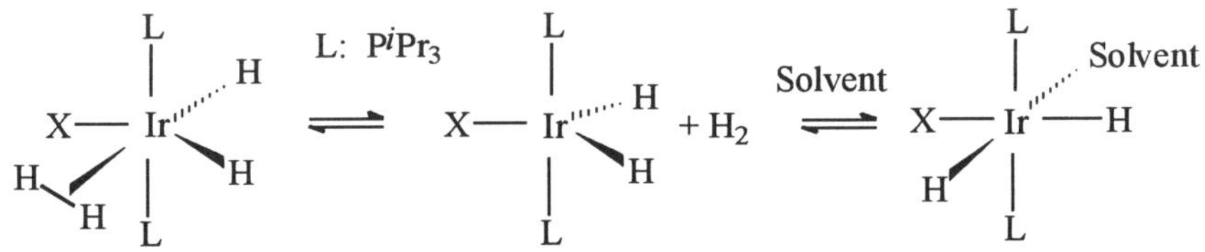

The values of ΔH and ΔS, determined for the chloride and bromide analogs, were higher in alkane solvents than in toluene (Table VI.1). On the basis of these results it has been concluded that Cl and Br complexes lose an H_2 molecule with little or no complexation of the alkane, and that solvent coordination to the resulting five-coordinate complex occurs in toluene solution. The low values obtained for the iodo complex's thermodynamic parameters indicate that this five-coordinate complex adds a solvent molecule to produce the adduct $IrIH_2$-$(P^iPr_3)_2$(Alkane). Influence of arene and alkane coordination on reversible hydrogen elimination from these complexes has been also investigated [17b].

Table VI.1. Thermodynamic parameters for the loss of H_2 from $IrXH_2$-$(H_2)(P^iPr_3)_2$.

X	Solvent	ΔH, kcal mol^{-1}	ΔS, e.u.
Cl	Toluene-d_8	8.6	15
	Methylcyclohexane-d_{14}	12.4	35
	n-Hexane-d_{14}	12.3	36
Br	Toluene-d_8	8.0	10
	Methylcyclohexane-d_{14}	12.1	31
	n-Hexane-d_{14}	12.0	32
I	Toluene-d_8	[a]	[a]
	Methylcyclohexane-d_{14}	10.5	22
	n-Hexane-d_{14}	10.2	22

[a] Equilibrium shifted too far toward $IrIH_2(H_2)(P^iPr_3)_2$ for accurate determination of the equilibrium constants.

The adducts of alkanes with various metal complexes formed at low temperature can be detected by IR spectroscopy [18]. EPR spectroscopy has revealed that the RhH_2 molecule is strongly complexed with methane in argon matrices at 4 K [19].

The investigation [20a] of the photoinitiated reaction of $Cp^*Rh(CO)_2$ with neopentane in liquid krypton by low-temperature IR flash kinetic spectroscopy gave the results that are consistent with a pre-equilibrium mechanism. According to this mechanism, an initially formed transient krypton complex Cp^*Rh-(CO)(Kr) is in rapid equilibrium with a transient (uninserted) alkane complex

Cp*Rh(CO)(Me_4C). The last complex forms the neopentyl hydride in a unimolecular step. It is interesting that rhodium is bound an order of magnitude more strongly to $(CD_3)_4C$ than to $(CH_3)_4C$. Recently, the reactive intermediate in a cyclohexane C–H bond activation reaction with $CpRh(CO)_2$ has been identified as the cyclohexane solvate η^5-CpRh(CO)(*c*-C_6H_{12}) by sub-picosecond IR spectroscopy [20b].

The reaction between cobalt atoms and diazomethane proceeds spontaneously to yield methane and MeCoH when these reactants and dihydrogen are co-condensed with argon onto a rhodium-plated copper surface at 12 K. Wavelength-dependent photolysis of this reaction gave evidence for a cobalt–methane complex formation [15]:

$$CH_3CoH \underset{UV}{\overset{\lambda > 400\ nm}{\rightleftharpoons}} Co(CH_4)$$

Intermediates (Alkane)$M(CO)_5$ are produced in predominant concentration in solution after flash photolysis of the metal carbonyls $M(CO)_6$ (M = Cr, Mo, W). It has been proposed that the alkane (used as a solvent) is coordinated to the metal via a C–H $\rightharpoonup$ M agostic interaction [21a,b]:

$$M(CO)_6 \xrightarrow{h\nu} [M(CO)_5] \xrightarrow{Alkane} (Alkane)M(CO)_5$$

The complex $CpRe(CO)_2$(*n*-heptane) has been characterized at or above room temperature in *n*-heptane by fast time-resolved infrared (TRIR) spectroscopy [21c]. The reactivity of this complex with carbon monoxide (k = 2.1×10^3 mol^{-1} dm^3 s^{-1}) demonstrates that $CpRe(CO)_2$(*n*-heptane) is the least reactive of the reported organometallic alkane complexes. It is interesting that an analogous xenone derivative $CpRe(CO)_2(Xe)$ is less reactive toward CO than all of the reported alkane complexes except $CpRe(CO)_2$(*n*-heptane) [21c]. More recently, direct observation of a transition metal alkane complex using NMR spectroscopy has been reported [21d]. When a supersaturated solution of the complex $CpRe(CO)_3$ (~ 4×10^{-3} mol dm^{-3}) in neat cyclopentane was cooled to –80 °C or below and photolyzed, a new singlet peak at δ 4.92 in the Cp region of the ^{1}H NMR spectrum, as well as another resonance at δ –2.32, appeared at the expense of the peak at δ 5.23 due to the starting complex. The authors attributed these new peaks to the alkane complex $CpRe(CO)_2$(cyclopentane) with the peak at δ –2.32 apparently due to one methylene unit of the cyclopentane ring being coordinated to the metal center:

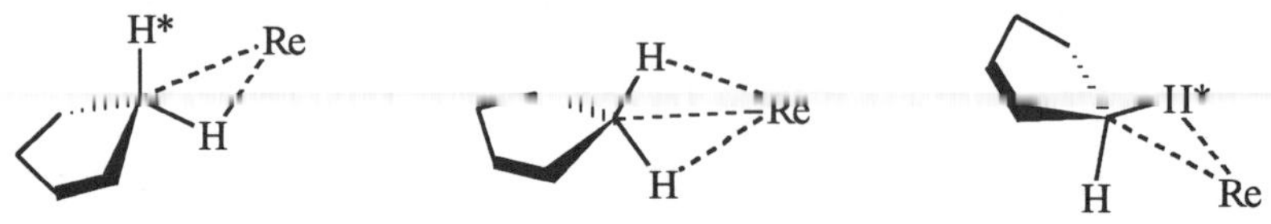

η^2-*C,H* η^2-*H,H* η^2-*C,H**

Results of nanosecond and microsecond transient absorption spectroscopic experiments on the photolysis of the complex (η^2-dfepe)$Cr(CO)_4$ [dfepe = $(C_2F_5)_2PCH_2CH_2P(C_2F_5)_2$] in hexane and cyclohexane solvents indicated the formation of alkane complexes of the type (η^2-dfepe)$Cr(CO)_3$(Alkane) and (η^1-dfepe)$Cr(CO)_4$(Alkane) [22]. The rearrangement of the silane complex CpMn-$(CO)_2(HSiEt_3)$ has been studied by photoacoustic calorimetry [23].

It has been established that methane inhibits the reaction chain free-radical autooxidation of dialkylcadmium [24a] and the reaction of butyllithium with butyl iodide [24b]. It has been proposed that in the second case this effect is due to the coordination of methane to the lithium alkyl in the reaction or to the prereaction complex. The investigation [25] of the kinetics of the reductive elimination of methane from the complex $Cp^*_2W(CH_3)H$ or $Cp_2W(CH_3)H$ led to the proposal of the formation of the so called methane "σ-complex", i.e., a complex with η^2-coordination of the C–H σ bond. The existence of an analogous intermediate has been postulated in the reaction involving the elimination of CH_4 from the methylrhenium hydride derivative [26a]. Rapid exchange of hydrogen atoms between hydride and methyl ligands in the complex [$Cp^*Os(dmpm)$-$(CH_3)H^+$], where dmpm is bis(dimethylphosphino)methane, can proceed by means of α-hydrogen elimination to form an osmium methylene dihydride intermediate or via reversible deprotonation of the Os–CH_3 group by the $HC(SO_2CF_3)_2^-$ anion present in the solution. However, the authors' opinion suggests that this reaction most likely takes place by formation of a coordinated methane ligand (Scheme VI.) [26b].

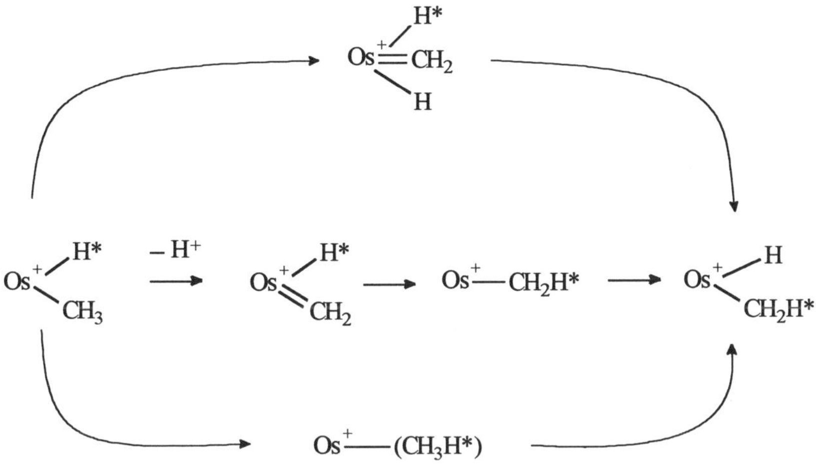

Scheme VI.4. Possible intermediates in the exchange of hydrogen atoms between hydride and methyl ligands in [Cp*Os(dmpm)(CH_3)H.

In the literature [25, 26a,c–g], inverse kinetic isotope effects for the reductive elimination of alkanes from metal centers, which is the microscopic reverse of alkane activation by oxidative addition, have been explained by the presence of an σ alkane intermediate. Recently, thermolysis of the diastereomerically pure complexes (*RS*),(*SR*)-[2,2-dimethylcyclopropyl) (Cp*)-(PMe_3)IrH] and (*RR*),(*SS*)-[2,2-dimethylcyclopropyl)(Cp*)(PMe_3)IrH] (see Scheme VI.5) in C_6D_6 has been shown [26h] to result in its interconversion to the other diastereomer. The analogous reaction of the deuterium-labeled complexes resulted additionally in scrambling of the deuterium from the α-position of the dimethylcyclopropyl ring to the metal hydride position. Diastereomer interconversion and isotopic scrambling occurred at similar rates and have been discussed in terms of a common intermediate mechanism involving a metal alkane complex (Scheme VI.5).

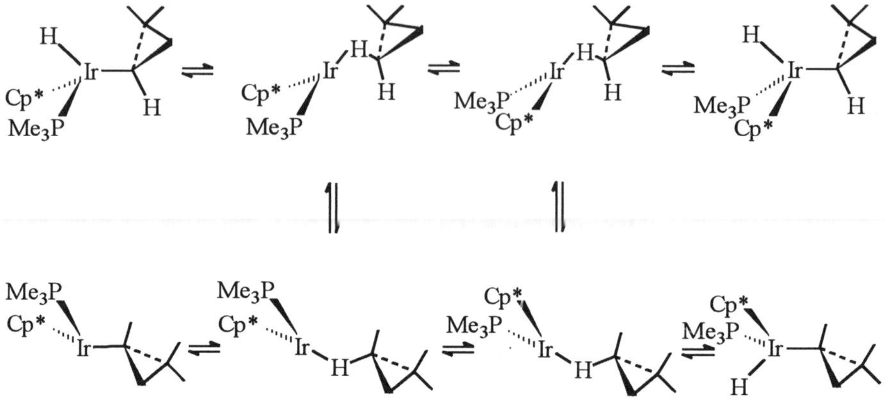

Scheme VI.5. The alkane-complex intermediate mechanism proposed for deuterium-scrambling and diastereomer interconversion in the iridium complexes.

Alkane σ-complexes have also been detected in the gas phase. The study by time-resolved IR spectroscopy of the interaction of a range of open-chain and cyclic alkanes with the 16-electron species $W(CO)_5$ has shown that reversible complexes with all the unsubstituted alkanes, L, are formed except with methane [27a]. The equilibrium constant

$$K_p = \frac{[W(CO)_3L]}{[W(CO)_3]p(L)}$$

at 300 K, increases with carbon number from 610 atm^{-1} in ethane to 5200 atm^{-1} in *n*-hexane, and from 1300 atm^{-1} in cyclopropane to 7300 atm^{-1} in cyclohexane. Binding energies are in the range of 7–11 kcal mol^{-1}, again increasing with the size of the alkane. The kinetic and activation parameters have been determined for the reactions of complexes (η^6-arene)$Mo(CO)_2$(Sol) and $M(CO)_5$(Sol) (M = Cr, Mo, W) with CO [27b]; for M = Cr the $\Delta H^{\ddagger}$ is constant (~ 5 kcal mol^{-1}), while the $\Delta S^{\ddagger}$ term becomes less negative as the alkane chain length increases. In high-pressure mass spectrometry studies of organometallic species, a [Mn, C_6, O_5, H_4]$^+$ ion was observed when $Mn_2(CO)_{10}$ was used as a precursor and methane as a chemical ionization agent [28a]. The authors proposed that this ion is a $(OC)_5Mn^+–CH_4$ complex formed in the high-pressure ion source by the interaction between $(OC)_5Mn^+$ and methane.

Reacting mass-selected species $Fe(CO_2)^+$ with the mixtures $N_2–CH_4$, $CO_2–CH_4$ and others provided an internally consistent set of equilibrium constants and free energies $\Delta\Delta G$ (298 K) (Table VI.2) [28b].

Table VI.2. Equilibrium constants for ligand-exchange reactions with $Fe(L)^+$ and $\Delta\Delta G$ (298 K) values.

No	Reaction	K_{eq}	$\Delta\Delta G_R$ kcal mol^{-1}
1	$Fe(N_2)^+ + CH_4 \longrightarrow Fe(CH_4)^+ + N_2$	5.4±0.7	– 1.0±0.1
2	$Fe(CO_2)^+ + CH_4 \longrightarrow Fe(CH_4)^+ + CO_2$	1.5±0.3	– 0.2±0.1
3	$Fe(Xe)^+ + CO_2 \longrightarrow Fe(CO_2)^+ + Xe$	2.7±1.3	– 0.6±0.4
4	$Fe(N_2)^+ + CO_2 \longrightarrow Fe(CO_2)^+ + N_2$	3.8±0.8	– 0.8±0.2
5	$Fe(N_2)^+ + Xe \longrightarrow Fe(Xe)^+ + N_2$	1.5±0.3	– 0.2±0.1

The η^2-coordination of the Si–H bond in complexes **VI-5** [29a] and **VI-6** [29b] and the C–H bonds of the *meso*-octaethyltetraoxaporphyrinogen [29c] has been confirmed by X-ray analysis.

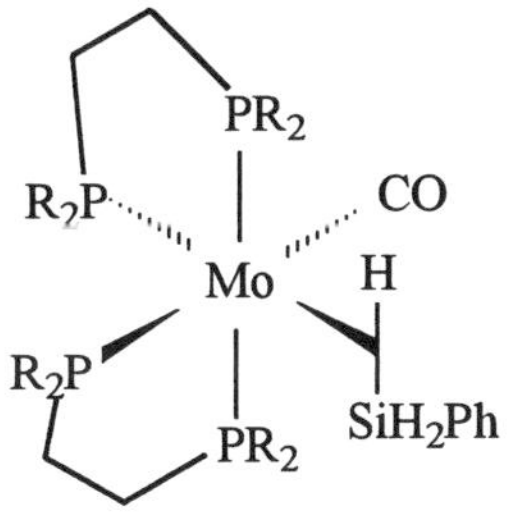

VI-5

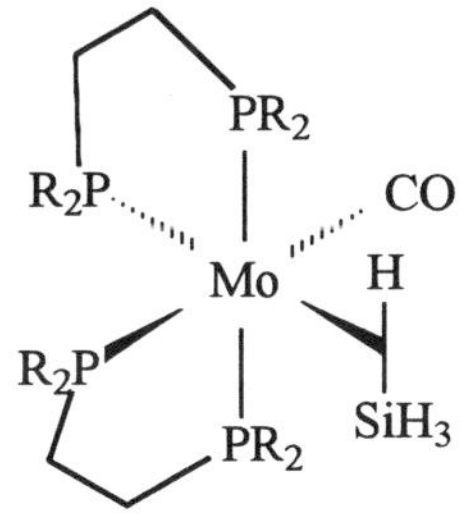

VI-6

The concept of the transient formation of a carbon–hydrogen σ bond complex has been used in the discussion [30] of the possible mechanism of equilibration between diastereomeric chiral rhenium alkene complexes [CpRe-(NO)(PPh$_3$)(H$_2$C=CHR)]$^+$.

Quantum-Chemical Calculations

The problem of the coordination of alkanes to metal atoms, ions and complexes has attracted the attention of theoretical chemists. It has been suggested that the interaction of saturated C–H bonds with transition metals is due to the overlap of the diffuse outer orbitals of the transition metal with the localized orbitals of the saturated bond [31].

The simplest system for modeling the interaction of the C–H bond with an unoccupied diffuse orbital of the metal is a methane molecule coordinated to a palladium atom. A calculation has been carried out for this system by the nonempirical SCF MO method [32]. Three possible symmetrical structures of Pd···CH$_4$, differing in the number of hydrogen atoms coordinated directly to the palladium atom, were examined: two structures with C_{3v} symmetry (η^1-*H* and η^3-*H,H,H* coordination, see Scheme VI.3) and one with the C_{2v} symmetry (η^2-*H,H* coordination). The calculated CH$_4$···Pd bond energies decrease in the sequence η^1-*H* > η^2-*H,H* > η^3-*H,H,H*, although they are similar for all three structures (8.4 and 7.3 kcal mol^{-1} for η^1-*H* and η^3-*H,H,H*, respectively).

Coordination leads to a slight transfer of electron density from methane (mainly from the hydrogen atoms) to the 5s and 5p orbitals of palladium via a donor–acceptor mechanism; this entails some redistribution of electron density in the individual AO of both components. Thus, the electron density in the $4d_{z^2}$ AO of the palladium atom decreases in all three structures. There is a simultaneous increase of electron density in the 5s, $5p_\pi$, and $5p_\sigma$ AOs, but the overall negative effective charge on palladium increases. Changes in the populations of the AOs of methane (much smaller) show that the electron density is to some extent transferred also via a dative mechanism. Since a carbon and not a hydrogen atom plays the main role in the structure with η^3-*H,H,H* coordination, where the methane molecule is oriented towards the palladium atom via three hydrogen atoms (the equilibrium Pd···H distance in this structure is 2.2 Å against 1.7 Å in the η^1-*H* structure), the dative transfer of electron density from palladium mainly to the carbon atom is most clearly expressed precisely in this structure.

It is of interest that the bond energy in the $Pd \cdots H_2$ system, calculated by the same method, is almost twice as high as the bond energy in the methane complex. The donor–acceptor transfer in the alkane complex is much smaller than in the complex with dihydrogen. On the other hand, in general, since the HOMO energy in methane is higher than in hydrogen (the ionization potentials are 12.7 and 15.4 eV respectively), methane should be a better electron donor. However, according to the calculation, the opposite behavior takes place. One can therefore conclude that the interaction is in this instance determined by the overlap of the corresponding orbitals and not by the difference between the energy levels.

The non-empirical MO method has been also used for the calculation of complexing copper(I) ion with molecules of hydrogen, methane and ethane [33]. The potential surfaces for the addition reactions of H_2, CH_4 and C_2H_6 from infinity up to the equilibrium state are of bonding character. There is only one minimum on these surfaces corresponding to the complexes $Cu^+ \cdot H_2$ (energy of complexing ~ 17 kcal mol^{-1}), $Cu^+ \cdot CH_4$ (~ 10 kcal mol^{-1}) and $Cu^+ \cdot C_2H_6$ (~ 13 kcal mol^{-1}). The H_2 molecule is more strongly coordinated to the central Cu^+ ion by an η^2 mode.

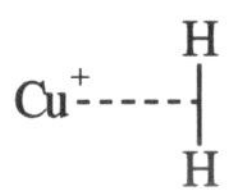

η^2

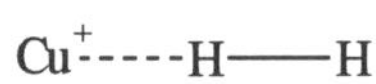

η^1

The structure with η^1 coordination lies 3.7 kcal mol^{-1} higher and transforms into the η^2 configuration without any barrier.

In the case of methane, the tridentate coordination η^3-*H,H,H* to the Cu^+ ion is more advantageous. Bidentate (η^2-*H,H*) and monodentate (η^1-*H*) coordination structures are situated 3–4 kcal mol^{-1} higher than the tridentate configuration. These differences are small and at low temperatures, the Cu^+ cation may apparently migrate around the CH_4 molecule, i.e., all possible modes of coordination can alternate. It is important that the geometrical parameters of CH_4 when it coordinates to the Cu^+ ion change only negligibly, so the rupture of the C–H bond in methane by this cation seems to be impossible. Two modes of motion can be discussed for the complex $Cu^+ \cdot C_2H_6$: migration (Cu^+ around C_2H_6)

and rotation (CH_3 groups around the C–C bond). There are two minimums on the potential surface corresponding to isomers σ and π (Figure VI.1). All complexes of the Cu^+ ion are stable in respect to any channel of monomolecular decay and can exist in the gas phase or in matrices of inert gases.

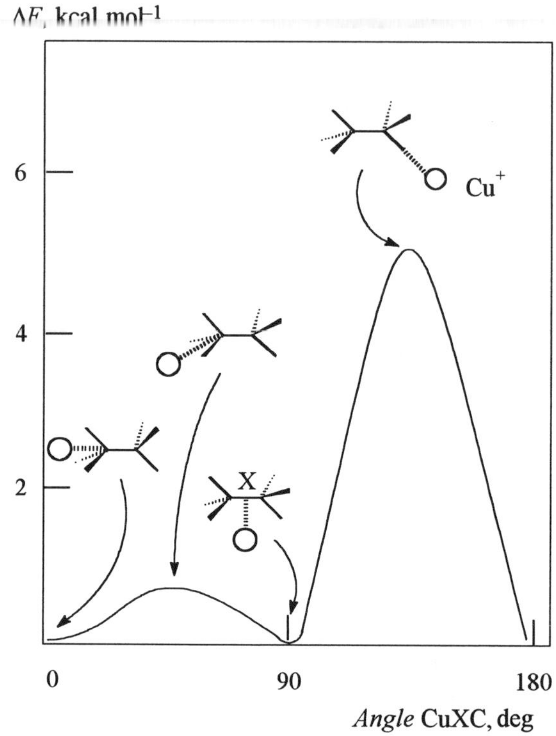

Figure VI.1. Coordination of the ethane molecule with ion Cu^+ according to the nonempirical MO caclulations.

Ab initio considerations [34a] of methane adducts of pyramidal complexes $M(NH_2)(=NH)_2$ (where M = Nb, Ta) and $M(=NH)_3$ (where M = Mo, W) have shown that these adducts have appreciable calculated binding enthalpies (~ 11–16 kcal mol^{-1}) (see, however, [34b]). Planar imidos $(NH_2)_2M=NH$, where M = Ti, Zr, and $(HO)_2Ti=NH$ have much smaller binding enthalpies. A significant covalent contribution to the bonding between the substrate and the formally d^0 complex has been evaluated by the calculations. Upon coordination there is a weakening of a methane C–H bond, charge transfer from methane to metal, and increased $C^{\delta-}$–$H^{\delta+}$ polarization; all indicative of a role for the adduct in the all-important C–H scission step to follow. In a recent publication, it has been shown that the reactivity of primary alkyl halides with $[(^tBu_3SiN=)_3WH]K$ provides evidence of alkane complexation and d^0 alkane complexes $(^tBu_3SiN=)_3W(RH)$ precede C–H activation [34c]. Although the results obtained were also consistent with a solvent cage comprised of $(^tBu_3SiN=)_3W$ and RH, calculational studies strongly supported the proposed alkane adducts (also see [34d]).

Calculations of high accuracy were used for the theoretical consideration of molecular precursor complexes for the reaction between methane and a selected set of second-row transition metal complexes [35]. It turned out that the electronic structure requirements are quite different to form a strong precursor and to obtain a low barrier for the oxidative addition reaction. Thus for the formation of a strong precursor, a ground state singlet is important and the precursor binding energy will be larger if the complex does not have π-bonding ligands. For example, the precursor binding energy for the complex RhH(CO) (10.8 kcal mol^{-1}) is smaller than the precursor binding energy for the complex $RhH(NH_3)$ (14.5 kcal mol^{-1}). A low barrier for the oxidative addition reaction with methane requires a low-lying triplet state of the reactant. Complexes of this type are RhCl(CO) and RuH_2, which are ground state triplets.

An *ab initio* study [36a] demonstrated that the complexes *cis*-Mo(CO)-$(PH_3)_4(HSiH_3)$ (compare to structure **VI-6**) and $CpMn(CO)_2(HSiR_3)$ can be regarded as pseudooctahedral d^6-ML_6, which defines a particularly stable class of complexes. Finally, theoretical analysis of the molecular complexes $(\eta^2$-$H_2)$Os-$Cl_2(PH_3)_2$ and $(\eta$-$CH_4)OsCl_2(PH_3)_2$ by RHF, MP2 and DFT methods has been recently reported [36b].

VI.2. Mechanistic Studies

In a series of recent publications, Bergman, Harris, Frei, Bromberg and their co-workers investigated the mechanism of alkane activation by complexes of rhodium and iridium [37, 38]. The study of ultrafast (femtosecond–picosecond)

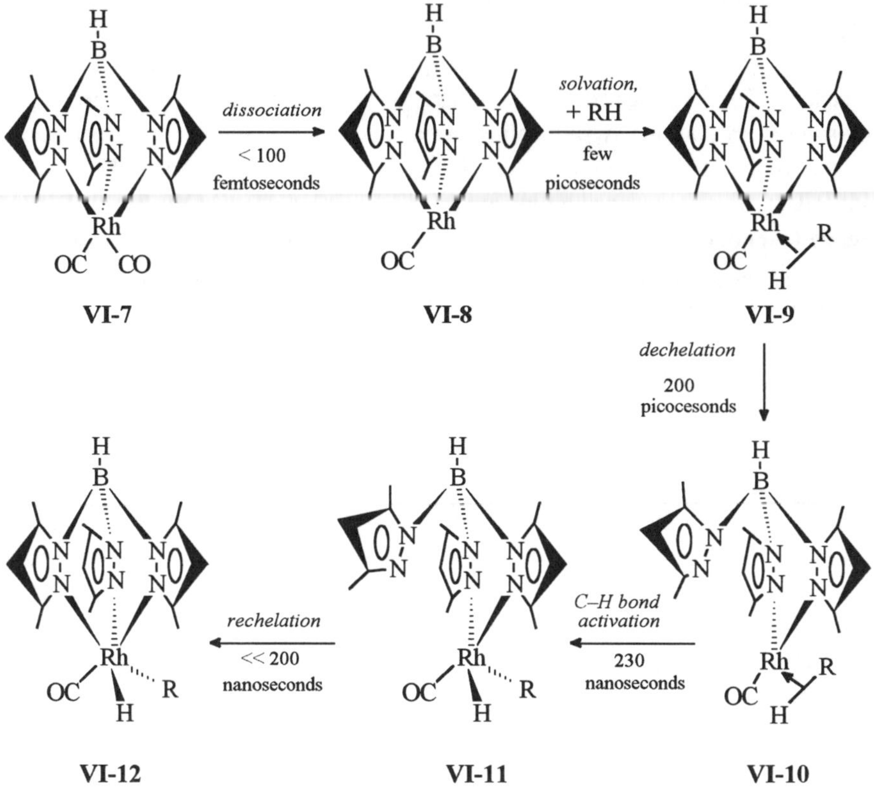

Scheme VI.6. Proposed mechanism for the C–H activation reaction of $Tp^*Rh(CO)_2$ in alkane solution.

excited state dynamics of the complexes $Cp^*M(CO)_2$, where M = Rh, Ir, and $Tp^*Rh(CO)_2$, where Tp* is H-tris(3,5-dimethylpyrazolyl)borate, in alkane solutions led to the conclusion that the primary photoprocess of the highly photoactive $Tp^*Rh(CO)_2$ is the dissociation of one carbon monoxide ligand to generate the vibrationally excited coordinatively unsaturated species $Tp^*Rh(CO)$. After the formation of solvated $Tp^*Rh(CO)$, activation of the C–H bonds in the alkanes occurs.

The authors established directly the time scale for activation of C–H bonds in solutions at room temperature by monitoring the C–H bond activation reaction in the nanosecond regime with infrared detection. In the first stage of the process, loss of one carbon monoxide ligand (reaction **VI-7** ⟶ **VI-8** in Scheme VI.6) substantially reduces back-bonding from the rhodium ion and increases the electron density at the metal center. Formed after the solvation stage, complex **VI-9** traverses a 4.2 kcal mol^{-1} barrier ($k = 5.0 \times 10^9$ s^{-1}) and forms the η^2-Tp* complex **VI-10** which is more reactive toward C–H oxidative addition.

The bond-breaking stage occurs with a barrier of ~ 8.3 kcal mol^{-1} and produces the unstable η^2 C–H activated complex **VI-11** containing the metal center in a formal oxidation state of III. After rechelation, which reduces the electron density on the metal center, the final product, **VI-12**, is formed.

Bergman *et al.* have suggested [39a,b] two mechanisms for the σ-bond metathesis reaction by cationic iridium(III) complexes (see review [39c]), i.e., a two-step addition–elimination mechanism:

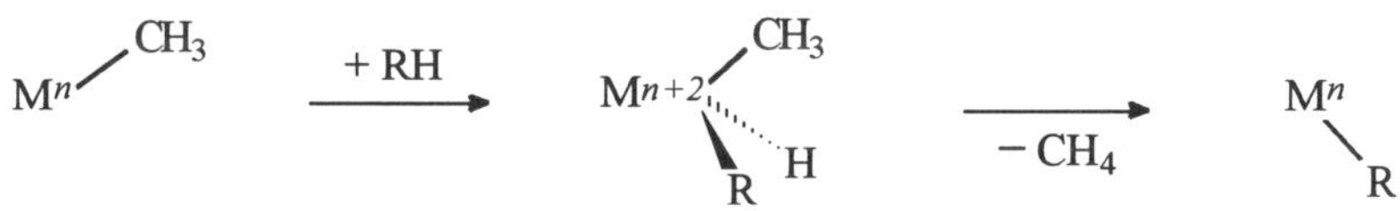

and concerted σ-bond metathesis mechanism:

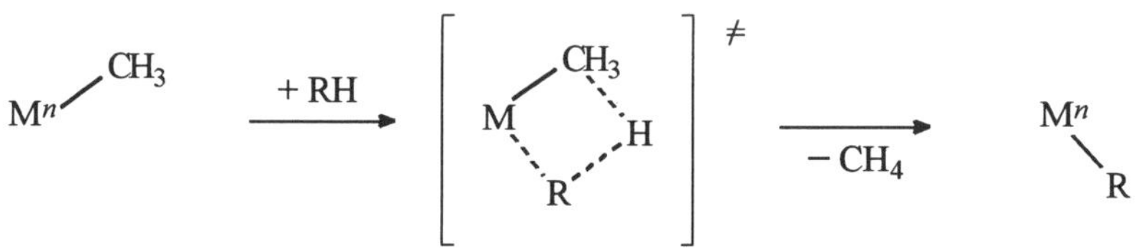

A similar pattern of reactivity has been reported for platinum(II) complexes [39d]. *Ab initio* calculations demonstrated that an oxidative addition–reductive elimination mechanism is more likely than the concerted metathesis [39e].

The C–H activation reactions of complexes Cp*(PMe$_3$)Ir(CH$_3$)$^+$ and Cp(PMe$_3$)Ir(CH$_3$)$^+$ has been studied by a combination of electrospray ionization MS/MS techniques, isotopic labeling experiments in the gas phase and in solution, and by *ab initio* calculations [40a,b]. The authors concluded that their experimental and computational results demonstrate the C–H activation in the gas phase proceeds by a third mechanism, namely the elimination of a methane molecule with the formation of an intermediate metallaphosphacyclopropane. Although the experimental results obtained did not distinguish decisively between the two possibilities shown in Scheme VI.7, the *ab initio* calculations found structure **VI-13** to be a minimum.

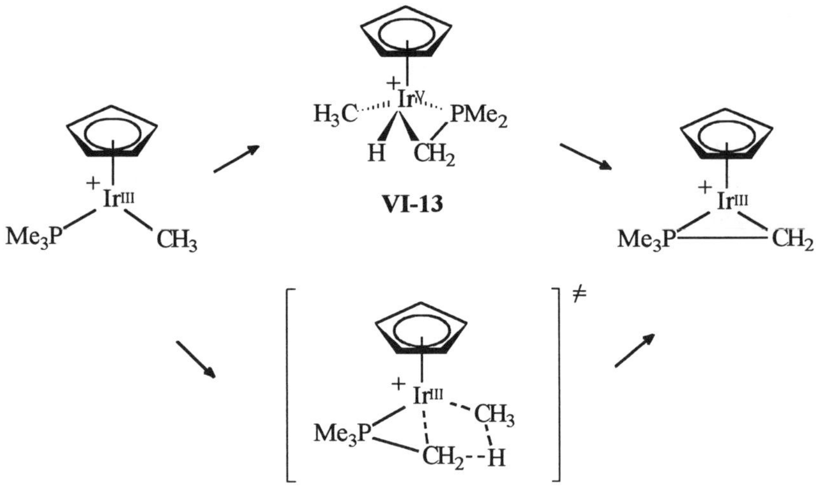

Scheme VI.7. Two possible mechanisms of C–H activation by the complex $Cp(PMe_3)Ir(CH_3)^+$.

A further investigation of C–H activation reactions of cyclometalated complexes of iridium(III) showed that the complex $Cp^*(PMe_3)Ir(CH_3)^+$ does not undergo intermolecular C–H activation in solution via a cyclometalated intermediate [40c] (see also below).

In a work which is relative to the problem of C–H bond addition, Goldman and coworkers have investigated the thermolysis of *trans*-$Ir(^iPr_3)_2(CO)$-$Cl(H)(C_6H_5)$ affording benzene and the Vaska-type complex $IrL_2(CO)Cl$ [41]. Two pathways for the elimination have been elucidated with no evidence for a direct elimination pathway (estimated upper limit for direct elimination, 1.5×10^{-6} s^{-1} at 120 °C, corresponds to a high kinetic barrier, $\Delta G^{\ddagger} \geq 34$ kcal mol^{-1}). The first pathway involves a catalysis by molecular hydrogen, and the authors assumed that it may be possible to use H_2 to catalyze oxidative addition of C–H bonds when such additions are thermodynamically favorable. The second pathway involves an isomerization proceeding via reversible loss of carbon monoxide.

VI.3. THERMODYNAMICS OF OXIDATIVE ADDITION

Thermodynamics of C–H oxidative addition to various metal complexes, as well as to metal ions and atoms, have been considered in a number of experimental and theoretical works (see, e.g., [42]). The energy of the oxidative addition of hydrocarbons, RH, to complexes L_nM is determined by the equation [43]:

$$E(\text{M–R}) + E(\text{M–H}) - E(\text{R–H}) \approx E(\text{M–M})$$

and hence can be estimated from E(M–M). The highest value of E(M–M) has been obtained for platinum and in addition it has been shown that the reaction involving Group VIII and I transition metals is energetically favorable.

An estimation [44a] of the heat of oxidative addition via the mechanism

$$L_nM + \text{R–H} \longrightarrow \text{R–}L_n\text{M–H}$$

demonstrated that this reaction is in many cases endothermic with $\Delta H \cong +10$ kcal mol^{-1}. Oxidative addition with cleavage of the C–C bond via the mechanism

$$L_nM + \text{R–R} \longrightarrow \text{R–}L_n\text{M–R}$$

should be thermodynamically even less favorable. The above considerations refer to complexes of the first transition series of metals. The alkyl-metal bonds can be stronger in the case of heavy metals, although, for example, the values D(M–CH_3) for the M^+–CH_3 ions in the gas phase do not support this trend when passing from light to heavy metals within a given Group. The calculated [44b] values D(Pt–C) = 36 kcal mol^{-1} and D(Pt–H) = 60 kcal mol^{-1} made it possible to estimate the enthalpies of the oxidative addition to $Pt(PH_3)_2$ of dihydrogen (–16 kcal mol^{-1}; exothermic reaction), methane via the CH_3–H bond (+9 kcal mol^{-1}; endothermic reaction), and ethane via the CH_3–CH_3 bond (+19 kcal mol^{-1}; strongly endothermic reaction). The complex *cis*-$OsHMe(CO)_4$ is apparently thermodynamically stable and does not exhibit a tendency towards intermolecular reductive elimination of CH_4. It has been suggested [44a] that the $Os(CO)_4$ species, generated photochemically, is capable of combining oxidatively with alkanes, this addition process being thermodynamically favorable. Thus the kinetic barrier to the oxidative addition of RH to low-valent metal complexes is fairly low and the inability of many 16-electron complexes to activate alkanes can be associated with the thermodynamic causes. Indeed, the rate of addition of methane to Cp*IrCO depends only very slightly on temperature [45]. The nature of ligand L has little influence on the reactivity of the complexes Cp*IrL in relation to alkanes, which can also indicate that the activation barrier is low.

Calculated (B3LYP) enthalpies of addition and the effects of phosphine methylation (substitution of PMe_3 for PH_3) in reactions of some iridium complexes with C–H compounds (as well as with molecular hydrogen) are summarized in Table VI.3 [46]. It can be seen that the addition of aryl and especially acetylene C–H bonds is thermodynamically more favorable than the addition of simple alkyl C–H bonds. Addition of an aryl C–H bond has been found to be at least 16 kcal mol^{-1} less exothermic than H_2 addition. However, on the basis of the Bryndza–Bercaw relationship [47]

$$\mathrm{BDE(M{-}X)} - \mathrm{BDE(M{-}Y)} = \mathrm{BDE(H{-}X)} - \mathrm{BDE(H{-}Y)}$$

(where BDE's are bond dissociation enthalpies), the enthalpy of H–C_6H_5 addition is predicted to be equal to that of H_2 addition.

The insertion of a neutral transition metal atom, M, into methane according to the equation

$$\mathrm{M} + \mathrm{CH_4} \longrightarrow \mathrm{H{-}M{-}CH_3}$$

is exothermic for a number of ground state transition metal atoms [48]. Decay to $MH+CH_3$ or MCH_3+H is highly endothermic ($\Delta H \sim +60$ kcal mol^{-1}) and cannot occur [49]. It is important that elimination of H_2 with formation of a metal carbene

$$M + CH_4 \longrightarrow H\text{–}M\text{–}CH_3 \longrightarrow M{=}CH_2 + H_2$$

is thermodynamically possible for several metals [48a]. We will also be concerned with the thermodynamics of oxidative addition in the next section while considering quantum-chemical calculations.

Table VI.3. Addition of some compounds to iridium complexes.

		ΔH, kcal mol^{-1}	
Metal center	C–H Compound	L = PH_3	L = PMe_3
IrL_2Cl	H–H	– 51.2	– 57.2
	H–CH_3	– 27.2	– 37.6
	H–C_6H_5	– 32.9 [a]	– 47.5 [a]
	H–CCH	– 49.8	– 64.4
$IrL_2Cl(CO)$	H–H	– 3.8	– 11.1
	H–CH_3	+ 20.4	+ 14.9
	H–C_6H_5	+ 15.1 [a]	+ 7.8 [a]
	H–CCH	+ 0.7	– 7.5

[a] ΔE value (ΔE were used in lieu of ΔH for the H–C_6H_5 additions).

VI.4. Quantum-Chemical Calculations

Processes involving the oxidative addition of saturated H–H or C–H bonds to metal complexes or metal atoms as well as the reductive elimination of RH have been investigated by quantum-chemical methods and published in numerous articles (see, for example, recent papers [50]). Some of the recent theoretical works have been used in the previous sections, others will be discussed below.

VI.4.A. ACTIVATION BY BARE METAL IONS AND ATOMS

The reaction of Ta^+ with methane in the gas phase has been computationally investigated using approximate density functional theory [51]. The overall reaction sequence

$$Ta^+ + CH_4 \longrightarrow Ta{=}CH_2^+ + H_2$$

is exothermic by ca. 14 kcal mol^{-1} and the Ta–C binding energy in the carbene complex is 119 kcal mol^{-1}. The bond activation process consists of several steps, commencing with the formation of an electrostatically bound encounter complex. Then the insertion of the Ta^+ ion into a C–H bond and the generation of a $H_2TaCH_2^+$ dihydrido species occur. The latter intermediate rearranges into a complex between molecular hydrogen and a carbene cation, which eliminates H_2 in the final step.

Theoretical studies [52] of the complete reaction profile for the dehydrogenation of methane by gaseous iridium ions have shown that three salient factors are responsible for the high reactivity of these ions: the ability of Ir^+ to change spin easily, the strength of the Ir–C and Ir–H bonds, and the ability of Ir^+ to form up to four covalent bonds. Iridium ions are unique in possessing all three characteristics. The reaction steps for the reaction are as follows:

(a) initial formation of an η^2-molecular complex, $Ir(CH_4)^+$;

(b) oxidative addition of a single C–H bond to form the hydridomethyliridium complex, $Ir(H)(CH_3)^+$;

(c) insertion into a second C–H bond to form the pyramidal dihydridomethylideneiridium complex, $Ir(H)_2(CH_2)^+$;

(d) coupling of the H–H bond to form the planar (dihydrogen)methylideneiridium complex, $Ir(H_2)(CH_2)^+$;

(e) elimination of H_2.

There is a global minimum (~ 63 kcal mol^{-1}) for the singlet $Ir(H)_2(CH_2)^+$ structure, which plays an important role in the activation. The overall exothermicity of the reaction is calculated to be 3 kcal mol^{-1}. On the basis of these calculations, the authors suggested solution-phase analogues that may also activate methane. Indeed, in order to obtain a solution-phase complex with chemistry analogous to the gas-phase iridium ion, it is necessary to remove two electrons. This suggests Re^+ in the form of complex $XReL_3$.

Calculations for the oxidative addition reactions between methane and the whole sequence of second row transition metal atoms from yttrium to palladium have been carried out [53]. The lowest barrier for the C–H insertion has been found for the rhodium atom. Palladium has the lowest methane elimination barrier. In another paper [54], the formation of complexes $[CH_4–Fe]^q$ (q = +1, 0, –1) and the oxidative addition of methane to Fe^q has been studied by using the MINDO/SR-UHF method. The potential energy curves for the oxidative addition were calculated for a C_S symmetry (structure **VI-14**).

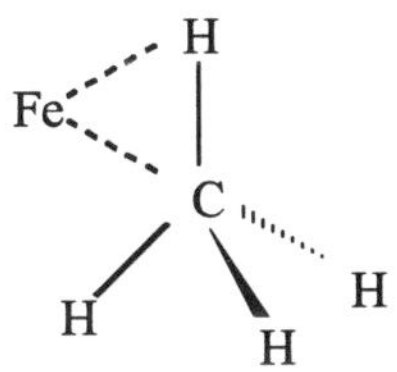

VI-14

It turned out that the formation of alkane complexes and oxidative addition were favored as the system became negatively charged, suggesting that an electronic transfer from the metal to the methane molecule promotes the C–H bond activation. An increase in the *p* character of the metal center favors the C–H bond splitting.

Both the C–C and C–H bond activation branches of the potential energy surface for the reaction between Fe^+ and ethane (calculations by B3LYP method) are characterized by a low barrier for the first step (the insertion of the Fe^+ into a C–C or C–H bond) [55]. The second step is the rate determining one. In the C–C bond activation this is a [1,3]-H shift leading to a complex between $Fe{=}CH_2^+$ and methane. Calculations using high-accuracy quantum chemical methods (B3LYP and PCI-80) for the endothermic reaction

$$Rh^+ + CH_4 \longrightarrow RhCH_2^+ + H_2$$

have demonstrated [56] that there is no barrier in excess of the endothermicity of the H_2 elimination reaction.

An investigation of the oxidative addition of CH_4 and CD_4 [57a], as well as ethane [57b], to a bare palladium atom has demonstrated that quantum tunneling plays a very important role in the process. The barrier of insertion of different transition metal atoms into a C–C bond has been found to be 14–20 kcal mol^{-1} higher than the barrier for insertion into a C–H bond [57c]. Calculations for the activation of the C–H bond in ethylene by second row transition metal atoms showed that the oxidative addition barrier is lowest for the atoms to the right (for rhodium there is no barrier and for palladium the barrier is almost zero) [57d]. The activation energy for B_2 insertion into methane has been predicted to be 4.1 kcal mol^{-1}, while this value increases to 16.2 kcal mol^{-1} for insertion of B [58].

Two mechanisms have been considered by the SCF CNDO/S method for the oxidative addition of methane to the palladium cluster Pd_2 [59a]. In the first possible reaction, the C–H bond oxidatively adds to different palladium atoms:

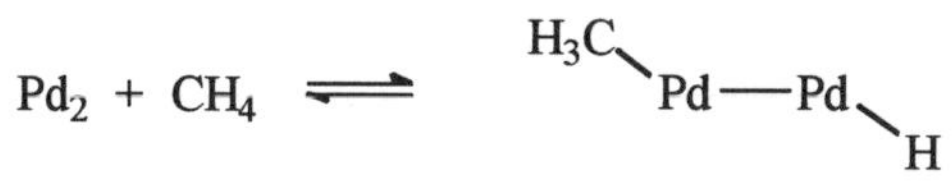

The second mechanism involves the transition state with simultaneous coordination of the methyl group and hydrogen atom to both palladium atoms:

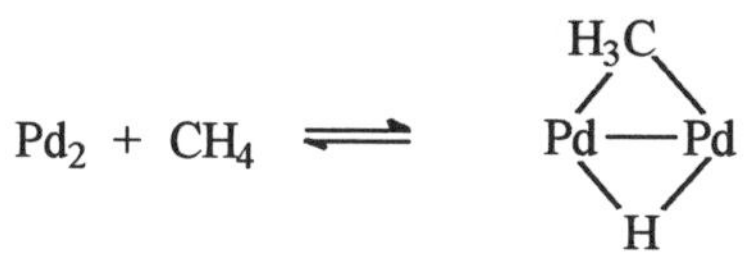

The calculations have elucidated the higher catalytic activity of the cluster compared to a bare palladium atom.

In the reaction of molecular hydrogen or methane with the platinum cluster Pt_2, activation of H–H or C–H bonds takes place first on one metal atom via structures far from planar [59b]. Then one of the hydrogen atoms migrates to the other platinum atom with a negligible activation barrier. In the system "Pd_3+H_2", Pd_3, like Pd_2, activates molecular hydrogen without an activation barrier [59c]. For the activation of the C–H bond in methane, the activation barriers on Pd_3 are much higher than those on Pd_2. Systems "Pt_3+H_2" and "Pt_3+CH_4" are similar to analogous systems containing Pt_2 clusters.

VI.4.B. OXIDATIVE ADDITION OF C–H AND C–C BONDS TO METAL COMPLEXES

In the most simple system "metal complex (ML_n) + H_2", when the bonding σ orbitals of hydrogen interact with the unoccupied acceptor orbital of the complex and with the antibonding σ^* orbital at the occupied donor orbital, four orbitals of a new hydride derivative are generated [60]. The two low-lying orbitals are filled with electrons, one being symmetrical and the other antisymmetrical relative to the twofold axis or the mirror symmetry plane. A symmetrical combination is formed, for the M–H bond generated, from the occupied orbital of the hydrogen molecule and such electron transfer from $\sigma_g(H_2)$ to ML_n weakens the H–H bond and strengthens the M–H bond. The electron transfer in the opposite direction, namely "donor orbital → σ_u^*", also weakens the H–H bond and strengthens the M–H bond. One can assume that the occupied a_1-orbitals and the three degenerate t_2-orbitals in the methane molecule correspond to the σ_g orbital of the hydrogen molecule. Since the energy of the t_2-orbital is greater by 2 eV than the energy of the $\sigma_g(H_2)$ orbitals, the methane molecule is a somewhat stronger donor.

The total energy of the system $H_2 \cdots Cr(CO)_5$ has been calculated by the extended Hückel method [60]. The interaction energy is negative for both the perpendicular orientation (structure **VI-15**) and the parallel orientation (structure **VI-16**) of the H_2 molecule approaching the $Cr(CO)_5$ fragment, which has the C_{4v} symmetry.

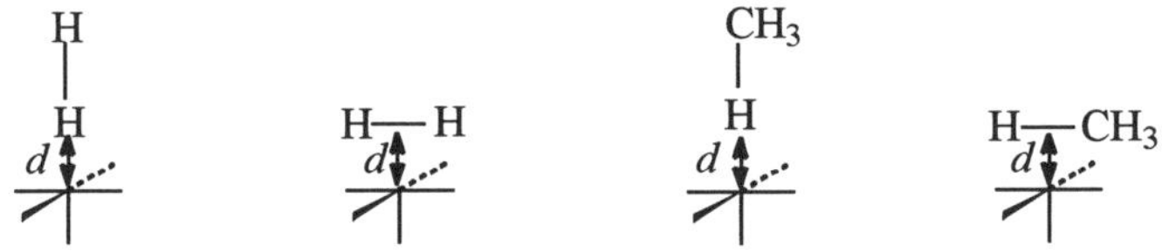

VI-15 **VI-16** **VI-17** **VI-18**

However, in the case of the parallel approach of H_2, the energy of the system has a much more distinct minimum. When H_2 is replaced by CH_4, the curve for the perpendicular approach (structure **VI-17**) has the same form but the parallel approach (structure **VI-18**) becomes extremely unfavorable. Whereas the d_{xz}-orbital (which functions as the donor orbital in the complex) is lowered for **VI-16**, in the case of CH_4 (structure **VI-18**) the energies of both d_{xz}- and d_{yz}-orbitals increase (Figure VI.2). If the methane molecule approaches $Cr(CO)_5$ in such a way that the $H–CH_3$ axis forms an angle θ with the height of the pyramid, the structure is stable for $d(M–H) = 2.0$ Å and $\theta \leq 130°$. However, the perpendicular orientation ($\theta = 180°$) remains the most favorable.

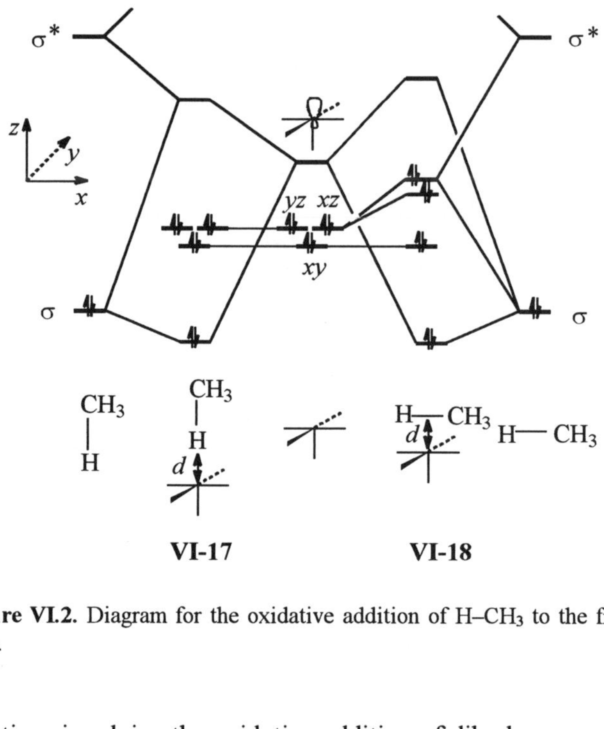

VI-17 **VI-18**

Figure VI.2. Diagram for the oxidative addition of $H–CH_3$ to the fragment $Cr(CO)_5$.

The reactions involving the oxidative addition of dihydrogen and methane to the fragment $Rh(CO)_4^+$, or the corresponding isolobal complex CpRh(CO), or to the metallic surfaces of nickel and titanium have also been analyzed [60].

When an η^2-C–H coordinated adduct $X(PH_3)_2Ir\cdots HCH_3$ is formed, considerable weakening of the coordinated methane C–H bond occurs [61a]. The calculated enthalpy for the reaction

$$Ir(PH_3)_2H + CH_4 \longrightarrow Ir(PH_3)_2(H)_2CH_3$$

is −12.8 kcal mol^{-1}. It has been shown in this work that although donation of electron density from methane to the metal is essential for adduct formation, it is not until back donation to the σ^*_{CH} increases that the C–H bond is activated and cleaved. Calculations [61b] by the *ab initio* restricted Hartree-Fock method of the reaction

$$\text{“}(H_3P)_2Ni\text{”} + CH_4 \longrightarrow (H_3P)_2NiH(CH_3)$$

showed that the oxidative addition is endothermic by 7.0 kcal mol^{-1} and the planar *trans* product is the most stable, being lower in energy than the *cis* isomer by 3.4 kcal mol^{-1}.

The oxidative addition reactions of C–H and Si–H σ-bonds of CH_4 and $HSiR_3$ to $Pt(PH_3)_2$ occur through a planar transition state structure, but the reactions of CH_3–CH_3 and SiH_3–CH_3 take place through a nonplanar transition state [62a]. The oxidative addition of a σ-C–H bond to $Pt(PH_3)_2$ and $Pd(PH_3)_2$ requires a considerably high activation energy (30 and 37 kcal mol^{-1} for Pt and Pd complexes, respectively) [62b]. It is interesting that if two monodentate phosphine ligands are replaced by one chelate phosphine, the reaction proceeds with a much lower activation energy (20.0 and 3.8 kcal mol^{-1} for Pd and Pt complexes with diphosphinomethane, respectively).

A theoretical study of the reactions between methane and ML_2 complexes (where M = Pd, Pt; L = PH_3, CO, L_2 = $PH'_2CH_2CH_2PH_2$), led to the conclusion that for 14-electron ML_2 complexes, a smaller L–M–L angle, a better electron-donating ligand, and a heavier transition metal center should be a potential model for the oxidative addition of C–H bonds [63a].

The theoretical calculations using density functional theory showed that the intermolecular C–H activation of alkanes by the complex $CpIr(PMe_3)(CH_3)^+$ (described by Bergman, *vide supra*) is a lower-energy process and that both *inter*- and *intra*molecular C–H activation proceed only through an oxidative-addition mechanism (Scheme VI.8) [64] (compare Scheme VI.7).

Calculations [65] carried out with the MP2 technique predicted that for the reaction

$$CH_4 + CpRh(CO) \longrightarrow CH_4\cdots RhCp(CO) \longrightarrow CH_3Rh(H)Cp(CO)$$

the intermediate alkane complex is stabilized by 14.8 kcal mol^{-1}, a transition state lies 4.1 kcal mol^{-1} higher than the intermediate and the reaction is exothermic (30.6 kcal mol^{-1}).

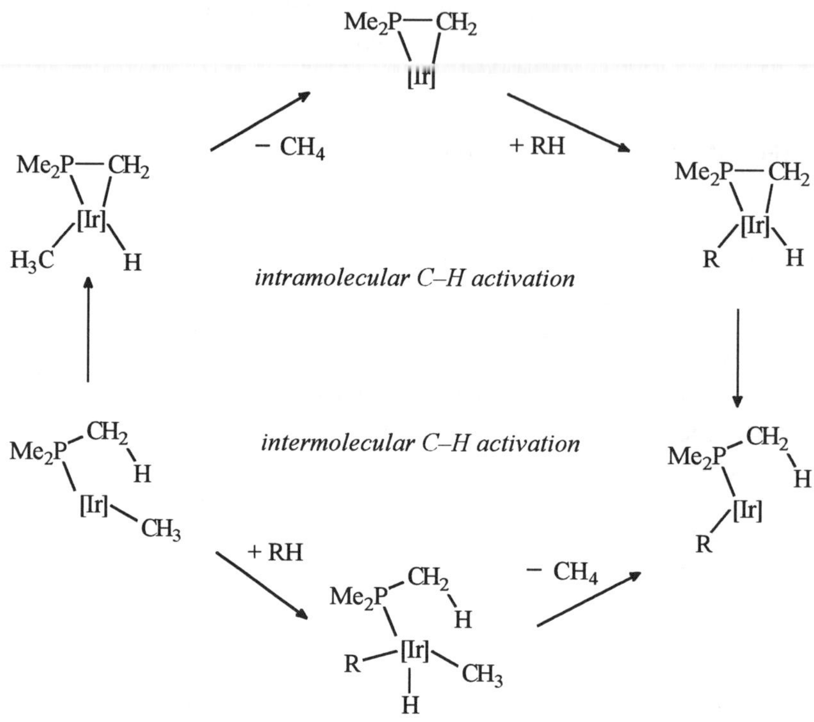

Scheme VI.8. Intra- and intermolecular C–H activation by the complex $Cp(PMe_3)Ir(CH_3)^+$

The effects of ligands in the activation of C–H bonds by transition metal complexes were discussed in certain works. It has been suggested [66a] that a combination of hydride and lone-pair ligands with a minimum of π-bonding should be an optimal combination for the reaction between methane and some model Rh(I) and Ru(II) complexes. For example, it should be advantageous to have methyl and lone-pair ligands with as little covalent bonding as possible. As for the methane activation by a Ru(II) complex, which has not yet been done experimentally, halide ligands and lone-pair ligands with strong π-bonding should be avoided. Only one ligand is needed to improve the exothermicity of the reaction of Rh^+ with methane and the position of this ligand is critical [66b]. Covalent ligand effects in the oxidative addition of methane to second-row transition-metal complexes have been evaluated [66c]. The theoretical study of methane activation by complexes $M(=NH)_2(NH_2)$ (M = V, Nb and Ta) demonstrated that an activation pathway involving C–H addition across the metal-amido bond (σ-bond metathesis) is disfavored relative to a pathway involving C–H addition across the metal-imido bond ($[2_\sigma+2_\pi]$ addition) [67].

A comparison of the activation of C–H and Si–H bonds has been carried out. It has been shown that while C–H bond activation by the complex $RhCl(PH_3)_2$ passes through an η^2-CH_4 adduct and a three-centered transition state, Si–H bond activation is downhill and the η^2-SiH_4 complex is a transition state for intramolecular rearrangement connecting two silyl hydride complexes [68]. The *ab initio* MO study of the mechanism of activation by transient species CpRh(CO) not only of C–H but also H–H, N–H, O–H and Si–H bonds has also been carried out [69].

VI.4.C. OTHER PROCESSES

Bare Ta^+ cations induce the coupling of methane and carbon dioxide in the gas phase [70a] according to the sequence:

$$Ta^+ + CH_4 \longrightarrow TaCH_2^+ + H_2$$

$$TaCH_2^+ + CO_2 \longrightarrow O{=}TaCH_2^+ + CO$$

$$O{=}TaCH_2^+ + CO_2 \longrightarrow TaO_2^+ + O{=}C{=}CH_2$$

The overall reaction equation is as follows:

$$Ta^+ + CH_4 + 2\,CO_2 \longrightarrow TaO_2^+ + H_2 + CO + O{=}C{=}CH_2$$

Density functional calculations employing the B3LYP hybrid functional method combined with an adequate one-particle description [70b] gave an overall computed exothermicity of –55.9 kcal mol^{-1} (experimental value is –53 kcal mol^{-1}). The formation of the final products, TaO_2^+ and the ketene O=C=CH_2, proceeds as a concerted metathesis reaction in which Ta–CH_2 and a C–O bond are broken while simultaneously a Ta–O and C–C bond are newly formed.

Calculations have been carried out using density functional theory [71a] on the hydrogen exchange

and methane elimination

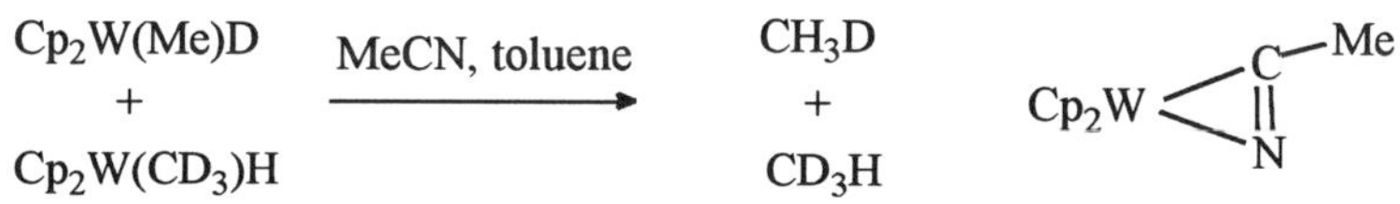

reactions of the complex $Cp_2W(Me)H$ [71b,c]. The hydrogen exchange proceeds along a reaction path on which lies a η^1-*H* σ-complex. The midpoint of the reaction is an η^2-*H,H* σ-complex and the transition state lies 19.5 kcal mol^{-1} above the ground state. Elimination of methane is calculated to have an overall reaction energy of 4.7 kcal mol^{-1} and an activation energy of ca. 19 kcal mol^{-1}.

A computational study of ethane dehydrogenation by the 14-electron complex $Ir(PH_3)_2H$ has been carried out [72]. The first step of the reaction is the formation of the ethane adduct

$$(H_3P)_2(H)Ir + H{-}CH_2{-}CH_3 \longrightarrow (H_3P)_2(H)Ir(H{-}CH_2{-}CH_3)$$

and oxidative addition of ethane to the catalyst:

$$(H_3P)_2(H)Ir(H{-}CH_2{-}CH_3) \longrightarrow (H_3P)_2(H)Ir(H)(CH_2{-}CH_3)$$

The oxidative addition transition state has a three-centered M···H···C geometry with an Ir–H–C angle of 93°. The activation barrier to C–H oxidative addition starting from the ethane adduct is only 1 kcal mol^{-1} and $\Delta H_{rxn} = -18$ kcal mol^{-1}. The ethyl(hydride) is then converted to an olefin complex through β-hydrogen transfer:

$$(H_3P)_2(H)Ir(H)(CH_2{-}CH_2{-}H) \longrightarrow (H_3P)_2(H)_3Ir(CH_2{=}CH_2)$$

Calculations of the intrinsic reaction coordinate showed the reactant in β-H transfer to be the agostic ethyl complex:

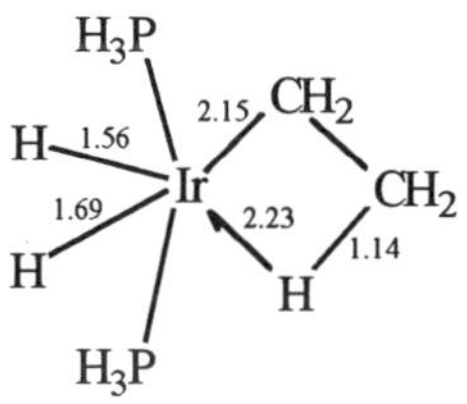

Ir–H–C = 97°
Ir–C–C = 89°
C–C = 1.55

with an activation barrier for β-H transfer of 12 kcal mol^{-1}. Then olefin dissociation occurs (47 kcal mol^{-1} uphill):

$$(H_3P)_2(H)_3Ir(CH_2{=}CH_2) \longrightarrow (H_3P)_2(H)_3Ir + CH_2{=}CH_2$$

and the Ir-trihydride regenerates the catalyst through reductive elimination of dihydrogen:

$$(H_3P)_2(H)_3Ir \longrightarrow (H_3P)_2(H)Ir + H_2$$

A dynamic density functional study of methane photo-carbonylation by the $Rh(PH_3)_2Cl$ catalyst has been reported [73]. It involves C–H bond activation to produce $Rh(PH_3)_2Cl(H)(CH_3)(CO)$ followed by migratory insertion of CO into the Rh–CH_3 bond to generate $Rh(PH_3)_2Cl(H)(CH_3CO)$ and, finally, the elimination of acetaldehyde to form $Rh(PH_3)_2Cl(CO)(H)(CO)(CH_3CO)$.

The following reaction has been recently studied by molecular modeling methods [74]:

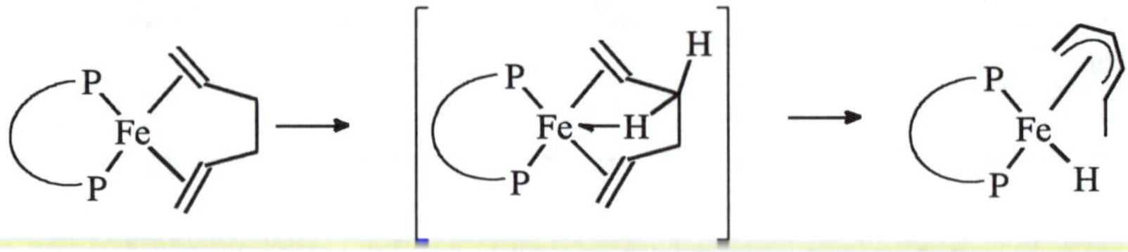

It has been concluded that sterically demanding ligands in the fragment $[R_2P(CH_2)_nPR_2]Fe$ lead to a small accessible molecular surface at the metal atom and further reaction of the coordinated hexadiene is hindered. On the contrary, less demanding ligands lead to a larger accessible molecular surface, which enables the metals atom to interact with a methylenic hydrogen atom with resulting hydrogen transfer and formation of a hydride species.

References for Chapter VI

1. (a) Siegbahn, P. E. M.; Blomberg, M. R. A. In *Theoretical Aspects of Homogeneous Catalysis*; van Leeuwen, W. N. M.; Morokuma, K.; van Lenthe, J. H., Eds.; Kluwer: Dordrecht, 1995, p. 15. (b) Folga, E.; Woo, T.; Ziegler, T. In *Theoretical Aspects of Homogeneous Catalysis*; van Leeuwen, W. N. M.; Morokuma, K.; van Lenthe, J. H., Eds.; Kluwer: Dordrecht, 1995, p. 115.
2. (a) Tsipis, C. A. *Coord. Chem. Rev.* **1991**, *108*, 163. (b) Mamaev, V. M.; Topaler, M. S.; Ustynyuk, Yu. A. *Metalloorg. Khim.* **1992**, *5*, 35 (in Russian).
3. (a) Gunnoe, T. B.; Sabat, M.; Harman, W. D. *J. Am. Chem. Soc.* **1998**, *120*, 8747. (b) Harman, W. D. *Chem. Rev.* **1997**, *97*, 1953.
4. (a) Crabtree, R. H.; Siegbahn, P. E. M.; Eisenstein, O.; Rheingold, A. L.; Koetzle, T. F. *Acc. Chem. Res.* **1996**, *29*, 348. (b) Yao, W.; Eisenstein, O.; Crabtree, R. H. *Inorg. Chim. Acta* **1997**, *254*, 105. (c) Tomaszewski, R.; Hyla-Kryspin, I.; Mayne, C. L.; Arif, A. M.; Gleiter, R.; Ernst, R. D. *J. Am. Chem. Soc.* **1998**, *120*, 2959. (d) Klooster, W. T.; Lu, R. S.; Anwander, R.; Evans, W. J.; Koetzle, T. F.; Bau, R. *Angew. Chem., Int. Ed. Engl.* **1998**, *37*, 1268. (e) Crabtree, R. H.; Eisenstein, O.; Sini, G.; Peres, E. *J. Organometal. Chem.* **1998**, *567*, 7.
5. (a) Crabtree, R. H.; Hamilton, D. G. *Adv. Organomet. Chem.* **1988**, *28*, 299. (b) Ginzburg, A. G.; Bagatur'yants, A. A. *Metalloorg. Khim.* **1989**, *2*, 249 (in Russian). (c) Crabtree, R. H. *Acc. Chem. Res.* **1990**, *23*, 95. (d) Crabtree, R. H. *Angew. Chem., Int. Ed. Engl.* **1993**, *32*, 789. (e) Heinekey, D. M.; Oldham, W. J., Jr. *Chem. Rev.* **1993**, *93*, 913. (f) Ayllon, J. A.; Gervaux, C.; Sabo-Etienne, S.; Chaudret, B. *Organometallics* **1997**, *16*, 2000. (g) Bender, B. B.; Kubas, G. J.; Jones, L. H.; Swanson, B. I.; Eckert, J.; Capps, K. B.; Hoff, C. D. *J. Am. Chem. Soc.* **1997**, *119*, 9179. (h) Tomas, J.; Lledos, A.; Jean, Y. *Organometallics* **1998**, *17*, 190. (i) Braga, D.; De Leonardis, P.; Grespioni, F.; Tedesco, E. *Inorg. Chim. Acta* **1998**, *273*, 116.
6. Buchachenko, A. L. *Complexes of Radicals and Molecular Oxygen with Organic Molecules*; Nauka: Moscow, 1984 (in Russian).
7. (a) Brookhart, M.; Green, M. L. H. *J. Organomet. Chem.* **1983**, *250*, 395. (b) Ginzburg, A. G. *Usp. Khim.* **1988**, *57*, 2046 (in Russian). (c) Brookhart, M.; Green, M. L. H.; Wong, L. *Prog. Inorg. Chem.* **1988**, *36*, 1.
8. Ankianiec, B. C.; Christou, V.; Hardy, D. T.; Thomson, S. K.; Young, G. B. *J. Am. Chem. Soc.* **1994**, *116*, 9963.

9. McLoughlin, M. A.; Flesher, R. J.; Kaska, W. C.; Mayer, H. A. *Organometallics* **1994**, *13*, 3816.

10. Adams, K. J.; McGrath, T. D.; Thomas, R. L.; Weller, A. S.; Welch, A. J. *J. Organometal. Chem.* **1997**, *527*, 283.

11. (a) Knappe, P.; Rösch, N. *J. Organometal. Chem.* **1989**, *359*, C5. (b) Goddard, R. J.; Hoffmann, R.; Jemmis, E. D. *J. Am. Chem. Soc.* **1980**, *102*, 7667. (c) Eisenstein, O.; Jeanne, Y. *J. Am. Chem. Soc.* **1985**, *107*, 1177.

12. Chen, M.-B.; Li, J.; Xu, G.-X. *Acta Chim. Sinica*, **1989**, 317.

13. Ujaque, G.; Cooper, A. C.; Maseras, F.; Eisenstein, O.; Caulton, K. G. *J. Am. Chem. Soc.* **1998**, *120*, 361.

14. (a) Perutz, R. N.; Belt, S. T. McCamley, A.; Whittlesey, M. K. *Pure Appl. Chem.* **1990**, *62*, 1539. (b) Perutz, R. N. *Chem. Soc. Rev.* **1993**, *22*, 361. (c) Hall, C. Perutz, R. N. *Chem. Rev.* **1996**, *96*, 3125. (d) Schneider, J. J. *Angew. Chem., Int. Ed. Engl.* **1996**, *35*, 1068.

15. Billups, W. E.; Chang, S.-C.; Hauge, R. H.; Margrave, J. L. *J. Am. Chem. Soc.* **1993**, *115*, 2039.

16. (a) Kitaygorodskiy, A. N. *Izv. Akad. Nauk SSSR, Ser. Khim.* **1985**, 1518 (in Russian). (b) Kitaygorodskiy, A. N.; Stepanov, A. G.; Nekipelov, V. M. *Magn. Reson. Chem.* **1986**, *24*, 705. (c) Kitaygorodskiy, A. N.; Belyaev, A. N. *Z. Naturforsch.* **1985**, *40a*, 1271. (d) Kitaygorodskiy, A. N. *Izv. Akad. Nauk SSSR, Ser. Khim.* **1987**, 2183 (in Russian). (e) Evans, D. R.; Drovetskaya, T.; Bau, R.; Reed, C. A.; Boyd, P. D. W. *J. Am. Chem. Soc.* **1997**, *119*, 3633.

17. (a) Lee, D. W.; Jensen, C. M. *J. Am. Chem. Soc.* **1996**, *118*, 8749. (a) Lee, D. W.; Jensen, C. M. *Inorg. Chim. Acta* **1997**, *259*, 359.

18. Lokshin, B. V.; Grinval'd, I. I. *Metalloorg. Khim.* **1989**, *2*, 118 (in Russian).

19. Van Zee, R. J.; Li, S.; Weltner, W., Jr. *J. Am. Chem. Soc.* **1993**, *115*, 2976.

20. (a) Bengali, A. A.; Schultz, R. H.; Moore, C. B.; Bergman, R. G. *J. Am. Chem. Soc.* **1994**, *116*, 9585. (b) Asbury, J. B.; Ghosh, H. N.; Yeston, J. S.; Bergman, R. G.; Lian, T. *Organometallics* **1998**, *17*, 3417.

21. (a) Dobson, G. R.; Asali, K. J.; Cate, C. D.; Cate, C. W. *Inorg. Chem.* **1991**, *30*, 4471. (b) Zhang, S.; Dobson, G. R. *Organometallics* **1992**, *11*, 2447. (c) Sun, Z.-Z.; Grills, D. C.; Nikiforov, S. M.; Poliakoff, M.; George, M. W. *J. Am. Chem. Soc.* **1997**, *119*, 7521. (d) Geftakis, S.; Ball, G. E. *J. Am. Chem. Soc.* **1998**, *120*, 9953.

22. Brookhart, M.; Chandler, W.; Kessler, R. J.; Liu, Y.; Pienta, N. J.; Santini, C. C.; Hall, C.; Perutz, R. N.; Timney, J. A. *J. Am. Chem. Soc.* **1992**, *114*, 3802.

23. Hu, S.; Farrell, G. J.; Cook, C.; Johnston, R.; Burkey, T. J. *Organometallics* **1994**, *13*, 4127.

24. (a) Aleksandrov, Yu. A.; Kuznetsova, N. V.; Lyakhmanov, K. S. *Metalloorg. Khim.* **1990**, *3*, 953 (in Russian). (b) Aleksandrov, Yu. A.; Baryshnikov, Yu. N.; Vesnovskaya, G. I. *et al. Zh. Obshch. Khim.* **1988**, *58*, 1683 (in Russian).

25. (a) Parkin, G.; Bercaw, J. E. *Organometallics* **1989**, *8*, 1172. (b) Bullock, R. M.; Headford, C. E. L.; Hennessy, K. M.; Kegley, S. E.; Norton, J. R. *J. Am. Chem. Soc.* **1989**, *111*, 3897.

26. (a) Gould, G. L.; Heinekey, D. M. *J. Am. Chem. Soc.* **1989**, *111*, 5502. (b) Gross, C. L.; Girolami, G. S. *J. Am. Chem. Soc.* **1998**, *120*, 6605. (c) Buchanan, J. M.; Stryker, J. M.; Bergman, R. G. *J. Am. Chem. Soc.* **1986**, *108*, 1537. (d) Periana, R. A.; Bergman, R. G. *J. Am. Chem. Soc.* **1986**, *108*, 7332. (e) Schultz, R. H.; Bengali, A. A.; Tauber, M. J.; Weiller, B. H.; Wasserman, E. P.; Kyle, K. R.; Moore, C. B.; Bergman, R. G. *J. Am. Chem. Soc.* **1994**, *116*, 7369. (f) Bengali, A. A.; Schultz, R. H.; Moore, C. B.; Bergman, R. G. *J. Am. Chem. Soc.* **1994**, *116*, 9585. (g) Wang, C. M.; Ziller, J. W.; Flood, T. C. *J. Am. Chem. Soc.* **1995**, *117*, 1647. (h) Mobley, T. A.; Schade, C.; Bergman, R. G. *Organometallics* **1998**, *17*, 3574.

27. (a) Brown, C. E.; Ishikawa, Y.; Hackett, R. A.; Rayner, D. M. *J. Am. Chem. Soc.* **1990**, *112*, 2530. (b) Breheby, C. J.; Kelly, J. M.; Long, C.; O'Keeffe, S.; Pryce, M. T.; Russell, G.; Walsh, M. M. *Organometallics* **1998**, *17*, 3690.

28. (a) Hop, C. E. C. A.; McMahon, T. B. *J. Am. Chem. Soc.* **1991**, *113*, 355. (b) Dieterle, M.; Harvey, J. N.; Heinemann, C.; Schwarz, J.; Schröder, D.; Schwarz, H. *Chem. Phys. Lett.* **1997**, *277*, 399.

29. (a) Luo, X.-L.; Kubas, G. J.; Bryan, J. C.; Burns, C. J.; Unkefer, C. J. *J. Am. Chem. Soc.* **1994**, *116*, 10312. (b) Luo, X.-L.; Kubas, G. J.; Burns, C. J.; Bryan, J. C.; Unkefer, C. J. *J. Am. Chem. Soc.* **1995**, *117*, 1159. (c) Kretz, C. M.; Gallo, E.; Solari, E.; Floriani, C.; Chiesi-Villa, A.; Rizzoli, C. *J. Am. Chem. Soc.* **1994**, *116*, 10775.

30. Peng, T.-S.; Gladysz, J. A. *J. Am. Chem. Soc.* **1992**, *114*, 4174.

31. (a) Bagatur'yants, A. A.; Gritsenko, O. V.; Zhidomirov, G. M. *Zh. Fiz. Khim.* **1980**, *54*, 2993 (in Russian). (b) Bagatur'yants, A. A. *Zh. Fiz. Khim.* **1983**, *57*, 1100 (in Russian).

32. Anikin, N. A.; Bagatur'yants, A. A.; Zhidomirov, G. M.; Kazanskiy, V. B. *Zh. Fiz. Khim.* **1982**, *56*, 3003 (in Russian).

33. Musaev, D. G.; Charkin, O. P. *Zh. Neorg. Khim.* **1990**, *35*, 689 (in Russian).

34. (a) Cundari, T. R. *Organometallics* **1993**, *12*, 1998. (b) Schaller, C. P.; Bonanno, J. B.; Wolczanski, P. T. *J. Am. Chem. Soc.* **1994**, *116*, 4133. (c)

Schafer, D. F., II; Wolczanski, P. T. *J. Am. Chem. Soc.* **1998**, *120*, 4881. (d) Wolczanski, P. T.; Schafer, D. F., II; Lobkovsky, E. Presented at the *217th Am. Chem. Soc. Natl. Meeting*; Anaheim, March **1999**; paper INOR No. 433.

35. Siegbahn, P. E. M.; Svensson, M. *J. Am. Chem. Soc.* **1994**, *116*, 10124.

36. (a) Fan, M.-F.; Jia, G.; Lin, Z. *J. Am. Chem. Soc.* **1996**, *118*, 9915. (b) Avdeev, V. I.; Zhidomirov, G. M. *Catalysis Today* **1998**, *42*, 247.

37. (a) Lian, T.; Bromberg, S. E.; Yang, H.; Proulx, G.; Bergman, R. G.; Harris, C. B. *J. Am. Chem. Soc.* **1996**, *118*, 3769. (b) Bromberg. S. E.; Lian, T.; Bergman, R. G.; Harris, C. *J. Am. Chem. Soc.* **1996**, *118*, 2069. (c) Vlček, A., Jr. *Chemtracts – Inorg. Chem.* **1997**, *10*, 112.

38. (a) Bromberg. S. E.; Yang, H.; Asplund, M. C.; Lian, T.; McNamara, B. K.; Kotz, K. T.; Yeston, J. S.; Wilkens, M.; Frei, H.; Bergman, R. G.; Harris, C. *Science* **1997**, *278*, 260. (b) Rouhi, M. *C&EN* October 13, **1997**, 5.

39. (a) Burger, P.; Bergman, R. G. *J. Am. Chem. Soc.* **1993**, *115*, 10462. (b) Arndtsen, B. A.; Bergman, R. G. *Science* **1995**, *270*, 1970. (c) Lohrenz, J. C. W.; Jacobsen, H. *Angew. Chem., Int. Ed. Engl.* **1996**, *35*, 1305. (d) Holtcamp, M. W.; Labinger, J. A.; Bercaw, J. E. *J. Am. Chem. Soc.* **1997**, *119*, 848. (e) Strout, D. L.; Zaric, S.; Niu, S.; Hall, M. B. *J. Am. Chem. Soc.* **1996**, *118*, 6068.

40. (a) Hinderling, C.; Plattner, D. A.; Chen, P. *Angew. Chem., Int. Ed. Engl.* **1997**, *36*, 243. (b) Hinderling, C.; Feichtinger, D.; Plattner, D. A.; Chen, P. *J. Am. Chem. Soc.* **1997**, *119*, 10793. (c) Luecke, H. F.; Bergman, R. G. *J. Am. Chem. Soc.* **1997**, *119*, 11538.

41. Rosini, G. P.; Wang, K.; Patel, B.; Goldman, A. S. *Inorg. Chim. Acta* **1998**, *270*, 537.

42. (a) Jones, W. D.; Feher, F. J. *J. Am. Chem. Soc.* **1984**, *106*, 1650. (b) Jones, W. D.; Feher, F. J. *J. Am. Chem. Soc.* **1985**, *107*, 620. (c) Smith, G. M.; Carpenter, J. D.; Marks, T. J. *J. Am. Chem. Soc.* **1986**, *108*, 6805. (d) Stoutland, P. O.; Bergman, R. G.; Nolan, S. P.; Hoff, C. D. *Polyhedron* **1988**, *7*, 1429. (e) Halpern, J. *Polyhedron* **1988**, *7*, 1483. (f) Jones, W. D. Hessel, E. T. *J. Am. Chem. Soc.* **1993**, *115*, 554. (g) Bennett, J. L.; Wolczanski, P. T. *J. Am. Chem. Soc.* **1997**, *119*, 10696.

43. Shestakov, A. F.; Shilov, A. E. *Khim. Fiz.* **1984**, *3*, 1591 (in Russian).

44. (a) Halpern, J. *Inorg. Chim. Acta* **1985**, *100*, 41. (b) Low, J. J.; Goddard, W. A., III *J. Am. Chem. Soc.* **1984**, *106*, 6928.

45. Rest, A. J.; Whitwell, I.; Graham, W. A. G.; Hoyano, J. K.; McMaster, D. *J. Chem. Soc., Chem. Commun.* **1984**, 624.

46. Rosini, G. P.; Liu, F.; Krogh-Jespersen, K.; Goldman, A. S.; Li, C.; Nolan, S. P. *J. Am. Chem. Soc.* **1998**, *120*, 9256.

47. Bryndza, H. E.; Fong, L. K.; Paciello, R. A.; Tam, W.; Bercaw, J. E. *J. Am. Chem. Soc.* **1987**, *109*, 1444.

48. (a) Carroll, J. J.; Haug, K. L.; Weisshaar, J. C.; Blomberg, M. R. A.; Siegbahn, P. E. M.; Svensson, M. *J. Phys. Chem.* **1995**, *99*, 13955. (b) Carroll, J. J.; Weisshaar, J. C.; Siegbahn, P. E. M.; Wittborn, C. A. M.; Blomberg, M. R. A. *J. Phys. Chem.* **1995**, *99*, 14388. (c) Willis, P. A.; Stauffer, H. U.; Hinrichs, R. Z.; Davis, H. F. *J. Chem. Phys.* **1998**, *108*, 2665.

49. Armentrout, P. B.; Georgiadis, R. *Polyhedron* **1988**, *7*, 1573.

50. (a) Koga, K.; Morokuma, K. *Chem. Rev.* **1991**, *91*, 823. (b) Siegbahn, P. E. M.; Blomberg, M. R. A. In *Theoretical Aspects of Homogeneous Catalysts, Application of Ab Initio Molecular Orbital Theory;* Leeuwen, v. P. W. N. M., Lenthe, v. J. H., Morokuma, K., Eds.; Kluwer Academic Publishers: Hingham, 1995. (c) Musaev, D. G.; Morokuma, H. In *Advances in Chemical Physics*, Volume XCV; Prigogine, I., Rice, S. A., Eds.; John Wiley & Sons: New York, 1996, p. 61. (a) Siegbahn, P. E. M. *J. Am. Chem. Soc.* **1996**, *118*, 1487. (b) Thomas, J. L. C.; Hall, M. B. *Organometallics* **1997**, *16*, 2318. (c) Su, M.-D.; Chu, S.-Y. *J. Am. Chem. Soc.* **1997**, *119*, 5373. (d) Espinosa-Garcia, J.; Corchado, J. C.; Truhlar, D. G. *J. Am. Chem. Soc.* **1997**, *119*, 9891. (e) Sargent, A. L.; Titus, E. P. *Organometallics* **1998**, *17*, 65. (f) Yi, S. S.; Blomberg, M. R. A.; Siegbahn, P. E. M.; Weisshaar, J. C. *J. Phys. Chem. A* **1998**, *102*, 395.

51. Sändig, N.; Koch, W. *Organometallics* **1997**, *16*, 5244.

52. Perry, J. K.; Ohanessian, G.; Goddard, W.A., III. *Organometallics* **1994**, *13*, 1870.

53. Blomberg, M. R. A.; Siegbahn, P. E. M.; Svensson, M. *J. Am. Chem. Soc.* **1992,** *114***,** 6095.

54. Rodríguez-Arias, E. N.; Rincõn, L.; Ruette, F. *Organometallics* **1992**, *11*, 3677.

55. Holthausen, M. C.; Fiedler, A.; Schwarz, H.; Koch, W. *J. Phys. Chem.* **1996**, *100*, 6236.

56. Westerberg, J.; Blomberg, M. R. A. *J. Phys. Chem. A* **1998**, *102*, 7303.

57. (a) Mamaev, V. M.; Gloriozov, I. P.; Ischenko, S. Ya.; Simonyan, V. V.; Myshakin, E. M.; Prisyajnyuk, A. V.; Ustynyuk, Yu. A. *Mendeleev Commun.* **1995**, 137. (b) Mamaev, V. M.; Gloriozov, I. P.; Prisyajnyuk, A. V.; Ustynyuk, Yu. A. *Mendeleev Commun.* **1996**, 203. (c) Blomberg, M. R. A.; Siegbahn, P.

E. M.; Nagashima, U.; Wennerberg, J. *J. Am. Chem. Soc.* **1991**, *113*, 424. (d) Siegbahn, P. E. M.; Blomberg, M. R. A.; Svensson, M. *J. Am. Chem. Soc.* **1993**, *115*, 1952.

58. Wang, Z.-X.; Huang, M.-B. *J. Am. Chem. Soc.* **1998**, *120*, 6758.

59. (a) Mamaev, V. M.; Gloriozov, I. P.; Simonyan, V. V.; Prisyajnyuk, A. V.; Ustynyuk, Yu. A. *Mendeleev Commun.* **1996**, 146. (b) Cui, Q.; Musaev, D. G.; Morokuma, K. *J. Chem. Phys.* **1998**, *108*, 8418. (c) Cui, Q.; Musaev, D. G.; Morokuma, K. *J. Phys. Chem. A* **1998**, *102*, 6373.

60. Saillard, J.-Y.; Hoffmann, R. *J. Am. Chem. Soc.* **1984**, *106*, 2006.

61. (a) Cundari, T. R. *J. Am. Chem. Soc.* **1994**, *116*, 340. (b) Re, N.; Rosi, M.; Sgamellotti, A.; Floriani, C.; Guest, M. F. *J. Chem. Soc., Dalton Trans.* **1992**, 1821.

62. (a) Sakaki, S.; Mizoe, N.; Musashi, Y.; Biswas, B.; Sugimoto, M. *J. Phys. Chem. A* **1998**, *102*, 8027. (b) Sakaki, S.; Biswas, B.; Sugimoto, M. *J. Chem. Soc., Dalton Trans.* **1997**, 803.

63. (a) Su, M.-D.; Chu, S. Y. *Inorg. Chem.* **1998**, *37*, 3400. (b) Su, M.-D.; Chu, S. Y. *Organometallics* **1997**, *16*, 1621.

64. Niu, S.; Hall, M. B. *J. Am. Chem. Soc.* **1998**, *120*, 6169.

65. Song, J.; Hall, M. B. *Organometallics* **1993**, *12*, 3118.

66. (a) Siegbahn, P. E. M. *Organometallics* **1994**, *13*, 2833. (b) Blomberg, M. R. A.; Siegbahn, P. E. M.; Svensson, M. *New J. Chem.* **1991**, *15*, 727. (c) Siegbahn, P. E M.; Blomberg, M. R. A.; Svensson, M. *J. Am. Chem. Soc.* **1993**, *115*, 4191.

67. Cundari, T. R. *Organometallics* **1994**, *13*, 2987.

68. Koga, N.; Morokuma, K. *J. Am. Chem. Soc.* **1993**, *115*, 6883.

69. Musaev, D. G.; Morokuma, K. *J. Am. Chem. Soc.* **1995**, *117*, 799.

70. (a) Wesendrup, R.; Schwarz, H. *Angew. Chem.* **1995**, *107*, 2176. (b) Sännnndig, N.; Koch, W. *Organometallics* **1998**, *17*, 2344.

71. (a) Green, J. C.; Jardine, C. N. *J. Chem. Soc., Dalton Trans.* **1998**, 1057. (b) Bullock, R. M.; Headford, C. E. L.; Kegley, S. E.; Norton, J. R. *J. Am. Chem. Soc.* **1985**, *107*, 727. (c) Bullock, R. M.; Headford, C. E. L.; Hennessy, K. M.; Kegley, S. E.; Norton, J. R. *Am. Chem. Soc.* **1989**, *111*, 3897.

72. Benson, M. T.; Cundari, T. R. *Inorg. Chim. Acta* **1997**, *259*, 91.

73. Margl, P.; Ziegler, T.; Blöchl, P. E. *J. Am. Chem. Soc.* **1996**, *118*, 5412.

74. Angermund, K.; Geier, S.; Jolly, P. W.; Kessler, M.; Krüger, C.; Lutz, F. *Organometallics* **1998**, *17*, 2399.

CHAPTER VII

ACTIVATION OF HYDROCARBONS BY PLATINUM COMPLEXES

In this chapter, we will consider the reactions of C–H compounds, such as alkanes, arenes as well as some others, with platinum complexes containing mainly chloride ligands. The reactions of alkanes with platinum(II) complexes have been the first examples of "true" homogeneous activation of saturated hydrocarbons in solution. Complexes of Pt(II) exhibit both nucleophilic and electrophilic properties, they do not react with alkanes via a typical oxidative addition mechanism nor can they be regarded as typical oxidants. Due to this, it is reasonable to discuss their reactions in a special chapter which is a bridge between previous chapters (devoted to the low-valent complexes) and further sections of the book that consider mainly complexes in a high oxidation state. Chloride complexes of platinum(IV) are oxidants and electrophiles and they will constitute the first subjects in our discussion of processes of electrophilic substitution in arenes and alkanes as well as their oxidation.

VII.1. NON-OXIDATIVE REACTIONS IN THE PRESENCE OF Pt(II)

The exchange of aromatic hydrocarbons with D_2O was discovered and studied by Garnett and Hodges [1]. The isotope exchange of alkanes in the same system was discovered two years later [2] and investigated by several groups. In many respects the behavior of arenes and alkanes happened to be unexpectedly similar to each other. Hence, we will review the results obtained for both types of hydrocarbons together, taking into consideration the peculiar behavior of aromatic molecules among the other effects exerted by the nature of organic molecules on the reactivity of C–H bonds. H–D Exchange is usually observed at temperatures of 80–120 °C. It takes place with hydrocarbons of the methane homologous series, their various derivatives (alkyl halides, carboxylic acids) and unsubstituted and substituted aromatic molecules.

Bearing in mind that Pt(II) solutions can give precipitates of metallic platinum on heating and, particularly, in the presence of reducing agents, and also that metallic platinum is a catalyst for various reactions, including H–D exchange, the suspicion may arise that the exchange reaction actually proceeds by a heterogeneous catalytic mechanism on the surface of the metallic platinum present, e.g., as colloid particles. However, a number of experimental results leaves no doubt that the H–D exchange is actually a homogeneous reaction. The addition of mineral acids stabilizes Pt(II) solutions by making the following disproportionation reaction difficult

$$2\ PtCl_4^{2-} \rightleftharpoons Pt + PtCl_6^{2-} + 2\ Cl^-$$

but does not decrease the rate of exchange; on the contrary, precisely under those conditions, where $PtCl_4^{2-}$ solutions remain homogeneous, reproducible results are obtained. Molecular oxygen and the addition of aromatic compounds (e.g., pyrene) also stabilize homogeneous platinum(II) solutions and make the results more reproducible, preventing the formation of metallic platinum. On the other hand, on the precipitation of metallic platinum, the exchange slows down or stops completely.

It must be noted that no metallic platinum is precipitated at all in the $PtCl_4^{2-}$–$PtCl_6^{2-}$ system, which oxidizes hydrocarbons at least in the early stages where sufficient amounts of Pt(IV) are present. As will be shown later, both H–D exchange and oxidation involve the same initial step of a reaction of Pt(II) with the hydrocarbon. Thus, there is no doubt at this time that homogeneous Pt(II) solutions react with hydrocarbons.

Solutions of the salt K_2PtCl_4 in aqueous acetic acid are more stable upon heating than aqueous solutions. A 50% aqueous solution of acetic acid is very convenient to use as a solvent. Furthermore, in aqueous solutions, the solubility of hydrocarbons is rather low. Besides, even with a consideration of a different solubility of the hydrocarbons, the rate of alkane isotope exchange with D_2O in 50% acetic acid turns out to be about 30 times greater than in water. Since a methyl group of acetic acid also exchanges with D_2O in the presence of Pt(II), a mixture of D_2O + CD_3COOD is used. The activation role of acetic acid can be attributed to the formation of a weak chelate complex. The ability of the acetate ion to act as a bidentate ligand is known. This has been demonstrated in the investigation of the molecular structure of the complex $RuH(CH_3COO)(PPh_3)_3$ [3]. The H–D exchange rate is lower in trifluoroacetic acid, which has a lower chelating capacity than acetic acid. By using H–D exchange catalyzed by $PtCl_4^{2-}$, we can obtain practically completely deuterated hydrocarbons if the reaction is carried out for a sufficiently long time. However, even in the initial stages of the reaction, di-, tri-, and polydeuterium-substituted molecules are formed. Thus several hydrogen atoms can be exchanged during a single contact between the hydrocarbon molecule and the platinum complex, i.e., a so-called multiple exchange takes place.

VII.1.A. SOME PECULIARITIES OF H–D EXCHANGE

For the case of alkanes in the presence of $HClO_4$, the H–D exchange rate within the $HClO_4$ concentration range of 0.2–1.0 mol dm^{-3} was shown to be virtually independent of the acidity and ionic strength. With $HClO_4$ concentrations lower that 0.2 mol dm^{-3}, the reaction rate decreases slightly. The kinetics of the exchange reaction is first order with respect to hydrocarbon concentration and of fractional order (<1) with respect to the K_2PtCl_4 concentration. As the Cl^- concentration increases, the order with respect to the chloride-ion concentration changes from 0 to −1. This dependence permits us to draw conclusions on the nature of the species reacting with the alkane molecule. It is known that, in aqueous solution, $PtCl_4^{2-}$ ions undergo an equilibrium dissociation (S is the solvent):

$$PtCl_4^{2-} \xrightleftharpoons{K_1\,(S)} SPtCl_3^- + Cl^-$$

$$SPtCl_3^- \xrightleftharpoons{K_2\,(S)} S_2PtCl_2 + Cl^-$$

The negative dependence on the Cl^- concentration shows that the species produced by $PtCl_4^{2-}$ dissociation are more active that $PtCl_4^{2-}$ itself. The dependence of the rate corresponds approximately to the change in the concentration of uncharged species (S_2PtCl_2) in the system. If for S_2PtCl_2, $SPtCl_3^-$, and $PtCl_4^{2-}$ the rates constants are, respectively k_1, k_2, and k_3, then the dependence of the effective rate constant (k) on the Cl^- concentration may be expressed by the equation:

$$k = \{k_1 + k_2K_2^{-1}[Cl^{-1}] + k_3K_1^{-1}K_2^{-1}[Cl^-]^2\}/\{1 + K_2^{-1}[Cl^-] + K_1^{-1}K_2^{-1}[Cl^-]^2\}$$

This equation is in good agreement with experimental results. An analysis of this dependence results in the following ratio of constants: $k_1 : k_2 : k_3 = 100 : 14 : 0.6$ (for cyclohexane at 100 °C), i.e., in effect, the uncharged complex S_2PtCl_2 is the most active. This may be an indication of the importance of the electrophilic properties of the platinum complex in the reaction with hydrocarbons, although an appreciable value of the rate constant of such a moiety as $PtCl_3^-$, in its reaction with alkanes in aqueous solution, does not fit the usual concepts of electrophiles very well. It has been shown that the particular Pt(II) complexes obtained, which should form positive ions upon solution in water, $[PtCl(H_2O)_3]^+$ and $[Pt(H_2O)_4]^{2+}$, react more slowly with alkanes than $PtCl_2$. This demonstrates a more complicated nature of interaction of the bivalent platinum and the hydrocarbon molecule than that of the interaction of an electrophile with a nucleophile.

This conclusion is supported by the study of the effect of the other ligands on the H–D exchange rate [4]. In particular, the effect of the ligand's nature in the complexes $PtLCl_3^-$ (L = H_2O, Cl^-, NO_2^-, DMSO, NH_3, py), $PtCl_2L_2$ (L = H_2O, Cl^-, Br^-, I^-, NO_2^-, CN^-, PPh_3), and others, has been investigated. Depending on the nature of the ligand, the rate constant changes by three orders of magnitude, decreasing as a result of the introduction of more basic and also of more highly polarizable ligands and of ligands with an increased tendency towards double bond formation.

The rate constants of H–D exchange catalyzed by the complexes formed in the system

$$PtCl_2S_2 + 2\,X^- \rightleftharpoons PtCl_2X_2^{2-} + 2\,S$$

decrease in the following order of ligands X:

$F > SO_4^- > Cl > Br > I > NO_2 > CN$

which correlates with the change of the PtX_4^{2-} complex stability constant and also with the overlap integral of the 6pσ-orbital of platinum and the p_σ-orbital of the ligand and the parameter (σ_χ) of the ligands, which characterizes their "softness". The ligand will obviously effect both the concentration of the PtX_2 complex in the solution and the electronic properties of the complex. The effect of the ligand nature on the catalytic properties of the complex may be followed by a comparison of the activity of complexes of the type $PtCl_3L$ where L = Cl^-, H_2O, NH_3, and DMSO.

The stability constants of the aquo-complexes and DMSO–platinum complexes are similar, i.e., the concentrations of the active complexes $PtCl_2$– (DMSO) and $PtCl_2(H_2O)S$ in solution are close, whereas the rate constants of H–D exchange with cyclohexane for those complexes differ by a factor of over ten. Dimethyl sulfoxide, which is a strong π-acceptor, should give rise to a higher positive charge on the central ion which, in agreement with the concept of the electrophilic character of its interaction with the hydrocarbon, should have resulted in a higher rate of H–D exchange (as in the replacement of NH_3 by H_2O). The slower rate of H–D exchange in the case of DMSO, as compared with H_2O, which is actually observed, is apparently connected with an increase (in comparison with H_2O) of the softness of the sulfur-containing ligand.

Soft ligands are well-known to stabilize Pt(II) complexes against oxidation to Pt(IV). All the ligands tested can be arranged in the following order, reflecting their influence on the rate of deuterium exchange:

$PPh_3 \cong py < DMSO < CN^- < NO_2^- < NH_3 < I^- < Br^- < Cl^- < F^- \cong H_2O$

This order is opposite to the known order for the *trans*-effect of ligands in substitution reactions of square platinum complexes. Therefore, the result obtained does not agree with the mechanism of usual electrophilic H^+ substitution in hydrocarbons, where the hydrocarbon molecule acts as a nucleophile, since a normal order of *trans*-effect would have been expected for a mechanism of this type. The order in the ligand effect is apparently determined by the strength of the covalent Pt–C bond formed and, possibly, of the Pt–H bond as well, both of which are weaker for softer ligands. Thus, the vibrational frequency of the Pt–C bond, which apparently changes approximately linearly with the energy of the bond dissociation, decreases in the order:

NO_3^-, NCS^- >> Cl^- > Br^- > NO_2^- > I^- > CN^-

A similar order of ligand effects is observed for vibrational frequencies of the Pt–H bond in PtA_2LH complexes [5]:

NO_3^- > Cl^- > Br^- > I^- >NO_2^- > SCN^- > CN^-

In both series, the order is close to opposite to that of the ligand *trans*-effects in substitution.

The isotope effect, though moderate, observed in the reaction of H–D exchange, indicates the participation of the C–H bond cleavage in the rate-determining step. In the systems CD_4–H_2O and CH_4–D_2O, the ratio of the constants is $k(CH_4)/k(CD_4) = 3.0 \pm 0.5$ (100 °C) [6], for cyclohexane $k(C_6H_{12})/k(C_6D_{12}) = 1.7 \pm 0.1$ (100 °C) [7], for benzene $k(C_6H_6)/k(C_6D_6) = 1.65$ [1b].

n-Alkanes are more reactive than branched alkanes in H–D exchange in the presence of $PtCl_4^{2-}$. In normal and branched alkanes the highest activity is observed for isolated methyl groups. For example, according to [7a], in normal pentane the rate of isotope exchange in two methyl groups is 4.5 ± 0.8 times greater than the rate of exchange in the three methylene groups. In alkanes containing primary, secondary and tertiary C–H bonds the reactivity changes in the order

1° > 2° > 3°

which is opposite to the "normal" orders of reactivity (e.g., with respect to free radical reactions). Both the methyl and methylene groups show a particularly low reactivity when they are bonded to a quaternary carbon atom (e.g., in 2,2-dimethylpropane and 2,2-dimethylbutane). Deuterium atoms are predominantly exchanged with hydrogenes in an equatorial position of 2-alkyl indanes [7b]:

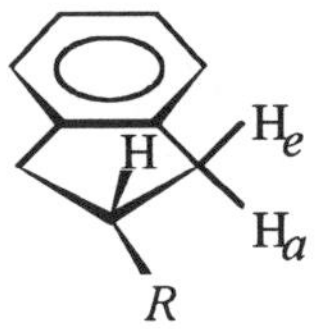

	D content, %	
	R = Me	*R* = Et
H_e	**33.5**	**21.0**
H_a	**6.5**	**4.0**

All these results are most naturally explained by the important role played by steric factors.

Arenes show higher reactivity than alkanes, but this difference is not so pronounced as in many other cases: benzene, which is the least reactive arene, exchanges its hydrogen atoms only twice as fast as the most reactive of the alkanes, cyclohexane. In the case of substituted aromatic molecules the steric effect is again very pronounced.

In an H–D exchange of monosubstituted benzenes, only *meta*- and *para*-hydrogen atoms of the ring are exchanged (with apparently similar reactivity). Thus, in the exchange only molecules containing a maximum of free deuterium atoms are formed from bromobenzene and chlorobenzene. *para*-Disubstituted benzenes do not exchange ring-hydrogen atoms at all. For example, in *p*-xylene H–D exchange occurs only in the side chain of the ring. Naphthalene exchanges hydrogen atoms only in the β-position [8].

The reactivity of C–H bonds in a side chain in the α-position to the benzene ring is higher than in other positions (which is connected apparently with a low value of the C–H bond energy). However, it is almost the same for the CH_2 group next to a phenyl as for the methyl group, where the energy of the C–H bond dissociation in the side chain is the highest. The importance of the electron donor properties of hydrocarbons in reactions of H–D exchange is stressed by Hodges *et al.* [7a], who reported a linear correlation between the logarithm of the exchange rate constant and the ionization potential of respective hydrocarbon. It is of interest that points corresponding to alkanes and aromatic hydrocarbons are located on the same straight line, which is evidence for a similar activation mechanism for the two types of hydrocarbons. However, branched hydrocarbons naturally do not show the above mentioned dependence, due to the steric effect discussed above.

In the case of halogen-substituted alkanes the relation between the rate of exchange and the ionization potential is reported to have a different slope [9]. The reaction rates of cyclic hydrocarbons (cycloheptane and cyclooctane) also do not correlate with the ionization potentials. Obviously, there is no reason to expect a simple (e.g., linear) dependence of log k on the ionization potential for all hydrocarbons and their derivatives for the case of Pt(II)-catalyzed H–D exchange with a solvent. The rate determining step must involve both the C–H bond cleavage and the formation of the M–C bond, and, probably, of the M–H bond. The facilitation of the reaction in the case of C–H bonds located in an α-position to the benzene ring as compared with simple alkanes is related to a decrease of dissociation energy of these bonds. At the same time, for aromatic hydrocarbons, higher rates of the reaction (as compared with alkanes) might be caused mainly by a more stable M–aryl bond as compared with the M–alkyl bond.

These effects are not expected to correlate with only one parameter, such as the ionization potential, even if this is considered as the potential of an electron of the orbital which is affected in the reaction with the platinum complexes. The apparent linearity of the observed log k relation with the ionization potential might be connected to a considerable extent with the fact that the effect of C–H bond differences on the rate of H–D exchange in the case of platinum complexes is relatively small. When we go from the most inert alkanes of the methane series to the most active aromatic hydrocarbons (such as phenanthrene and pyrene), the rate differs by slightly more than two orders of magnitude, whereas the ionization potential changes by over 5 eV. This would have changed the reaction rate by a factor of approximately 10^{64} at 120 °C if the latter were determined entirely by the energy of electron transfer from the hydrocarbon molecule.

Another approach to the quantitative correlation between the structure and the reactivity of hydrocarbons may be the use of the Taft correlation equation. A consideration of the dependence of the exchange rate on the constant σ* for a number of functionally substituted methane and ethane derivatives has shown [10] that the two-parameter equation is obeyed:

$$\log (k_S/k_0) = \rho^*\sigma^* + n\psi$$

Besides the dependence on the polar parameter σ* (which characterizes the inductive effect of a substituent) the reaction rate depends also on the resonance term ψ which characterizes the "conjugation" of the α-substituent with the radical reaction center or with the π-system (n is equal to the number of such substituents). The value of ρ^* has been found to be equal to -1.4, and shows that Pt(II) acts as a moderate electron acceptor with respect to hydrocarbon, which acts as a donor in the reaction. The presence of a resonance term in the correlation equation indicates the stabilizing effect of the electron pair of the substituent in the cleavage of the C–H bond in the interaction with the platinum complex. This could have taken place in the homolytic cleavage of the C–H bond to produce a quasi-free radical in the transition state.

VII.1.B. MULTIPLE H–D EXCHANGE

The H–D exchange between alkanes and the solvent catalyzed by platinum(II) is multiple, i.e., the hydrocarbon molecule entering into the reaction may exchange several hydrogen atoms with deuterium without leaving the coordination sphere of a catalyst. Table VII.1 presents the value of the parameter *M* which shows the average number of hydrogen atoms replaced by deuterium during a contact of a hydrocarbon molecule with the catalyst [7a].

For alkanes this parameter has a value of 1.4–1.7, for benzene it increases to 3.5, and for naphthalene and phenanthrene it drops down to 1. The value of *M* increases only slightly with decreasing temperature.

The phenomenon of multiple exchange is well known for heterogeneous catalysis and is usually explained by the intermediate formation of surface π-complexes from alkyl metal derivatives. A similar mechanism may be proposed for the homogeneous exchange of ethane and other hydrocarbons (Scheme VII.1).

Each **VII-1** → **VII-2** → **VII-1** cycle amounts to the introduction of one deuterium atom into the alkane molecule in the deuterated solvent ($[D^+] >> [H^+]$). If the rates of conversion of **VII-2** → **VII-1** and **VII-2** → **VII-3** are comparable, then every cycle **VII-1** → **VII-2** → **VII-1** may involve several cycles of **VII-2** → **VII-3** → **VII-2** which leads to the insertion of several deuterium atoms during the contact of one molecule with the platinum complex.

Multiple exchange in the case of methane may be accounted for in the same way assuming the intermediate formation of a methylene–platinum complex, $Pt{=}CH_2$. The distribution of deuterohydrocarbons, which does not depend on time

Table VII.1. Parameters of multiple exchange catalyzed by Pt(II) in CH_3COOD–D_2O mixture at 80–100 °C.

Hydrocarbon	% D	*M*
Methane	7.80	1.48
Ethane	5.56	1.70
Propane	7.16	1.63
Butane	3.33	1.62
Pentane	6.39	1.65
Hexane	6.65	1.50
Cyclopentane	6.45	1.68
Cyclohexane	6.31	1.42
Benzene	21.18	3.50
Naphthalene	10.92	1.02
Phenanthrene	13.3	0.95

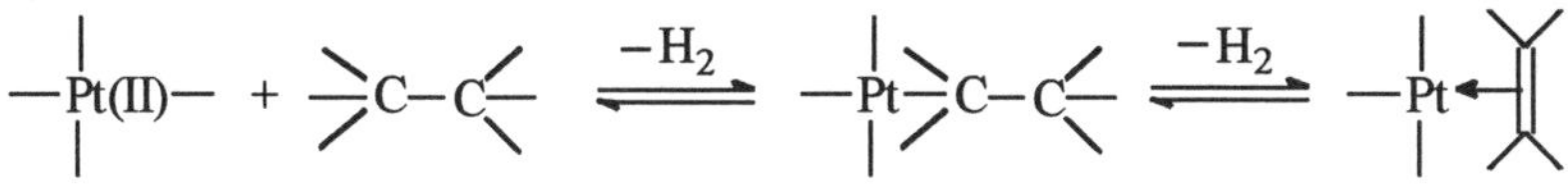

VII-1 **VII-2** **VII-3**

Scheme VII.1. Mechanism of multiple H–D exchange in higher alkanes.

during the initial reaction stages, can be evaluated by a method of steady-state concentrations (Scheme VII.2). The amount of each of the deuteromethane present in the mixture

$$\alpha_i = \frac{d_i}{\Sigma d_i}$$

is completely determined by the factor k_3/k_2. A calculation of deuterocarbon distribution (α_i) for methane and ethane in a 1 : 1 CH_3COOD–D_2O mixture shows that Scheme VII.2 adequately describes the experimental results [11].

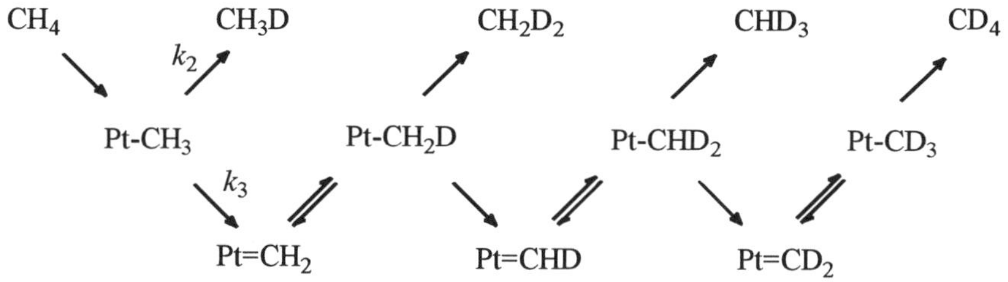

Scheme VII.2. Mechanism of multiple H–D exchange in methane.

The multiple exchange in alkanes, which is very sensitive to reaction conditions, may be used to prove the formation of alkyl platinum derivatives as intermediate products of the exchange. Alkyl platinum derivatives can be produced using organomercuric compounds via the exchange reaction

$$RHgBr + PtCl_2 \longrightarrow RPtCl + HgClBr$$

The alkanes produced in this interaction are certainly formed as a result of a hydrolytic cleavage of alkyl platinum derivatives. The close agreement between the ratio of methanes-d_n formed in platinum(II) methylation and in direct hydrogen-deuterium exchange with alkanes [12] is a sound support for the formation of alkyl platinum derivatives in the homogeneous activation of alkanes.

A distribution of deuterohydrocarbons in the case of methane and ethane in a 1 : 1 mixture of CH_3COOD–D_2O is smoothly descending, i.e., the greater the number of deuterium atoms in the hydrocarbon, the lower will be the proportion of the corresponding hydrocarbon. Experiments on the hydrogen-deuterium exchange of ethane in pure D_2O have demonstrated, however, an unusual stepwise distribution of deuteroethanes which may be visualized in the form of two groups, α_1–α_3 and α_4–α_6. The distribution of each of these group ascending, although the total content of ethane in the group α_4–α_6 is considerably smaller than that in the group α_1–α_3. It was shown by mass-spectrometry of the deuteroethanes that the asymmetrical molecules CHD_2CH_3, CD_3CH_3, and CD_3CH_2D prevail in the d_2–d_4 deuteroethanes. Such a distribution for the H–D exchange in water (predominance of d_1–d_3 over d_4–d_6 and the same ascending pattern in each group) suggests that the exchange takes place first in one of methyl groups of ethane and then in the other, which is thus not equivalent to the first.

Obviously, such a distribution is inconsistent with the mechanism involving an intermediate π-complex formation in which both carbon atoms should be equivalent. There is a logical explanation of a stepwise distribution if it is assumed that the multiple exchange in ethane is similar to that in methane involving the intermediate formation of complexes with the carbene, Pt=$CHCH_3$, formed as a result of α-elimination of a proton. In this case, not more than three atoms of one methyl group will initially be exchanged and then, provided there is an elementary act of rotation of the alkyl group to form a platinum–carbon bond with another carbon atom, exchange in another methyl group will produce polydeuterated molecules with up to six atoms exchanged, i.e., C_2D_6. The isomerization will probably proceed via the intermediate formation of a π-complex (Scheme VII.3). A calculation of the deuteroethane distribution by the method of steady-state concentrations according to Scheme VII.3 is in good agreement with the experimental data obtained for an aqueous solution.

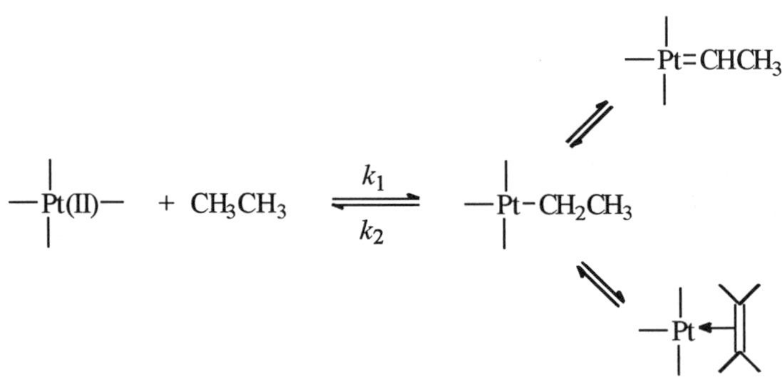

Scheme VII.3. The interaction of ethane with a platinum(II) complex.

The smoothly descending distribution in deuteroethanes (obtained for solutions in 50% acetic acid) may be explained by both mechanisms, the one involving the ethylidene complex and the other consistent with only π-complex formation. In effect, if we accept that the conversion

$$Pt{-}CH_2CH_3 \longrightarrow Pt{=}CHCH_3$$

occurs more slowly than the turnover of the alkyl group

$$Pt{-}CH_2{}^{1}CH_3{}^{2} \rightleftharpoons Pt{-}CH_2{}^{2}CH_3{}^{1}$$

the carbene mechanism and that with the participation of only π-complexes will give the same distribution.

However, it is more natural to assume that the carbene derivative is present both in aqueous and in acetic acid solutions, and that it is the ratio of rate constants that changes, than to assume that the mechanism changes completely with a change of the solvent composition. In the case of other alkanes, a stepwise distribution is also observed with a clearly distinguishable group α_1–α_3, which is an indication of a predominant exchange occurring in one of the methyl groups. This shows that platinum is more likely to be bounded with one (terminal) carbon atom with a transition 1-(alkyl) → 1,1-(carbene) (α-elimination) than to form a 1,2-(olefin)-bonded fragment (β-elimination). The isomerization proceeding via 1,2-bonded fragments (or π-complexes) leads to an extensive multiple exchange to produce polydeuterated alkanes via the mechanism 1,1 → 1,2 → 2,2 or 1,1 → 1 → 1,2 → 2,2. In the case of propane it is necessary to take into account the possibility of the formation of the *n*- and *i*-propyl platinum complexes. An analysis of deuteropropane distribution makes it possible to determine the relative rates of platinum(II) reactions with primary and secondary carbon atoms. For this purpose the distribution of deuteropropanes in the case of platinum alkylation with *n*-PrHgBr and *i*-PrHgBr was investigated. The distribution of the deuteropropanes obtained by the H–D exchange of propane coincides with that detected for *n*-PrPt- and *i*-PrPt-derivatives if we accept that they are initially formed with a relative probability of 0.75 and 0.25, respectively. Taking into consideration that the propane molecule contains six primary and two secondary C–H bonds, it may be concluded that the selectivity in the attack of primary and secondary bonds in propane is in accordance with the ratio 1° : 2° = 1 : 1. This demonstrates a much higher relative reactivity of a secondary C–H bond in propane as compared with the reactivity of secondary C–H bonds in pentane where the probability of attack of a primary C–H bond is 4.5 times as great as that at a secondary bond. This can be caused by the smaller steric hindrances of the secondary C–H bond in propane bound to two methyl groups in comparison with secondary C–H bonds in pentane bound to, at least, one long alkyl group. Similarly, in linear alkanes, the methylene groups of the second carbon atoms ($\omega-1$ position) should show a higher reactivity than all the other CH_2 groups on analogy with some cases of biological oxidation (see below).

The presence of a second step α_4–α_5 in the mass-spectrum of deuteropropanes in the platinum alkylation by *n*-PrHgBr is consistent with the 1,2-shift of the platinum along the carbon atoms. However, the considerable proportion of α_6 in the case of platinum alkylation using both *n*-PrHgBr and *i*-PrHgBr confirms the strong probability of a 1,3-shift of platinum. This shift may include a four-membered metallacycle with platinum participation [13]. Intermediate metallacycles are apparently formed in the multiple exchange of such hydrocarbons as isobutane and isooctane, for which a 1,2-shift should be sterically hindered. A 1,3-shift in the case of these hydrocarbons is supported by the triplet groups in the mass-spectra of the corresponding hydrocarbons, and the considerable contribution to the distribution of the α_4–α_6 group shows the ease of the 1,3-shift. The presence of a group of lines α_7–α_9 for isobutane, and absence of an α_{10} line, in contrast to *n*-butane, shows that the tertiary C–H bond does not participate in the exchange in either the primary attack or in the 1,2-shift. The presence of a group of lines α_7–α_9 in the isooctane exchange shows the possibility of a 1,4-shift. This result also demonstrates that a NMR-observed small exchange in methyl groups bonded to a quaternary carbon atom is at least partially due to a multiple exchange of the molecule, the rate of the initial platinum(II) interaction with these groups being even lower than follows from the NMR data.

The differences in the factor of multiple exchange for aromatic molecules could be explained to a considerable extent by steric factors. The large value of the factor *M* for benzene testifies for a high probability of the 1,2-shift (possibly via the intermediate formation of dehydrobenzene):

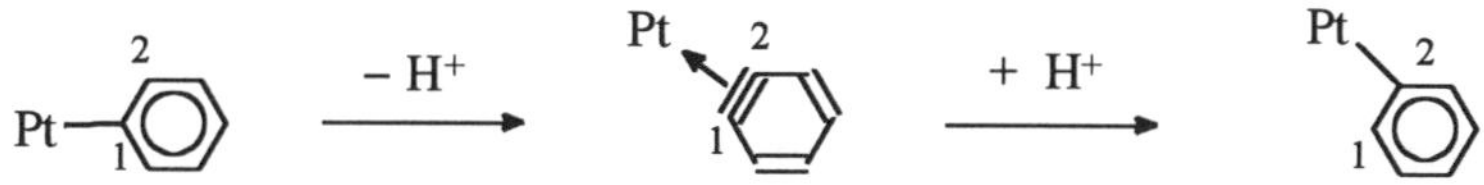

The hydrogen-deuterium exchange in naphthalene occurs selectively in the β-position, which is accounted for by the intermediate formation of β-naphthyl derivatives. The α-naphthyl platinum derivatives are apparently sterically hindered, retarding the 2 → 1 shift.

It may be concluded that the data on multiple exchange indicate the intermediate formation of complexes with carbenes and metallacycles together with alkyl platinum complexes (Scheme VII.4).

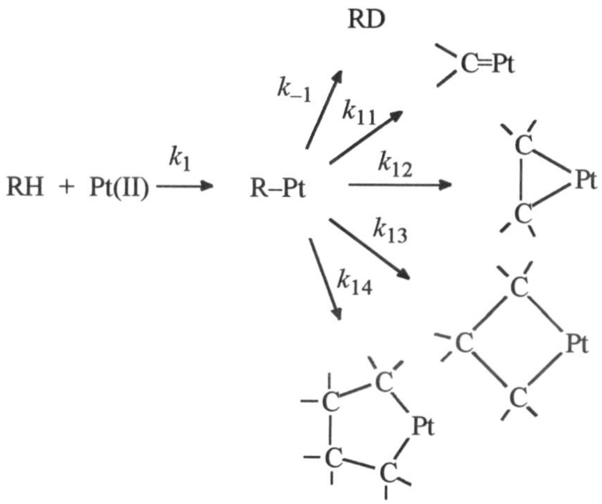

Scheme VII.4. The interaction of alkanes, RH, with a platinum(II) complex.

An accurate calculation of the relative rates of the elementary reactions from the data on multiple exchange for hydrocarbons containing more than two carbon atoms is difficult. However, the analysis of the distribution in the absence of steric hindrance shows that the rate constants decrease in the order:

$k_{11} >> k_{12} > k_{13} > k_{14}.$

An analysis of ascending distribution in the case of methane H–D exchange in aqueous solution, from the viewpoint of several possible mechanisms, demonstrates that the reaction proceeds with the participation of the methylene complex, but no evidence for the formation of the Pt≡CH methyne complex was obtained.

Rudakov and coworkers have proposed another mechanism for multiple exchange [14]. The authors found that the multiple exchange factor M for H–D exchange of cyclohexane increases from 1 to 1.8 with an increase of acidity. Proceeding from the assumption that the dissociation rate

$$\text{Pt–CHR}_2 \longrightarrow \text{Pt=CR}_2 + \text{H}^+$$

does not depend on H^+ concentration, the authors have come to the conclusion that the multiple exchange is due to the formation of the complex of platinum(II) with an alkane **VII-4**:

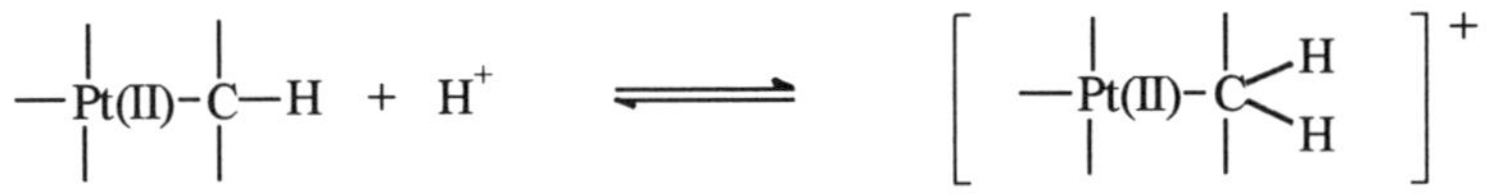

VII-4

According to this hypothesis, the rate of the complex decomposition to form free alkane must be comparable with its dissociation rate into the alkyl platinum complex and a proton:

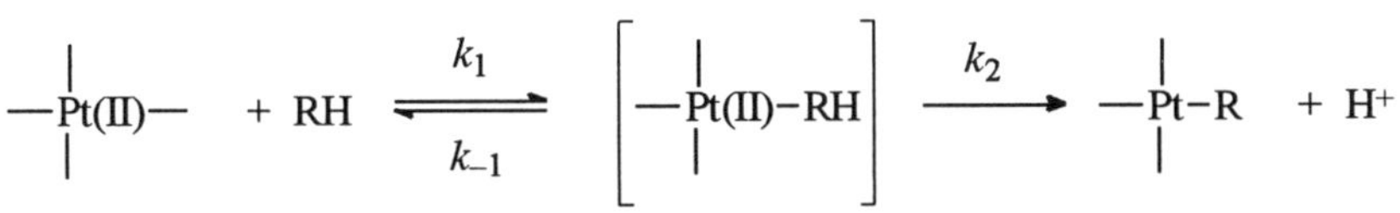

In steady-state conditions:

$$k_2\,[\text{Pt(II)RH}] + k_{-1}\,[\text{Pt(II)RH}] = k_1\,[\text{Pt(II)}]\,[\text{RH}]$$

and, since both terms of the left side of the equation are comparable, the rate of the complex formation k_1 [Pt(II)] [RH] must be very slow and close to the total rate of H–D exchange. Moreover, the rate of the complex decomposition must be also rather slow, implying that it is comparatively stable. This seems unlikely for the reaction of alkanes with platinum(II) complexes. It looks more plausible that the rate of the carbene-platinum complex formation increases with acidity, e.g., as a result of protonation:

$$X_2Pt(OH)\text{–}R + H^+ \rightleftharpoons [X_2Pt(OH_2)\text{–}R]^+$$

An increase of positive (or decrease of negative) charge of the complex could increase the rate of the carbene formation in the reaction of the C–H bond dissociation in the Pt(II) coordination sphere to form a proton and a carbene.

Concluding this section which is devoted to multiple exchange in alkanes, we should note that more recently, peculiarities of this phenomenon have been explained by Zamaschikov, Labinger and co-workers. Thus the experiments with methyl platinum(IV) complexes demonstrated that reversible deprotonation of the alkylhydridoplatinum(IV) complexes can lead to the multiple H–D exchange with the solvent (see below, particularly, Section VII.5.B, page 298).

VII.2. OXIDATION OF ALKANES BY Pt(IV) IN THE PRESENCE OF Pt(II)

It has been shown by Hodges and Garnett in a study of hydrogen-deuterium exchange in benzene catalyzed by platinum(II) that when hexachloroplatinic acid is added, a homogeneous reduction of platinum(IV) by benzene takes place [1b]. The formation of platinum(II), chlorobenzene, and small amounts of diphenyl has been detected. The study of the H–D exchange of alkanes showed that added Pt(IV) also oxidized acetic acid (present as a solvent) to chloroacetic acid [15]. It also has been demonstrated that if an aqueous solution of H_2PtCl_6 and Na_2PtCl_4 is heated in the presence of alkanes, a mixture of isomeric alkyl chlorides, alcohols and ketones is formed while platinum(IV) is reduced to platinum(II).

VII.2.A. MAIN PECULIARITIES OF ALKANE OXIDATION BY THE SYSTEM “Pt(IV)+Pt(II)”

Alkanes reduce platinum(IV) to platinum(II) under conditions similar to those of hydrogen–deuterium exchange with the solvent (at 100–120 °C), platinum(II) being the necessary catalyst for the reaction [15]. Products of the reactions are mainly chloroalkanes together with alcohols, ethers, ketones, and acids for linear hydrocarbons. When cyclic alkanes (cyclohexane and decalin) are oxidized, aromatic hydrocarbons (benzene, naphthalene) are obtained. The π-complex of hex-1-ene with platinum(II) has been isolated in 1% yield from the mixture of products formed in the reaction with *n*-hexane [16]:

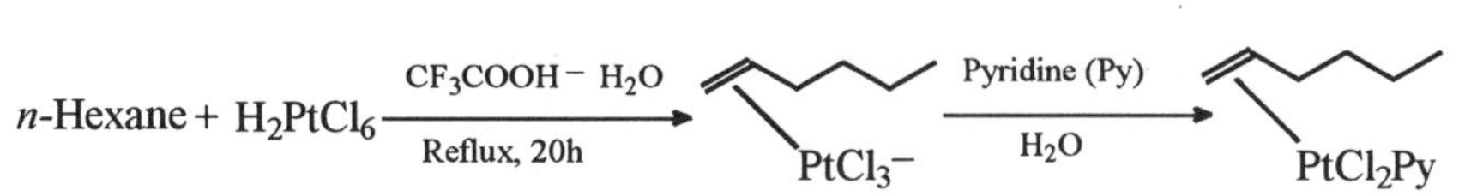

The same complex also has been obtained by the reaction between $PtCl_6^{2-}$ and hexanol-1 or hex-1-ene in CF_3COOH–H_2O. More recently it has been shown [17a] that Pt^{II}–C_2H_5 and $Pt^{II}(CH_2{=}CH_2)$ species are viable intermediates in the conversion of ethane to ethanol and ethane-1,2-diol.

Platinum(II) salts as well as the system Pt(II) + Pt(IV) have been used as catalysts in the oxidation of C–H compounds with various strong oxidants. Thus the reaction of methane with chlorine in water at 125 °C in the presence of platinum chlorides affords methyl chloride which is partially hydrolyzed in situ to methanol [17b]. A combination of Pt(II) and metallic platinum oxidized ethane in the presence of oxygen to a mixture of acetic and glycolic acids [17c,d].

VII.2.B. KINETICS OF THE OXIDATION

Valuable information can be derived from the comparison of reaction rates of various hydrocarbons, measured by the decrease of their concentration due to oxidation [18]. It is essential to note the important role of steric factors which manifest themselves in a very similar way to the case of H–D exchange. Thus methyl and methylene groups adjacent to the quaternary carbon atom practically do not participate in the oxidation. 2,2-Dimethylpropane and 2,2,3,3-tetramethylbutane were shown to be oxidized by platinum(IV) much more slowly than alkanes with the same number of carbon atoms but with no quaternary carbon. According to [18], the number of carbon atoms (n_0) accessible to attack by the platinum compounds can be approximately determined by the formula

$$n_0 = n - n_{tert} - 5n_{quat}$$

where n is the total number of carbon atoms, and n_{tert} and n_{quat} are the numbers of tertiary and quaternary carbon atoms in the molecule, respectively.

The value k/n decreases for normal alkanes with increasing n. This effect reflects the growth of steric barriers as a result of the hydrophobic molecule rolling itself up into a "coil". For n-octane, almost half the carbon atoms do not participate in the reaction. It should be noted that the rate constant of the oxidation remains practically unchanged despite the increase of the number of carbon atoms from pentane to octane. This may mean that only the C–H bonds of the primary and secondary carbon atoms from the hydrocarbon are reactive, i.e., ω- and ω–1-carbon atoms. If this is so, then steric factors in the alkane oxidation by platinum complexes produce the same effect as they do in the selective oxidation under the action of monooxygenase with participation of cytochrome P450 and methane monooxygenase.

Only the primary methyl groups (if not directly attached to the quaternary carbon atom) and the secondary methylene group adjacent to the methyl group are active in the oxidation. The remaining methylene groups, bonded to longer alkyl groups other then methyl, are practically inactive. There is comparatively little difference between the rate constants relating to one C–H bond in the reactive methyl or methylene groups of different alkane molecules. Only the C–H bond in methane is somewhat less active than that in other hydrocarbons but, in this case also, the difference is only a factor of two.

Similar to hydrogen–deuterium exchange, neither $[Pt(H_2O)]_4^{2+}$ nor $PtCl_4^{2-}$ is active in the oxidation of alkanes. The values of k_1, k_2, k_3 are not very different from each other. The positive doubly-charged ion shows an adequate activity for benzene whereas the maximum activity is found for the singly-charged $[PtCl(H_2O)_3]^+$ ion. This evidently reflects the more classical electrophilic nature of the reaction displayed in the case of positively charged platinum complexes interacting with aromatic compounds. In the case of alkanes, the Pt–C bond formation and C–H bond cleavage should be a synchronous process (i.e., the "electrophilic" addition of platinum to the carbon atom should be accompanied by "nucleophilic" addition of platinum or one of its ligand to the proton). Hence, the positive charge is not of practical importance here.

The reaction rate constants of the less electrophilic species – the uncharged platinum complex $PtCl_2(H_2O)$ and the negative ion $[PtCl_3(H_2O)]^-$ – are practically identical for benzene and cyclohexane and that the ratio of the rate constants of the reactions of cyclohexane and isobutane remains constant for the reaction with different platinum complexes present in solution.

A number of similarities existing in the reactions of alkane hydrogen-deuterium exchange with the solvent and of alkane oxidation under the action of platinum(IV) (e.g., the similarity in the selectivities of attack in reactions with various C–H bonds) leads us to the conclusion that the same stage is involved in both cases. This stage must include the alkane's reaction with the platinum(II) complex acting as a catalyst. It is natural to assume that the common stage must be the formation of alkyl platinum derivatives, which are intermediates in the oxidation and exchange reactions. The kinetic data confirm this assumption. A comparison of the rate constants of H–D exchange and hydrocarbon oxidation reveals the direct correlation between them [18]. Under certain conditions the rate constants of oxidation and exchange and their activation energies may be practically the same (Table VII.2).

Table VII.2. Parameters for exchange and oxidation of various C–H compounds in CH_3COOD–D_2O.

	$k \times 10^4$, dm mol^{-1} s^{-1} [a]		E_a, kcal mol^{-1}	
C–H Compound	Exchange	Oxidation	Exchange	Oxidation
Hexane	18	16	–	–
Benzene	60	55	25.7	25
Acetic acid	0.9	1.25	23.7	22.5

[a] At 100 °C.

Direct evidence for a common stage in the oxidation and exchange of alkanes in the presence of platinum(II) complexes has been obtained [19], where both the reactions were studied jointly at different concentrations of acid and platinum(IV) complexes. The total of the exchange and oxidation rates was shown to remain constant and independent of the concentration of acid and platinum(IV), but their ratio changes with a change of its concentration. With increasing Pt(IV) concentration and constant acid concentration, the rate of the oxidation process increases, whereas with increasing acid concentration (and constant Pt(IV) concentration) there is an increase in the rate of the exchange reaction. These data are in agreement with the common mechanism of the formation of alkyl platinum derivatives in the first step:

$$Cl_2Pt(II) + RH \longrightarrow R\text{–}PtCl + H^+ + Cl^-$$

followed by two competing reactions:

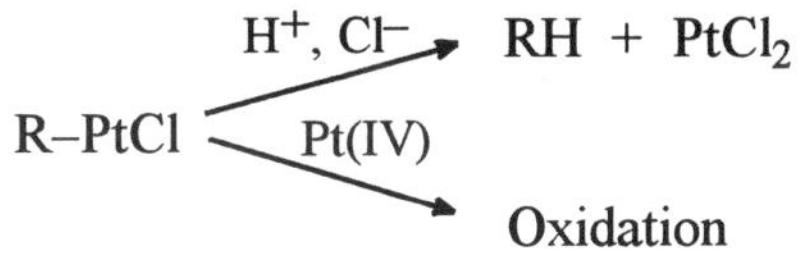

The kinetics of the catalytic chlorination of the C–H bond have been studied extensively for the case of the $PtCl_6^{2-}$ reaction of acetic acid [20]. Similar to the hydrocarbon reactions, the reaction is catalyzed by platinum(II) chloride complexes and proceeds almost quantitatively according to the scheme:

$$PtCl_6^{2-} + CH_3COOH \xrightarrow{PtCl_4^{2-}} PtCl_4^{2-} + ClCH_2COOH + HCl$$

In the absence of platinum(II), the Pt(IV) concentration decreases autocatalytically, the observed induction period being removed by the addition of bivalent platinum. The reaction is convenient for kinetic study (because of the absence of by-products) and obviously proceeds according to a mechanism similar to that for the oxidation of alkanes. The reaction is first order with respect to the platinum(II) and acetic acid concentrations, and is retarded on the addition of acid and Cl^- ions, its rate being inversely proportional to the square of chloride-ion concentration at high Cl^- concentrations. The order of reaction with respect to Pt(IV) changes from 0 to 1. The mechanism suggested for this reaction is evidently common for various C–H bond containing compounds including saturated hydrocarbons:

$$PtCl_4^{2-} \rightleftharpoons PtCl_3^- + Cl^- \quad \text{(VII.1)}$$

$$PtCl_3^- \overset{k}{\rightleftharpoons} PtCl_2 + Cl^- \quad \text{(VII.2)}$$

$$PtCl_2 + RH \underset{k_{-1}}{\overset{k_1}{\rightleftharpoons}} RPtCl + H^+ + Cl^- \quad \text{(VII.3)}$$

$$RPtCl + PtCl_6^{2-} \xrightarrow{k_2} RCl + 2\,PtCl_3^- \quad \text{(VII.4)}$$

The rate-determining steps may be (VII.3) or (VII.4), depending on the conditions, since reactions (VII.1) and (VII.2) proceed rapidly to reach equilibrium. At temperatures below 100 °C and at sufficiently high platinum concentrations, the reaction rate depends neither on Pt(IV) concentration nor on acidity and is apparently determined by the reaction (VII.3) of the substrate interaction with platinum(II). At higher temperatures and low Pt(IV) concentrations, there appears to be a dependence on Pt(IV) concentration (up to direct proportionality) and on the acidity (inverse proportionality). In this case the rate determining step is stage (VII.4), the platinum(II) alkyl derivative interaction with platinum(IV).

The following kinetic equation, based on the scheme above, can be obtained using the steady-state concentrations method and assuming reaction (VII.1) shifted to the right:

$$v = k_1 k_2 K[\mathrm{RH}](\mathrm{Pt(II)})[\mathrm{Pt(IV)}] \,/\, \{([\mathrm{Cl}^-] + K)(k_{-1}[\mathrm{H}^+][\mathrm{Cl}^-] + k_2[\mathrm{Pt(IV)}])\}$$

This equation is in good agreement with the experimental data. The temperature dependence of the reaction rate corresponds to two activation energies. Over the temperature range of 80–100 °C the effective activation energy is $E_{\mathrm{eff}} = 22.5$ kcal mol^{-1}, and over the range of 100–120 °C $E_{\mathrm{eff}} = 8.6$ kcal mol^{-1}. At a temperature below 100 °C and low chloride concentration, when the reaction rate is practically independent of [Cl$^-$], the equilibrium (VII.2) seems to be almost completely shifted to the right and we can disregard [Cl$^-$] in comparison with K

and $k_1[H^+]$ [Pt(II)] as compared with k_2[Pt(IV)]. In this case, the kinetic equation corresponds to

$$v = k_1[RH][[Pt(II)]$$

In this temperature range the activation energy of the oxidation is close to that of the exchange. At temperatures above 100 °C, the reaction of Pt(II) alkyl derivatives with platinum(IV) becomes the rate-determining step. The low values of both the activation energy and the preexponential factor correspond to the nature of process (VII.4), which is evidently complicated and involves, as a first stage, the formation of a platinum(IV) alkyl derivative with a subsequent reductive elimination.

$$RPtCl + PtCl_6^{2-} \longrightarrow [RPtCl\text{–}PtCl_6^{2-}] \longrightarrow RPtCl_3 + PtCl_4^{2-}$$

$$RPtCl_3 \longrightarrow RCl + PtCl_2$$

This oxidation process – at least in the case when R = CH_2CH_2OH – involves electron transfer (see below).

The kinetics of the reaction with participation of methane have been investigated in detail [21a]. The study of methane (pressure ca. 100 atm.) oxidation at 120 °C under the action of H_2PtCl_6 (in the presence of $PtCl_4^{2-}$) in a mixture of water and trifluoroacetic acid showed the solution to contain a platinum complex producing methane if decomposed in the presence of a reducing agent, such as hydrazine hydrate or sodium boron hydride. When decomposing this complex by DCl in D_2O, CH_3D is produced. Upon heating in water the complex produces methyl chloride and methanol, i.e., the same products as those formed in methane oxidation. The kinetics of the methyl platinum(IV) complex concentration change were followed together with the decrease of platinum(IV) concentration and the increase of the products' (methyl chloride and methanol) concentration in the reaction of $PtCl_6^{2-}$ + $PtCl_4^{2-}$ with methane in water. The reaction rate

$$-\,d[Pt(IV)]/dt = d([CH_3Cl] + [CH_3OH])/dt$$

was found to be proportional to the methyl platinum(IV) complex concentration, in agreement with the suggestion that the latter is the intermediate. However, the most direct proof that the Pt(IV)–CH_3 complex is the sole intermediate of the reaction came out of measurements of its decomposition, which produces methyl chloride and methanol. The rate of the decomposition measured at different temperatures, when extrapolated to the temperatures of the reaction of $PtCl_6^{2-}$ with methane catalyzed by Pt(II), exactly coincides with the total reaction rate for the concentration of the Pt(IV)–CH_3 complex observed in the reaction mixture. Thus, knowing the decomposition rate constant measured separately and the concentration of the Pt(IV)–CH_3 complex, the reaction rate of methane oxidation by $PtCl_6^{2-}$ can be correctly predicted, leaving no doubt that the Pt(IV)–CH_3 complex is the sole intermediate. The activation energy for the decomposition of the methyl complex σ-$CH_3PtCl_5^{2-}$ was found to be 26 kcal mol^{-1}, the decomposition rate constant at 120 °C is 0.14 s^{-1}, and the maximum concentration of the complex is 7.7×10^{-5} mol dm^{-3}. Hence it is possible to calculate the rate of decomposition of the complex at its maximum concentration ($W_{dec} = 1.1 \times 10^{-5}$ mol $dm^{-3} \cdot s^{-1}$). This calculated rate is almost exactly equal to the observed rate of platinum(IV) reduction (and methane oxidation) at 120° C.

It is interesting that dimethylplatinum(II) complexes can be oxidized to platinum(IV) alkoxides by dioxygen (or peroxides) [21b]. This finding suggests that these reactions constitute a promising approach to the catalytic oxidation of alkanes.

VII.3. ACTIVATION OF SOME OTHER C–H COMPOUNDS

Water-soluble organic compounds are selectively oxidized by aqueous solutions of platinum salts [22]. For example, *p*-toluenesulfonic acid undergoes stepwise hydroxylation to the corresponding alcohol and aldehyde while *p*-ethylbenzenesulfonic acid is functionalized at both the benzylic and methyl positions. Ethanol affords a variety of products (about 50% of the substrate is converted into products):

$$CH_3CH_2OH \longrightarrow \underset{3\%}{HOCH_2CH_2OH} + \underset{5\%}{ClCH_2CH_2OH} +$$

$$\underset{1\%}{ClCH_2CH(OH)_2} + \underset{21\%}{CH(OH)_2CH(OH)_2} + \underset{4\%}{(\pi\text{-}CH_2{=}CH_2)PtCl_3^-} +$$

$$CH_3COOH + HOCH_2COOH + CH(OH)_2COOH +$$

6% 4% trace

$$ClCH_2COOH + CO_2$$

trace 8%

1-Propanol is also significantly attacked at the methyl position. In contrast to most alkane conversion systems, the reactivity of a C–H bond of a methyl group is at least as high as that of the C–H bond in the α-position to an oxygen atom. The relative rate of C–H bond activation by the Pt(II) ion decreases in the order [23a]:

$$H–CH_2CH_3 > H–CH_2CH_2OH > H–CH(OH)CH_3$$

The platinum(II) ion, in the presence of a Pt(IV) complex, catalyzes the hydroxylation of unactivated C–H bonds of aliphatic carboxylic acids in water [23b]. The following order of reactivity has been evaluated:

$$\alpha\text{-C-H} \ll \beta\text{-C-H} < \gamma\text{-C-H} \geq \delta\text{-C-H} \approx \varepsilon\text{-C-H}$$

This reaction formed lactones, together with hydroxyacids (yields are given relative to Pt(II)):

39% 140% 19%

A system consisting of aqueous $PtCl_4^{2-}$ as the catalyst and phosphomolybdic acid as the redox mediator in a carbon cloth anode electrochemical cell, electrocatalytically hydroxylated *p*-toluenesulfonic acid according to the equation [24]:

$$R–H + H_2O \longrightarrow R–OH + 2\,e^- + 2\,H^+$$

Neutral five-coordinated platinum(II) complexes, $Pt(SnCl_3)_2[P(OR)_3]_3$ containing π-acceptor ligands (and hence having enhanced cationic character of the metal ion) have been used for thermal catalytic dehydrogenation of cyclo-octane [25]. The authors attributed the high activity of these complexes in the dehydrogenation reaction to both enhanced electrophilicity and stability against reduction.

VII.4. PHOTOCHEMICAL REACTIONS OF $PtCl_6^{2-}$ WITH ALKANES

The reaction between the $PtCl_6^{2-}$ ion and alkanes can be induced by irradiation [26]. When a solution of hexachloroplatinic acid and *n*-hexane in acetic acid is irradiated by light ($\lambda > 300$ nm) [26a,c,d] or γ-quanta [26b], a π-complex of hex-1-ene with platinum(II) is formed in addition to isomeric chlorohexanes. This complex has been isolated in the form of the pyridine adduct, (hex-1-ene)$PtCl_2Py$. The yield of the π-complex in the γ-induced reaction reaches 17% based on Pt. The photochemical reaction is of first order with respect to hexane. It is interesting that the photochemical reaction with hex-2-ene affords the π-complex of hex-1-ene (Scheme VII.5). This complex can be also prepared by the reaction of hex-1-ene with $PtCl_6^{2-}$ under irradiation. It should be noted that the photoinduced reaction of $PtCl_6^{2-}$ or $PtCl_4^{2-}$ with olefins is a convenient method for the synthesis of various π-olefin complexes of platinum(II) [27]. The photostimulated reaction of $PtCl_6^{2-}$ with stilbene does not occur, possibly due to steric restrictions.

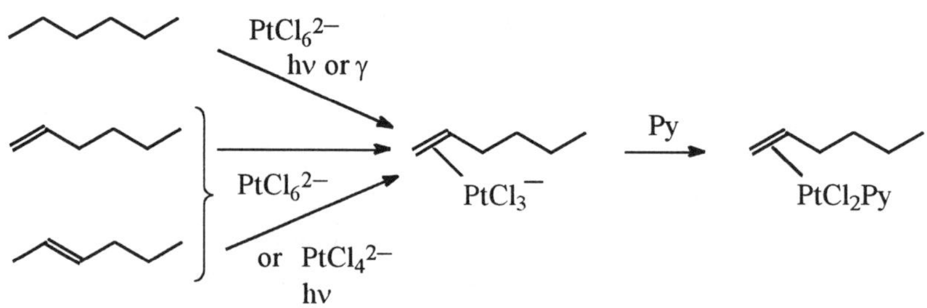

Scheme VII.5. Photochemical reactions of hexane and hexenes with platinum(IV) and platinum(II) complexes.

The following mechanism has been proposed for the photo- and γ-induced dehydrogenation of *n*-hexane [26c,d]:

$$Pt^{IV}Cl_6^{2-} \xrightarrow{h\nu} \{Pt^{III}Cl_5^{2-} + Cl\bullet\}$$

$$n\text{-}C_6H_{14} + Cl\bullet \longrightarrow n\text{-}C_6H_{13}\bullet + HCl$$

$$n\text{-}C_6H_{13}\bullet + Pt(IV) \longrightarrow n\text{-}C_6H_{13}^+ + Pt(III)$$

$$n\text{-}C_6H_{13}^+ + Cl^- \longrightarrow n\text{-}C_6H_{13}Cl$$

$$n\text{-}C_6H_{13}\bullet + ClPt(IV) \longrightarrow n\text{-}C_6H_{13}Cl + Pt(III)$$

$$2\ Pt(III) \longrightarrow Pt(II) + Pt(IV)$$

$$n\text{-}C_6H_{13}^+ \longrightarrow n\text{-}C_4H_9CH{=}CH_2 + H^+$$

$$2\ n\text{-}C_6H_{13}\bullet \longrightarrow n\text{-}C_4H_9CH{=}CH_2 + n\text{-}C_6H_{14}$$

$$n\text{-}C_4H_9CH{=}CH_2 + Pt(II) \longrightarrow \pi\text{–}(n\text{-}C_4H_9CH{=}CH_2)Pt(II)$$

$$n\text{-}C_6H_{13}\bullet + Pt(III) \longrightarrow \sigma\text{-}(n\text{-}C_6H_{13})\text{–}Pt(IV)$$

$$\sigma\text{-}(n\text{-}C_6H_{13})\text{–}Pt(IV) \longrightarrow \pi\text{-}(n\text{-}C_4H_9CH{=}CH_2)Pt(II) + H^+$$

The formation of the radicals Pt(III) and R• has been detected by EPR through modeling using the photoreaction between $PtCl_6^{2-}$ and acetone:

$$Pt^{IV}Cl_6^{2-} + CH_3COCH_3 \xrightarrow{h\nu} Pt^{III}Cl_5^{2-} + CH_3COCH_2\bullet$$

VII.5. ON THE MECHANISM OF ALKANE ACTIVATION

VII.5.A. STAGES OF THE PROCESS. FORMATION OF ORGANOMETALLICS

Oxidation and H–D exchange reactions apparently begin with the attack by the reactive form of the platinum(II) complex on the hydrocarbon molecule, RH, which results in the formation of an alkylplatinum(II) σ-complex:

$$PtCl_4^{2-} \rightleftharpoons PtCl_3^- + Cl^-$$

$$PtCl_3^- \rightleftharpoons PtCl_2 + Cl^-$$

$$PtCl_2 + RH \rightleftharpoons \sigma\text{-}RPtCl + H^+ + Cl^-$$

In the absence of platinum(IV), the alkyl σ-derivative undergoes electrophilic attack by the D^+ ion:

$$\sigma\text{-}RPtCl + D^+ + Cl^- \longrightarrow RD + PtCl_2$$

If a platinum(IV) derivative is present in the system, it interacts with the platinum(II) σ-complex and converts it into an alkylplatinum(IV) σ-complex:

$$\sigma\text{-}RPt(II) + Pt(IV) \longrightarrow \sigma\text{-}RPt(IV) + Pt(II)$$

Experiments with complexes with isotopically enriched ^{195}Pt demonstrated that such an oxidation process (at least in the case when R = CH_2CH_2OH) involves electron not alkyl transfer (step *a* rather than *b*) [29a,b]:

$$\sigma\text{-}R\text{–}Pt^{II}Cl_3^{2-} + {}^*Pt^{IV}Cl_6^{2-} \xrightarrow{a} \sigma\text{-}R\text{–}Pt^{IV}Cl_5^{2-} + {}^*Pt^{II}Cl_4^{2-}$$

$$\not\downarrow\ b$$

$$\sigma\text{-}R\text{–}{}^*Pt^{II}Cl_3^{2-} + Pt^{IV}Cl_6^{2-}$$

The kinetics and mechanism for oxidation and reduction of platinum(II) and platinum(IV) complexes with some reagents have been recently reported [29c].

The σ-alkylplatinum(IV) complex reacts with nucleophiles X (H_2O, Cl^-) to afford an alcohol ROH and an alkyl chloride RCl, respectively. Complexes **VI-5** and **VII-7** (R = CH_3 or CH_2CH_2OH) in aqueous solutions containing excess chloride have been shown [29b] to exist in equilibrium with $K = 0.9$ mol dm^{-3} at 25 °C. Rapid dissociative exchange of ligand *trans* to the alkyl substituent was proposed to occur via the five-coordinate intermediate **VII-6**.

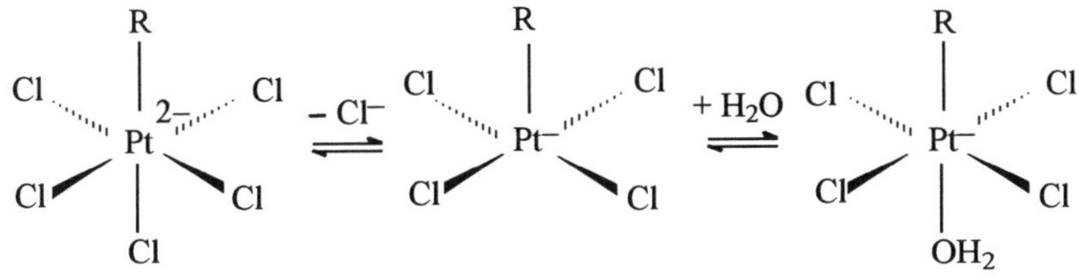

VII-5 **VII-6** **VII-7**

The following rate law for nucleophilic displacement of Pt(II) by water and chloride has been established:

$$-d[\mathbf{VII\text{-}5} + \mathbf{VII\text{-}7}]/dt = k_{obs}[\mathbf{VII\text{-}5} + \mathbf{VII\text{-}7}]$$

where

$$k_{obs} = (k_1[Cl^-] + k_2[H_2O])/(1 + K^{-1}[Cl^-])$$

The authors proposed an S_N2 mechanism for the formation of the final products of the oxidation reaction.

It should be noted that complexes of types **VII-5** or **VII-7** can be easily prepared by the method initially described in [30a–d]. The chloride complex $Pt^{II}Cl_4^{2-}$ reacts at room temperature in aqueous solution with alkyl iodide to produce a σ-alkyl complex of platinum(IV). Methyl [30a], ethyl [30b,c], and acetonyl [30d] complexes of Pt(IV) have been prepared by this method. An interesting feature of the reaction is that the sixth coordination site in the octahedral molecule of the product is occupied by water, while the I^- ion is bound to another molecule of the platinum(II) complex (PtI_2 and analogous products are formed and precipitate):

$$Pt^{II}Cl_4^{2-} + RI + H_2O \longrightarrow \sigma\text{-}RPt^{IV}Cl_4(H_2O)^- + I^-$$

The acetonyl complex turned out to be the most stable, which is apparently due to the absence of hydrogen atoms in a β-position of the alkane chain [30b,c]. Indeed, the σ-ethyl derivative readily decomposes to produce a π-ethylene complex of Pt(II) in addition to the product of the reaction with Cl^- and water. It should be also noted that the reaction between $PtCl_4^{2-}$ and *n*-hexyl iodide gives rise to the formation of the hex-1-ene π-complex of platinum(II) [30c]. More recently, generation of the complex $MePtCl_3^{2-}$ has been achieved by reduction of $Na_2[MePtCl_5]$ in THF with cobaltocene [30e]. This reaction gave the monomethyl dianion, not in pure form, but combined with $Me_2PtCl_2^{2-}$ and $PtCl_4^{2-}$ as the mixed cobalticenium salts.

σ-Alkyl complexes of platinum(IV) can be obtained in the reaction of $PtCl_6^{2-}$ with certain alkyl derivatives of non-transition metals. Thus, $PtCl_6^{2-}$ reacts slowly [31] with $(CH_3)_4Sn$ in CD_3COOD at room temperature resulting in the formation of σ-CH_3–$PtCl_5^{2-}$. The reaction is accelerated by the addition of $PtCl_4^{2-}$ and is of first order with respect to platinum(II):

$$(CH_3)_4Sn + PtCl_6^{2-} \xrightarrow{PtCl_4^{2-}} \sigma\text{-}CH_3\text{–}PtCl_5^{2-} + (CH_3)_3SnCl$$

In the absence of $PtCl_6^{2-}$, tetramethyltin reacts with Na_2PtCl_4 in aqueous acetone, also forming a methylplatinum(IV) complex. The precipitation of metallic platinum is then observed. The reaction partly proceeds as oxidative addition of the components, involved in the Me_3Sn–Me bond, to platinum(II). As in the case of alkyl iodides, one component adds to the platinum(II) ion while the other is bound by another platinum species (reducing the latter):

$$(CH_3)_4Sn + 2\,PtCl_4^{2-} + H_2O \longrightarrow$$
$$CH_3PtCl_4(H_2O)^- + Pt + (CH_3)_3SnCl + 3Cl^-$$

Both reactions, with participation of tetramethyltin, to a certain extent model the interaction of the Pt(IV) + Pt(II) system with saturated hydrocarbons. The cleavage of the Sn–CH_3 bond by the $PtCl_6^{2-}$ complex is accelerated by irradiation [31]. The formation of the complex σ-C_2H_5–$PtCl_5^{2-}$ has been observed by ^1H NMR in the photochemical reaction in CD_3COOD. Light apparently accelerates the conversion of the ethyl σ-complex into the π-ethylene complex [31]:

$$(CH_3)_2Sn(CH_2CH_3)_2 + PtCl_6^{2-} \longrightarrow$$

$$\longrightarrow \sigma\text{-}CH_3CH_2\text{–}PtCl_5^{2-} + \sigma\text{-}CH_3\text{–}PtCl_5^{2-}$$

$$\downarrow$$

$$(\pi\text{-}CH_2{=}CH_2)PtCl_3^{-}$$

Thus, it can be concluded that reactions, in which there is a possibility of the formation of platinum(IV) derivatives with alkyl fragments containing a hydrogen bond in the β-position, lead to olefin π-complexes of platinum(II). These reactions are the interactions of platinum chlorides with alkyl iodides and alkyl derivatives of tin as well as the thermal and photochemical dehydrogenation of *n*-hexane by $PtCl_6^{2-}$ (Scheme VI.5). One possible mechanism for the dehydrogenation of *n*-hexane is the formation of a σ-hexyl complex of platinum(IV) and its subsequent transformation into a π-olefin complex of platinum(II) via β-elimination of hydrogen [16]:

$$\sigma\text{-}RCH_2CH_2\text{–}PtCl_5^{2-} \xrightarrow{-Cl^-} (\pi\text{-}RCH{=}CH_2)\overset{H}{\overset{|}{Pt}}Cl_4^{2-} \xrightarrow{-HCl} (\pi\text{-}RCH{=}CH_2)PtCl_3^{-}$$

VII.5.B. INTERACTION BETWEEN Pt(II) AND ALKANES

The intimate mechanism of the reaction deserves special attention not only because it was the first example of alkane activation by a metal complex. Activation of alkanes by platinum(II) complexes remains unique in many aspects. The reaction takes place in neutral water solution with conventional chloride ligands at the metal without special ways to form coordinatively unsaturated species (e.g., by irradiation). A number of works were directed towards the elucidation of the nature of the interaction between an alkane and a platinum(II) complex. The unique feature of platinum(II) complexes is to exhibit both nucleophilic and electrophilic properties.

Two main mechanisms may be proposed for the first step of alkane interaction with platinum(II) complexes:

(i) oxidative addition

$$R\text{–}H + Pt(II) \longrightarrow \sigma\text{-}R\text{–}Pt(IV)\text{–}H$$

followed by proton reversible elimination

$$\sigma\text{-R–Pt(IV)(Cl)–H} \rightleftharpoons \sigma\text{-R–Pt(II)} + H^+ + Cl^-$$

and (ii) electrophilic substitution with simultaneous proton abstraction

$$\text{R–H} + \text{Pt(II)–Cl} \longrightarrow \sigma\text{-R–Pt(II)} + H^+ + Cl^-$$

The oxidative addition mechanism was proposed for the first step of the reaction in the original publication [2]. It looked more probable in particular because of the lack of strong rate dependence on polar factors and on the acidity of the medium. Later, however, the mechanism of electrophilic substitution was proposed in some publications.

A few other possible mechanisms have been suggested on the basis of quantum-chemical calculations for the interaction between Pt(II) and the alkane C–H bond in earlier publications. Let us consider briefly some of these theoretical as well as kinetic investigations.

The electronic structures of complexes activating alkanes – square planar and pseudo-square chloride and aquachloride derivatives of platinum(II) – as well as the interaction of these complexes with methane have been studied by the semiempirical MO LCAO SCF method in the CNDO approximation [32]. Calculation for the complexes $[PtCl_{4-n}(H_2O)_n]^{n-2}$ with $n = 0, 1, 2, 3, 4$ and *trans*-$[PtCl_2(H_2O)(OH)]^-$ showed that the changes in their quantum-chemical characteristics are monotonous and are independent of the symmetry (thus the positive charge on platinum increases monotonously with the increase of n). It was concluded that, when the interaction of such complexes with an axial ligand is controlled by the LUMO, then this is an interaction with a σ AO and to a lesser extent a d_{z^2} AO of platinum. The methane molecule was regarded as the ligand and the calculation was performed for the following structures:

$CH_4 \cdots PtCl_4^{2-}$,

$CH_4 \cdots$ *trans*-$PtCl_2(H_2O)(OH)^-$,

$CH_4 \cdots$ *trans*-$PtCl_2(H_2O)_2$,

$CH_4 \cdots Pt(H_2O)_4^{2+}$.

It was found that coordination involves a redistribution of electron density in the methane molecule and as a result, the polarity of the C–H bonds increases and the bonds (particularly those coordinated to platinum ion or an equatorial acido-ligand) are weakened. Furthermore, an increase in the free valence of the carbon atom of methane was noted. This actually results in the activation of the methane molecule, facilitating the attack on the latter by polar species. Analysis of the distribution of electron density in methane complexes showed that the alkane is bound as a result of the interaction of the 1s AO of the hydrogen atoms and the 2s and $2p_z$ AO of the carbon atom with the 6s and $6p_z$ AO of platinum. Thus the involvement of the d AO of the metal in such complex formation is smaller than the involvement of the outer s and p AO. For virtually any mode of coordination of CH_4 to the platinum(II) complex (coordination via the vertex, edge, and face of the methane tetrahedron to platinum and also simultaneously to platinum and the acido-ligand was considered), the resulting transfer of electron density from methane entails its decrease at the hydrogen atoms. However, owing to the contribution by the dative interaction, the electron density on the carbon atom increases.

It is important that the acceleration of the H–D exchange, observed experimentally on the replacement of the $PtCl_4^{2-}$ or $Pt(H_2O)_4^{2+}$ catalysts by $PtCl_2(H_2O)_2$, could not be related to any calculated quantum-chemical parameters. It was therefore postulated that an optimum combination of the donor–acceptor and dative mechanisms of the transfer of electron density is essential for the effective alkane-platinum(II) complex interaction. Since the contribution of the former increases and that of the latter diminishes with the increase of the positive charge on the complex, the optimum combination can be achieved for the complex with zero charge. Somewhat earlier, the presence of a catalytic activity maximum for the complex $PtCl_2(H_2O)_2$ was attributed to the influence of the symmetry of the ligand environment [32]. On the basis of the hypothesis that the formation of a bond with the activated alkane takes place mainly on interaction with the d_{z2} AO of the complex, the rates of reaction were correlated with the contribution by this orbital to the LUMO of the platinum(II) complex. Symmetry considerations led to the conclusion that, for $n = 0$ and $n = 4$, the LUMO of the complex is formed solely by the d_{x2-y2} AO and cannot contain an admixture of the d_{z2} state. As a consequence of the decrease of symmetry for $n = 1, 2, 3$, such an admixture exists but the $n = 2$ complex occupies a special place (thus the contribution of the d_{z2} AO for the *trans*-isomer with $n = 2$ should be greatest). It is believed, however [33], that this approach, based on crystal field theory, is insufficiently complete, since, according to MO theory, the d_{z2} AO of platinum participates in the binding of the equatorial ligands even in highly symmetrical complexes, i.e., in these complexes this orbital, being partly unoccupied, can be involved in the donor-acceptor interaction with the alkane. In another paper [34], the total energies have been calculated by the extended Huckel method for alkanes coordinated to the $PtCl_3(H_2O)^-$ species and it has been shown that the coordination of the terminal atom, for example, in pentane, is most preferred.

It may be assumed, that the reaction of the chloroaquoplatinum(II) complex with alkanes begins as oxidative addition, proceeds through a three-center transition state, and terminates by the synchronous formation of a platinum–carbon bond with elimination of a proton (which can be transferred to a molecule of water). A similar mechanism has been proposed [35a] for the cyclometalation of 8-alkylquinolines by palladium(II). It has also been suggested that the reverse process (the protolysis of the platinum-carbon bond in alkyl complexes) involves a three-center transition state [35b], and the concerted oxidative addition to Ir(I) complex has been proposed [35c].

The cleavage of the C–H bond in methane by the *trans*-$PtCl_2(H_2O)_2$ complex has been subjected to quantum-chemical analysis by the extended Hückel method [36]. It was shown that the approach of the methane molecule to the square planar platinum(II) complex (structure **VII-8**) leads to strong four-electron repulsion, the main cause of which is the interaction between t_2 orbitals of CH_4 and the occupied orbitals of the metal complex localized in the region of the free coordination site (HOMO consisting mainly of the d_{z2} orbital and also the low-lying d_{yz} orbital) (Figure VII.1).

When the distance between the platinum atom and the center of the C–H bond reaches $d = 2.8$ Å, the distortion of the metal complex (structure **VII-9**) lowers the repulsion energy. The composition of the HOMO then changes, namely the contribution of the orbital diminishes and that of the d_{x2-y2} orbital increases, which entails a decrease of the electron density in the region of the free coordination site. The d_{x2-y2} orbital is hybridized with the s, p_z and d_{z2} orbitals of the metal, as a result of which the low acceptor MO of the complex becomes extended in the axial direction. The overlap of the acceptor orbital with the σ orbital of the C–H bond increases and hence the contribution of the transfer of electron density via the donor–acceptor mechanism also increases. This is accompanied by the simultaneous increase of the dative interaction, the main contribution to which is made by the charge transfer from the d_{z^2} and d_{yz} orbitals to the σ* orbitals of methane.

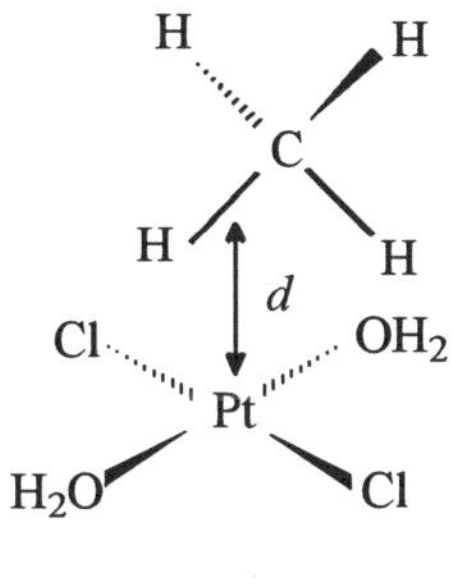

VII-8

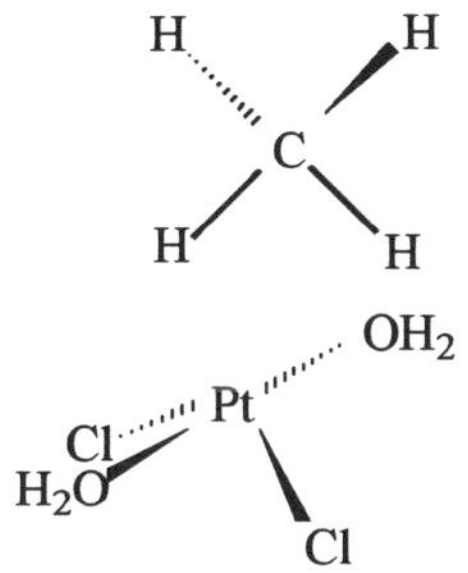

VII-9

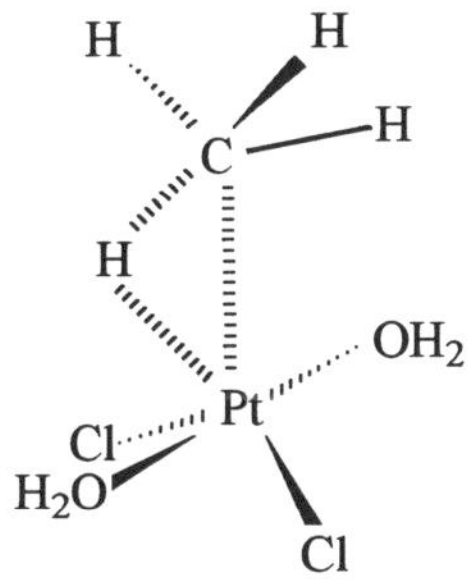

VII-10

In the second stage, the deformation distortion of the methane molecule occurs starting with $d = 2.4$ Å and the repulsion energy diminishes still further. At the same time, the angle of the displacement of the ligand from the plane of the complex increases from 10° to 36°. The populations of the Pt–C and Pt–H bonds reach 30–40% and the C–H bond is loosened to the extent of 20–30%. Under these conditions, the overlap of the d_{yz} and t_2 orbitals diminishes. Since an analogue of an unshared pair oriented towards the complex appears at the carbon atom, namely the *3a′* component at the t_2 level (see diagram in Figure VII.1), the donor–acceptor interaction is enhanced. The transition state **(VII-10)** arises for $d = 1.8$ Å , whereupon the Pt–C and Pt–H bond lengths reach the usual values and the angle of deviation of the ligand in the complex is 36°. When the C–H bond is extended by 0.28 Å, a maximum appears on the curve describing the energy of the system and the C–H, Pt–C, and Pt–H bonds become approximately equivalent. The extension of the C–H bond enhances both the donor–acceptor and dative interactions; the main contribution to the latter comes in this case from the transfer of electron density to the $3a_1$ component of the t^*_2 level. This transfer plays a decisive role in the rupture of the C–H bond. It is noteworthy that, when the C–H bond is extended, the four-electron repulsion increases faster than the interactions of both types and only when CH_4 and the complex are close together do the repulsion and dative interaction energies become equalized.

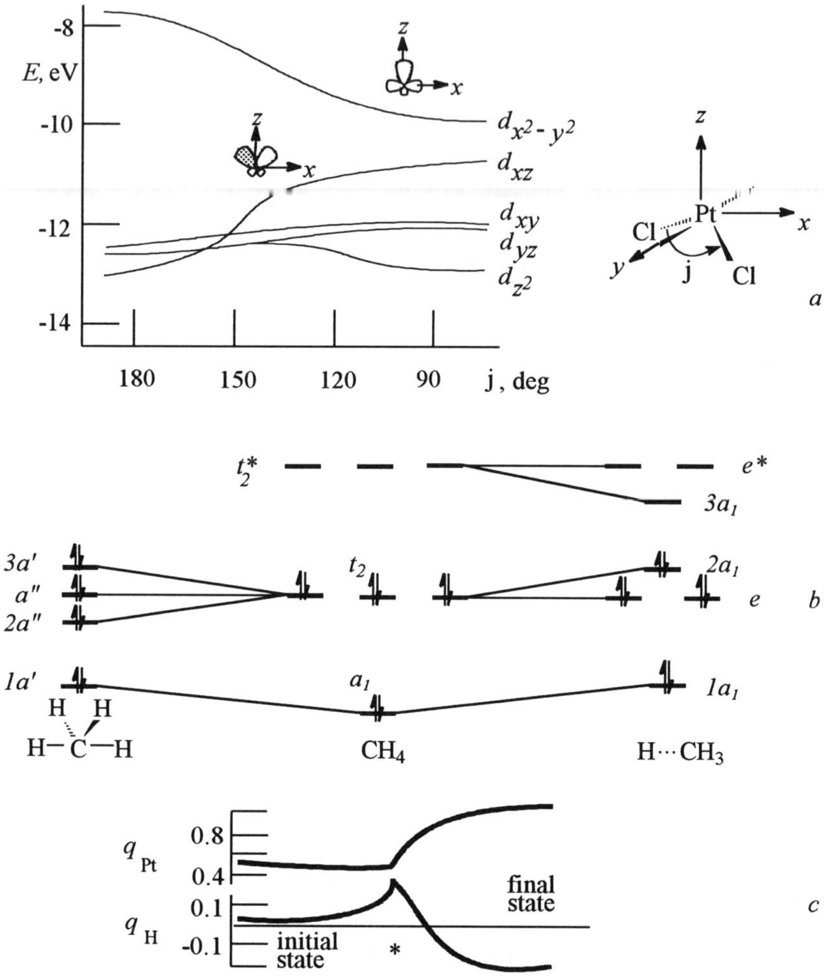

Figure VII.1. The reaction of CH_4 with *trans*-$PtCl_2(H_2O)_2$. The changes are shown (*a*) in the energies of the Pt(II) complex orbitals on deformation of the angle Cl–Pt–Cl, (*b*) in the orbitals of the methane molecule on deformation of the angles and following an increase in the C–H bond length, (*c*) in the charges *q* on the Pt and H atoms along the reaction coordinate (the transition state is denoted by an asterisk).

Calculation has shown that, on passing from *trans*-$PtCl_2(H_2O)_2$ to $PtCl_4^{2-}$, the activation energy for the reaction with the alkane should increase by 0.24 eV, which agrees with the experimental data. In the first stages of the process, up to the appearance of the transition state, the C–H bond is polarized, increasing the acidity of the hydrogen atom. This is apparently why, in the presence of bases stronger than the platinum(II) complex, it is easier for the system to transfer hydrogen in the form of a proton to the molecule of the base (e.g., water) than to continue the process of oxidative addition. Thus the reaction may be completed by the synchronous formation of the Pt–C bond and the elimination of a proton, i.e., can proceed as "soft" electrophilic substitution:

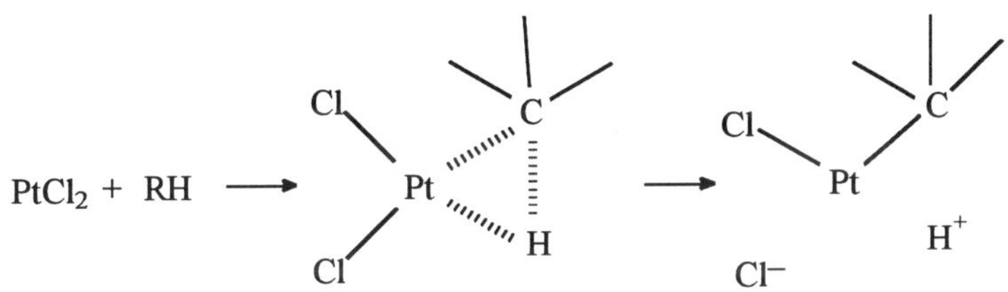

A few other mechanisms have been proposed to explain the kinetics of the reaction and especially the multiple H–D exchange. The multiple exchange may be due to the formation of carbene complexes:

$$\mathrm{Pt(II)} + \mathrm{RH} \underset{+D^+}{\overset{-H^+}{\rightleftharpoons}} \mathrm{Pt{-}R} \underset{+D^+}{\overset{-H^+}{\rightleftharpoons}} \mathrm{Pt{=}C}\!<$$

or π-olefin complexes and metallacycles:

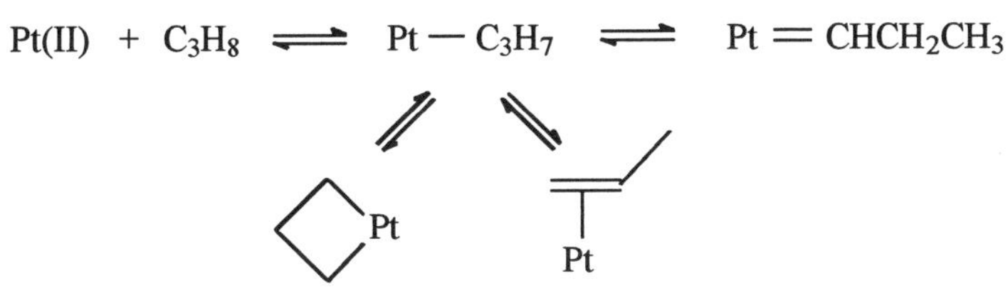

The formation of the Pt(II)–alkane complex as the first kinetically significant intermediate in the reaction has been proposed in the so called "alkane–alkyl" mechanism (see above) [14]. This intermediate is analogous to alkonium ions of the type CH_5^+, whose formation is associated with the deformation of both the planar complex and of the tetrahedral hydrocarbon molecule.

In the another hypothesis, it has been assumed [37] that the rotation of the alkyl group R in the coordination sphere of platinum entails an appreciable intensification of the nucleophilic properties of the carbon atom, which should facilitate the electrophilic attack on the latter by the deuterium cation D^+. The alkyl group rotates synchronously with the approach of D^+ and the agostic coordination between the C–H bond and platinum ion is formed. Since it is evident that several hydrogen atoms can be substituted by deuterium in the alkyl group before it is split off from platinum, the proposed mechanism explains satisfactorily the multiplicity of different types of H–D exchange in the presence of platinum(II) complexes.

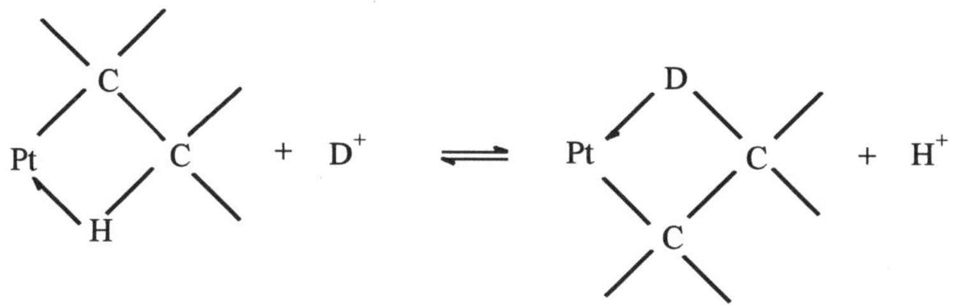

The electropilic substitution of H^+ by D^+ possibly proceeds via the three-center transition state:

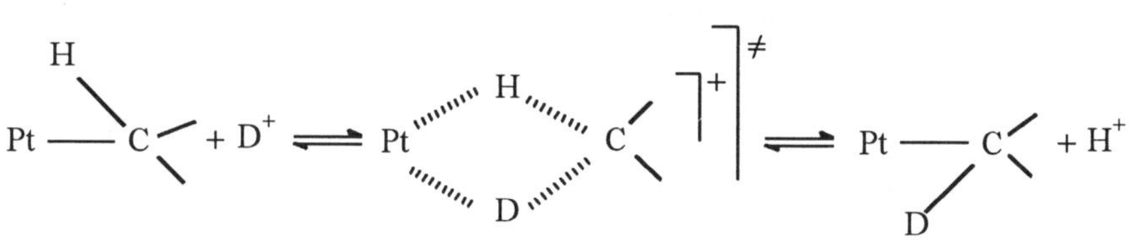

It is believed that the "alkyl" mechanism agrees better with certain features of the exchange reaction than the carbene and alkane–alkyl mechanisms.

More recently, the mechanism of oxidative addition was confirmed in investigations of decomposition and protonolysis of alkylplatinum complexes, i.e., in reactions reverse to the activation of alkanes. Zamashchikov *et al.* [38] investigated the decomposition of the dimethyl platinum(IV) complex $Cs_2Pt(CH_3)_2Cl_4$. There are two routes in this process. The first leads to reductive elimination under the action of Cl^- and produces methyl chloride and methane. The second leads to the formation of ethane. There is strong kinetic evidence that ethane is the product of decomposition of the ethylhydridoplatinum(IV) complex formed from the initial dimethyl platinum(IV) complex. In D_2O–DCl, ethane formed contains several D atoms with practically the same parameter of multiple exchange and similar distribution as in H–D exchange of ethane with D_2O catalyzed by platinum(II) complexes. Moreover, ethyl chloride is formed in the presence of platinum(IV) in the reaction competitive with H–D exchange. From the principle of microscopic reversibility, it follows that the same ethylhydridoplatinum(IV) complex is the intermediate in the reaction of ethane with platinum(II). Assuming the first step of the reaction of methane with Pt(II) as oxidative addition according to equation

$$\text{R–H} + \text{Pt}^{\text{II}} \longrightarrow \sigma\text{-R–Pt}^{\text{IV}}\text{–H}$$

allows the calculation of the deuterium distribution in methane. Indeed, if the mechanism of H–D exchange is accepted as shown in Scheme VII.6 [intermediates $CH_{3-i}D_iPt(II)$ (i = 0–3) and product Pt(II) are omitted for simplicity] the calculated parameters for D distribution in mixture of products of the reaction between $PtCl_4^{2-}$ and CH_4 in D_2O at 373 K turned out in close agreement with experimental data (Figure VII.2).

Important results were obtained by Labinger and Bercaw [39] in the investigation of the protonolysis mechanism of several alkylplatinum(II) complexes at low temperatures, i.e., the reactions which could model the microscopic reverse of C–H activation by platinum(II) complexes. Alkylhydridoplatinum(IV) complexeswere observed as intermediates in certain cases, e.g., under treatment of the complexes (tmeda)$Pt(CH_2Ph)Cl$ or (tmeda)$PtMe_2$ (tmeda = *N,N,N′,N′*-tetramethylethylenediamine) by HCl in CD_2Cl_2 and CD_3OD respectively. In some cases H–D exchange takes place between methyl groups on platinum and CD_3OD prior to methane loss.

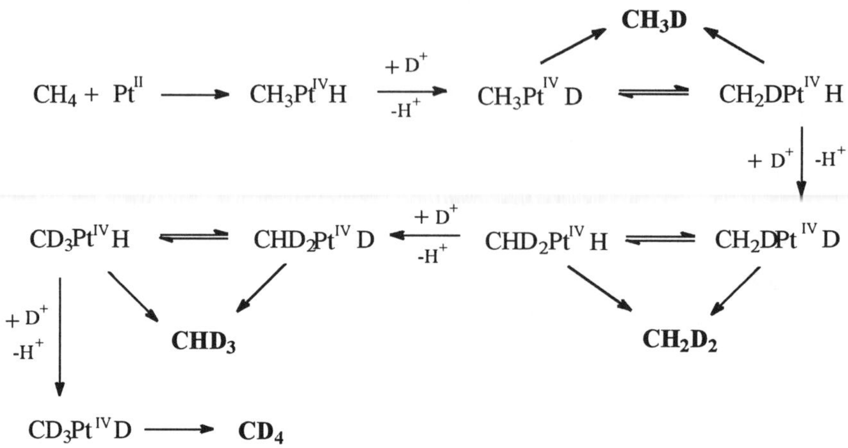

Scheme VII.6. H–D Exchange in methane catalyzed by Pt(II).

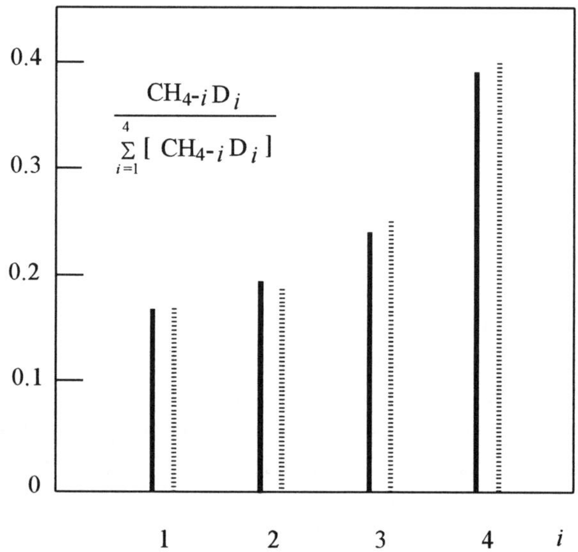

Figure VII.2. Distribution of deuterated methanes in the mixture of products of the reaction between CH_4 and $PtCl_4^{2-}$ in D_2O at 373 K (black bars, experimental results; dotted bars, calculated values).

Based on the kinetic results a common mechanism was proposed to operate in all the reactions:

(1) protonation of Pt(II) to generate alkylhydridoplatinum(IV) intermediate,

(2) dissociation of solvent or chloride to generate a cationic, five-coordinate platinum(IV) species,

(3) reductive C–H bond formation producing a platinum(II) alkane σ-complex, and

(4) loss of alkane either through associative or dissociative substitution pathway.

These results implicate the presence of both alkane σ-complexes and alkylhydridoplatinum(IV) as intermediates in the C–H activation reactions by Pt(II) complexes. The first step is the formation of a σ-complex with the alkane which then undergoes oxidative addition to produce the alkylhydrido complex. Reversible interconversion of these intermediates, together with reversible deprotonation of the alkylhydridoplatinum(IV) complexes, lead to the multiple H–D exchange with the solvent.

It has been shown [40] that heating a solution of $[(tmeda)Pt(CH_3)(NC_5F_5)]^+$ in pentafluoropyridine under 30 atm of $^{13}CH_4$ at 85 °C gives rise to the slow growth of a Pt–$^{13}CH_3$ resonance in the ^{13}C NMR spectrum due to methyl exchange. The reaction is accompanied by the formation of CH_4 and by deposition of metallic platinum, which is associated with the C–H activation process. The authors note a very close similarity between this reaction and the activation of alkane C–H bonds observed with $[Cp^*Ir(PMe_3)(CH_3)(ClCH_2Cl)]^+$ and $Cp^*Ir(PMe_3)(CH_3)(OTf)$ (see Chapter VI). When both Ir(III) and Pt(II) participate in the C–H activation process, a σ-bond metathesis mechanism cannot be excluded [40] as an alternative to oxidative addition (however, see Chapter VI). A proposed mechanism for H–D exchange between $[(tmeda)Pt(CH_3)(NC_5F_5)]^+$ and CD_4 [40b] is shown in Scheme VII.7.

Puddephatt *et al.* [41] have studied the C–H or C–C bond activation in the alkane complexes $[PtMe(CH_4)L_2]^+$ or $[PtMe(CH_3CH_3)L_2]^+$ ($L = NH_3$ or PH_3) as well as the reductive elimination of methane or ethane from the five-coordinate model complexes $[PtHMe_2L_2]^+$ or $[PtMe_3L_2]^+$, respectively, by carrying out extended Hückel molecular orbital calculations and density functional theory. The oxidative addition and reductive elimination reactions occur by a concerted mechanism, probably with a pinched trigonal-bipyramidal complex on the reaction coordinate. The methane remains coordinated to platinum through the C–H σ complex in the C–H reductive elimination, and for C–C reductive elimination, the transition state is a C–C σ complex. The activation energy for reductive elimination from the complexes with L = NH_3 and PH_3 was the same at only 3 kcal mol^{-1}, but the activation energy for methane activation was calculated to be 12 and 19 kcal mol^{-1} for complexes containing NH_3 and PH_3, respectively (Figure VII.3).

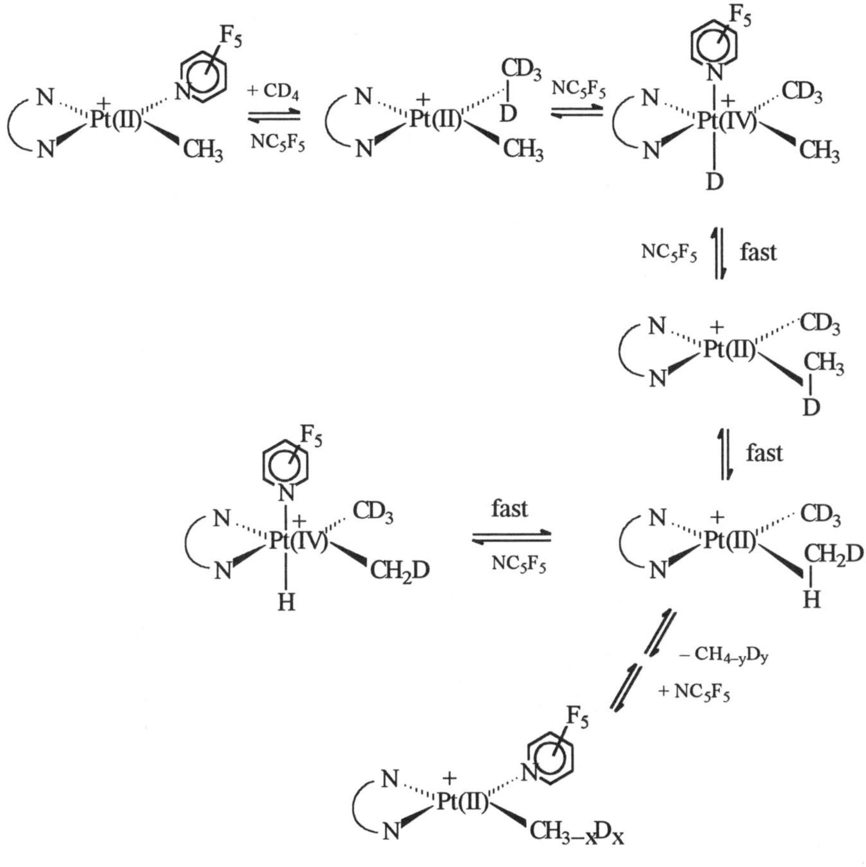

Scheme VII.7. H–D Exchange between Pt(II) complex and CD_4.

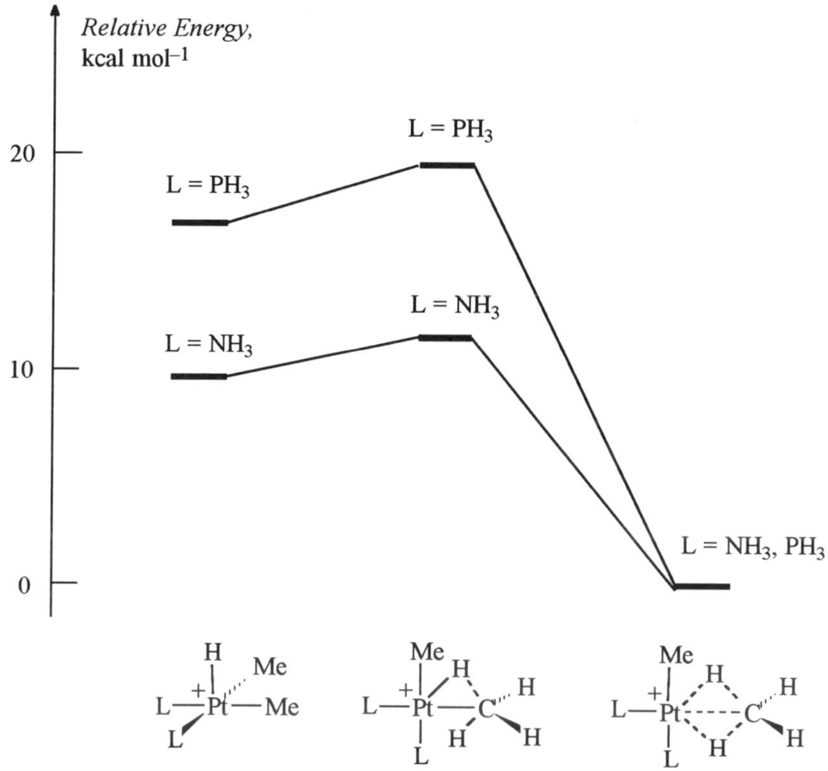

Figure VII.3. Energies of the Pt(II), Pt(IV) complexes and transition states for the reductive elimination of methane from $[PtHMe_2L_2]^+$ and the corresponding C–H bond activation in the alkane complex $[PtMe(CH_4)L_2]^+$.

Mechanistic studies of the C–H activation by aqueous platinum complexes and related topics have been also described in recent publications [42]. Some reactions with participation of platinum derivatives and their mechanisms are discussed also in other chapters of this book.

VII.6. σ-Aryl Complexes of Pt(IV) Formed in Thermal and Photochemical Reactions of Aromatic Hydrocarbons with Pt(IV) Halide Complexes

The experimental data reported in the previous sections leave no doubt of the intermediate formation of aryl and alkyl platinum derivatives in the reactions of Pt(II) and Pt(IV) chloride complexes with arenes and alkanes in aqueous solutions. In the 1980s, it was discovered that the reaction of H_2PtCl_6 with arenes in aqueous carboxylic acids produces stable σ-aryl complexes of platinum(IV) which can be isolated in crystalline form (see, for example, [43]).

VII.6.A. THE THERMAL REACTION OF ARENES WITH $PtCl_6^{2-}$

Heating a solution of H_2PtCl_6 and an aromatic compound ArH in the $CF_3COOH–H_2O$ mixture or in CH_3COOH leads to the formation of fairly stable σ-aryl complexes of platinum(IV) in yields up to 95%, which can be isolated in the form of anionic adducts with ammonia after chromatography on silica gel containing ammonia [43, 44]:

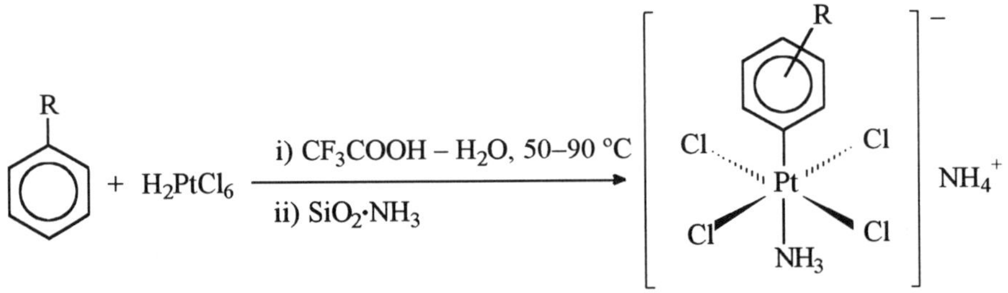

(VII.5)

Before their isolation in the form of the ammonia adduct, the complexes apparently exist as σ-$ArPtCl_5^{2-}$ and σ-$ArPtCl_4(H_2O)^-$. The complexes stabilized with an ammonia ligand have been characterized by X-ray diffraction [45a] and ^{195}Pt NMR spectra [45b]. The same complexes can be prepared by the reaction of H_2PtCl_6 with mercury, tin, lead or boron aryls [46].

On prolonged heating, the complexes decompose [43, 44d] to produce biaryls, chlorinated arenes and platinum(II) derivatives. The process of formation of the σ-tolyl complex of platinum(IV) is accompanied by its *para–meta* isomerization [43, 44e, 47]. If in the initial instant the substitution takes place mainly (to extent of ~ 90%) in the *para*-position in toluene, then the statistical distribution *meta:para* = 2 : 1 is gradually attained (Figure VII.4).

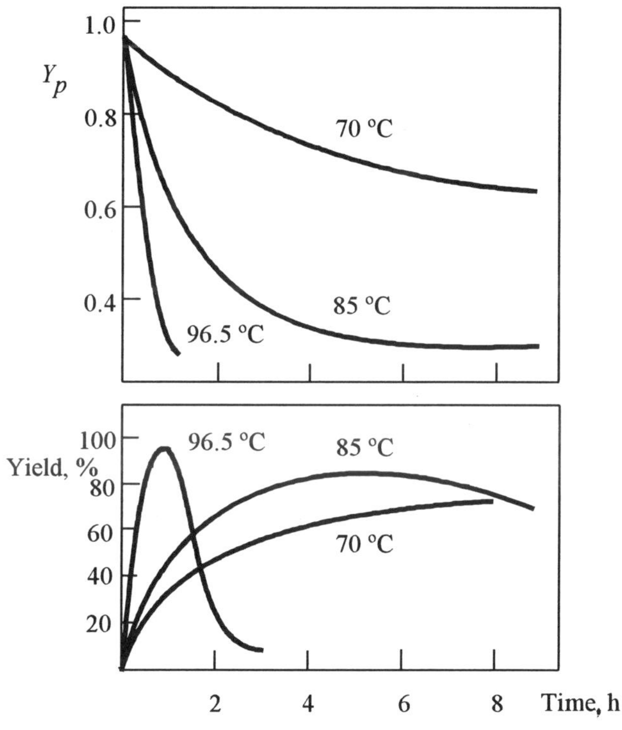

Figure VII.4. The reaction of H_2PtCl_6 with toluene in CF_3COOH–H_2O to afford the σ-tolyl complex of platinum(IV). Accumulation and decomposition of the σ-tolyl complex (the yield of a sum of *para* and *meta* isomers is given at the bottom) and *para–meta* isomerization of this complex (Y_p is the content of *para* isomer in the mixture *para* + *meta*).

The substituent does not enter the *ortho*-position for steric reasons. The relative content of *para*-isomer Y_p can be calculated from the equation

$$Y_p = \frac{[p\text{-Tolyl-Pt}^{IV}]}{[p\text{-Tolyl-Pt}^{IV}] + [m\text{-Tolyl-Pt}^{IV}]} = \frac{1}{3} + \frac{3k_p - k}{3(k - 1.5k_i)} \times \frac{e^{-1.5k_i t} - e^{-kt}}{1 - e^{-kt}}$$

which corresponds to the following scheme:

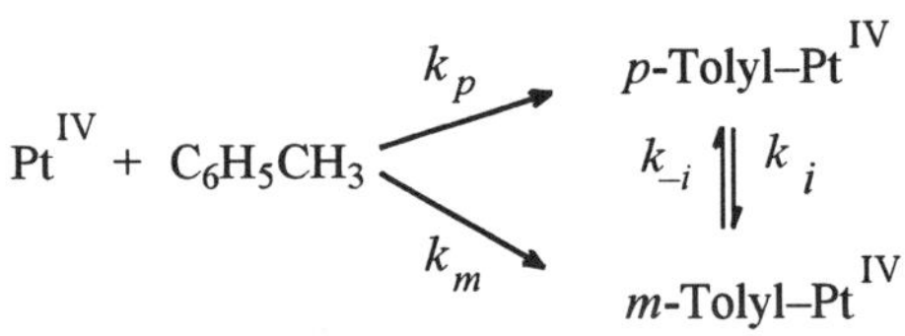

$$k = k_p + k_m$$
$$k_i = 2k_{-i}$$

The activation energies of the formation and *para–meta* isomerization are ca. 25 kcal mol^{-1}.

It has been established [43, 44e] by the method of competing reactions that the relative rates of reactions (given in parenthesis) with participation of the arenes C_6H_5X decrease in the following sequence of substituent X:

OH (16), OCH_3 (8.5), CH_3 (3), C_2H_5 (2.7), OC_6H_5 (2.0), $CH(CH_3)_2$ (1.9), H (1.0), C_6H_5 (0.9), F (0.3), $COCH_3$ (0.1), COOH (0.09), Cl (0.08), NO_2 (0.04).

The logarithms of these values are correlated with the Hammett constant σ and the Brown constant σ^+ (with the parameters $\rho = -3.0$ and $\rho^+ = -1.5$). The kinetic isotope effect of the reaction is small (~ 3 for benzene and ~ 2.3 for toluene).

In a study of the products of the decomposition of the *meta*- and *para*-isomers of the σ-tolylplatinum(IV) complex, formed in the reaction between H_2PtCl_6 and toluene in $CF_3COOH–H_2O$ (4.5 : 1) at 90 °C, it was observed that, together with the "expected" 3,3′-, 4,4′-, and 3,4′-bitolyl isomers, a large amount of the "surprising" 2,3′- and 2,4′-bitolyls is formed after a long induction period (Figure VII.5) [48a].

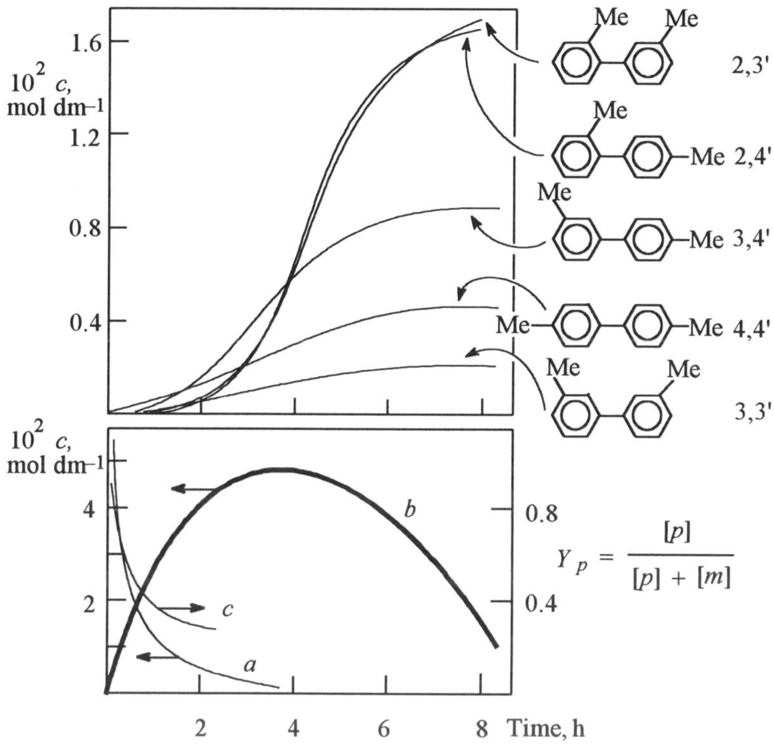

Figure VII.5. Decomposition of the σ-tolyl complex of platinum(IV) to produce isomers of bitolyls. *a* – Consumption of $PtCl_6^{2-}$; *b* – accumulation and decomposition of the σ-tolyl complex (sum of *para* and *meta* isomers); *c* – *para–meta* isomerization of the σ-tolyl complex (Y_p is the content of para isomer in a mixture *para* + *meta*).

The reaction of H_2PtCl_6 with anisole in $CF_3COOH–H_2O$ (9 : 1) at 85 °C produces the σ-methoxyphenyl complex of platinum(IV) which decomposes to give 4,4′-dimethoxybiphenyl and, after an induction period, a smaller amount of 2,4′-dimethoxydiphenyl (Figure VII.6) [48b].

Such products of *ortho*-substitution relative to the methyl group are apparently formed on interaction of *meta*- and *para*-platinated toluenes with free toluene present in solution. Indeed, if toluene is removed from the reaction mixture when the concentration of the σ-tolyl complex of platinum(IV) is close to a maximum (yield ca. 90%) the rate of decomposition of this complex diminishes [48c].

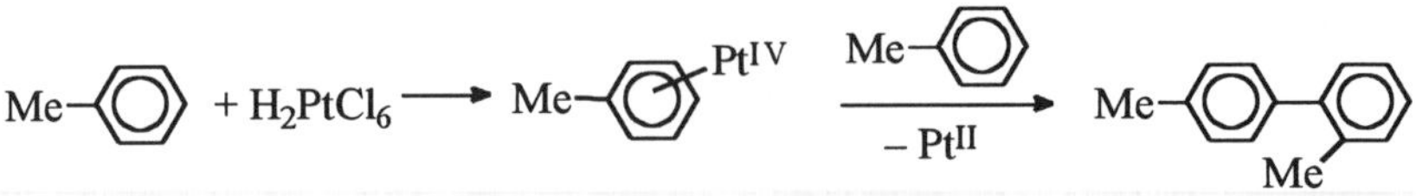

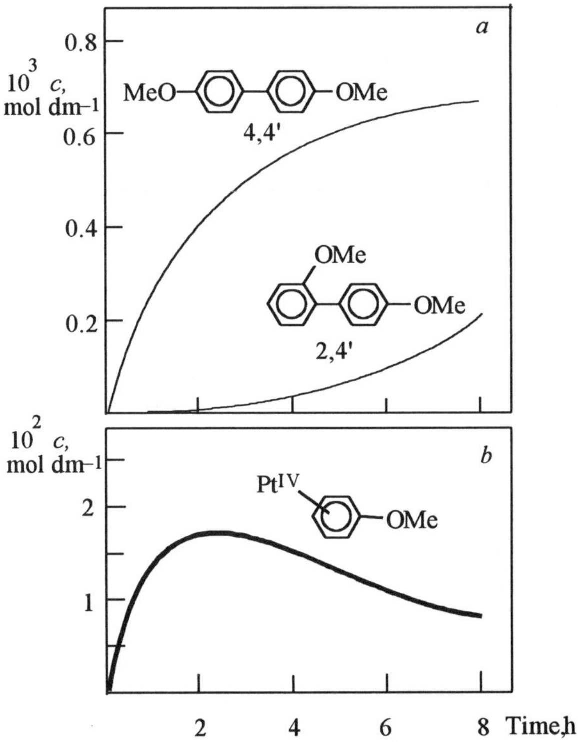

Figure VII.6. The reaction of H_2PtCl_6 with anisole. *a*) Formation of dimethoxybiphenyls; *b*) accumulation and decomposition of the σ-methoxybiphenyl complex (sum of *para* and *meta* isomers).

The platinum(IV)-containing fragment can be easily transferred from one arene to another. For example, when a solution of the σ-tolyl complex of platinum(IV) is heated in aqueous trifluoroacetic acid with an excess of anisole or ethylbenzene, transarylation (transmetalation) occures [48c]:

$$\text{Ar–Pt}^{\text{IV}} + \text{Ar}'\text{–H} \longrightarrow \text{Ar}'\text{–Pt}^{\text{IV}} + \text{Ar–H} \qquad \text{(VII.6)}$$

Upon prolonged heating with anisole, the σ-tolyl complex decomposes mainly to isomers of dimethoxybiphenyl and lesser amounts of ditolyl and tolylmethoxyphenyl. Heating a solution of the σ-tolyl complex with acrylic acid yield the product of olefin arylation:

$$\sigma\text{-}CH_3C_6H_4PtCl_4(H_2O)^- + CH_2{=}CHCOOH \longrightarrow$$
$$CH_3C_6H_4CH{=}CHCOOH + Pt(II) \qquad \text{(VII.7)}$$

The proposed [43, 44e] mechanism of the reaction between $PtCl_6^{2-}$ and the arene to afford *para* and *meta* isomers of the σ-arylplatinum(IV) complex (**VII-14*p*** and **VII-14*m*** in Scheme VII.8) involves the formation of a weak π-arene complex of platinum(IV) **VII-11** which is transformed into an intermediate Wheland-type complex. The isomerization of the Wheland complexes **VII-12*p*** and **VII-12*m*** possibly proceeds though the transition state **VII-13**, and/or is due to the reversibility, to some extent, of the stage of **VII-12** formation. The transarylation mentioned above (equation VII.6) is perhaps also due to the reversibility of the Wheland complex formation.

Heating a solution of an aromatic compound (for example, benzene or toluene) with Na_2PtCl_4 in aqueous trifluoroacetic acid affords a σ-aryl complex of platinum(II which is much less stable in comparison with a corresponding complex of platinum(IV). However this complex was identified after the oxidation with H_2PtCl_6 (at room temperature) to produce the latter derivative [48d]:

$$\text{Ar–H} + PtCl_4^{2-} \longrightarrow \text{Ar–Pt}^{\text{II}} \xrightarrow{PtCl_6^{2-}} \text{Ar–Pt}^{\text{IV}}$$

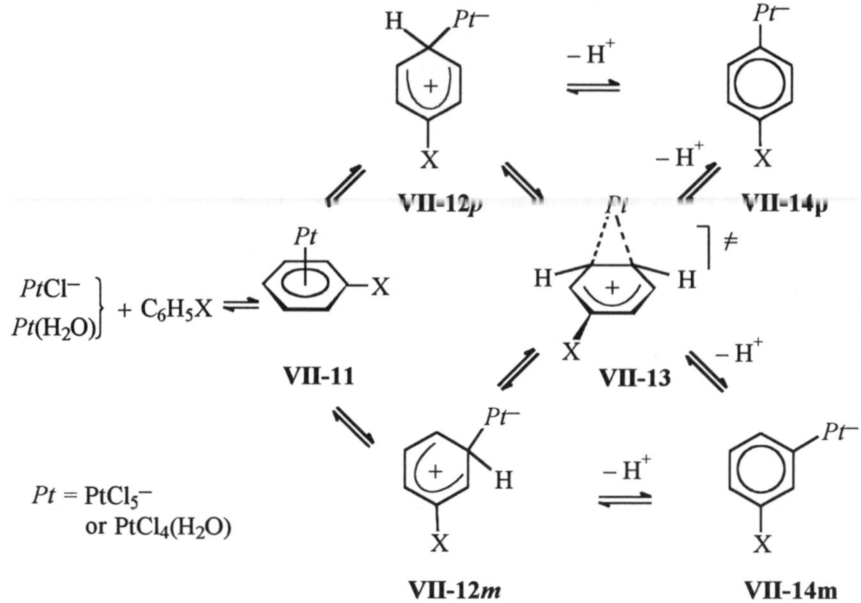

Scheme VII.8. The mechanism proposed for the reaction of $PtCl_6^{2-}$ with monosubstituted benzene C_6H_5X.

The reaction of $PtBr_6^{2-}$ with toluene in aqueous CF_3COOH apparently yields the unstable σ-tolyl complexes of platinum(IV), which rapidly decompose to generate 3,3′-, 3,4′-, and 4,4′- isomers of bitolyls with the ratio 2.5 : 27.5 : 70.0, respectively. Unlike the reaction of $PtCl_6^{2-}$ (*vide supra*), the oxidation with $PtBr_6^{2-}$ gives no *ortho*-substituted bitolyls. An induction period preceding the accumulation of these products was noticed [48e].

VII.6.B. PHOTOELECTROPHILIC SUBSTITUTION IN ARENES

The reaction (VII.5) can be also carried out at room temperature if the reaction solution is irradiated with light (full light of high-pressure mercury arc, in a Pyrex vessel, $\lambda > 300$ nm) [44e, 49a–c] or γ-quanta (^{60}Co source, with a nominal dose rate of 6 Mrad h^{-1}) [44e, 49d]. Aqueous trifluoroacetic acid [44e, 49a,b,d] or methylene chloride (in this case $PtCl_6^{2-}$ has been introduced into the reaction in the form of the tetrabutylammonium salt) [49c] have been used as solvents. The quantum yield of the σ-aryl complex in the reaction with anisole was ca. 0.08 for $\lambda = 365$ nm. Unlike the thermal reaction, the process induced by irradiation apparently does not depend on the acidity of the reaction solution. Another distinguishing peculiarity of the irradiation-stimulated reaction is the formation of the pure *para*-isomer in the cases of interaction with mono-substituted benzenes. No *para–meta* isomerization occurred at room temperature even under irradiation. The relative rates of the photoinduced reaction decrease in the following sequence:

$$OH > OC_2H_5 > OCH_3 > CH_3.$$

The correlation with σ^+ parameters gives the value $\rho^+ = -1.5$. The same sequence of rates and ρ^+ value have been obtained for the γ-induced process. The isotope effect for the photoreaction was $k_H/k_D \cong 2$. The effective activation energy of the photo-induced reaction is ca. 5 kcal mol^{-1}.

When a frozen solution of $PtCl_6^{2-}$ and arene in acetic acid (or H_2PtCl_6 and phenol in water) is irradiated at 77 K, the EPR spectrum is observed (Figure VII.7) containing the characteristic signals of platinum(III) complexes in

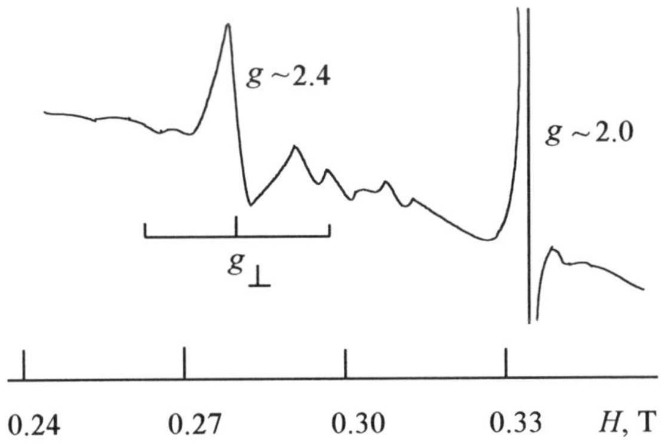

Figure VII.7. The EPR spectrum of the frozen solution of H_2PtCl_6 and phenol in water after light irradiation at 77 K.

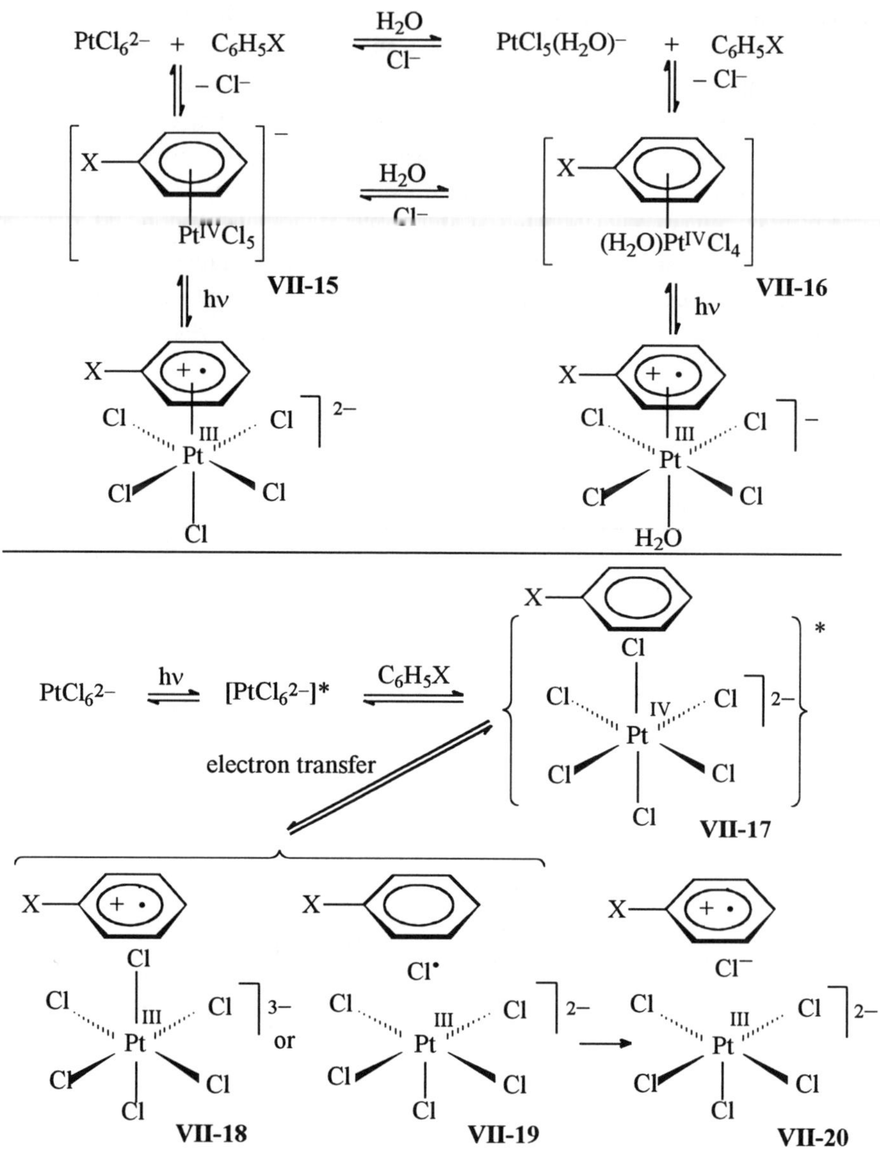

Scheme VII.9. Proposed electron transfer stage in the photoinduced reaction between $PtCl_6^{2-}$ and arene C_6H_5X.

perpendicular orientation in the region $g \cong 2.4$. These spectrum consists of an intensive central signal due to the non-magnetic ^{196}Pt isotope and two satellites due to the splitting on the ^{195}Pt isotope. Intensive narrow singlet resonances due to organic free radicals are located in the region $g \cong 2.0$. These signals can be assigned to cation-radicals or radicals derived from aromatics. The EPR spectrum also contains signals at $g \cong 4$ which may indicate the existence of dimers of paramagnetic species of radical pairs in these systems.

The proposed mechanism of the reaction involves electron transfer from an arene to a platinum(IV) compound to give an intermediate ion-radical pair, $[ArH]^{+\bullet}$ $[Pt^{III}Cl_5^{2-}]$ or $[ArH]^{+\bullet}$ $[Pt^{III}Cl_4(H_2O)^-]$ (Scheme VII.9) (for the photochemistry of $PtCl_6^{2-}$, see, e.g., [50]). The route to this ion-radical pair may be conceived ether as an electron transfer within the π-complex of ArH with $PtCl_5^{2-}$ (**VII-15**) or $PtCl_4(H_2O)^-$ (**VII-16**), or as an outer-sphere electron transfer to the excited complex of platinum(IV) with participation of a chlorine ligand (the transformation of structure **VII-17** into structures **VII-18** or **VII-19**). Subsequent extrusion of Cl^- from structure **VII-20** gives rise to the same ion-radical pair $[ArH]^{+\bullet}$ $[Pt^{III}Cl_5^{2-}]$.

The collapse of the ion–radical pair may produce the Wheland intermediate which after elimination of the proton is transformed into the σ-aryl complex of platinum(IV):

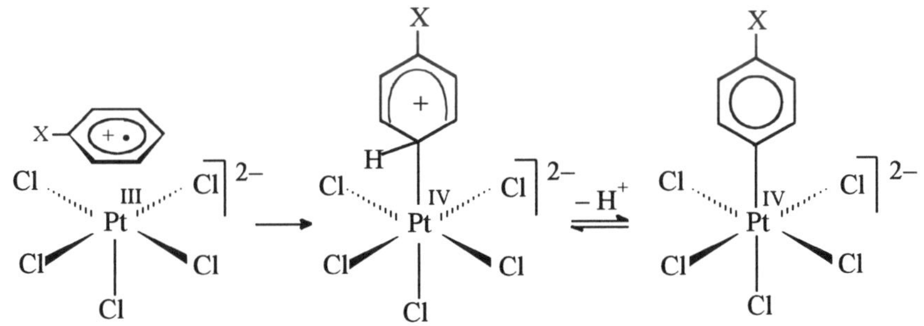

Para–meta isomerization in this case is impossible since the activation energy of the isomerization is ca. 25 kcal mol^{-1} and its rate at room temperature is too low.

The mechanism described above is analogous to that proposed for the nitration of arenes by tetranitromethane [51]. It is known that an arene and mercury trifluoroacetate form a charge-transfer complex that undergoes photoinduced electron transfer. When certain arenes are oxidized with this mercury derivative in trifluoroacetic acid, the EPR spectra of the arene radical cations can be observed [52]. It may be assumed that some thermal reactions of metal compounds with arenes also involve the electron transfer step. As for the mechanisms of both thermal and radiation-induced reaction of $PtCl_6^{2-}$ and arenes (particularly as concerned with the possibility of electron transfer stage under thermal conditions), they are not quite clear in detail and additional investigations are necessary [53].

References for Chapter VII

1. (a) Garnett, J. L.; Hodges, R. J. *J. Am. Chem. Soc.* **1967**, *89*, 4546. (b) Hodges, R. J.; Garnett, J. L. *J. Phys. Chem.* **1968**, *72*, 1673.
2. Gol'dshleger, N. F.; Tyabin, M. B.; Shilov, A. E.; Shteinman, A. A. *Zh. Fiz. Khim.* **1969**, *43*, 2174 (in Russian).
3. Skapski, A. C.; Stephens, F. A. *J. Chem. Soc., Chem. Commun.* **1969**, 1008.
4. Gol'dshleger, N. F.; Shteinman, A. A. *React. Kinet. Catal. Lett.* **1977**, *6*, 43.
5. Chatt, J.; Shaw, B. L. *J. Chem. Soc.* **1962**, 5075.
6. Tyabin, M. B.; Shilov, A. E.; Shteinman, A. A. *Dokl. Akad. Nauk SSSR* **1971**, *198*, 380 (in Russian).
7. (a) Hodges, R. J. Webster, D. E.; Wells, P. B. *J. Chem. Soc. A* **1971**, 3230. (b) Sakhabutdinov, A. G.; Zinchenko, I. S.; Zinchenko, S. V.; Shmidt, F. K. *Metalloorg. Khim.* **1990**, *3*, 1426 (in Russian).
8. Garnet, J. L.; West, J. C. *Aust. J. Chem.* **1974**, *27*, 129.
9. Hodges, R. J. Webster, D. E.; Wells, P. B. *J. Chem. Soc., Dalton Trans.* **1972**, 2577.
10. Repka L. F.; Shteinman, A. A. *Kinet. Katal.* **1974**, *15*, 805 (in Russian).
11. Rudakov, E. S.; Shteinman, A. A. *Kinet. Katal.* **1973**, *14*, 1346 (in Russian).
12. Gol'dshleger, N. F.; Moiseev, I. I.; Khidekel', M. L.; Shteinman, A. A. *Dokl. Akad. Nauk SSSR* **1972**, *206*, 106 (in Russian).
13. Rudakov, E. S. *Dokl. Akad. Nauk SSSR* **1976**, *229*, 149 (in Russian).
14. Rudakov, E. S.; Tretyakov, V. P.; Mitchenko, S. A.; Bogdanov, A. V. *Dokl. Akad. Nauk SSSR* **1981**, *259*, 899 (in Russian).
15. (a) Gol'dshleger, N. F.; Es'kova, V. V.; Shilov, A. E.; Shteinman, A. A. *Zh. Fiz. Khim.* **1972**, *46*, 1353 (in Russian). (b) Es'kova, V. V.; Shilov, A. E.; Shteinman, A. A. *Kinet. Katal.* **1972**, *13*, 534 (in Russian).
16. Nizova, G. V.; Zcilc-Krcvor, J. V.; Kitaygorodskiy, A. N.; Shul'pin, G. B. *Bull. Acad. Sci. USSR, Div. Chem. Sci.* **1982**, 2480.
17. (a) Basickes, N.; Hutson, A. C.; Sen, A.; Yap, G. P. A.; Rheingold, A. L. *Organometallics* **1996**, *15*, 4116. (b) Horváth, I.T.; Cook, R.A.; Millar, J.M.; Kiss, G. *Organometallics* **1993**, *12*, 8. (c) Sen, A.; Lin, M. *J. Chem. Soc., Chem. Commun.* **1992**, 508. (d) Sen, A.; Lin, M.; Kao, L.-C.; Hutson, A. C. *J. Am. Chem. Soc.* **1992**, *114*, 6385.

18. Tretyakov, V. P.; Rudakov, E. S.; Galenin, A. A.; Rudakova, R. I. *Dokl. Akad. Nauk SSSR* **1975**, *225*, 583 (in Russian).

19. Tretyakov, V. P.; Rudakov, E. S.; Bogdanov, A.V.; Zimtseva, G. P.; Kozhevina, L. I. *Dokl. Akad. Nauk SSSR* **1979**, *249*, 878 (in Russian).

20. Lavrushko, V. V.; Shilov, A. E.; Shteinman, A. A. *Kinet. Katal.* **1975**, *16*, 1479 (in Russian).

21. (a) Kushch, L .A., Lavrushko, V. V., Misharin, Yu. S., Moravskiy, A. P., Shilov, A. E. *Nouv. J. Chim.* **1983**, *7*, 729. (b) Rostovtsev, V. V.; Labinger, J. A.; Bercaw, J. E.; Lasseter, T. L.; Goldberg, K. I. *Organometallics* **1998**, *17*, 4530.

22. (a) Labinger, J. A.; Herring, A. M.; Bercaw, J. E. In *Homogeneous Transition Metal Catalyzed Reactions*; Moser, W. R., Slocum, D. W., Eds.; ACS: Washington, 1992, p. 221. (b) Labinger, J. A.; Herring, A. M.; Lyon, D. K.; Luinstra, G. A.; Bercaw, J. E.; Horváth, I. T., Eller, K. *Organometallics* **1993**, *12*, 895.

23. (a) Hutson, A. C.; Lin, M.; Basickes, N.; Sen, A. *J. Organometal. Chem.* **1995**, *504*, 69. (b) Kao, L-C.; Sen, A. *J. Chem. Soc., Chem. Commun.* **1991**, 1242.

24. Freund, M. S.; Labinger, J. A.; Lewis, N. S.; Bercaw, J. E. *J. Mol. Catal.* **1994**, *87*, L11.

25. Yamakawa, T.; Fujita, T.; Shinoda, S. *Chemistry Lett.* **1992**, 905.

26. (a) Shul'pin, G. B.; Nizova, G. V.; Shilov, A. E. *J. Chem. Soc., Chem. Commun.* **1983**, 671. (b) Serdobov, M. V.; Nizova, G. V.; Shul'pin, G. B. *J. Organometal. Chem.* **1984**, *265*, C12. (c) Shul'pin, G. B.; Nizova, G. V.; Geletiy Yu. V. *J. Gen. Chem. USSR* **1987**, *57*, 548. (d) Shul'pin, G. B.; Nizova, G. V.; Nikitaev, A. T. *J. Gen. Chem. USSR* **1985**, *55*, 1251.

27. (a) Shul'pin G. B.; Nizova, G. V. *Izv. Akad. Nauk SSSR, Ser. Khim.* **1983**, 669 (in Russian). (b) Shul'pin G. B.; Nizova, G. V.; Lederer, P. *J. Organometal. Chem.* **1984**, *275*, 283.

28. Nizova, G. V.; Serdobov, M. V.; Nikitaev, A. T.; Shul'pin, G. B. *J. Organometal. Chem.* **1984**, *275*, 139.

29. (a) Luinstra, G. A.; Wang, L.; Stahl, S. S.; Labinger, J. A.; Bercaw, J. E. *Organometallics* **1994**, *13*, 755. (b) Luinstra, G. A.; Wang, L.; Stahl, S. S.; Labinger, J. A.; Bercaw, J. E. *J. Organometal. Chem.* **1995**, *504*, 75. (c) Hindmarsh, K.; House, D. A.; van Eldik, R. *Inorg. Chim. Acta* **1998**, *278*, 32.

30. (a) Zamashchikov, V. V.; Rudakov, E. S.; Mitchenko, S. A.; Litvinenko, S. L. *Teor. Eksper. Khim.* **1982**, *18*, 510 (in Russian). (b) Zamashchikov, V. V.;

Rudakov, E. S.; Mitchenko, S. A.; Kitaygorodskiy, A. N.; Shul'pin, G. B. *Dokl. Akad. Nauk Ukrain. SSR, Ser. B* **1983** (No. 6), 33 (in Russian). (c) Zamashchikov, V. V.; Rudakov, E. S.; Mitchenko, S. A.; Nizova, G. V.; Kitaygorodskiy, A. N.; Shul'pin, G. B. *Bull. Acad. Sci. USSR, Div. Chem. Sci..* **1986**, 175. (d) Zamashchikov, V. V.; Rudakov, E. S.; Mitchenko, S. A.; Nizova, G. V.; Kitaygorodskiy, A. N.; Shul'pin, G. B. *Izv. Akad. Nauk SSSR, Ser. Khim.* **1984**, 1657 (in Russian). (e) Wang, L.; Stahl, S. S.; Labinger, J. A.; Bercaw, J. E. *J. Mol. Catal. A: Chem.* **1997**, *116*, 269.

31. Shul'pin, G. B.; Nizova, G. V.; Kitaygorodskiy, A. N.; Serdobov, M. V. *J. Organometal. Chem.* **1984**, *275*, 273.

32. Shestakov, A. F., *Koord. Khim.* **1980**, *6*, 117 (in Russian).

33. Lebedev, V. L.; Bagatur'yants, A. A.; Zhidomirov, G. M.; Kazanskiy, V. B. *Zh. Fiz. Khim.* **1983**, *57*, 1057 (in Russian).

34. Li, G.; Ding,; F.; Zhang, L. *Huaxue Xuebao* **1983**, *41*, 993 [*Chem. Abstr.* **1984**, *100*, 102417].

35. (a) Deeming, A. J.; Rothwell, I. P. *J. Organometal. Chem.* **1981**, *205*, 117. (b) Romeo, R.; Minniti, D.; Lanza, S. *J. Organometal. Chem.* **1979**, *165*, C36. (c) Harper, T. G. P.; Desrosiers, P. J.; Flood, T. C. *Organometallics* **1990**, *9*, 2523.

36. (a) Shestakov, A. F.; Shilov, A. E. *Khim. Fiz.* **1984**, *3*, 1591 (in Russian). (b) Vinogradova, S. M.; Shestakov, A. F. *Khim. Fiz.* **1984**, *3*, 371 (in Russian). (c) Shestakov, A. F.; Vinogradova, S. M. *Koord. Khim.* **1983**, *9*, 248(in Russian).

37. Shestakov, A. F. *Kinet. Katal.* **1990**, *31*, 586 (in Russian).

38. Zamashchikov, V. V.; Popov, V. G.; Rudakov, E. S. *Kinet. Katal.* **1994**, *35*, 372 (in Russian).

39. (a) Stahl, S. S.; Labinger, J. A.; Bercaw, J. E. *J. Am. Chem. Soc.* **1996**, *118*, 5961. (b) Holtcamp, M. W.; Labinger, J. A.; Bercaw, J. E. *Inorg. Chim. Acta* **1997**, *265*, 117.

40. (a) Holtcamp, M. W.; Labinger, J. A.; Bercaw, J. E. *J. Am. Chem. Soc.* **1997**, *119*, 848. (b) Holtcamp, M. W.; Henling, L. M.; Day, M. W.; Labinger, J. A.; Bercaw, J. E. *Inorg. Chim. Acta* **1998**, *270*, 467.

41. Hill, G. S.; Puddephatt, R. J. *Organometallics* **1998**, *17*, 1478.

42. (a) Siegbahn, P. E. M.; Crabtree, R. H. *J. Am. Chem. Soc.* **1996**, *118*, 4442. (b) Wick, D. D.; Goldberg, K. I. *J. Am. Chem. Soc.* **1997**, *119*, 10235. (c) Stahl, S. S.; Labinger, J. A.; Bercaw, J. E. *Inorg. Chem.* **1998**, *37*, 2422. (d) Goldberg, K. I. Presented at the *217th Am. Chem. Soc. Natl. Meeting*; Anaheim, March **1999**;

paper INOR No. 349. (e) Labinger, J. A.; Rostovtsev, V.; Scollard, J.; Wong-Foy, A.; Bercaw, J. E. Presented at the *217th Am. Chem. Soc. Natl. Meeting*; Anaheim, March **1999**; paper INOR No. 351. (f) Martin, R. L.; Hay, P. J. Presented at the *217th Am. Chem. Soc. Natl. Meeting*; Anaheim, March **1999**; paper INOR No. 435.

43. Shul'pin, G. B. *Zh. Obshch. Khim.* **1981**, *51*, 2100 (in Russian).

44. (a) Shul'pin, G. B.; Shilov, A. E.; Kitaigorodskii, A. N.; Zelle-Krevor, J. V. *J. Organometal. Chem.* **1980**, *201*, 319. (b) Shul'pin, G. B. *J. Organometal. Chem.* **1981**, *212*, 267. (c) Shul'pin, G. B.; Kitaygorodskiy, A. N. *J. Organometal. Chem.* **1981**, *212*, 275. (d) Shul'pin, G. B. *Kinet. Katal.* **1981**, *22*, 520 (in Russian). (e) Shul'pin, G. B.; Nizova, G. V.; Nikitaev, A. T. *J. Organometal. Chem.* **1984**, *276*, 115.

45. (a) Shibaeva, R. P.; Rozenberg, L. P.; Lobkovskaya, R. M.; Shilov, A. E.; Shul'pin, G. B. *J. Organometal. Chem.* **1981**, *220*, 271. (b) Kitaigorodskii, A. N.; Nekipelov, V. M.; Nikitaev, A. T.; Shul'pin, G. B. *J. Organometal. Chem.* **1984**, *275*, 295.

46. (a) Shul'pin, G. B. *Zh. Obshch. Khim.* **1980**, *50*, 2628 (in Russian). (b) Shul'pin, G. B.; Lederer, P.; Nizova, G. V. *J. Gen. Chem USSR*, **1982**, *52*, 1262. (c) Shul'pin, G. B.; Nizova, G. V. *J. Organometal. Chem.* **1984**, *276*, 109.

47. (a) Shul'pin, G. B.; Nikitaev, A. T. *Izv. Akad. Nauk SSSR, Ser. Khim.* **1981**, 1416 (in Russian). (b) Nikitaev, A. T.; Lontsov, V. V.; Nizova, G. V.; Shul'pin, G. B. *Zh. Organich. Khim.* **1984**, *20*, 2570 (in Russian).

48. (a) Shul'pin, G. B.; Shilov, A. E.; Nizova, G. V.; Yatsimirskiy, A. K.; Deyko, S. A.; Lederer, P. *Bull. Acad. Sci. USSR, Div. Chem. Sci.* **1986**, 2177. (b) Shul'pin, G. B.; Skripnik, S. Yu.; Deyko, S. A.; Yatsimirskiy, A. K. *Metalloorg. Khim.* **1989**, *2*, 1301 (in Russian). (c) Shul'pin, G. B.; Nizova, G. V. *Izv. Akad. Nauk SSSR, Ser. Khim.* **1982**, 1172 (in Russian). (d) Shul'pin, G. B.; Nizova, G. V. *Kinet. Katal.* **1981**, *22*, 1061 (in Russian). (e) Shul'pin, G. B.; Deyko, A. S.; Yatsimirskiy, A. K. *J. Gen. Chem. USSR* **1991**, *61*, 742.

49. (a) Shul'pin, G. B.; Nizova, G. V.; Shilov, A. E. *J. Chem. Soc., Chem. Commun.* **1983**, 671. (b) Shul'pin, G. B.; Nizova, G. V.; Shilov, A. E.; Nikitaev, A. T.; Serdobov, M. V. *Izv. Akad. Nauk SSSR, Ser. Khim.* **1984**, 2681 (in Russian). (c) Nizova, G. V.; Shul'pin, G. B. *Metalloorg. Khim.* **1990**, *3*, 463 (in Russian). (d) Serdobov, M. V.; Nizova, G. V.; Shul'pin, G. B. *J. Organometal. Chem.* **1984**, *265*, C12.

50. (a) Grivin, V. P.; Khmelinski, I. V.; Plyusnin, V. F. *J. Photochem. Photobiol. A* **1990**, *51*, 379. (b) Monreal, O.; Esmaeili, T.; Hoggard, P. E. *Inorg. Chim. Acta* **1997**, *265*, 279.

51. (a) Kochi, J. K. *Acta Chem. Scand.* **1990**, *44*, 409. (b) Kochi, J. K. *Acc. Chem. Res.* **1992**, *25*, 39. (c) Eberson, L.; Hartshorn, M P.; Radner, F.; Robinson, W. T. *J. Chem. Soc., Chem. Commun.* **1992**, 566.

52. Davies, A. G.; McGuchan, D. C. *Organometallics* **1991**, *10*, 329.

53. Tret'yakov, V. P.; Min'ko, L. A.; Rudakov, E. S. *Kinet. Katal.* **1990**, *31*, 763 (in Russian).

CHAPTER VIII

HYDROCARBON REACTIONS WITH HIGH-VALENT METAL COMPLEXES

This chapter is devoted to reactions of hydrocarbons and other C–H containing compounds with complexes of metals in a high oxidation state [1]. Section VIII.1 describes reactions which lead to isolable or detectable organometallic compounds. However, many known processes of hydrocarbon oxidation by high-valent metal complexes either do not involve a step of σ-organyl derivative formation at all or the formation of such intermediates is only suspected. High-valent metal intermediates have been proposed to take part in certain biological oxidation processes (see Chapter XI).

VIII.1. ELECTROPHILIC METALATION OF C–H BONDS ("ORGANOMETALLIC ACTIVATION")

Reactions of aromatic and saturated hydrocarbons with high-valent metal complexes may involve direct metalation (with or without subsequent decomposition of σ-organyl complex formed).

VIII.1.A. METALATION OF AROMATIC COMPOUNDS

The electrophilic substitution of hydrogen in arenes is known both for non-transition – Hg(II), Tl(III), Pb(IV) – and transition metals, such as Au(III), Pd(II), Pt(IV), Rh(III). All these reactions apparently proceed with the intermediate formation of Wheland complexes. Some parameters for these reactions are summarized in Table VIII.1 [2].

Table VIII.1. Metalation of arenes by metal compounds.

Parameter		Hg(II)	Tl(III)	Pb(IV)	Au(III)	Pd(II)	Pt(IV)
ρ (σ)		?	–12.8	?	?	?	–3.0
ρ^+ (σ^+)		–6.3	–6.9[a] –7.4[a] –8.3[a]	–9	?	–0.4 [a] –0.7 [a] –1.4 [a]	–1.5
k_H/k_D	Benzene	5.6	4.3		?	4.5 [a] 5.1 [a]	3.0
	Toluene	7.0	5.0		?	3.5	2.3
	Anisole	?	?	4.5	?	?	?
$S_f = \log(f_p^{Me}/f_m^{Me})$		1.3	1.5 [a]	1.7 [a] 1.8 [a]	? 2.0 [a]	–0.03	1.2
ortho-meta-para- Isomerization		yes	yes	?	?	no	yes
Transmetalation		yes	yes	?	?	yes	yes
Acceleration by light		yes	?	?	?	no	yes

[a] The data are given from different works.

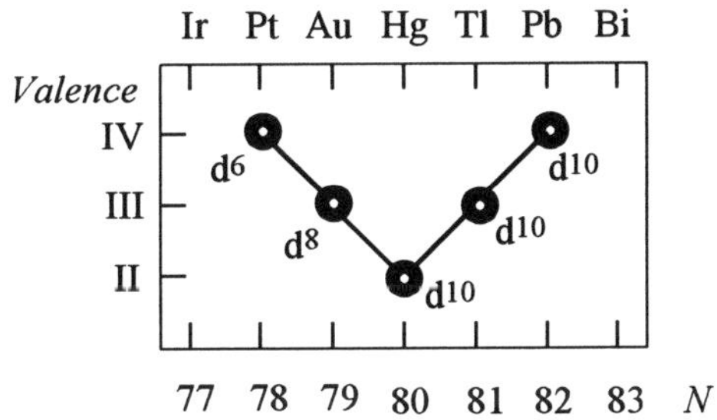

Figure VIII.1. Metal ions which are known to metalate arenes electrophilically to afford stable σ-aryl compound.

It is noteworthy that all the heavy metals capable of arene metalation to produce stable compounds arrange in the sequence shown in Figure VIII.1. All these metals belong to the same row of the Periodic Table. The valence of the metal active in the metalation decreases from IV to II on going from platinum to mercury, and then increases to IV again on going to lead [2].

Orthometalation

Great number of publications has been devoted to cyclometalation, especially to the reactions with C–H bonds of aromatic compounds (orthometalation) [3] which proceed according to the general equation:

Examples of orthometalation reactions [4] are presented in Scheme VIII.1.

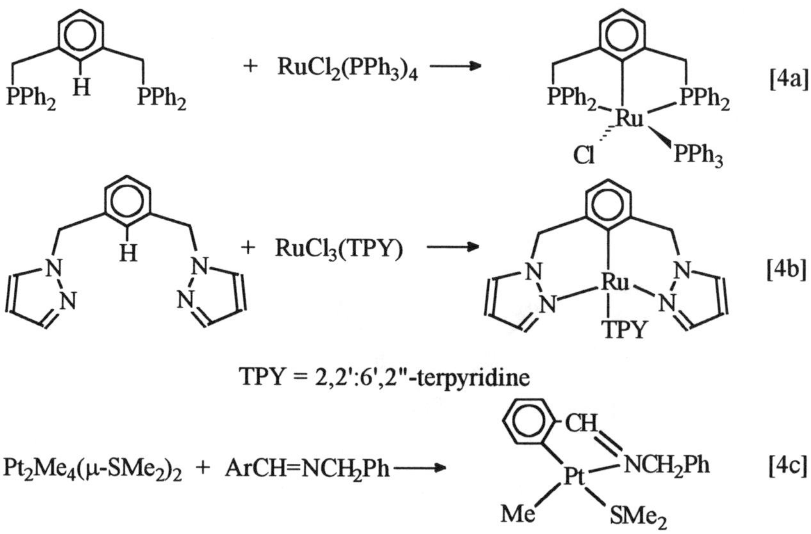

Scheme VIII.1. Examples of orthometalation reactions. Reference numbers are in brackets.

Palladium(II) derivatives also are commonly used in these transformations (orthopalladation) [5, 6] (Scheme VIII.2).

The cyclometalation reactions of benzylidenebenzylamines, -anilines, and -propylamine with palladium acetate occur via C–H electrophilic bond activation [7a]. The formation of a highly ordered four-centered transition state involving the C–H and Pd–O(acetato) bonds has been proposed:

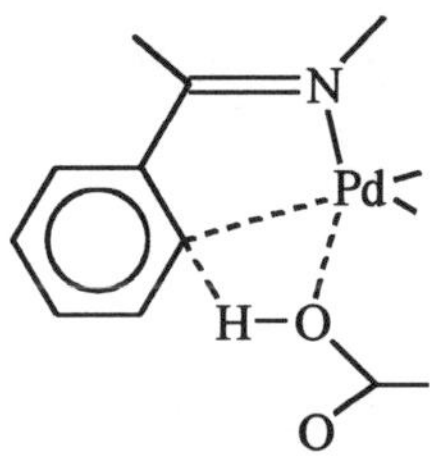

In this transition state, the neighboring terminal acetato group acts as a proton acceptor to produce acetic acid as a leaving group.

The activation parameters for the cyclopaladation in neat solvent

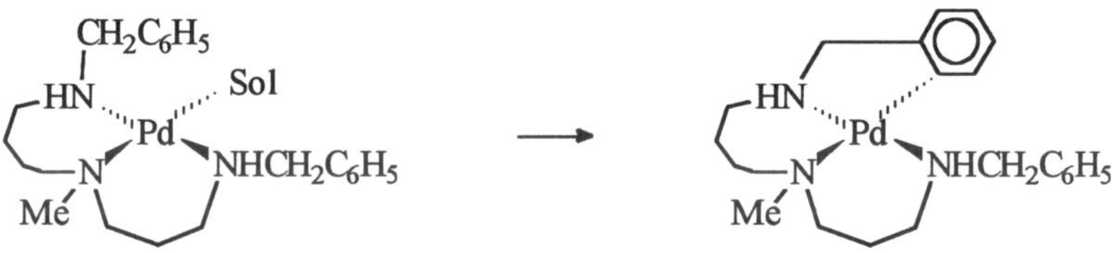

have been obtained as follows: $k^{298} = 5.74 \times 10^{-6}$ s^{-1}, $\Delta H^{\ddagger} = 24.8$ kcal mol^{-1} in DMF and $k^{298} = 3.13 \times 10^{-4}$ s^{-1}, $\Delta H^{\ddagger} = 20.0$ kcal mol^{-1} in DMSO [7b].

Intermolecular Reactions of Arenes with Palladium(II) Compounds

Despite the numerous works devoted to the oxidation of aromatics by palladium(II) compounds, the mechanism of these processes is still not clear in detail. There are three main types of such reactions.

1) Oxidative coupling of aromatic nuclei, as discovered by van Helden [8]:

$$2\,\mathrm{ArH} + \mathrm{Pd}^{2+} \longrightarrow \mathrm{Ar{-}Ar} + \mathrm{Pd}^0 + 2\,\mathrm{H}^+$$

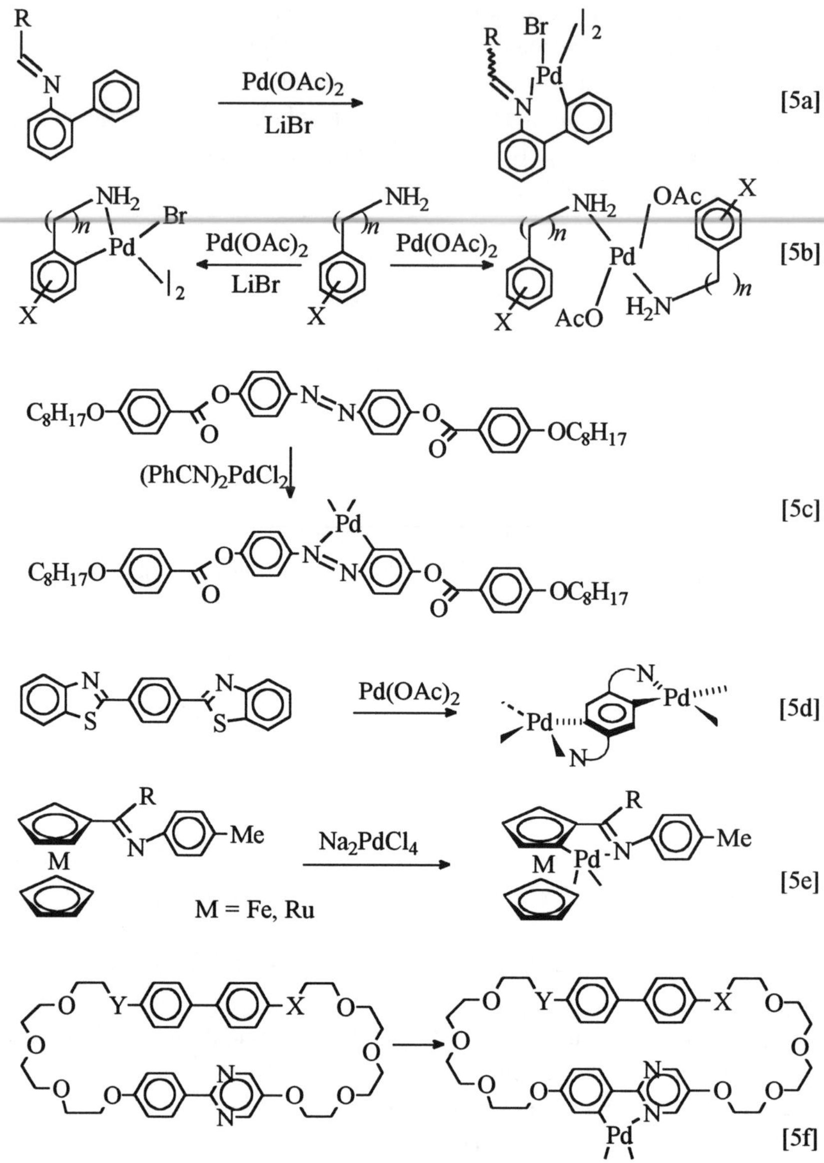

Scheme VIII.2. Examples of orthopalladation reactions. Reference numbers are in brackets.

The mechanism of this process has been investigated [9]. The reaction exhibits the features of electrophilic substitution: electron-releasing substituents in the aromatic nucleus accelerate the metalation. Usually σ-aryl complexes of palladium(II) are not stable and cannot be isolated. However, dialkyl sulfides stabilize the complexes which can be isolated as yellow crystals [10]:

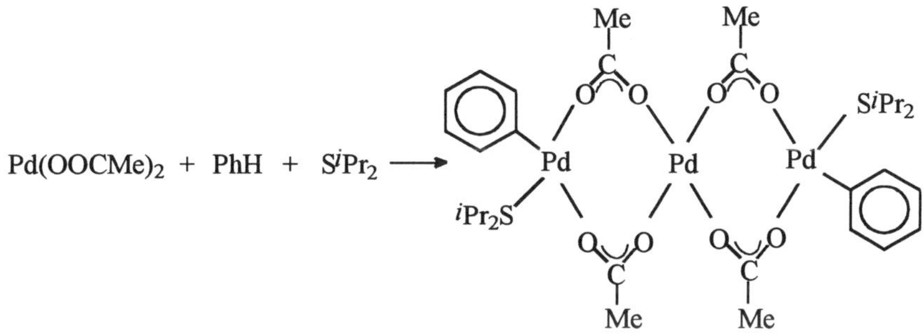

The kinetics of the oxidative coupling of benzene under the action of the system $PdCl_2$–NaOAc in acetic acid follows the equation:

$$v = k[\mathrm{Pd(II)}][(\mathrm{C_6H_6}],$$

sodium acetate being an obligatory reagent though it is not reflected in the equation. The first step of the oxidative coupling is proposed to be electrophilic substitution of a proton. The role of acetate in the coordination sphere of palladium is apparently to facilitate the abstraction of a proton. The reaction appears to be irreversible since there is no H–D exchange with the solvent.

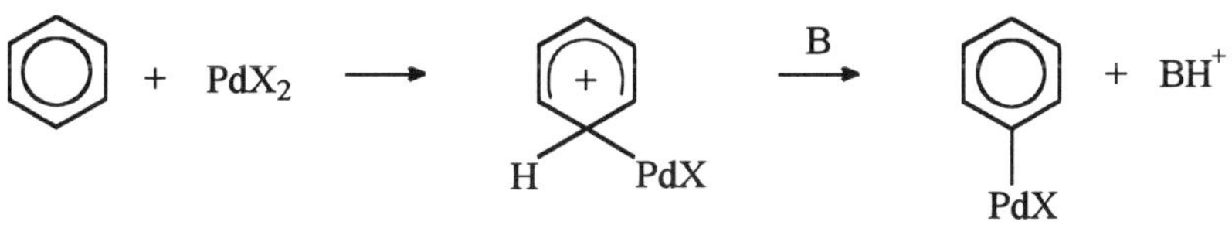

Then the disproportionation followed by the biaryl formation is assumed:

$$2\,ArPdX \rightleftharpoons Ar_2Pd + PdX_2$$
$$Ar_2Pd \longrightarrow ArAr + Pd$$

2) Oxidative coupling of arenes and olefins, as discovered by Fujiwara and co-workes [11]:

$$ArH + CH_2{=}CHR + Pd^{2+} \longrightarrow ArCH{=}CHR + Pd(0) + 2\,H^+$$

The first step of this reaction is the palladation of the arene. Olefin insertion into the Ar–Pd bond is then possible [12]:

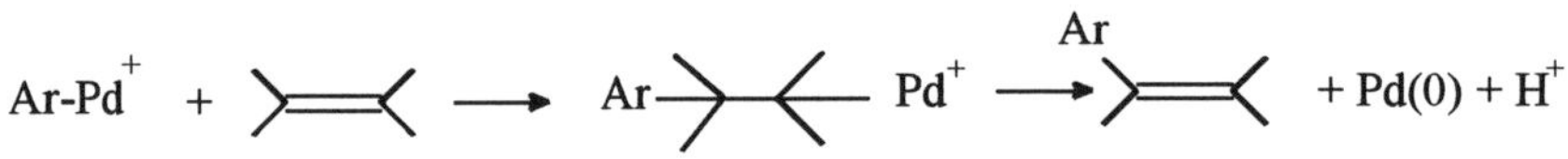

An alternative mechanism involves the formation of an aryl hydride intermediate followed by the insertion of the olefin into the Ar–Pd bond:

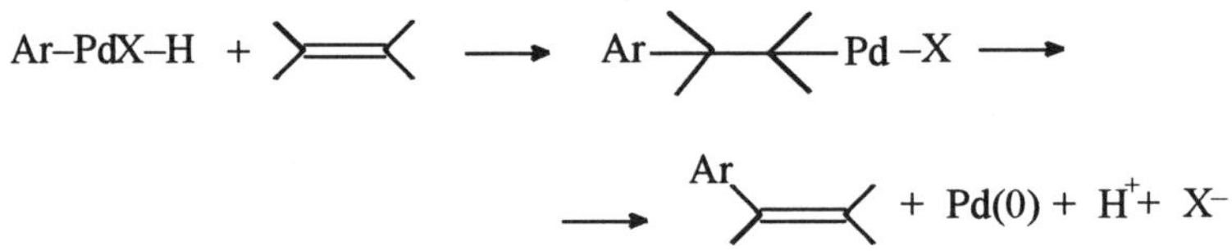

3) Acetoxylation:

$$C_6H_5CH_3 + CH_3COOH + Pd^{2+} \longrightarrow$$
$$AcOC_6H_4CH_3 + C_6H_5CH_2OAc + Pd(0) \qquad [13a]$$

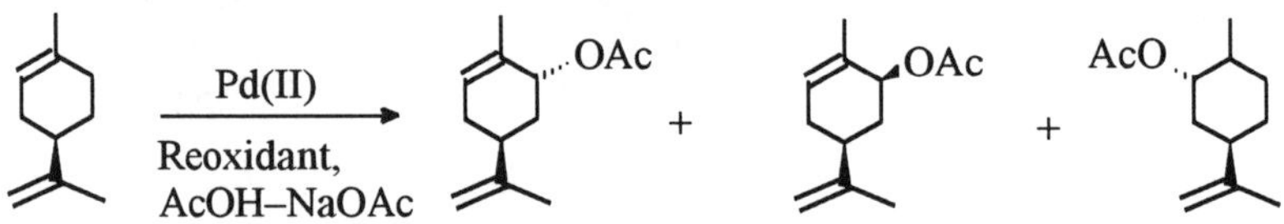

[13b]

4) Carbonylation by CO [14]:

$$ArH + CO + CH_3COOH + Pd^{2+} \longrightarrow ArCOOAc + Pd(0)$$

and by CO_2 [14, 15]:

$$ArH + CO_2 + Pd^{2+} \longrightarrow ArCOOH + Pd(0)$$

The addition of an oxidizing reagent makes reactions 1–4 catalytic with respect to palladium(II). Palladium compounds catalyze many other reactions invol-ving C–H bond activation. For example, benzaldehyde and benzoic acid can be produced by partial oxidation of toluene applying the fuel cell reaction in the gas phase using palladium black as the anode. The authors proposed a π-allyl–benzyl–Pd^{2+} complex as the reactive intermediate [16].

Intermolecular Reactions of Aromatic Compounds with Complexes of Other Metals

A few reactions of electrophilic metalation of an aromatic nucleus with transition metal complexes are known. The porphyrin complex of rhodium(III) PorphRhCl reacts with benzene and its derivatives in the presence of a silver salt to give the metalated arene [17]:

$$PorphRhCl + C_6H_5X \longrightarrow PorphRh(p\text{-}C_6H_4X)$$

Exclusively the *para*-metalated product is formed. There is a correlation between logarithms of relative rates of metalation and Hammett σ constants of the substituent X (ρ = –5.43).

A ruthenium(II) complex in the presence of a methyl derivative of aluminum metalates arenes [18]:

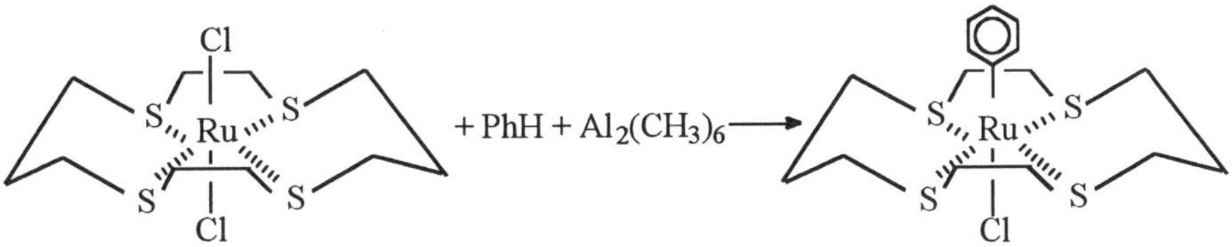

An interesting C-metalation of a coordinated pyrrol ligand in a rhenium complex has been described [19a]. Methylated aromatic hydrocarbons have been activated by $(ZrCl_4)_n$ [19b]. A novel mode of electrophilic activation of an aliphatic C–H bond, assisted by porphyrin complexes of Zr(IV) and achieved by the use of hydrides of lithium, sodium or potassium, has been reported [19c]. It has been shown that the ligands 2,3,5,6-tetraphenylphenoxide and 3,5-dimethyl-2,6-diphenylphenoxide undergo intramolecular activation by tantalum alkylidene groups at rates 20 and 100 times slower than that of the simple 2,6-diphenylphenoxide ligand [19d].

VIII.1.B. METALATION OF sp^3-C–H BONDS

Transition metal complexes can react with aliphatic amines, phosphines and analogous compounds to cleave the C–H bonds at sp^3-hybridized carbon atoms and produce cyclometalated derivatives [20] (Scheme VIII.3). If in the starting complex the metal ion is in a high oxidation state, the metalation occurs apparently as electrophilic substitution.

Another type of reaction which occurs between a saturated C–H bond and a high-valent metal complexes is metathesis of the C–H bond according to equation:

$$\text{M–R} + \text{R}'\text{–H} \rightleftharpoons \text{M–R}' + \text{R–H}$$

For example, the exchange between a methyl complex of lutetium or yttrium and labeled methane proceeds as metathesis [21]:

$$\text{Cp}^*{}_2\text{M–CH}_3 + {}^{13}\text{CH}_4 \rightleftharpoons \text{Cp}^*\text{M}^{13}\text{–CH}_3 + \text{CH}_4$$

Analogously, the reaction with other C–H compounds RH gives a σ-organyl derivative of scandium [22]:

$$\text{Cp}^*\text{Sc–CH}_3 + \text{RH} \longrightarrow \text{Cp}^*\text{Sc–R} + \text{CH}_4$$

Here RH is $^{13}CH_4$, arenes, or alkynes.

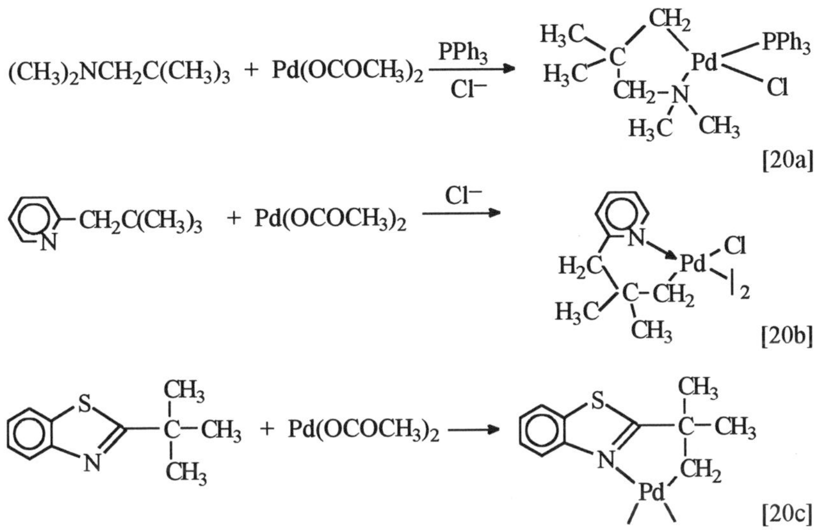

Scheme VIII.3. Cyclopalladation of sp^3-hybridized carbon atoms. Reference numbers are in brackets.

The exchange reaction of methane with a lutetium complex apparently proceeds via the transition state **VIII-1** rather than via oxidative addition involving intermediate **VIII-2**. Quantum-chemical calculations have been carried out [23] for the mechanism including *inter alia* a four-electron four-center transition state of the type **VIII-1**.

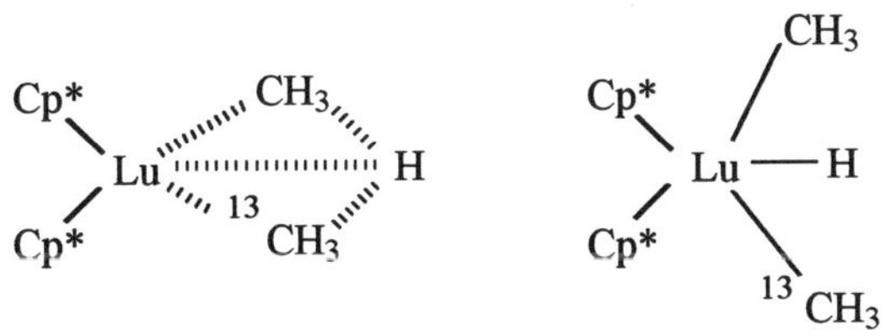

VIII-1 **VIII-2**

A calculated [23b] barrier for methane exchange is 28.0 kcal mol^{-1} for the reaction

$$Cl_2ScCH_3 + CH_4$$

and 61.7 kcal mol^{-1} for the interaction

$$Cl_2AlCH_3 + CH_4.$$

Examples of some other reactions between alkanes and electrophilic metal complexes are shown in Scheme VIII.4. The electrophilic "Cp^*Ru^+" fragment generated by protonation of $[Cp^*Ru(OMe)]_2$ with CF_3SO_3H is capable of activating the bonds C–H, C–O, and C–C in various organic compounds.

Irradiation (λ 180–360 nm) of an aqueous solution of an alkane C_2–C_7 and mercury(II) sulfate in air gives rise to the alkane oxidation products (predominantly carbon dioxide) [25]. The authors tentatively assumed that the key step of the reaction is the interaction of the hydrocarbon with the photoexcited Hg(II) species. This process is considered as electrophilic substitution:

$$RH + [Hg(II)]^* \longrightarrow RHg(II) + H^+$$

The alkyl mercury derivative thus formed can be then photolized to produce Hg(I) and radicals R• which react further with molecular oxygen. However, an alternative explanation of the reaction could be proposed.

A novel mode of electrophilic activation of aliphatic C–H bonds also has been reported. This reaction is induced by Zr(IV) porphyrin complexes and achieved through the use of lithium, sodium or potassium hydrides [26].

VIII.1.C. SOME SPECIAL CASES

Numerous cases of C–H bond activation are known, for which the reaction mechanism cannot be definitely attributed to either typical oxidative addition to a nucleophilic metal center or to an electrophilic substitution at an electron-deficient metal ion. In some cases the mechanism is not at all clear.

The intramolecular activation of a C–D bond in d_6-benzene by the complex $(Me_3P)_2Pt(CH_2CMe_3)(SO_3CF_3)$ apparently begins with the formation of the coor-

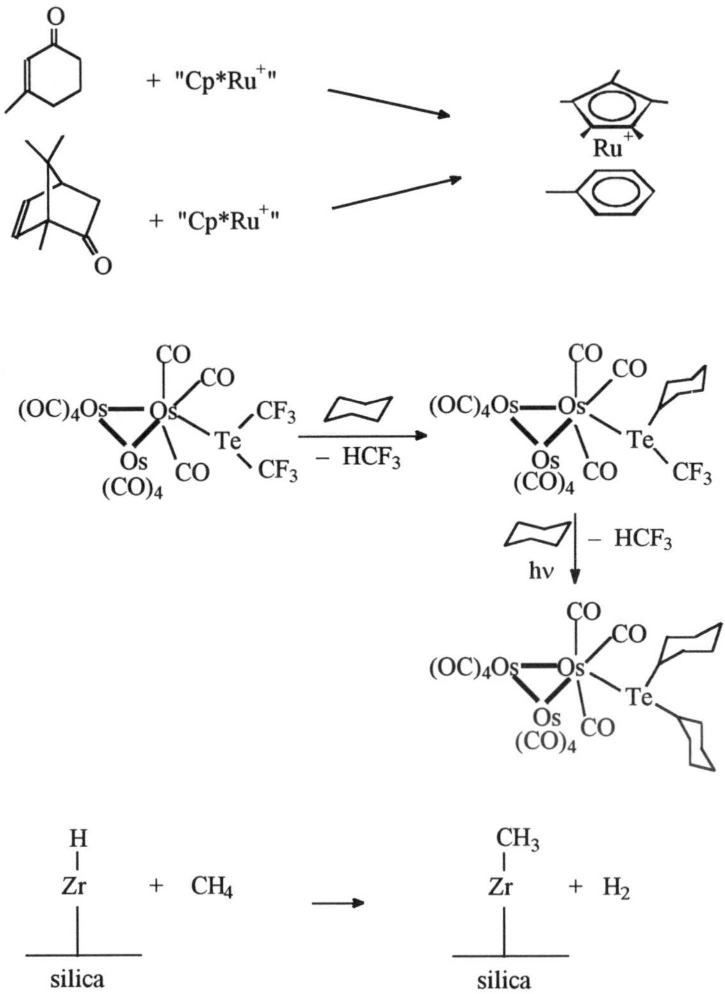

[24a]

[24b]

[24c]

[24d]

Scheme VIII.4. Activation of sp^3-hybridized C–H bonds. Reference numbers are in brackets.

dinatively unsaturated species $[(PMe_3)_2Pt(CH_2CMe_3)]^+$, which gives rise to an arylplatinum hydride complex with d_6-benzene as a result of oxidative addition [27]. The product $(Me_3P)_2Pt(C_6D_5)(SO_3CF_3)$ is formed after the elimination of

CH_2DCMe_3. However, the authors do not rule out the possibility that the positively charged species carries out an electrophilic attack on the benzene ring.

Thermolysis of complex **VIII-3** in d_6-benzene leads to the appearance of the intramolecular metalation product **VIII-5** [28]. The reaction takes place in several steps and the intermediate complex **VIII-4** has been isolated. The introduction of several substituents into the benzene ring has virtually no effect on the rate of metalation, which enables the authors to regard the transformation **VIII-4** ⟶ **VIII-5** not as a typical electrophilic metalation by the Hf(IV) ion but as a process which proceeds via a four-membered transition state.

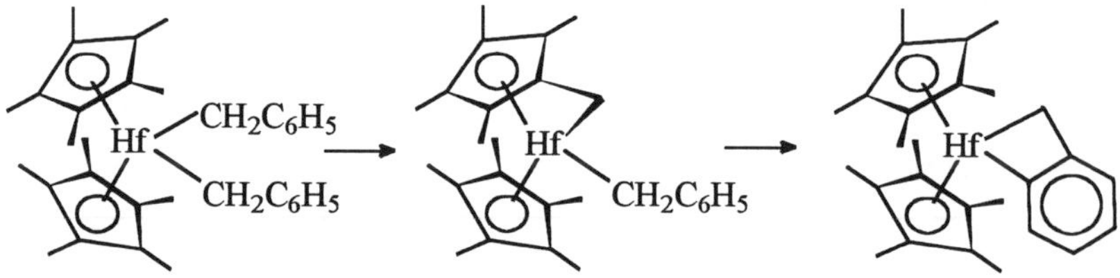

VIII-3 **VIII-4** **VIII-5**

The mechanism of the reaction between tetraarylrhenium and phosphines is not completely clear [29]:

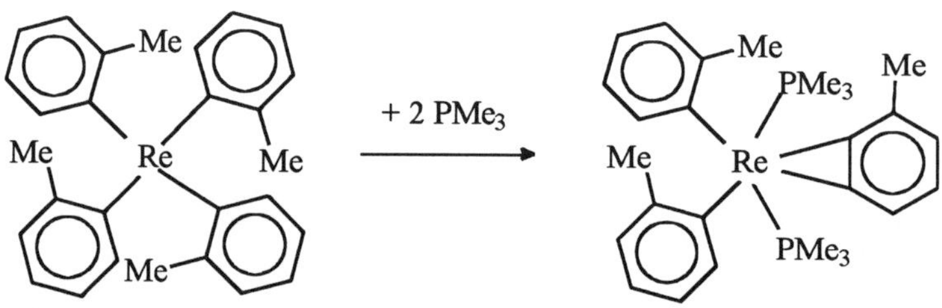

According to the authors' proposal, one step of the reaction is oxidative addition of the C–H bond in the *ortho*-position to the metal atom.

The reduction of the complex $(NH_3)_5Os$(trifluoromethylsulfonate) with cobaltocene in the presence of *N*-methylpyridinium forms the π-complex which rearranges into the σ-aryl complex of osmium(II) [30a]:

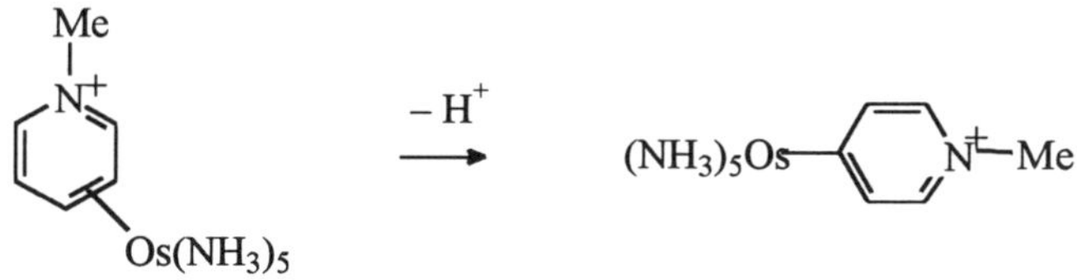

The mechanism of this activation of the C–H bond is unknown although the reaction may proceed by an oxidative addition. Generally, the pentaammine-osmium(II) system is known to activate phenols, anilines, and anisoles toward electrophilic addition and substitution reactions by binding the aromatic ligand in an η^2-fashion. Protonation, for example, results in the formation of a heterotriene system [30b]:

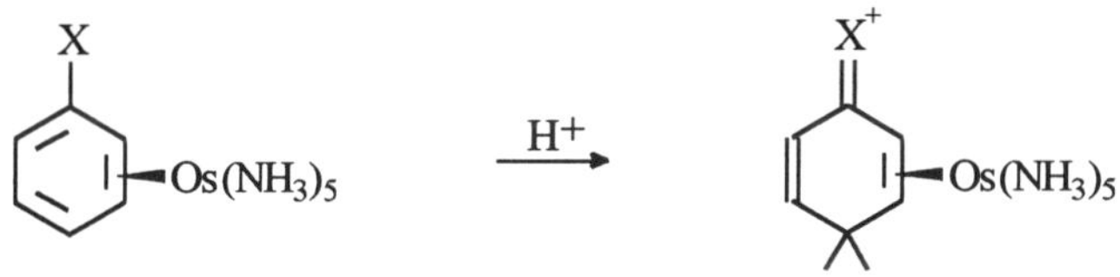

The interaction between the complex $Cp^*Ir(CH_3)_2(DMSO)$ and arene C_6H_5X leads to the formation of σ-aryl derivatives $Cp^*Ir(CH_3)(C_6H_4X)(DMSO)$ and the evolution of methane [31]. The rate of metalation decreases in the following sequence of X:

$I > NO_2 > CF_3 > CN > COOCH_3 > COCH_3 > Br > H >$
$OCH_3 > Cl > F > NH_2$

The metal atom substitutes only the *meta* and *para* positions of the benzene ring. In the case of X = NH_2, only the *meta* isomer is formed. The authors propose that the reaction proceeds as DMSO is replaced by the arene with subsequent oxidative addition of the C–H bond to iridium to produce an iridium(IV) derivative. Then reductive elimination of methane and addition of DMSO gives the final product. A synchronous elimination of methane and addition of the arene cannot be ruled out however.

Dihalogeno*bis*(triphenylphosphine)palladium(II) complexes react [32] with alkanes and arenes to form unusual hydride complexes $Pd(H)(R)X_2(PPh_3)_2$. The authors propose that the reaction involves the oxidative addition of the hydrocarbon to palladium(II), well-known "electrophilic" center. However, other authors have not reproduced these results.

The pentamethylcyclopentadienyl complex of iridium relative to the iridium complexes that activate alkanes via typical oxidative addition, also reacts [33] with arenes by an unknown mechanism:

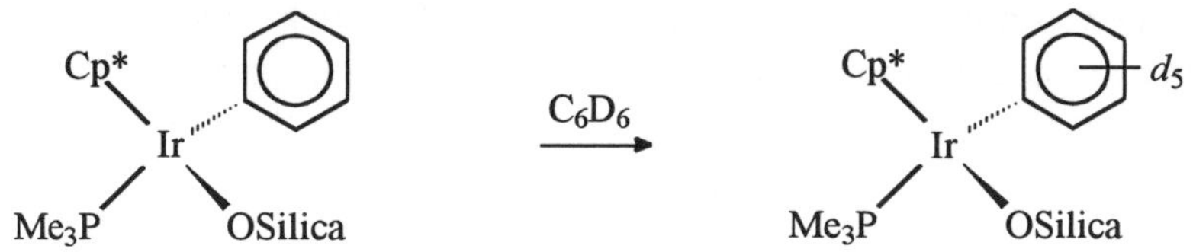

The mechanism proposed for the formation of the complex $W(PMe_3)_4(\eta^2\text{-}CH_2PMe_2)H$ from $W(PMe_3)_6$ involving a concerted C–H bond metalation via the intermediate or transition state **VIII-6** has been discounted by the lack of an observable kinetic isotope effect [34].

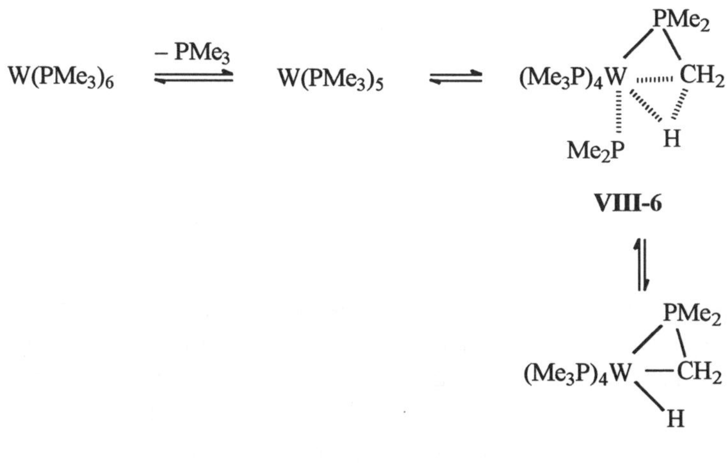

VIII-6

The formation of the oxametallacycles in the reactions of the complex $W(PMe_3)_4(\eta^2\text{-}CH_2PMe_2)H$ with phenols does not proceed via oxidative addition of the OH group at the tungsten center, but rather by direct reaction at the W–C bond of $W(PMe_3)_4(\eta^2\text{-}CH_2PMe_2)H$ [34].

In the reaction between a tantalum complex and benzene, thermolysis of the initial complex produces an electrophilic three-coordinate imido intermediate which is capable of metalating benzene (R: t-Bu_3Si) [35a]:

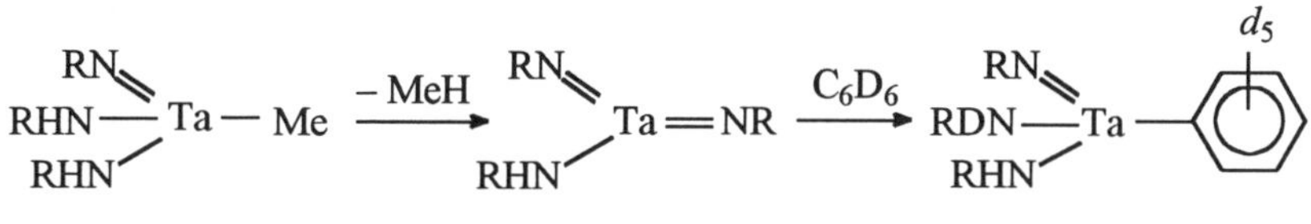

It has been concluded that the interaction of an analogous vanadium complex with hydrocarbons does not occur by a simple σ bond metathesis mechanism [35b]. The alkylidene tantalum(V) complexes undergo intermolecular cyclometalation of the aryloxide ligand [35c] (Scheme VIII.5).

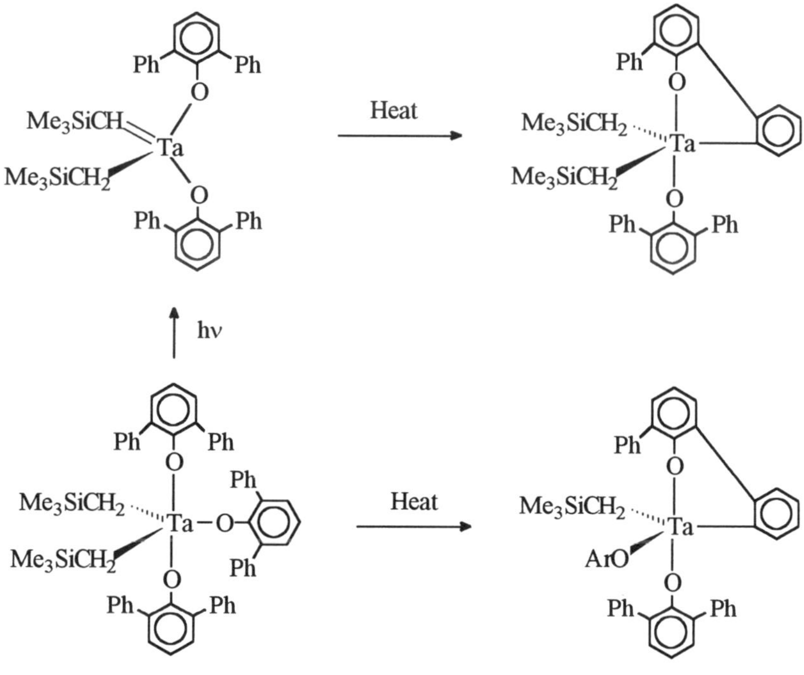

Scheme VIII.5. Transformations of tantalum(V) complexes.

The mechanism proposed for the complex $[Pt(PPh_3)_2(\eta^3\text{-}C_3H_3)](BF_4)$ for hydropyrrolylation [36] with pyrrole:

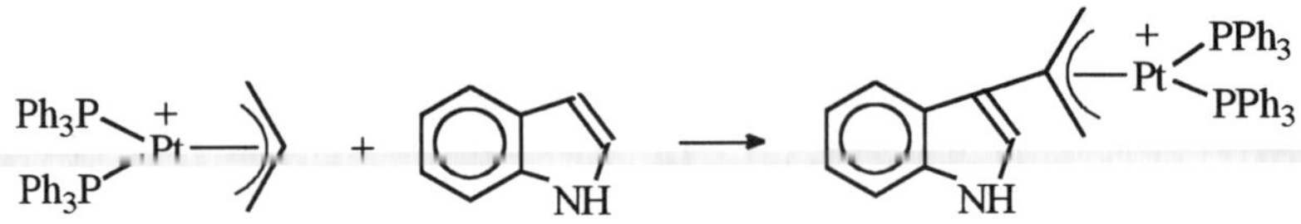

involves electrophilic aromatic substitution:

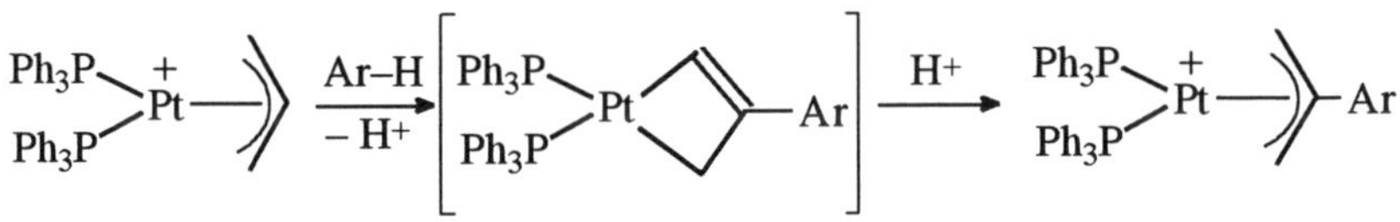

Reaction of the rhenium(V) complex $[Re(O)Cl_2(C_5Me_5)]$ with H_2S leads to the products of activation of a ring methyl substituent [37]:

$$[Re(O)Cl_2(C_5Me_5)] + H_2S \longrightarrow [ReO\{(SCH_2)C_5Me_4\}(C_5Me_5)]$$

Cyclopalladation of the racemic *N*-methyl-α-*tert*-butylbenzylamine **VIII-7** gave the products of activation of the *ortho*-C–H bond in the phenyl ring of **VIII-8** and of the sp^3-C–H bond of the methyl group of **VIII-9**. Additional heating of complexes **VIII-8** and **VIII-9** led to the palladium mediated cleavage of an sp^3-C–sp^3-C bond between two non-activated carbons to produce derivative **VIII-10** [38] (Scheme VIII.6).

Finally, photolysis of the Re(V) oxo–idodide compound $(HBpz_3)ReO(I)Cl$ in arene (ArH) solvents gives the aryl complexes $(HBpz_3)ReO(Ar)Cl$ [39]. Substituted arenes react with electrophilic selectivity and the authors propose that the reactive species is a rhenium(V) oxo complex which can add aromatic C–H bonds across the Re=O linkage.

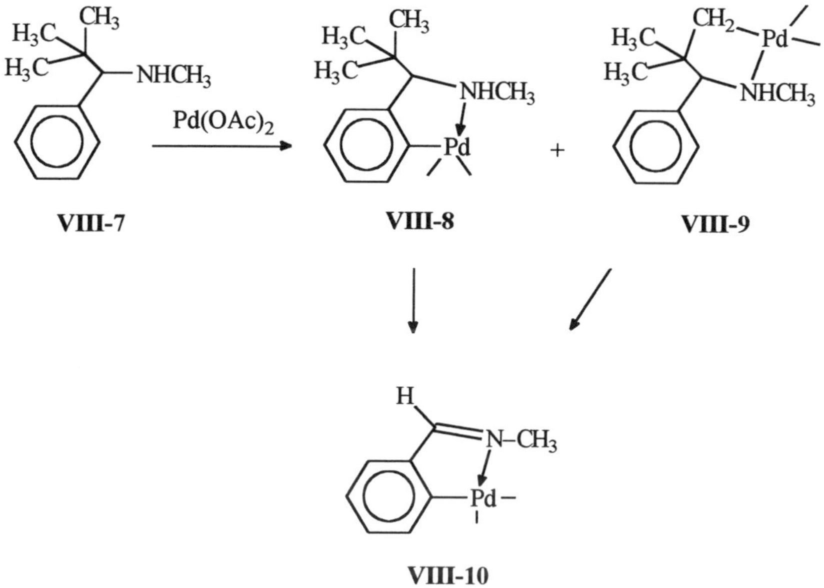

Scheme VIII.6. Activation of *ortho*-Ar–H, sp^3-C–H, and sp^3-C–sp^3-C bonds in the cyclopalladation reaction.

VIII.2. OXIDATIONS IN AQUEOUS AND ACIDIC MEDIA PROMOTED BY METAL CATIONS AND COMPLEXES

Alkanes can be oxidized in the presence of some transition metal complexes in aqueous and acidic media For example, in concentrated sulfuric acid the oxidative properties of the complexes are enhanced. Solutions of derivatives of palladium(II), platinum(III), manganese(III) and mercury(II) as well as some other compounds (hydrogen peroxide, ammonium persulfate, nitric acid and even concentrated sulfuric acid itself) can be used as oxidants. In the cases of metal-free oxidants the active species are apparently electrophiles such as NO_2^+ or SO_3H^+ (for nitration of aromatics, see, for example, recent publication [40] and references therein).

VIII.2.A. KINETICS AND FEATURES OF THE OXIDATION IN AQUEOUS AND ACIDIC MEDIA

According to Rudakov and co-workers (see reviews [41] and original papers [42]), due to the difficulty of the C–H bond rupture, the alkane activation usually begins from the preactivation of a reagent M (step *a*), followed by the formation of a weak adduct alkane–reagent (step *b*) which then is transformed into the first product Y_1:

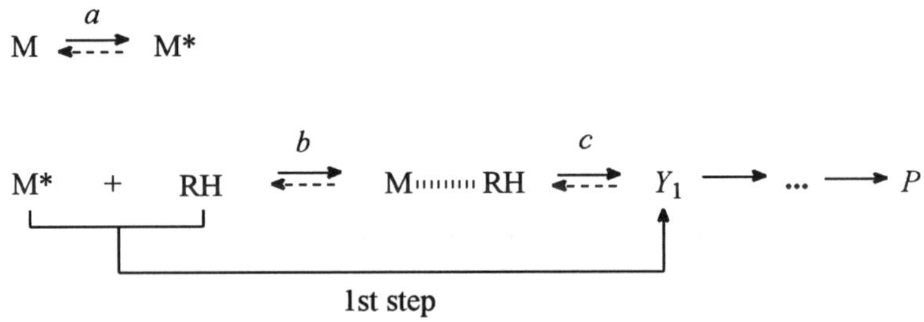

Here M* is the direct reagent which is chemically rebuilt and has the enhanced energy and necessary structural possibilities, including coordination vacancy, to participate in the reaction. A sum of elementary acts *b*+*c*, which cannot be easily discriminated, has been called "1st step" since in this stage of the reaction the alkane is involved in the interaction for the first time. A sequence of steps *a*+*b*+*c*

$$\text{M} + \text{RH} \longrightarrow Y_1$$

is the alkane activation in a broad sense. Steps *a*, *b*, and *c* can be catalyzed by acids, metal complexes or stimulated by light. The first product Y_1 in the sequence of the following transformations leads to the final products *P*. The activation in various systems was investigated kinetically by measurement of the substrate consumption. The rate of the reaction v follows the kinetic equation

$$v = -\frac{\text{d[RH]}}{\text{d}t} = k\text{[RH][M]} \qquad \text{when [RH]} \ll \text{[M]}$$

Two methods for the kinetic studies of the substrate activation have been developed. The first (kinetically distributive) method is based on the measurement of the effective pseudo-first order rate constants

$$k_1 = -\frac{1}{[\mathrm{RH}]_g}\frac{d[\mathrm{RH}]_g}{dt} \qquad \text{when } [\mathrm{RH}] \ll [\mathrm{Oxidant}]$$

for the reaction in a closed shaken reactor containing gas and solution. Values of k can be calculated from the equation

$$k = k_\lambda(1 + \alpha\lambda),$$

where

$$\alpha = [\mathrm{RH}]_0/[\mathrm{RH}]_g$$

is the thermodynamic coefficient of the substrate distribution between the gas phase and the solution, $\lambda = (V_g/V_S)$ is the ratio of values of gas and solution in the reactor, and k is the true (liquid-phase) constant of the rate at $\lambda = 0$.

The second (syringe-reactor) method permits the study of the kinetics of the activation in the absence of the gas phase. Though the solubility of alkanes in water or sulfuric acid is very low it is sufficient to follow the decreasing concentration of RH in the solution by GLC tool. The competitive versions of the method (when two substrates, R_AH and R_BH, are used in the same experiment), based on the equation

$$\lg[R_AH] = (k_A/k_B)\lg[R_BH] + \mathrm{const},$$

are the most effective for the investigation of selectivity and kinetic isotope effect (KIE).

The essential temperature dependence of KIE of alkane activation has been detected [43]. In all cases, the value of KIE is decreased when temperature rises, according to the equation

$$k_H/k_D = (A_H/A_D)\exp[(E_D - E_H)/RT],$$

the value $E_D - E_H$ being ca. 1–2 kcal mol^{-1}.

The 5/6 effect (i.e., the ratio of rates of the C–H bond splitting in cyclopentane and cyclohexane) turned out to be an even more informative test for the evaluation of the mechanism of the reaction than KIE. The dependence of the 5/6 effect on temperature has been found:

$$k_5/k_6 = (A_5/A_6)\exp[(E_6 - E_5)/RT]$$

Only for three reagents (NO_2^+, Pd^{2+} and SO_3H^+) is the value $E_6 - E_5 < 0$, all other systems investigated have exhibited $E_6 - E_5 > 0$.

The following common features of the oxidation reaction in sulfuric acid has been evaluated [41c].

1) All the reactions follow the kinetic equation of the second order:

$$-\,d[RH]/dt = k_2[RH][M^{2+}]$$

The only exception are manganese(III) complexes, since in this case, the active species is apparently a radical formed during the manganese complex decomposition, which occurs independently of the alkane.

2) The cleavage of the C–H bond takes place at the rate determining step of the reaction. KIE is ca. 2.0 ± 0.2 for almost all systems investigated and is the same for the cleavage of both tertiary and secondary C–H bonds. However, for CrO_3, MnO_3, and O_3 the values of KIE is higher (4 ± 1). It is interesting that for these oxidants the 5/6 effect is < 1, i.e., the rate of the oxidation in this case follows the decreasing energy of the C–H bond:

$$C_3 > C_4 > C_6 > C_5 > C_7 > C_8.$$

3) In all cases, the selectivity for the rupture of the C–H bond decreases in the sequence:

tertiary > secondary > primary.

This selectivity parameter is different for different reagents. The 3° : 2° and 2° : 1° ratios decrease in the sequence:

$$Hg(II) > Pt(III) > H_3SO_4^+ > NO_2^+ > Pd(II) > S_2O_8^{2-} \approx Mn(III).$$

The ratio 3°: 2° is 3000 for the most selective system, Hg(II)–H_2SO_4, and only 12 for the system exhibiting poor selectivity, Mn(III)–H_2SO_4.

4) For alkanes containing *tert*-C–H bonds, the rate of the oxidation rises in the sequence

$$Me_3C–H < Me_2EtC–H < MeEt_2C–H < Et_3C–H.$$

For all systems of the type Mn^+–H_2SO_4, the rates obey the Taft equation

$$\log k = const + \rho^*\Sigma\sigma^* + \delta\Sigma E_S^0$$

with the value ρ^* changing in the range from −3 to −1, which is typical for the abstraction of the hydrogen atom.

5) The rate of the reaction increases exponentially with an increase of the acidity function according to the Hammett equation

$$\log k = const - mH_0,$$

which shows that protonation is a mode of activation of the species reacting with the alkanes.

6) A common feature of the systems that are capable of oxidizing in sulfuric acid is the participation of bases in the process. For example, in the case of the system NO_2^+–H_2SO_4, the rate dependence passes through a maximum at a concentration of 92–94%.

VIII.2.B. ALKANE FUNCTIONALIZATIONS IN PROTIC MEDIA

Sen *et al.* have described several metal-catalyzed systems for the activation and functionalizations of alkanes in protic media (see reviews [44]). The main product of the methane oxidation in 98% sulfuric acid is CH_3OSO_3H [45a]. A variety of oxidants, such as $S_2O_8^{2-}$, Ce(IV), Pd(II) and Hg(II) have been employed [45b]. In the case of ethane, the observed products are CH_3OSO_3H and $HO_3SCH_2CH_2OSO_3H$. Proposed steps of the reaction are shown in Scheme VIII.7.

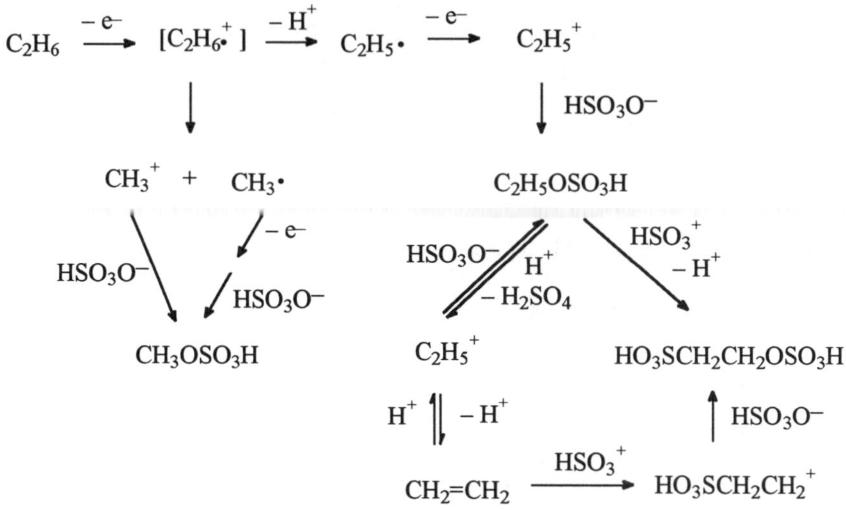

Scheme VIII.7. Ethane oxidation in H_2SO_4.

Remarkable similarity in rate constants for methane and methanol under the action of the platinum(II) ion has been observed: the ratio of rate constants for methane versus methanol oxidations is 0.17. The methyl group of ethanol is oxidized to produce 1,2-ethanediol as the predominant product. The electrophilic pathway of the activation of C–H compounds is presented in Scheme VIII.8 [45a].

The proposed [45b] radical mechanism for the formation of methansulfonic acid in the $S_2O_8^{2-}$ initiated reaction involves the steps of initiation

$$K_2S_2O_8 \longrightarrow 2\,KSO_4\bullet$$

$$KSO_4\bullet + CH_4 \longrightarrow KSO_4H + CH_3\bullet$$

and propagation

$$CH_3\bullet + SO_3 \longrightarrow CH_3SO_3\bullet$$

$$CH_3SO_3\bullet + CH_4 \longrightarrow CH_3SO_3H + CH_3\bullet$$

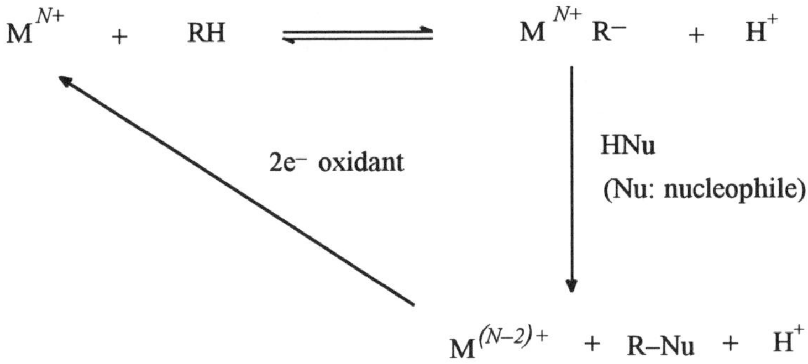

Scheme VIII.8. Electrophilic activation of C–H compounds.

Periana *et al.* [46] reported selective catalytic oxidation of methane by sulfuric acid to produce methyl bisulfate at 180 °C. The reaction is catalyzed by mercuric ions. Sulfur dioxide is the product of sulfuric acid reduction. At methane conversion of 50%, 85% selectivity to methyl bisulfate is observed. The major side product is carbon dioxide. The mercury turnover efficiency is $10^{-3}\ s^{-1}$. The Hg^{2+} ion reacts with methane as an electrophile substituting a proton and producing initially an intermediate methylated mercury complex, CH_3HgOSO_3H. The complex is formed in appreciable steady-state concentration and was observed directly by ^{13}C and ^{199}Hg NMR spectroscopy. Under the reaction conditions, methyl mercuric bisulfate decomposes to produce methyl bisulfate, CH_3OSO_3H, and the reduced mercurous species, Hg_2^{2+}. The catalytic cycle is completed by reoxidation of Hg_2^{2+} with H_2SO_4 to regenerate Hg^{2+} and to form other products, SO_2 and H_2O.

$$CH_4 + 2\,H_2SO_4 \xrightarrow{Hg(II)} CH_3OSO_3H + 2\,H_2O + SO_2$$

In the presence of D_2SO_4, the incorporation of deuterium into the methane was observed in the presence of mercuric salts under reaction conditions. Independent reaction of specially prepared methyl mercuric bisulfate confirmed that protolysis occurs at 180 °C and this provides the mechanism for the observed isotope exchange.

A few years later, platinum complexes derived from the bidiazine ligand family have been reported by the same authors [47] to catalyze the direct oxidative conversion of methane to a methanol derivative at grater than 70% one-pass yield based on methane (the term "one-pass yield" was used to describe the yield of product generated without separation and recycle of the starting material). The proposed system is catalytic in the platinum complex and a potentially practical scheme can be developed consisting of three steps:

$$CH_4 + H_2SO_4 + SO_3 \longrightarrow CH_3OSO_3H + H_2O + SO_2$$

$$CH_3OSO_3H + H_2O \longrightarrow CH_3OH + H_2SO_4$$

$$SO_2 + \tfrac{1}{2} O_2 \longrightarrow SO_3$$

with an overall equation:

$$CH_4 + \tfrac{1}{2} O_2 \longrightarrow CH_3OH$$

Dichloro(η-2-{2,2′-bipyrimidyl})platinum(II) **(VIII-11)** was among the most effective catalysts. The authors proposed that the process is electrophilic in character and occurs with the cationic, highly electrophilic, coordinatively unsaturated, 14-electron, T-shaped complex **(VIII-12)**, as shown in Scheme VIII.9.

Complex $RhCl_3$ catalyzed the direct formation of methanol and acetic acid from methane, CO and O_2 in a mixture of perfluorobutyric acid and water [48a]. At 80–85 °C the turnover rate was ca. 2.9 h^{-1} based on Rh. Ethane was more reactive, and under similar conditions gave ethanol, acetic acid and methanol (the rate was ca. 7.5 h^{-1}). The latter product arose from a C–C bond cleavage. It is interesting that for both methane and ethane, the product alcohols are less reactive than the starting alkanes. The authors assumed that the ratio of alcohol to the corresponding carboxylic acid is a function of the relative rates of nucleophilic attack versus CO insertion into a Rh–alkyl bond (Scheme VIII.10). It has been also shown that the formation of acetic acid from methane and O_2+CO in the presence of a $RhCl_3$–DCl–KI catalytic system is strongly accelerated when an aqueous medium is substituted by an aqueous–acetic one [48b].

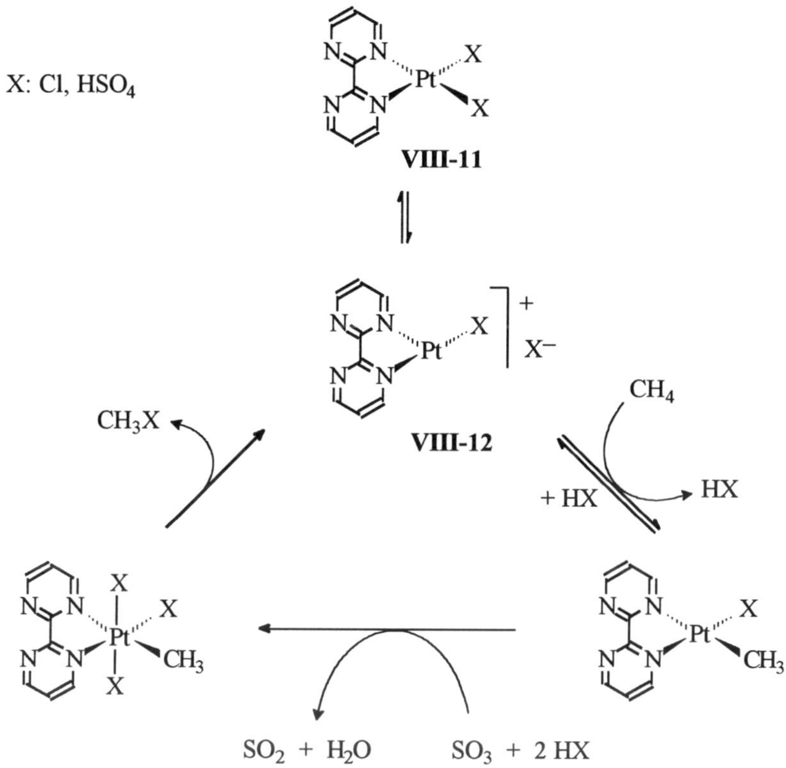

Scheme VIII.9. Proposed catalytic cycle for the oxidation of alkanes by SO_3 in H_2SO_4.

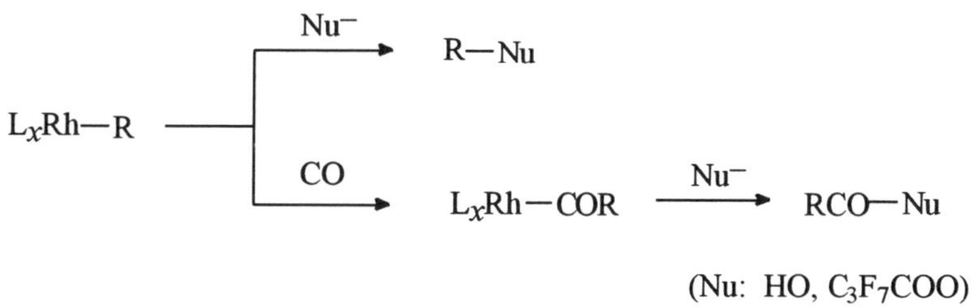

Scheme VIII.10. Transformations of a Rh–R complex formed from an alkane.

It is interesting that ethane can be converted to propionic acid and its mixed anhydride and trifluoromethyl ethyl ketone under the action of hydrogen peroxide in trifluoroacetic anhydride at 80 °C in the absence of a metal compound [49]. The addition of palladium trifluoroacetate resulted in simple oxidation to ethanol and acetaldehyde.

The catalytic system consisting of a mixture of copper chloride and metallic palladium, operating in a 3 : 1 mixture of $CF_3COOH–H_2O$ in the presence of O_2 and CO, converts methane selectively to methanol [50]. The rate of formation of methanol from methane was ca. 6.5×10^{-3} mol dm^{-3} min^{-1} at 145–150 °C. The first step of the reaction is assumed to be the water gas shift reaction involving the oxidation of CO to CO_2 with simultaneous formation of H_2 (Scheme VIII.11). The second step involves the combination of dihydrogen with dioxygen to give hydrogen peroxide. Finally, the third step involves the metal-catalyzed oxidation of the alkane by H_2O_2. It is important that in the absence of copper chloride formic and acetic acids are main products formed from methane and ethane, respectively, but the addition of $CuCl_2$ to the reaction mixture has a dramatic effect. In the presence of the copper salt, methanol and its ester become the preferred products.

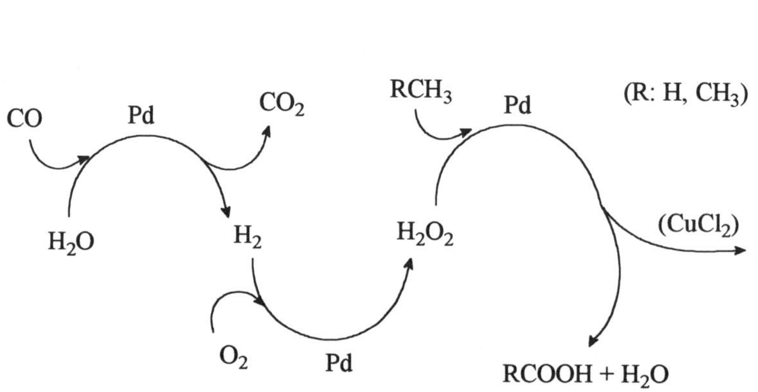

Scheme VIII.11. Catalytic cycles in the oxidation of alkanes by an O_2–CO mixture.

Fujiwara *et al.* have discovered that aromatic and saturated hydrocarbons can be carbonylated to carboxylic acids by carbon monoxide:

$$R\text{–}H + CO \longrightarrow R\text{–}COOH \qquad (R: CH_3, C_2H_5, C_3H_7 \text{ etc.})$$

and aminomethylated by *tert*-methylamine *N*-oxides:

$$C_nH_{2n+2} + CH_3R_2N{=}O \longrightarrow C_nH_{2n+1}CH_2NR_2 + H_2O$$

under the action of soluble palladium(II) derivatives [14, 51]. Copper(II) also catalyzes the reaction of gaseous alkanes with amine *N*-oxides in trifluoroacetic acid [52]. Recently, new systems for carboxylation of methane with carbon monoxide in the same solvent to produce acetic acid have been described: $VO(acac)_2/K_2S_2O_8$ [53], $Yb(OAc)_3/Mn(OAc)_2/NaClO/H_2O$ [54], and $Ln_2O_3/K_2S_2O_8$ (Table VIII.2) [55].

Other works on alkane oxidations in trifluoroacetic acid have been also described [56]. Finally, it has been found that heating an aqueous solution of sodium vanadate in the presence of methane, carbon monoxide and air gives rise to the formation of acetic acid, as well as methanol and formaldehyde in smaller amounts [57]. The yield of CH_3COOH attains 3700% based on vanadium after 50 h at 100 °C, the total turnover number being 49. The reaction is sensitive to the pH of the solution. For example, the yields of the products decrease noticeably if pH > 5. The reaction apparently involves hydrogen atom abstraction from methane by a radical or radical-like species which could be generated via the reduction of V(V) with carbon monoxide.

VIII.2.C. ON THE MECHANISMS OF ALKANE ACTIVATION IN AQUEOUS AND ACIDIC MEDIA

Kinetics and Mechanism of the Oxidation with Palladium Complexes

Rudakov and co-workers have studied the oxidation of alkanes with complexes of palladium in more detail. It was proposed [43b] that palladium(II) is an electrophilic reagent in these reactions. It attacks the C–H bond in a cyclic transition state simultaneously with the proton abstraction with a basic ligand (L: H_2O, HSO_4^-):

Table VIII.2. Methane carboxylation promoted by lanthanoid compounds [a].

	Yield of CH_3COOH	
Lanthanoid compound	μmol	% based on lanthanoid
None	6.0	–
Yb_2O_3	49.8	100
$Yb(OAc)_3$	44.0	88
$YbCl_3$	23.4	47
Yb_2S_3	5.5	11
$EuCl_3$	14.1	38
$TmCl_3$	14.0	28
$TbCl_3$	13.2	26
$LaCl_3$	12.0	24
$DyCl_3$	10.0	20
$HoCl_3$	8.4	17
$LuCl_3$	8.0	16
$PrCl_3$	6.6	13
$NdCl_3$	6.1	12
Ce_2S_3	7.8	16
Sm_2S_3	5.0	10
La_2S_3	4.4	8

[a] Conditions: autoclave (120 mL), CH_4 (40 atm), CO (20 atm), lanthanoid compound (50 μmol atom), $K_2S_2O_8$ (5 mmol), CF_3COOH (5 mL), 80 °C, 20h.

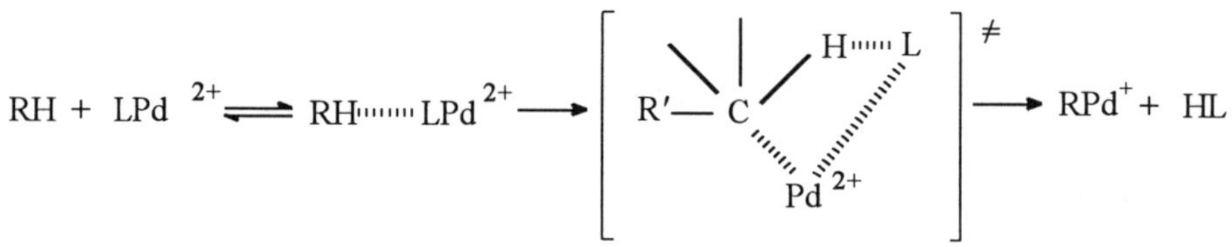

There is no H–D exchange with a solvent apparently due to the rapid irreversible transformation:

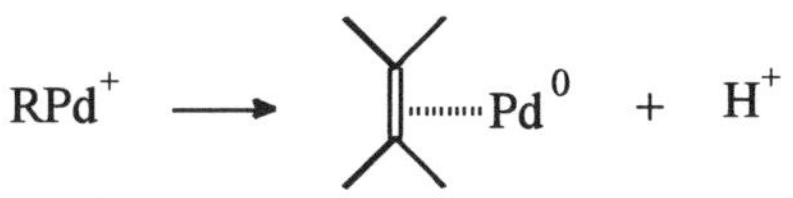

Other mechanisms of splitting the C–H bond are also possible: abstraction of the hydrogen atom in the oxidative homolysis (structure **VIII-13**), abstraction of a hydride ion in oxidative heterolysis (structure **VIII-14**), abstraction of two hydrogen atoms (structure **VIII-15**) or simultaneously of H^- and H^+ (structure **VIII-16**).

VIII-13

VIII-14

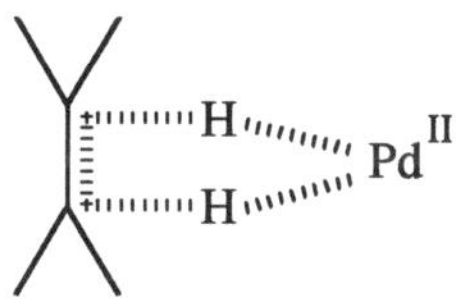

VIII-15

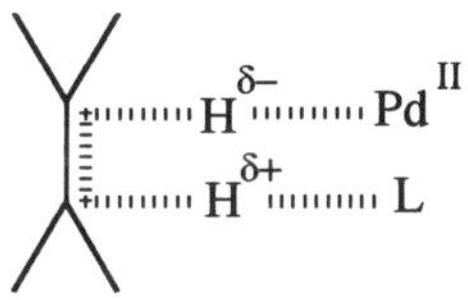

VIII-16

The products of these reactions are carbocations R^+ or olefins which are in equilibrium with the carbocations. Mechanisms involving structures **VIII-13–VIII-14** can be only hardly discriminated. The only evidence for the mechanism of electrophilic substitution might be the deep similarity in selectivity and KIE for the reactions between palladium(II) and alkanes on the one hand, and nitration of alkanes with NO_2^+ on the other hand. Indeed, such a similarity has been evaluated in the study of the temperature dependencies of KIE and 5/6 effects.

Main Types of Mechanisms

Based on experimental data, predominantly on the temperature dependency of KIE and 5/6 effect for various oxidizing reagents, Rudakov proposed the following main types of mechanisms of the reactions in aqueous and acidic media [43c].

a) Direct nucleophile-electrophile attack of the metal complex at the carbon alkane atom, for example:

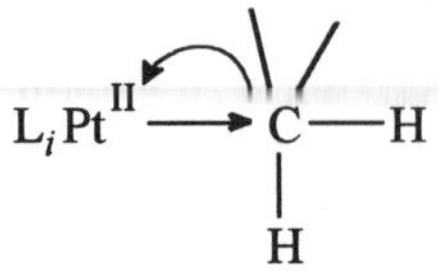

b) Electrophilic substitution of hydrogen, for example:

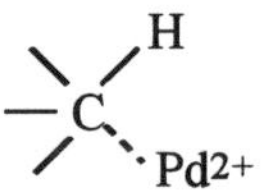

The same mechanism has been proposed for reactions with NO_2^+ and SO_3H^+.

c) Direct abstraction of a hydride, for example:

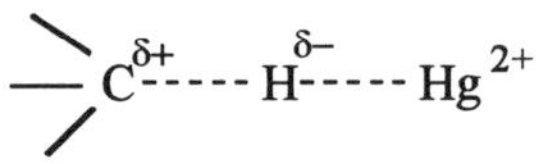

d) Synchronous insertion of an oxo or hydroxo reagent into the C–H bond in the cyclic transition state. Examples of such a mechanism are shown below:

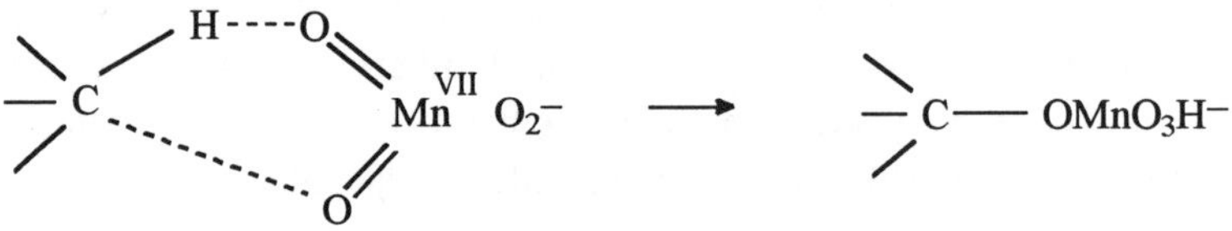

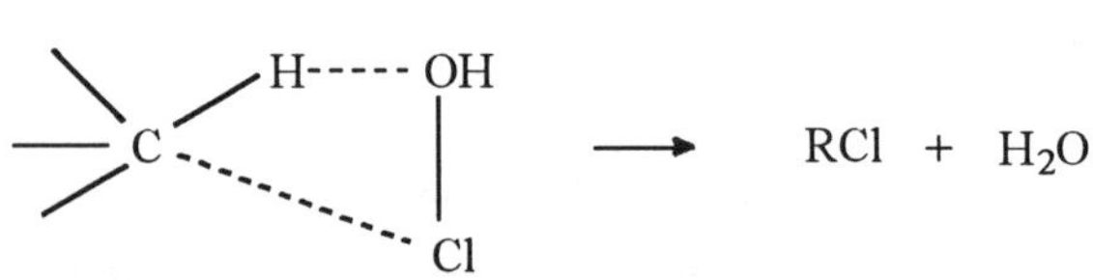

e) Outer-sphere oxidative homolysis of the C–H bond via the ligand oxidation with or without the entering ligand in the volume of solution, e.g.:

$$RH + LMn^{III} \longrightarrow \dot{R}\text{-----}\dot{H}\text{-----}\dot{L}\text{-----}Mn^{II} \longrightarrow R^{\bullet} + HL + Mn^{II}$$

As a result, a large group of electrophilic reagents have been divided into two types: proper electrophiles, such as Pd^{2+}, NO_2^+, SO_3H^+ (mechanism *b*), and acceptors of hydride ion, e.g., R^+, Hg^{2+} (mechanism *c*). The feature of mechanisms *b* and *c* is inner-sphere attack of the metal at the carbon atom and the formation of the metal alkyl as the first product.

VIII.2.D. OXIDATION OF ARYLALKANES BY METAL CATIONS

Derivatives of the following metal ions are capable of oxidizing saturated and alkylaromatic hydrocarbons:

d^0: Cr(VI), Mn(VII), Ce(IV), Pb(IV)

d^4: Mn(III)

d^6: Co(III)

d^7: Pt(III)

d^8: Pd(II), Cu(III), Ag(III)

The standard redox potential for the ions, which react with alkanes in aqueous solutions should apparently be > 1.5 V. In strongly acidic media, ions with lower redox potentials (measured in aqueous solutions) can be active relative the hydrocarbons (see preceding Section). In many cases the reactions between the alkane and metal ion give rise to the transient formation of free alkyl radicals. For branched hydrocarbons, "normal" selectivity (*tert* > *sec* > *prim*) is usual. It is interesting that there are some exceptions when the reactivity of the *tert*-C–H bond has been found very low.

Arylalkanes can react with electrophilic oxidants [58] to produce alkyl radicals via simultaneous electron transfer and deprotonation:

$$M^{n+} + C_6H_5CH_3 \longrightarrow C_6H_5CH_2\bullet + M^{(n-1)+} + H^+$$

The first step of the reaction can involve the formation of a cation radical [59] (this route is usual in the case of arenes possessing a low ionization potential, i.e., $I < 8$ eV):

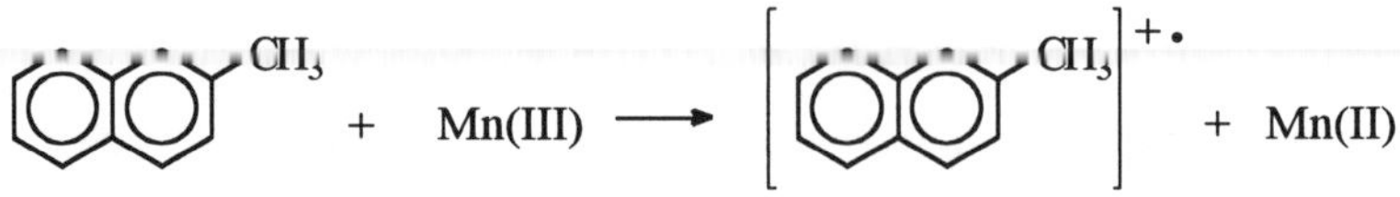

Further possible transformations of cation radicals are shown in Scheme VIII.12 [60].

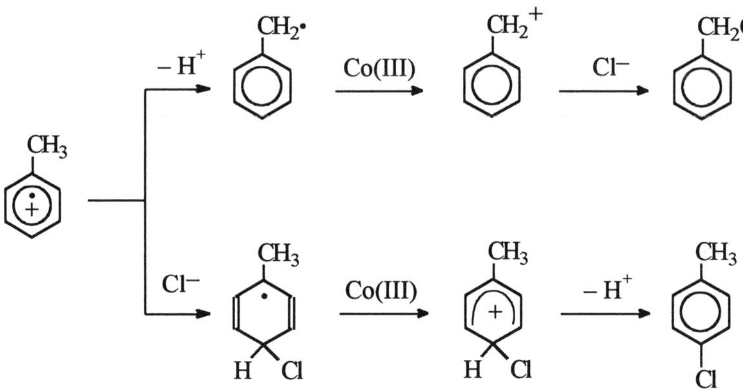

Scheme VIII.12. Transformations of the toluene cation radical.

It is interesting that the clorination of aromatic compounds using an acetyl chloride – manganese triacetate system is significantly accelerated when performed under sonication [61].

VIII.3 OXIDATION BY METAL OXO COMPLEXES

Reactions of hydrocarbons with metal complexes exhibit some common features. Abstraction of a hydrogen atom is a key step in most of the hydrocarbon oxidations by metal oxo complexes [62].

VIII.3.A. OXYGENATION OF ALKANES WITH Cr(VI) DERIVATIVES

Chromium(VI) oxo derivatives are strong oxidants which are also used as catalysts in organic synthesis [63]. The reaction of alkanes with chromium(VI) oxo compounds apparently takes place with the intermediate formation of radicals:

$$\text{RH} + \text{O=Cr(VI)} \longrightarrow \text{R}\bullet + \text{HO–Cr(V)}$$

It is important, however, that not all the radicals formed are liberated in the solution, since the oxidation of (+)-3-methylheptane by chromic acid involves the formation of (+)-3-methyl-3-heptanol with retention of configuration to the extent of 70–85%. The normal selectivity has been observed in the hydroxylation of branched alkanes.

Chromyl chloride, CrO_2Cl_2, reacts [64a] with cyclohexane to give a dark precipitate along with chlorocyclohexane and a small amount of cyclohexene. Hydrolysis of the precipitate yields cyclohexanone and chlorocylohexanone. Cyclohexene is readily oxidized by the chromium compound to give mostly ring-opened products with some 2-chlorocylohexanone and cyclohexanone (Scheme VIII.13).

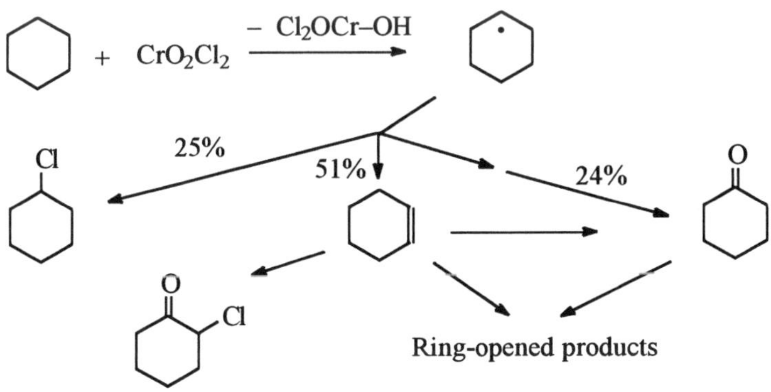

Scheme VIII.13. Oxidation of cyclohexane with chromyl chloride.

This reaction proceeds by an initial hydrogen atom transfer from cyclohexane to Cr(VI). The cyclohexyl radical is rapidly trapped by the chromium species via one of three pathways:

i) chlorine atom abstraction;
ii) formation of a C–O bond;
iii) transfer of a second hydrogen atom.

A correlation has been found between O–H bond strength and the hydrogen atom transfer rate (Figure VIII.2).

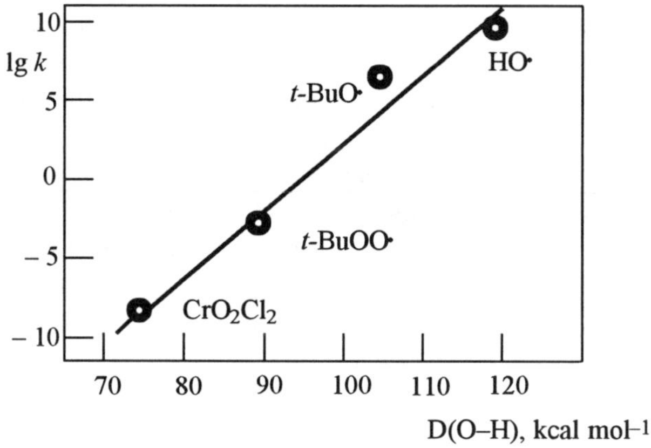

Figure VIII.2. Rates of hydrogen atom abstraction from cyclohexane by various radicals at 25–28° C vs the formed O–H bond strength.

The reaction of chromyl chloride with isopropylcyclopropane at 65 °C gives at least 20 organic products, among them 2-cyclopropyl-2-chloropropane, 2-cyclopropyl-2-propanol and ring-opened products [64b] (Scheme VIII.14). These products have been explained by a mechanism involving an initial hydrogen atom abstraction from the substrate. The resulting dimethylcyclopropylcarbinyl radical can either be trapped by CrO_2Cl_2 or ring-opened.

The addition of ruthenium(IV) or iridium(IV) chloride complexes increased the rate of the oxidation reaction leading to the formation of the chloroalkanes [65]. The selectivity of the oxidation in the presence of ruthenium(IV) (1° : 2° :

3° = 1 : 100 : 1000) is different from that in the oxidation catalyzed by iridium(IV) (1 : 30 : 250); in the latter case, the primary C–H bond is relatively more reactive. One of the possible causes of the greater oxidizing capacity of chromium(VI) compounds is the formation of the mixed complex $[Cl_4RuO\text{-}CrO_3]^{3-}$. Strong acids accelerate the reaction similarly (the rate of oxidation is proportional to the acidity of the medium), giving rise to protonated species, for example $O{=}Cr(OH)_3^+$ and $HCrO_3^+$.

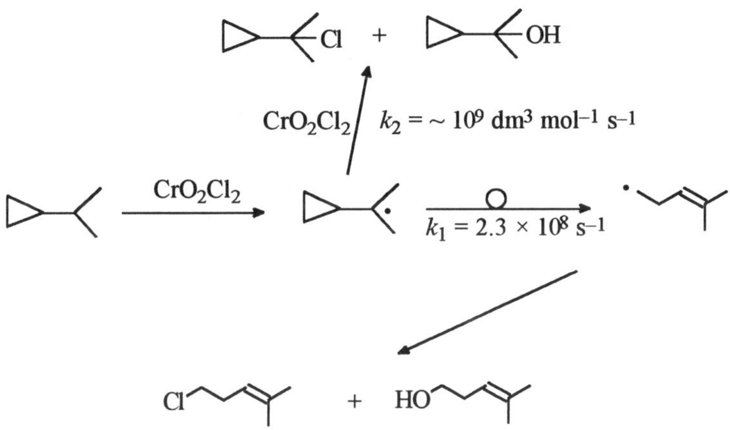

Scheme VIII.14. Oxidation of isopropylcyclopropane with chromyl chloride.

The oxidation of alkanes and aromatic alkyl-substituted compounds by oxo derivatives of chromium(VI) is greatly accelerated by light irradiation [66]. Acetic acid, acetonitrile and methylene chloride have been used as solvents with the products being an alcohol and a carbonyl compound. The reaction apparently begins with the abstraction of a hydrogen atom from the C–H bond of the oxo complex by an excited species. The logarithms of the relative rate constants for the oxidation of substituted toluenes $XC_6H_4CH_3$ by CrO_3 in acetic acid, both in the dark and under light ($\lambda > 300$ nm), correlate with the Brown constants σ^+ of substituent X (Figure VIII.3) [66c]. It is interesting that in both the dark and light reactions the points corresponding to the substituent p-OCH_3 do not lie on the straight line.

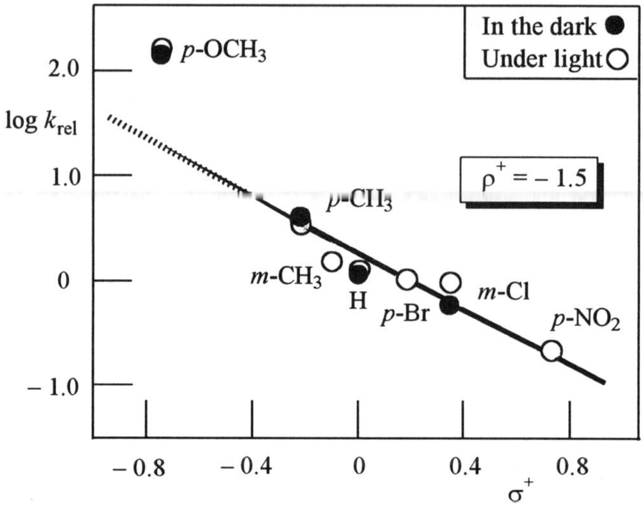

Figure VIII.3. The correlation between the logarithms of relative rate constants of the oxidation of substituted toluenes $XC_6H_4CH_3$ and the Brown constants σ^+ of substituent X.

The chromium trioxide–3,5-dimethylpyrazol complex has been found to be a mild and selective reagent for the oxidation of cyclopropyl hydrocarbons (reaction proceeds in the dark at –20 °C) [67].

VIII.3.B. OXYGENATION OF HYDROCARBONS WITH Mn(VII) COMPOUNDS

Permanganate anion oxidizes alkanes at room temperature in trifluoroacetic acid solution [68a]. The cation MnO_3^+ is apparently the active species. The selectivity of the reaction corresponds to 1° : 2° : 3° = 1 : 60 : 2100. In $CF_3COOH–H_2O$ solution, the active species are possibly MnO_4^-, $HMnO_4$, or $O_3MnOCOCF_3$ [68b]. The oxidation of benzene proceeds via either cation **VIII-17** or oxenoid **VIII-18**.

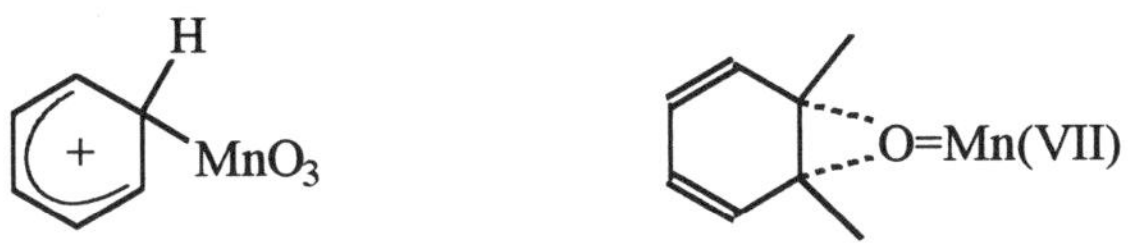

VIII-17 **VIII-18**

Oxidations of arylalkanes by $^{n}Bu_4NMnO_4$ in toluene solvent led to the conclusion that rate-limiting hydrogen atom transfer from the substrate to a permanganate oxo group is consistent with all of the experimental evidence [69a]:

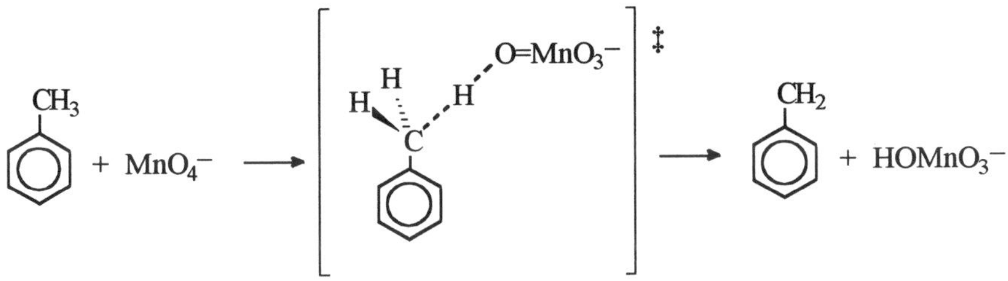

An alternative mechanism which involves hydride transfer giving a carbocation has been proposed for the oxidation of alkylbenzenes with permanganate adsorbed on a solid support [69b]:

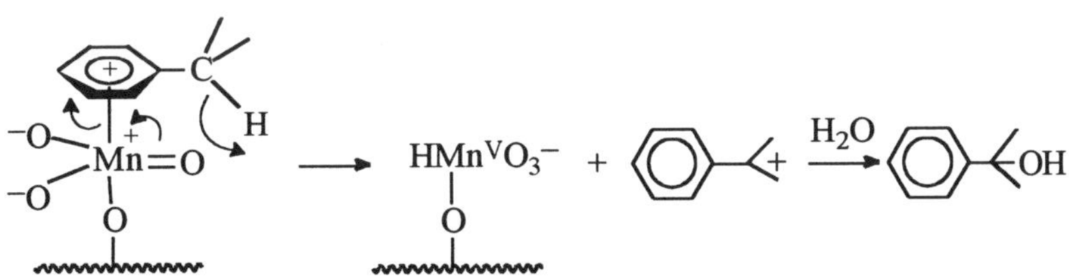

Potassium permanganate impregnated on alumina oxidizes arenes and alkylarenes to ketones under microwave activation [69c]. Potassium permanganate–triethylamine is a convenient reagent for the oxidation of benzylic methyl, methylene and methine groups [70a]. Aqueous solutions of permanganate oxidize methane at 40–100 °C to produce carbon dioxide as the sole product [70b]. Oxidation of hydrocarbons containing weak C–H bonds (9,10-dihydroanthracene, fluorene) by a mixed valent ($Mn^{III}Mn^{IV}$) complex $[(phen)_2Mn(\mu\text{-}O)_2Mn(phen)_2]^{3+}$ (phen: 1,10-phenanthroline) proceeds via hydrogen atom abstraction [71].

VIII.3.C. OXYGENATIONS BY OTHER COMPLEXES

Selective oxidation of the alkyl ligand in rhenium(V) oxo complexes $(HBpz_3)ReO(R)OTf$, where $HBpz_3$ = hydrotris(1-pyrazolyl)borate and OTf = OSO_2CF_3, has been reported to occur via transfer of an α-hydrogen from the alkyl group (R =Me, Et, *n*-Bu) to an oxo ligand [72]. An alkylidene complex is then trapped by SMe_2 or pyridine to give the observed intermediates:

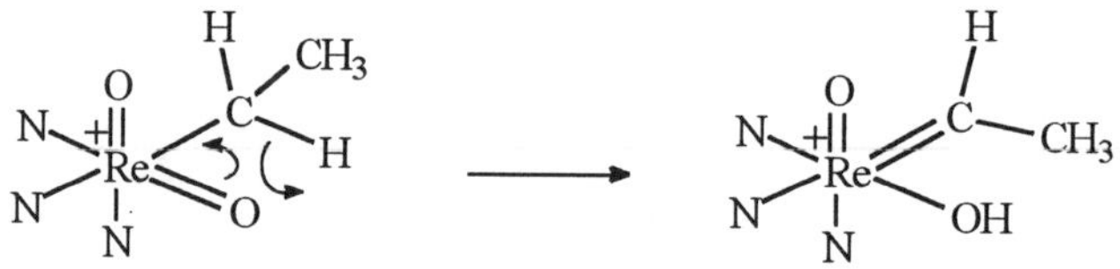

Arylalkanes have been oxidized by a molybdenum(VI) complex dioxodithiocyanato-4,4′-di-*tert*-butyl-2,2′-bipyridylmolybdenum [73]. The reaction can also be carried out under catalytic conditions using DMSO as the oxo-donor reactant.

Ruthenium tetroxide formed from $RuO_2 \cdot 2H_2O$ or other ruthenium derivatives and sodium periodate oxidize saturated hydrocarbons [74]:

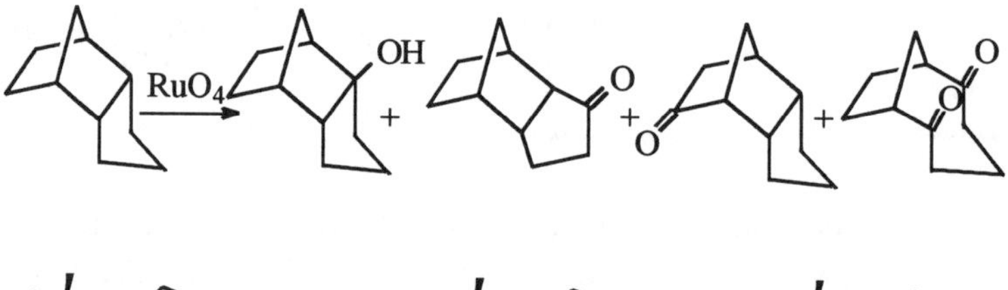

[74c]

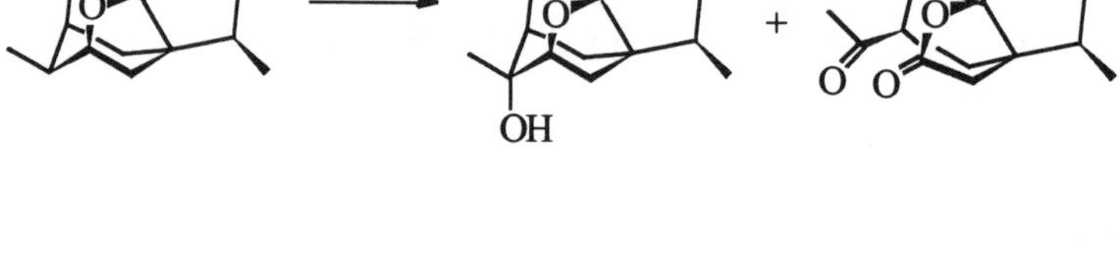

[74e]

Proposed mechanistic pathways for this oxidation are shown in Scheme VIII.15 [74e].

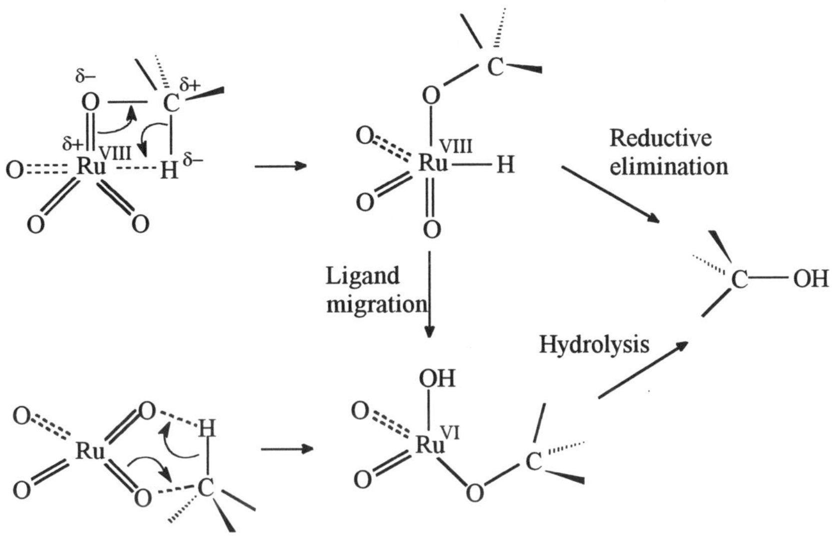

Scheme VIII.15. Mechanism proposed for alkane oxidation by RuO_4.

Ruthenium trichloride catalyzes the oxidation of 2-methylnaphthalene to 2-methylnaphthaquinone with ammonium dichromate [75]. Cyclohexane, adamantane, *n*-hexane, toluene and ethylbenzene can be oxidized at room temperature by barium ruthenate [76a]. Barium ruthenate in the presence of 2,2′-bipyridine generates a highly reactive ruthenium-oxo system that is capable of oxidizing ethane and propane at room temperature [76b]. Complexes of ruthenium containing chelating and some other ligands effectively oxidize alkanes both in the dark and under light irradiation [77]. A ruthenium(VI) complex containing two oxo ligands oxidizes alkyl chains in alkylaromatics [77h].

An oxidizing reagent based on potassium ferrate(VI) has been described [78]. This potassium ferrate, when used in conjunction with an appropriate heterogeneous catalyst such as K10 montmorillonite clay, is a strong oxidant which produces cycloalkanols and cycloalkanones from cycloalkanes, and benzyl alcohol and benzaldehyde from toluene.

A theoretical study has demonstrated that the enthalpies for the reactions of transition metal oxo complexes with aliphatic alcohols follows the order [79]:

OH hydrogen abstraction > CH hydrogen abstraction >
CH bond addition > OH bond addition.

Finally, sodium bismuthate in the presence of acetic acid oxidizes benzylic methylenes in polycyclic systems to the benzylic ketones (Table VIII.3) [80].

Table VIII.3. Oxidation by sodium bismuthate in the presence of acetic acid.

Substrate	Product	Yield, %
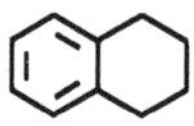	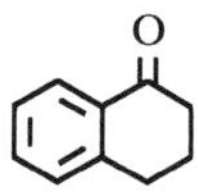	82
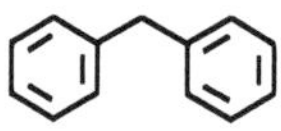	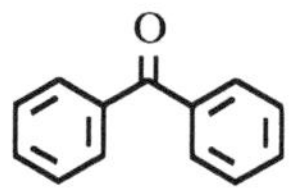	90
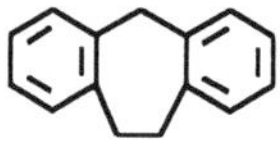	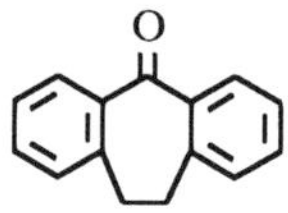	68
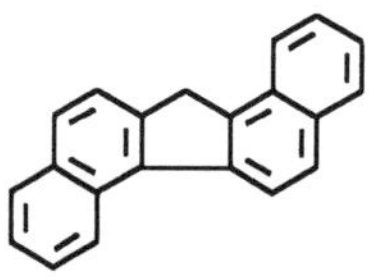	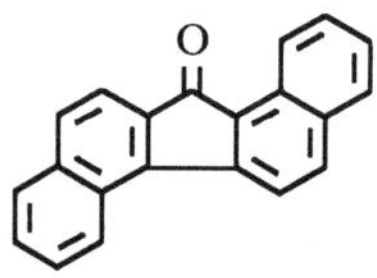	56

VIII.3.D. ALKANE FUNCTIONALIZATION UNDER THE ACTION OF POLYOXOMETALATES

Polyoxometalates (POMs) are widely used in various oxidations, particularly of saturated hydrocarbons [81]. In this section, we will discuss only POM-promoted transformations of alkanes which occur in the absence of oxygen (air). Hill and co-workers have reported [82] that in the absence of air, the light irradiation of an alkane solution in acetonitrile in the presence of polyoxometalates and metallic platinum gives rise to the formation of a variety of products, for example, as shown in the following equation:

$$\text{Alkane} + h\nu \xrightarrow{\text{Polyoxometalate}} \text{Oxidation product} + \text{Reduction product } (H_2)$$

Here the oxidation product is either an alkene, *N*-alkylacetamide, alkylmethylketone, or dialkyl. Thus, hydrocarbon **VIII-19** was acetylated by this method:

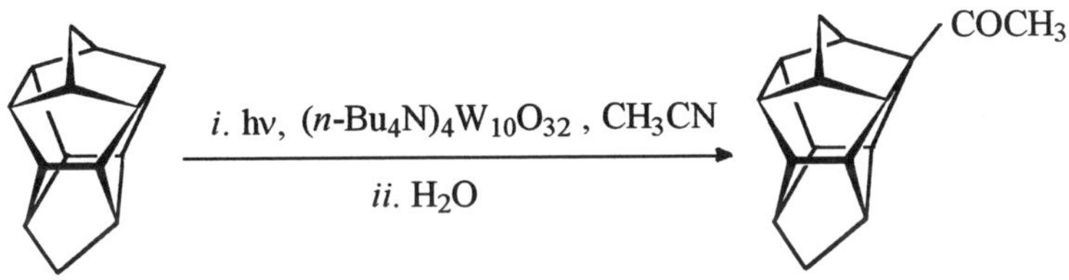

VIII-19

The following sequence of steps has been proposed. First, the photoexcited POM species M* abstracts hydrogen atoms from the substrate, SH_2. The reduced form of the metalate transforms the H^+ ions into molecular hydrogen:

$$PW_{12}O_{40}^{3-} + h\nu \longrightarrow [PW_{12}O_{40}^{3-}]^*$$
$$[PW_{12}O_{40}^{3-}]^* + SH_2 \longrightarrow PW_{12}O_{40}^{4-} + SH\bullet + H^+$$
$$PW_{12}O_{40}^{3-} + SH\bullet \longrightarrow PW_{12}O_{40}^{4-} + S + H^+$$
$$2\,PW_{12}O_{40}^{4-} + 2\,H^+ \longrightarrow 2\,PW_{12}O_{40}^{3-} + H_2$$

Alkane derivatives may be formed according to the following scheme:

$$RH + M^* \longrightarrow R\bullet + H^+ + M$$
$$R\bullet + NCCH_3 \rightleftharpoons RN{=}C^+CH_3 \xrightarrow{H_2O} RNHCOCH_3 + H^+$$

The relative rate constants for linear alkanes C_5–C_9 correlate with the ionization potentials of these hydrocarbons.

Calculations by the semiempirical theoretical method ASED-MO of the photodimerization of cyclohexene and methane by decatungstate anions have been carried out [84]. Calculated hydrogen abstraction activation energies for a C–H bond in methane are higher than for cyclohexene, but they are low enough that in the case of methane it can be suggested that dimerization could be looked for in future experiments.

In some cases POMs can oxygenate hydrocarbons. For example, the oxidation of alkanes and benzene by polyvanadate in CF_3COOH to produce alkyl (phenyl) trifluoroacetates and ketones is stimulated by light [85]. In fact, the alkane and benzene reactions with the vanadate virtually do not proceed in the absence of light. On the other hand, ethylbenzene and toluene are rapidly oxidized in the dark reaction, even at 10 °C.

VIII.4 Oxygenation by Peroxo Complexes

The peroxo complexes of transition metals [86] are capable of oxidizing various organic compounds including alkanes and aromatic hydrocarbons. Thus the peroxo complex of chromium $(Ph_3PO)Cr(O)(O_2)_2$ stoichiometrically oxygenates cyclohexane [87a]. The interaction of this complex with cyclohexane in CH_2Cl_2–Me_3COH (1 : 1) at room temperature gives 9.3% of cyclohexanol and 1.6% of cyclohexanone based on chromium. The vanadium complex $(pic)V(O)(O_2)$-$(H_2O)_2$, which contains picolinate (pic), is an effective oxidizing reagent capable of the oxygenation of both alkanes and arenes [87b–f]. Benzene is oxidized by this complex to produce phenol together with dioxygen. The reaction is a radical chain process. The initiation produces the actual oxidant which may be described as a radical anion derived from the peroxovanadium complex. In the propagation steps, such species react either with the original peroxo complex, giving dioxygen, or with the aromatic hydrocarbon yielding phenols [87d]:

$$2\ \mathrm{picVO(O_2)} \rightleftharpoons \mathrm{dimer}$$

$$\mathrm{dimer} \longrightarrow \mathrm{picVO(O_2)}^{\bullet} + \mathrm{picVO(O_2)}^{+\bullet}$$

$$\mathrm{picVO(O_2)}^{+\bullet} \xrightarrow{\text{fast}} \mathrm{picVO}^{+} + \mathrm{O_2}$$

$$\mathrm{picVO}^{+} + \mathrm{picVO(O_2)} \longrightarrow \mathrm{picVO(O_2)}^{-\bullet} + \mathrm{picVO}^{2+}$$

$$picVO(O_2)^{-}\bullet + ArH \rightleftharpoons [picVO(O_2)^{-}\bullet\ ArH]$$

$$[picVO(O_2)^{-}\bullet\ ArH] + picVO(O_2) \longrightarrow$$
$$\longrightarrow ArOH + picVO^{2+} + picVO(O_2)^{-}\bullet$$

$$[picVO(O_2)^{-}\bullet\ ArH] + picVO(O_2) \longrightarrow$$
$$\longrightarrow picVO_3^{2-} + picVO(O_2)^{+}\bullet + ArH$$

$$picVO(O_2)^{-}\bullet + picVO(O_2)^{+}\bullet \longrightarrow [picVO\text{–}O\text{–}]_2$$

$$picVO^{2+} + picVO_3^{2-} \xrightarrow{\text{fast}} [picVO\text{–}O\text{–}]_2$$

The oxidation of cyclohexane in acetonitrile affords cyclohexyl hydroperoxide, cyclohexanol and cyclohexanone in the ratio ca. 2 : 1 : 1. The reaction both with alkanes and arenes is accelerated upon irradiation with visible and, especially, UV light [87f].

When a solution of the complex $[(t\text{-BuOO})Pd(OCOCH_3)]_4$ and cyclohexane in methylene chloride was exposed to the light of a high-pressure mercury lamp in a glass vessel ($\lambda > 310$ nm), cyclohexanol and cyclohexanone were detected in the reaction mixture [88]. The oxidation of alkanes by the complex $[(t\text{-BuOO})\text{-}Pd(OCOCF_3)]_4$ occurs in a benzene solution both in the dark and under light irradiation. It is interesting that in the dark reaction a significant amount of cyclohexyl hydroperoxide is formed in addition to cyclohexanol and cyclohexanone [88].

There is evidence that metal peroxocomplexes can also oxidize alkanes by a non-radical mechanism. Peroxocomplexes, formed by the reaction of hydrogen peroxide with high-valent metal compounds in acid solutions, can react with alkanes, e.g., cyclohexane, and produce alcohols and ketones in a reaction which does not seem to involve free radicals. Particularly convincing results were obtained by Moiseev *et al.* [89]. Thus, the peroxovanadium complex produced in reaction of $VO(acac)_2$ or KVO_3 with H_2O_2 reacted with cyclohexane in acetic acid to form cyclohexanol and cyclohexanone. The evidence is against free radical mechanism, i.e., the reaction is not inhibited by typical inhibitors of chain reactions, the reaction rate and the yield of cyclohexanol is not influenced in the presence of isopropyl alcohol and carbon tetrachloride which would be easily attacked by free radicals if they were present. The reaction was much faster in trifluoroacetic acid than in acetic acid, and in CF_3COOH produces cyclohexanol and cyclohexyl trifluoroactate but not cyclohexanone. The conversion of the hydrocarbon reached 5–8% in acetic acid with a 10-fold excess of hydrogen peroxide and 95–98% in trifluoroacetic acid with only a 4–5 fold excess of H_2O_2. 100% selectivity for reacted cyclohexane is reached with low H_2O_2/C_6H_{12} ratios. In the absence of a substrate, ozone is detected in the products of hydrogen peroxide decomposition in the presence of acid; the ozone which was formed in CF_3COOH comprised 10–15% of all the gaseous products formed. This formation of ozone shows that the oxygen atom is easily transferred from the peroxo ligand of the active center to even such a weak acceptor as another peroxo ligand. Apparently, the C–H bond in an alkane molecule is a sufficiently strong acceptor to add and then to insert an O atom to form an alcohol. The situation is close to superacid solutions where hydrogen peroxide and ozone react with alkanes inserting an oxygen atom into C–H bonds in a non-radical fashion.

Peroxo derivatives of transition metals are apparently intermediates in the oxidations of organic compounds by hydrogen peroxide or alkyl hydroperoxides catalyzed by various metal complexes. These processes will be discussed in the following chapters.

References for Chapter VIII

1. (a) Stahl, S. S.; Labinger, J. A.; Bercaw, J. E. *Angew. Chem., Int. Ed. Engl.* **1998**, *37*, 2180. (b) Sen, A. In *Applied Homogeneous Catalysis with Organometallic Compounds*; Cornils, B.; Herrmann, W. A., Eds.; VCH: Weinheim, 1996, p. 1081. (c) Collins, T. J. *Acc. Chem. Res.* **1994**, *27*, 279.
2. (a) Shul'pin, G. B. *Organic Reactions Catalyzed by Metal Complexes*; Nauka: Moscow, 1988 (in Russian). (b) Shul'pin, G. B.; Nizova, G. V.; Nikitaev, A. T. *J. Organometal. Chem.* **1984**, *276*, 115.
3. (a) Omae, I. *Coord. Chem. Rev.* **1988**, *83*, 137. (b) Dunina, V. V.; Zalevskaya, O. A.; Potapov, V. M. *Usp. Khim.* **1988**, *57*, 434 (in Russian). (c) Ryabov, A. D. *Chem. Rev.* **1990**, *90*, 403.
4. (a) Steenwinkel, P.; Gossage, R. A.; van Koten, G. *Chem Eur. J.* **1998**, *4*, 759. (b) Hartshorn, C. M.; Steel, P. J. *Organometallics* **1998**, *17*, 3487. (c) Crespo, M.; Solans, X.; Font-Bardia, M. *Polyhedron* **1998**, *17*, 1651. (d) Ryabov, A. D.; Panyashkina, I. M.; Polyakov, V. A.; Howard, J. A. K.; Kuz'mina, L. G.; Datt, M. S.; Sacht, C. *Organometallics* **1998**, *17*, 3615.
5. (a) Albert, J.; Granell, J.; Luque, A.; Font-Bardia, M.; Solans, X. *J. Organometal. Chem.* **1997**, *545*, 131. (b) Vicente, J.; Saura-Llamas, I.; Palin, M. G.; Jones, P. G.; de Arellano, M. C. R. *Organometallics* **1997**, *17*, 826. (c) Ghedini, M.; Neve, F.; Pucci, D. *Eur. J. Inorg. Chem.* **1998**, 501. (d) O'Keefe, B. J.; Steel, P. J. *Organometallics* **1998**, *17*, 3621. (e) Troitskaya, L. L.; Ovseenko, S. T.; Sokolov, V. I.; Gruselle, M. *Izv. Akad. Nauk, Ser. Khim.* **1998**, 1421 (in Russian). (f) Neumann, B.; Hegmann, T.; Wolf, R.; Tschierke, C. *Chem. Comm.* **1998**, 105.
6. (a) Carina, R. F.; Williams, A. F.; Bernardinelli, G. *J. Organometal. Chem.* **1997**, *548*, 45. (b) Grellier, M.; Pfeffer, M. *J. Organometal. Chem.* **1997**, *548*, 301. (c) Crispini, A.; Ghedini, M. *J. Chem. Soc., Dalton Trans.* **1997**, 75. (d) Fuchita, Y.; Yoshinaga, K.; Ikeda, Y.; Kinoshita-Kawashima, J. *J. Chem. Soc., Dalton Trans.* **1997**, 2495. (e) Lopez, C.; Bosque, R.; Solans, X.; Font-Bardia, M. *J. Organometal. Chem.* **1997**, *547*, 309. (f) Hiraki, K.; Ibaraki, S.; Onishi, M.; Ono, Y.; Kawashima, J. K.; Ando, M. *J. Organometal. Chem.* **1997**, *547*, 199. (g) Vila, J. M.; Gayoso, M.; Pereira, M. T.; Ortigueira, J. M.; Torres, M. L.; Castiñeiras, A.; Suarez, A.; Fernandez, J. J.; Fernandez, A. *J. Organometal. Chem.* **1997**, *547*, 297. (h) Avis, M. W.; van der Boom, M. E.; Elsevier, C. J.; Smeets, W. J. J.; Spek, A. L. *J. Organometal. Chem.* **1997**, *527*, 263. (i)

Fedorov, B. S.; Golovina, N. I.; Strukov, G. V.; Kedrov, V. V.; Arakcheeva, V. V.; Trofimova, R. F.; Shilov, G. V.; Atovmyan, L. O. *Izv. Akad. Nauk, Ser. Khim.* **1997**, 1704 (in Russian). (j) Dunina, V. V.; Kazakova, M. Yu.; Grishin. Yu. K.; Malyshev, P. R.; Kazakova, E. I. *Izv. Akad. Nauk, Ser. Khim.* **1997**, 1375 (in Russian). (k) Zhao, G.; Xue, F.; Zhang, Z.-Y.; Mak, T. C. W. *Organometallics* **1997**, *17*, 4023. (l) Zhao, G.; Wang, Q.-G.; Mak, T. C. W. *J. Chem. Soc., Dalton Trans.* **1998**, 1241. (m) Lopez, C.; Bosque, R.; Salans, X.; Font-Bardia, M. *New J. Chem.* **1998**, 977. (n) Zhao, G.; Wang, Q.-G.; Mak, T. C. W. *Organometallics* **1998**, *17*, 3437. (o) Lopez, C.; Granell, J. *J. Organometal. Chem.* **1998**, *555*, 211. (p) Vila, J. M.; Pereira, M. T.; Ortigueira, J. M.; Torres, M. L.; Castiñeiras, A.; Lata, D.; Fernandez, J. J.; Fernandez, A. *J. Organometal. Chem.* **1998**, *556*, 21. (q) Navarro-Ranninger, C.; Montero, E. I.; Lopez-Solera, I.; Masaguer, J. R.; Lippert, B. *J. Organometal. Chem.* **1998**, *558*, 103. (r) Sokolov, V. I.; Bulygina, L. A. *Izv. Akad. Nauk, Ser. Khim.* **1998**, 1268 (in Russian).

7. (a) Gomez, M.; Granell, J.; Martinez, M. *Organometallics* **1997**, *16*, 2539. (b) Yagyu, T.; Aizawa, S.; Funabishi, S. *Bull. Chem. Soc. Jpn.* **1998**, *71*, 619.

8. van Helden, R.; Verberg, G. *Recl. Trav. Chim. Pays-Bas* **1965**, *84*, 1263.

9. (a) Kozhevnikov, I. V.; Kim, V. I.; Sidel'nikov, V. N. *Izv. Akad. Nauk SSSR, Ser. Khim.* **1986**, 1352 (in Russian). (b) Ignatenko, V. M.; Rudakov, E. S. *Kinet. Katal.* **1989**, *30*, 44 (in Russian). (c) Ryabov, A. D. *Zh. Obshch. Khim.* **1987**, *57*, 249 (in Russian).

10. (a) Fuchita, Y.; Kawakami, M.; Shimoke, K. *Polyhedron* **1991**, *10*, 2037. (b) Fuchita, Y.; Taga, M.; Kawakami, M.; Kawachi, F. *Bull. Chem. Soc. Jpn.* **1993**, *66*, 1294.

11. Fujiwara, Y.; Moritani, I.; Danno, S. *et al. J. Am. Chem. Soc.* **1969**, *91*, 7166.

12. (a) Kozhevnikov, I. V.; Sharapova, S. Ts.; Kurteeva, L. I. *Kinet. Katal.* **1982**, *23*, 352. (b) Ryabov, A. D.; Sakodynskaya, I. K.; Yatsimirskiy, A. K. *J. Chem. Soc., Perkin Trans. 2* **1983**, 1511.

13. (a) Benazzi, E.; Cameron, C. J.; Mimoun, H. *J. Mol. Catal.* **1991**, *69*, 299. (b) El Firdoussi, L.; Baqqa, A.; Allaoud, S.; Ait Allal, B.; Karim, A.; Castanet, Y.; Mortreux, A. *J. Mol. Catal. A: Chem.* **1998**, *135*, 11.

14. (a) Fujiwara, Y.; Jintoku, T.; Takaki, K. *CHEMTECH* **1990**, 636. (b) Fujiwara, Y.; Takaki, K.; Taniguchi, Y. *Synlett* **1996**, 591.

15. Stromnova, T. A.; Busygina I. N.; Tihonova, N. Yu.; Moiseev, I. I. *Mendeleev Commun.* **1991**, 58.

16. Otsuka, K.; Ishizuka, K.; Yamanaka, I.; Hatano, M. *Catalyst*, **1990**, *32*, 452.

17. Aoyama, Y.; Yoshida, T.; Sakurai, K.; Ogoshi, H. *Organometallics* **1986**, *5*, 168.

18. Ueda, T.; Yamanaka, H.; Adachi, T.; Yoshida, T. *Chem. Lett.* **1988**, 525.

19. (a) Johnson, T. J.; Alvey, L. J.; Brady, M.; Mayne, C. L.; Arif, A. M.; Gladysz, J.A. *Chem. Eur. J.* **1995**, *1*, 294. (b) Solari, E.; Musso, F.; Ferguson, R.; Floriani, C.; Chiesi-Villa, A.; Rizzoli, C. *Angew. Chem., Int. Ed. Engl.* **1995**, *34*, 1510. (c) Jacoby, D.; Isoz, S.; Floriani, C.; Chiesi-Villa, A.; Rizzoli, C. *J. Am. Chem. Soc.* **1995**, *117*, 2805. (d) Lockwood, M. A.; Clark, J. R.; Parkin, B. C.; Rothwell, I P. *Chem. Commun.* **1996**, 1973.

20. (a) Fuchita, Y.; Hiraki, K.; Matsumoto, Y. *J. Organometal. Chem.* **1985**, *280*, C51. (b) Fuchita, Y.; Hiraki, K.; Uchiyama, T. *J. Chem. Soc., Dalton Trans.* **1983**, 897. (c) Hiraki, K.; Fuchita, Y.; Nakashima, M.; Hiraki, H. *Bull. Chem. Soc. Jpn.* **1986**, *59*, 3073.

21. Watson, P. L. *J. Am. Chem. Soc.* **1983**, *105*, 6491.

22. Thompson, M. E.; Baxter, S. M.; Bulls, A. R. *et al. J. Am. Chem. Soc.* **1987**, *109*, 203.

23. (a) Rabaâ, H.; Saillard, J.-Y.; Hoffmann, R. *J. Am. Chem. Soc.* **1986**, *108*, 4327. (b) Rappé, A. K.; Upton, T. H. *J. Am. Chem. Soc.* **1992**, *114*, 7507. (c) Zigler, T.; Folga, E.; Berces, A. *J. Am. Chem. Soc.* **1993**, *115*, 636.

24. (a) Rondon, D.; Chaudret, B.; Xe, X.-D.; Labroue, D. *J. Am. Chem. Soc.* **1991**, *113*, 5671. (b) Carreno, R.; Urbanos, F.; Dahan, F.; Chaudret, B. *New J. Chem.* **1994**, *18*, 449. (c) Süss-Fink, G.; Langenbahn, M.; Stoeckli-Evans, H.; Naumann, D. *J. Chem. Soc., Chem. Commun.* **1991**, 447. (d) Quignard, F.; Choplin, A.; Basset, J.-M. *J. Chem. Soc., Chem. Commun.* **1991**, 1589.

25. Rudakov, E. S.; Mitchenko, S. A.; Miroshnichenko, N. A. *Kinet. Katal.* **1987**, *28*, 187 (in Russian).

26. Jacoby, D.; Isoz, S.; Floriani, C.; Chiesi-Villa, A.; Rizzoli, C. *J. Am. Chem. Soc.* **1995**, *117*, 2805.

27. Brainard, R. L.; Nutt, W. R.; Lee, T. R.; Whitesides, G. M. *Organometallics* **1988**, *7*, 2379.

28. Bulls, A. R.; Schaefer, W. P.; Serfas, M.; Bercaw, J. E. *Organometallics* **1987**, *6*, 1219.

29. Arnold, J.; Wilkinson, G.; Hussain, B.; Hursthouse, M. B. *Organometallics* **1989**, *8*, 415.

30. (a) Cordone, R.; Harman, W. D.; Taube, H. *J. Am. Chem. Soc.* **1989**, *111*, 2896. (b) Winemiller, M. D.; Kopach, M. E.; Harman, W. D. *J. Am. Chem. Soc.* **1997**, *119*, 2096.

31. Gómez, M.; Yarrow, P. I. W.; Robinson, D. J.; Maitlis, P. M. *J. Organometal. Chem.* **1985**, *279*, 115.

32. (a) Vedernikov, A. N.; Kuramshin, A. I.; Solomonov, B. N. *Russ. J. Org. Chem.* **1993**, *29*, 1767. (b) Vedernikov, A. N.; Kuramshin, A. I.; Solomonov, B. N. *J. Chem. Soc., Chem. Commun.* **1994**, 121.

33. Meyer, T. Y.; Woerpel, K. A.; Novak, B. M.; Bergman, R. G. *J. Am. Chem. Soc.* **1994**, *116*, 10290.

34. Rabinovich, D.; Zelman, R.; Parkin, G. *J. Am. Chem. Soc.* **1992**, *114*, 4611.

35. (a) Schaller, C. P.; Wolczanski, P. T. *Inorg. Chem.* **1993**, *32*, 131. (b) de With, J.; Horton, A.D. *Angew. Chem., Int. Ed. Engl.* **1993**, *32*, 903. (c) Vilardo, J. S.; Lockwood, M. A.; Hanson, L. G.; Clark, J. R.; Parkin, B. C.; Fanwick, P. E.; Rothwell, I. P. *J. Chem. Soc., Dalton Trans.* **1997**, 3353.

36. Chen, J.-T.; Hsu, R.-H.; Chen, A.-J. *J. Am. Chem. Soc.* **1998**, *120*, 3243.

37. Herberhold, M.; Jin, G.-X.; Rheingold, A. L. *Chem. Comm.* **1996**, 2645.

38. Dunina, V. V.; Golovan', E. B. *Inorg. Chem. Commun.* **1998**, *1*, 12.

39. Brown, S. N.; Myers, A. W.; Fulton, J. R.; Mayer, J. M. *Organometallics* **1998**, *17*, 3364.

40. Strazzolini, P.; Giumanini, A. G.; Runcio, A.; Scuccato, M. *J. Org. Chem.* **1998**, *63*, 952.

41. (a) Rudakov, E. S. *The Reactions of Alkanes with Oxidants, Metal Complexes, and Radicals in Solutions*; Naukova Dumka: Kiev, 1985 (in Russian). (b) Rudakov, E. S. *Izv. SO Akad. Nauk SSSR, Ser. Khim.* **1980,** No 3, 161 (in Russian). (c) Rudakov, E. S.; Lutsyk, A. I. *Neftekhimiya* **1980**, *20*, 163 (in Russian).

42. (a) Rudakov, E. S. *Dokl. Akad. Nauk SSSR* **1982**, *263*, 942 (in Russian). (b) Tishchenko, N. A.; Rudakov, E. S.; Lyubchik, S. B. *Kinet. Katal.* **1989**, *30*, 588 (in Russian).

43. (a) Tishchenko, N. A.; Rudakov, E. S. *Kinet. Katal.* **1989**, *30*, 1058 (in Russian). (b) Rudakov, E. S.; Lutsyk, A. I.; Yaroshenko, A. P. *Ukr. Khim. Zhurn.* **1983**, *49*, 1083 (in Russian). (c) Rudakov, E. S. *Kinet. Katal.* **1990**, *31*, 37 (in Russian).

44. (a) Sen. A. *Platinum Metals Rev.* **1991**, *35*, 126. (b) Sen, A. In *Applied Homogeneous Catalysis with Organometallic Compounds*; Herrmann, W. A., Cornils,

B, Eds.; VCH: Weinhein, 1996; Vol. 2, p. 987. (c) Sen, A. *Acc. Chem. Res.* **1998**, *31*, 550. (d) Sen, A. Presented at the *217th Am. Chem. Soc. Natl. Meeting*; Anaheim, March **1999**; paper INOR No. 434.

45. (a) Sen, A.; Benvenuto, M. A.; Lin, M.; Hutson, A. C.; Basickes, N. *J. Am. Chem. Soc.* **1994**, *116*, 998. (b) Basickes, N.; Hogan, T. E.; Sen, A. *J. Am. Chem. Soc.* **1996**, *118*, 13111.

46. Periana, R. A.; Taube, D. J.; Evitt, E. R.; Löffer, D. G.; Wentrecek, P. R.; Voss, G.; Masuda, T. *Science,* **1993,** *259*, 340.

47. (a) Periana, R. A.; Taube, D. J.; Gamble, S.; Taube, H.; Satoh, T.; Fujii, H. *Science,* **1998,** *280*, 560. (b) Rouhi, M. *Chem. Eng. News* **1998**, April 27, 8. (c) Periana, R. A. Presented at the *217th Am. Chem. Soc. Natl. Meeting*; Anaheim, March **1999**; paper INOR No. 348.

48. (a) Lin, M.; Hogan, T. E.; Sen, A. *J. Am. Chem. Soc.* **1996**, *118*, 4574. (b) Chepaikin, E. G.; Boyko, G. N.; Bezruchenko, A. P.; Lescheva, A. A.; Grigoryan, E. H. *J. Mol. Catal. A: Chem.* **1998**, *129*, 15.

49. Hogan, T. E.; Sen, A. *J. Am. Chem. Soc.* **1997**, *119*, 2642.

50. Lin, M.; Hogan, T. E.; Sen, A. *J. Am. Chem. Soc.* **1997**, *119*, 6048.

51. (a) Nakata, K.; Yamaoka, Y.; Miyata, T.; Taniguchi, Y.; Takaki, K.; Fujiwara, Y. *J. Organomet. Chem.* **1994**, *473*, 329. (b) Kurioka, M.; Nakata, K.; Jintoku, T.; Taniguchi, Y.; Takaki, K.; Fujiwara, Y. *Chem. Lett.* **1995**, 244. (c) Nishiguchi, T.; Nakata, K.; Takaki, K.; Fujiwara, Y. *Chem. Lett.* **1992**, 1141. (d) Nakata, K.; Miyata, T.; Taniguchi, Y.; Takaki, K.; Fujiwara, Y. *J. Organometal. Chem.* **1995**, *489*, 71. (e) Taniguchi, Y.; Horie, S.; Takaki, K.; Fujiwara, Y. *J. Organomet. Chem.* **1995**, *504*, 137.

52. Taniguchi, Y.; Kitamura, T.; Fujiwara, Y.; Horie, S.; Takaki, K. *Catalysis Today* **1997**, *36*, 85.

53. Fujiwara, Y. In *Organometallic Chemistry on the Eve of the 21st Century*, Abstracts of the International Workshop, INEOS: Moscow, 1998, p. L6.

54. Asadullah, M.; Taniguchi, Y.; Kitamura, T.; Fujiwara, Y. *Appl. Organometal. Chem.* **1998**, *12*, 277.

55. Asadullah, M.; Taniguchi, Y.; Kitamura, T.; Fujiwara, Y. *Sekiyu Gakkaishi* **1998**, *41*, 236.

56. (a) Gol'dshleger, N. F.; Moravskiy, A. P. *Usp. Khim.* **1994**, *63*, 130 (in Russian). (b) Sen, A.; Gretz, E.; Oliver, T. F.; Jiang, Z. *New J. Chem.* **1989**, *13*, 755. (c) Shishkin, D. I.; Gekhman, A. E.; Moiseev, I. I. *Kinet. Katal.* **1989**, *30*,

1513 (in Russian). (d) Stolarov, I. P.; Vargaftik, M. N.; Moiseev, I. I. *Kinet. Katal.* **1989**, *30*, 1513 (in Russian). (e) Vargaftik, M. N.; Stolarov, I. P.; Moiseev, I. I. *J. Chem. Soc., Chem. Commun.* **1990**, 1049. (f) Stolarov, I. P.; Vargaftik, M. N.; Shishkin, D. I.; Moiseev, I. I. *J. Chem. Soc., Chem. Commun.* **1991**, 938. (g) Gol'dshleger, N. F.; Kresova, E. I.; Moravskiy, A. P. *Kinet. Katal.* **1991**, *32*, 1023 (in Russian).

57. Nizova, G. V., Süss-Fink, G., Stanislas, S., Shul'pin, G. B. *Chem. Comm.* **1998**, 1885.

58. Sheldon, R. A.; Kochi, J. K. *Metal-Catalyzed Oxidations of Organic Compounds*, Academic Press: New York, 1981.

59. (a) Baciocchi, E. *Acta Chem. Scand.* **1990,** *44*, 645. (b) Iqbal, J.; Bhatia, B.; Nayyar *Chem. Rev*. **1994**, *94*, 519.

60. Rudakov, E. S.; Lobachev, B. L. *Kinet. Katal.* **1988**, *29*, 1056 (in Russian).

61. Prokes, I.; Toma, S.; Luche, J.-L. *J. Chem. Res. (S)* **1996**, 164.

62. (a) Mayer, J. M. *Acc. Chem. Res.* **1998**, *31*, 441. (b) Mayer, J. M. Presented at the *217th Am. Chem. Soc. Natl. Meeting*; Anaheim, March **1999**; paper INOR No. 432.

63. (a) Muzart, J. *Chem. Rev.* **1992**, *92*, 113. (b) Chisem, I. C.; Martin, K.; Shieh, M. T.; Chisem, J.; Clark, J. H.; Jachuck, R.; Macquarrie, D. J.; Rafelt, J.; Ramshaw, C.; Scott, K. *Org. Proc. Res. Dev.* **1997**, *1*, 365.

64. (a) Cook, G. K.; Mayer, J. M. *J. Am. Chem. Soc.* **1994**, *116*, 1855. (b) Wang, K.; Mayer, J. M. *J. Org. Chem.* **1997**, *62*, 4248.

65. Tret'yakov, V. P.; Rudakov, E. S. In *Complexes of Platinum Group Metals in Synthesis and Catalysis*, Chernogolovka, 1983, p. 71 (in Russian).

66. (a) Shul'pin, G. B.; Lederer, P.; Mácová, E. *Bull. Acad. Sci. USSR, Div. Chem. Sci.* **1986**, *35*, 2422. (b) Shul'pin, G. B.; Mácová, E.; Lederer, P. *J. Gen. Chem. USSR* **1989**, *59*, 2329. (c) Shul'pin, G. B.; Kitaygorodskiy, A. N. *J. Gen. Chem. USSR* **1989**, *59*, 2335. (d) Shul'pin, G. B.; Druzhinina, A. N.; Nizova, G. V. *Bull. Acad. Sci. USSR, Div. Chem. Sci.*, **1990**, *39*, 195. (e) Shul'pin, G. B.; Kats, M. M. *J. Gen. Chem. USSR* **1990**, *61*, 684.

67. Banwell, M. G.; Haddad, N.; Huglin, J. A.; MacKay, M. F.; Reum, M. E.; Ryan, J. H.; Turner, K. A. *J. Chem. Soc., Chem. Commun.* **1993**, 954.

68. (a) Stewart, R.; Spitzer, U. A. *Can. J. Chem.* **1978**, *56*, 1273. (b) Rudakov, E.S.; Lobachev, V.L.; Zaychuk, E.V. *Kinet. Katal.* **1996**, *37*, 180 (in Russian).

69. (a) Gardner, K. A.; Kuehnert, L. L.; Mayer, J. M. *Inorg. Chem.* **1997**, *36*, 2069. (b) Noureldin, N. A.; Zhao, D.; Lee, D. G. *J. Org. Chem.* **1997**, *62*, 8767. (c) Oussaid, A.; Loupy, A. *J. Chem. Res. (S)* **1997**, 342.

70. (a) Li, W.-S.; Liu, L. K. *Synthesis* **1989**, 293. (b) Belavin, B. V.; Kresova, E. I.; Moravskiy, A. P.; Shilov, A. E. *Kinet. Katal.* **1990**, *31*, 764 (in Russian).

71. Wang, K.; Mayer, J. M. *J. Am. Chem. Soc.* **1997**, *119*, 1470.

72. DuMez D. D.; Mayer, J. M. *J. Am. Chem. Soc.* **1996**, *118*, 12416.

73. (a) Arzoumanian, H.; Agrifoglio, G.; Krentzien, H.; Capparelli, M. *J. Chem. Soc., Chem. Commun.* **1995**, 655. (b) Arzoumanian, H.; Maurino, L.; Agrifoglio, G. *J. Mol. Catal. A: Chem.* **1997**, *117*, 471.

74. (a) Tenaglia, A.; Terranova, E.; Waegell, B. *Tetrahedron Lett.* **1989**, *30*, 5271. (b) Tenaglia, A.; Terranova, E.; Waegell, B. *J. Chem. Soc., Chem. Commun.* **1990**, 1344. (c) Bakke, J. M.; Brænden, J. E. *Acta Chem. Scand.* **1991**, *45*, 418. (d) Tenaglia, A.; Terranova, E.; Waegell, B. *Tetrahedron Lett.* **1991**, *32*, 1169. (e) Tenaglia, A.; Terranova, E.; Waegell, B. *J. Org. Chem.* **1992**, *57*, 5523.

75. Chocron, S.; Michman, M. *Applied Catalysis, A: General* **1990**, *62*, 119.

76. (a) Lau, T.-C.; Mak, C.-K. *J. Chem. Soc., Chem. Commun.* **1993**, 766. (b) Lau.T.-C.; Mak, C.-K. *J. Chem. Soc., Chem. Commun.* **1995**, 943.

77. (a) Yam, V. W.-W.; Che, C.-M. *New J. Chem.* **1989**, *13*, 707. (b) Che, C.-M.; Tang, W.-T.; Wong, W.-T.; Lai, T.-F. *J. Am. Chem. Soc.* **1989**, *111*, 9048. (c) Che, C.-M.; Yam, V. W.-W.; Mak, T. C. W. *J. Am. Chem. Soc.* **1990**, *112*, 2284. (d) Ho, C.; Leung, W.-H.; Che, C.-M. *J. Chem. Soc., Dalton Trans.* **1991**, 2933. (e) Che, C.-M.; Ho, C.; Lau, T.-C. *J. Chem. Soc., Dalton Trans.* **1991**, 1901. (f) Li, C.-K.; Che, C.-M.; Tong, W.-F.; Tang, W.-T.; Wong, K.-Y.; Lai, T.-F. *J. Chem. Soc., Dalton Trans.* **1992**, 2109. (g) Taqui Khan, M. M.; Chatterjee, D.; Merchant, R. R.; Bhatt, A.; Kumar, S. *J. Mol. Catal.* **1991**, *66*, 289. (h)) Cheng, W.-C.; Yu, W.-Y.; Li, C.-K.; Che, C.-M. *J. Org. Chem.* **1995**, *60*, 6840.

78. Delaude, L.; Laszlo, P. *J. Org. Chem.* **1996**, *61*, 6360.

79. Deng, L.; Ziegler, T. *Organometallics* **1996**, *15*, 3011.

80. Banik, B. K.; Venkatraman, M. S.; Mukhopadhyay, C.; Becker, F. F. *Tetrahedron Lett.* **1998**, *39*, 7247.

81. (a) Hill, C. L. *Chem. Rev.* **1998**, *98*, 1. (b) Weinstock, I. A. *Chem. Rev.* **1998**, *98*, 113. (c) Kozhevnikov, I. V. *Chem. Rev.* **1998**, *98*, 171.

82. (a) Renneke, R. F.; Hill, C. L. *Angew. Chem.* **1988,** *100*, 1583. (b) Hill, C. L. *Adv. Oxygenated Processes*, **1988**, *1*, 1. (c) Prosser-McCartha, C. M.; Hill, C. L. *J. Am. Chem. Soc.* **1990**, *112*, 3672. (d) Renneke, R. F.; Pasquali, M.; Hill, C. L. *J. Am. Chem. Soc.* **1990**, *112*, 6585. (e) Chambers R. C.; Hill, C. L. *J. Am. Chem. Soc.* **1990**, *112*, 8427. (f) Renneke, R. F.; Kadkhodayan, M.; Pasquali, M.; Hill, C. L. *J. Am. Chem. Soc.* **1991**, *113*, 8357. (g) Combs-Walker, L. A.; Hill, C. L. *J. Am. Chem. Soc.* **1992**, *114*, 938.

83. Kothe, T.; Martschke, R.; Fischer, H. *J. Chem. Soc., Perkin Trans. 2* **1998**, 503.

84. Awad, M. K.; Anderson, A. B. *J. Am. Chem. Soc.* **1990**, *112*, 1603.

85. Kats, M. M.; Shul'pin, G. B. *Bull. Acad. Sci. USSR, Div. Chem. Sci.* **1990**, *39*, 2233.

86. (a) Mimoun, H. *Angew. Chem., Int. Ed. Engl.* **1982**, *21*, 734. (b) Mimoun, H. *Isr. J. Chem.* **1983**, *23*, 451. (c) Gerasimova, S. O.; Kharitonov, Yu. Ya. *Koord. Khim.* **1988**, *14*, 1443 (in Russian). (d) Spirina, I. V.; Maslennikov, V. P.; Aleksandrov, Yu. A. *Usp. Khim.* **1987**, *56*, 1167 (in Russian). (e) Fomin, V. M.; Glushakova, V. N.; Aleksandrov, Yu. A. *Usp. Khim.* **1988**, *57*, 1170 (in Russian). (f) Butler, A.; Clague, M. J.; Meister, G. E. *Chem. Rev.* **1994**, *94*, 625.

87. (a) Daire, E.; Mimoun, H.; Saussine, L. *Nouv. J. Chim.* **1984**, *8*, 271. (b) Mimoun, H.; Saussine, L.; Daire, E.; Postel, M.; Fisher, J.; Weiss, R. *J. Am. Chem. Soc.* **1983**, *105*, 3101. (c) Bonchio, M.; Conte, V.; Di Furia, F.; Modena, G. *J. Org. Chem.* **1989**, *54*, 4368. (d) Bonchio, M.; Conte, V.; Di Furia, F.; Modena, G.; Moro, S. *J. Org. Chem.* **1994**, *59*, 6262. (e) Talsi, E. P.; Shalyaev, K. V. *J. Mol. Catal.* **1994**, *92*, 245. (f) Attanasio, D.; Suber, L.; Shul'pin, G. B. *Bull. Russ. Acad. Sci. Div. Chem. Sci.* **1992**, *41*, 1502.

88. Muzart, J.; Nizova, G. V.; Riahi, A.; Shul'pin, G. B. *J. Gen. Chem. (Russia)* **1992**, *62*, 964.

89. (a) Gekhman, A. E.; Moiseeva, N. I.; Moiseev, I. I. *Izv. Akad. Nauk, Ser. Khim.* **1995**, 605 (in Russian). (b) Gekhman, A. E.; Stalarov, I. P.; Moiseeva, N. I.; Rubajlo, V. L.; Vargaftik, M. N.; Moiseev, I. I. *Inorg. Chim. Acta* **1998**, *275–276*, 453.

CHAPTER IX

HOMOGENEOUS CATALYTIC OXIDATION OF HYDROCARBONS BY MOLECULAR OXYGEN

The oxidation of hydrocarbons by molecular oxygen in the absence of metal complexes has been discussed in Chapter II. The oxidation of hydrocarbons with molecular oxygen, as well as with donors of an oxygen atom (hydrogen peroxide, alkyl hydroperoxides and some other compounds), is a very important field since many industrial processes are based on these reactions [1]. In many cases, chain radical non-catalyzed autoxidation of saturated hydrocarbons is not very selective and the yields of valuable products are often low. The use of salts and complexes of transition metals creates great possibilities for solving problems of selective oxidation, as has been demonstrated for a number of important processes.

IX.1. TRANSITION METAL COMPLEXES IN THE THERMAL AUTOXIDATION OF HYDROCARBONS

IX.1.A. CLASSICAL RADICAL-CHAIN AUTOXIDATION

Alkanes and alkyl aromatic hydrocarbons can be oxidized when heated at rather high (usually above 100 °C) temperatures under oxygen. A long induction period can be reduced or eliminated if a donor of free radicals is present. The formation of oxygen-containing products from hydrocarbons and molecular oxygen is always thermodynamically allowed due to the high exothermicity of oxidation reactions. However, this same fact makes these processes usually unselective. The main problem is to prevent various parallel and consecutive oxidation reactions to produce numerous byproducts. Destruction of the carbon chain is one of the possible routes in the oxidation [2a]. For example, adipic acid is prepared commercially from cyclohexane by a two-step oxidation, first with air and then with nitric acid (Scheme IX.1) [2b].

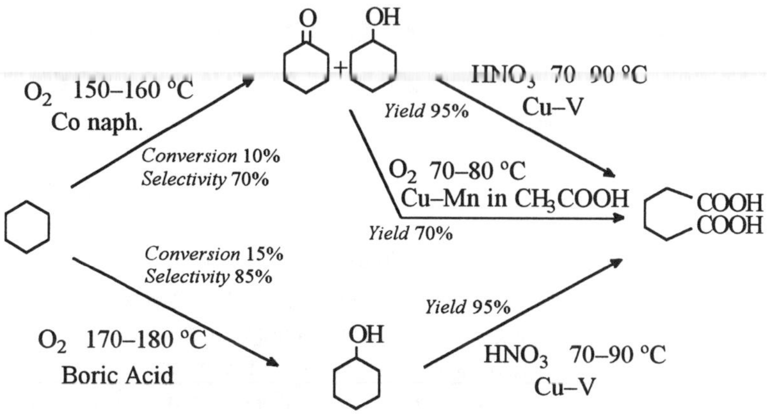

Scheme IX.1. Industrial process for the conversion of cyclohexane to adipic acid.

Autoxidation of saturated hydrocarbons or fragments occurs as a free radical-chain process. Azobis(isobutyronitrile) (AIBN, In–N=N–In) is frequently used as an initiator of radical-chain reactions [3]. For example, heptane (RH) is oxidized according to the following scheme [3c]:

$$\text{In–N=N–In} \longrightarrow 2\,\text{In}\bullet + N_2$$

$$2\,\text{In}\bullet \longrightarrow \text{InIn}$$

$$\text{In}\bullet + O_2 \longrightarrow \text{InOO}\bullet$$

$$2\,\text{InOO}\bullet \longrightarrow \text{InOOOOIn}$$

$$\text{InOOOOIn} \longrightarrow 2\,\text{InO}\bullet + O_2$$

$$\text{In}\bullet + \text{RH} \longrightarrow \text{InH} + \text{R}\bullet$$

$$\text{InOOH} \longrightarrow \text{InO}\bullet + \text{HO}\bullet$$

$$\text{InO}\bullet + \text{RH} \longrightarrow \text{InOH} + \text{R}\bullet$$

$$R\bullet + O_2 \longrightarrow ROO\bullet$$

$$ROO\bullet + RH \longrightarrow ROOH + R\bullet$$

$$InOO\bullet + ROO\bullet \longrightarrow InOOOOR$$

$$InOOOOR \longrightarrow InO\bullet + RO\bullet + O_2$$

$$InOOOOR \longrightarrow InOOR + O_2$$

$$2\,ROO\bullet \longrightarrow \text{products}$$

Any additive which could react with free radicals formed in such processes to yield stable adducts will inhibit the oxidation [4].

Mechanism of Catalysis by Transition Metal Salts

Ions of transition metals (homogeneously or in some cases supported on polymers [5]) also effectively catalyze the autoxidation. Salts of cobalt, manganese, iron, copper, chromium, lead, and nickel are used as catalysts that allow the reactions to be carried out at lower temperatures, therefore increasing the selectivity of the oxidation (see, for example, [6]). However, it is more important that the catalyst itself may regulate the selectivity of the process, leading to the formation of a particular product. The studies of the mechanism of the transition metal salt involvement have shown their role to consist, in most cases, of enhancing the formation of free radicals in the interaction with the initial and intermediate species.

The common mechanism of catalysis by transition metal salts was formulated at the beginning of the thirties by Haber and Willstäter [7]. They proposed a scheme involving a metal ion interaction with a molecule (A–B) involving a change of ion oxidation state and free radical formation:

$$M^{n+} + A\text{–}B \longrightarrow M^{(n+1)+} + A\bullet + B^-$$

$$M^{(n+1)+} + A\text{–}B \longrightarrow M^{n+} + A^+ + B\bullet$$

According to this scheme, a metal ion alternatively increasing and decreasing its valency can participate in the formation of a large number of free radicals, thus playing the role of a catalyst. The study of the participation of metal ions in the chain reactions of oxidation shows them to take part in all the stages of the reaction, i.e. chain initiation, branching, propagation, and termination. In effect, small additions of transition metal complexes increase the initial rate of chain reactions and decrease the induction period, which is characteristic of non-catalytic oxidation. For example, naphthenates of cobalt, chromium and manganese reduce the induction period in cyclohexane oxidation. The same action is displayed by cobalt stearate, the induction period being shorter if more catalyst is introduced.

Formation of free radicals in the presence of transition metal compounds may be the result of their interaction with hydrocarbon or dioxygen. Such catalysts as cobalt and manganese salts are usually introduced in the bivalent state. An increase of chain initiation rate here may be connected with the activation of dioxygen by a metal complex via the following reaction [8]:

$$M^{2+} + O_2 \rightleftharpoons M^{(2+\delta)+} \cdots O_2^{\delta-}$$

$$M^{(2+\delta)+} \cdots O_2^{\delta-} + RH \longrightarrow M^{2+} + R\bullet + HOO\bullet$$

If the catalyst is in a high-valent state, chain initiation can proceed in the reaction:

$$M^{3+} + RH \longrightarrow M^{2+} + R\bullet + H^+$$

This mechanism was unequally established for the case of benzaldehyde oxidation at room temperature in solutions of glacial acetic acid [9]. The kinetic equation obtained for the reaction is

$$v = k\,[RCHO]^{3/2}\,[catalyst]^{1/2}.$$

For the chain termination of the second order, the chain reaction rate is

$$v = const\,[RH]\,v_i^{1/2}$$

which leads to:

$$v_i = k_1\,[RCHO]\,[catalyst]$$

The rate constant of chain initiation was determined from the special experiments with the introduction of an inhibitor. It was found to be

$$k = 3 \times 10^9 \exp(-14800/RT) \text{ dm}^3 \text{ mol}^{-1} \text{ s}^{-1}.$$

The rate constant of the reaction of cobalt(III) acetate with benzaldehyde in the absence of dioxygen was determined in independent experiments. It turned out to be virtually the same as the rate constant of the chain initiation in the oxidation reaction in the presence of O_2. However, the contribution of chain initiation to the radical formation is insignificant in the developed oxidation process. The radicals are mainly formed in the reactions of the intermediates in the process of degenerate chain branching. These reactions are also catalyzed by transition metal ions. Especially well studied is the acceleration of radical decomposition of intermediately formed hydroperoxides (see, e.g., [10]).

As early as 1932, Haber and Weiss [11] suggested a mechanism of interaction between hydrogen peroxide and iron(II) ions where electron transfer to the peroxide molecule led to the formation of a free hydroxyl radical:

$$Fe^{2+} + H_2O_2 \longrightarrow Fe^{3+} + HO^- + HO\bullet$$

The reaction of iron(II) ions with hydroperoxides proceeds similarly [12]:

$$Fe^{2+} + ROOH \longrightarrow Fe^{3+} + HO^- + RO\bullet$$

Other ions, such as Co^{2+} and Mn^{2+} can react in the same way:

$$Co^{2+} + ROOH \longrightarrow Co^{3+} + HO^- + RO\bullet$$

The newly formed high-valent ions can further react with hydroperoxide, e.g.,

$$Co^{3+} + ROOH \longrightarrow ROO\bullet + Co^{2+} + H^+$$

Therefore transition metal salts, in particular those of cobalt, can catalyze the decomposition of hydroperoxides into free radicals:

$$2\,ROOH \xrightarrow{Co(II)-Co(III)} RO\bullet + ROO\bullet + H^+ + HO^-$$

In this case, the time taken to reach the maximum rate, as well as the steady-state concentration of hydroperoxide, will be reduced.

In non-polar hydrocarbon media the reaction of hydroperoxide proceeds with undissociated salt, for example cobalt(II) stearate, which can be present in solution in small concentrations:

$$St_2Co + ROOH \longrightarrow RO\bullet + St_2CoOH$$
$$St_2CoOH + ROOH \longrightarrow ROO\bullet + St_2Co + H_2O$$

Such a reaction obviously proceeds more slowly than in aqueous or other polar media, but much faster than does thermal hydroperoxide decomposition. This results in a sharp drop of the steady-state concentration of the hydroperoxide in the hydrocarbon oxidation in the presence of a transition metal salt. Here, a decrease of hydroperoxide concentration will be compensated by an increase of its specific rate of decomposition so that the maximum oxidation rate will remain the same as in non-catalytic oxidations. In effect, the maximum rate

$$v_{max} = (-\,d[RH]/dt)_{max} = f^{-1}\, k_2\, [ROO\bullet]_{max}\, [RH]$$

will be reached under the condition that

$$nk_2\, [ROO\bullet]\, [RH] = 2\, k_6\, [ROO\bullet]^2$$

that is [13]:

$$v_{max} = (n/f)(k_2^2/k_6)\, [RH]$$

The same expression was obtained for non-catalytic reactions, i.e., the catalyst does not increase the maximum rate but merely reduces the induction period. This seemingly paradoxical conclusion is based on the fact that in both cases the maximum rate is determined by the equality of catalyst-independent reaction rates of chain propagation which, in the maximum, are equal to the rate of chain branching and termination. The existence of a maximum rate was quantitatively confirmed for the oxidation of some hydrocarbons (tetralin, ethylbenzene, diphenylmethane, etc.) [14a].

The maximum rate of dioxygen consumption

$$- d[O_2]/dt = k_{eff}\,[RH]^2$$

was found to be independent of the catalyst concentration and the value k_{eff} was found to coincide quantitatively with the parameter $k_2^2/2k_6$, measured independently. Chemiluminescent techniques have shown the hydroperoxide formation rate in these systems to be equal to that of dioxygen consumption. This is the evidence that the process proceeds entirely via the hydroperoxide formation [14b].

Hydroperoxide decomposition in the presence of transition metal compounds may be, in effect, more complicated than the above simple radical scheme suggests. It may involve, for example, the formation of intermediate complexes between the catalyst and hydroperoxide. The reaction may result in the formation of molecular products without the participation of free radicals or proceed itself by a radical chain mechanism [15].

The classical radical-chain mechanism leads to a variety of products in hydrocarbon oxidation due to the high temperature of the process and low selectivity of the interacting radicals. In view of developing selective industrial processes based on these reactions for the production of various substances, their low selectivity is an important shortcoming. The rate of oxidation usually increases in the order

$$RCH_3 < RCH_2OH < RCHO.$$

Thus, oxidation products are oxidized in their turn faster than the hydrocarbons themselves, and therefore the yields of the required intermediate products are usually low. In many cases the catalyst merely decreases the induction period, changing neither the mechanism nor the maximum rate of the process. This is typical of low catalyst concentrations in hydrocarbon media when the metal salt concentration, due to low solubility, does not exceed several hundredths of one percent. The use of polar solvents, e.g., acetic acid, trifluoroacetic acid, etc., allows for a considerable increase (100–1000 times) of the catalyst concentration. In this case the reaction can acquire new features. Not only the ratio but also the kinetics of the process can be markedly changed. Especially important is the fact that the process very often becomes highly selective, which allows it to be used on the industrial scale.

Cobalt compounds are among the most efficient catalysts of alkane and aryl alkane autoxidation [16]. Toluene oxidation by dioxygen at 60 °C in acetic acid in the presence of Co(III) at the initial stages gives exclusively benzaldehyde [16d]. The reaction proceeds without an induction period, reaching the maximum rate at the beginning of the reaction. The kinetic equation for the oxidation is

$$-\,d[O_2]/dt = k\,[PhCH_3]\,[Co(III)]^2\,[Co(II)]^{-1}$$

It is noteworthy that the reaction rate increases sharply on the addition of sodium acetate, which may mean that the primary interaction of Co(III) with toluene, which the authors of [16d] represented as electron transfer from toluene to Co(III), actually proceeds through electron transfer with synchronous elimination of the proton with the participation of a base (in this case, acetate). Since Co^{3+} can rather effectively oxidize a number of hydrocarbons, the following catalytic pathway:

$$ROO\bullet + Co^{2+} + H^+ \rightleftharpoons ROOH + Co^{3+}$$

$$Co^{3+} + RH \longrightarrow R\bullet + H^+ + Co^{2+}$$

is presumably realized in the presence of high cobalt salt concentrations. It is important that in some reactions of alkanes, an unusual selectivity is observed in the presence of cobalt compounds with or without dioxygen. This is strikingly different from that observed in radical reactions and implicates the direct interaction of Co(III) with hydrocarbons.

Butane oxidation in the presence of cobalt(III) acetate in acetic acid occurs at temperatures of 100–125 °C. Acetic acid is the reaction product with 83% selectivity (at 80% conversion) [1j, 17]. These data are markedly different from those observed for butane autoxidation at low initiator concentrations, where temperatures up to 170 °C and higher are required and acetic acid is produced with 40% selectivity. Cyclohexane oxidation in the presence of cobalt(II) acetate in acetic acid gives adipic acid in one step as the main product with 75% selectivity at more than 80% cyclohexane conversion [2b]. The induction period which is observed in the reaction decreases on the addition of promoters, such as acetaldehyde, or disappears completely if the cobalt salt is introduced in the three-valent state.

The kinetics of dioxygen consumption are described by the equation which is common for reactions with Co(III) participation

$$-\,d[O_2]/dt = k\,[RH]\,[Co(III)]^2\,[Co(II)]^{-1}$$

It is noteworthy that benzene under these conditions is not oxidized and is an appropriate solvent for cyclohexane oxidation. However, in a mixture of benzene – acetic acid the rate is 40% higher than that in acetic acid alone.

The alkanes containing a tertiary C–H bond appear to be less reactive than normal alkanes (the same is observed in the absence of dioxygen) and the reactivity surprisingly decreases with the increase of the hydrocarbon chain length. Thus, under similar conditions, isobutane reacts slower than butane, *n*-butane is more reactive than *n*-pentane, and undecane is completely unreactive [17]. It is difficult, as yet, to find any plausible explanation for these results. Heptane oxidation by cobalt(III) acetate in trifluoroacetic acid proceeds selectively without dioxygen in position 2, and in the presence of dioxygen results in 2-heptanone with 83% selectivity [18]. Undoubtedly, the alkane in these reactions interacts with the cobalt(III) salt and the selectivity observed, as has been pointed out, is of special interest, indicating the significant role of steric factors. The kinetics of the reaction, however, seem to be complicated by the oxidation of substances produced from alkanes. These secondary reactions might proceed according to a mechanism different from the mechanism of alkane oxidation.

Cobalt-Bromide Catalysis

The oxidation of alkylaromatic hydrocarbons proceeds particularly easily in the presence of both cobalt and bromide ions (a so-called "cobalt-bromide catalysis"). Carboxylic acids are the final products of the reaction. For example, terephthalic acid is selectively formed from *p*-xylene, the whole process being used in the industrial production of the acid [1k, 19]. Despite the large number of works on cobalt-bromide catalysis, its mechanism has long remained speculative.

The process has interesting kinetic characteristics, specifically, in the presence of bromide ions the methylbenzenes oxidation rate is of second order with respect to cobalt. The rate does not depend on dioxygen concentration and depends only very insignificantly on the concentration of the compound being oxidized. To elucidate the "cobalt-bromide catalysis", a variety of schemes has been suggested. The reaction undoubtedly follows the chain mechanism (though some non-chain schemes were also suggested, some of them even without free radicals). The catalytic effect of cobalt and bromide ions is connected with their participation in chain propagation. Cobalt(II) ions can interact with hydroperoxide radicals:

$$ROO\bullet + Co^{2+} \longrightarrow Co^{3+} + ROO^{-}$$

Cobalt(III) ions are again reduced to Co(II) by Br^- ions, the latter converting into bromide atoms:

$$Co^{3+} + Br^{-} \longrightarrow Co^{2+} + Br\bullet$$

Bromine atoms further react with hydrocarbons causing H elimination and forming a free radical:

$$Br\bullet + RH \longrightarrow HBr + R\bullet$$

The bromide ions may enter the coordination sphere of Co(II) as ligands, facilitating the process of cobalt oxidation:

$$ROO\bullet + Co(II)Br^{-} \longrightarrow Co(III)Br^{-} + ROO^{-}$$

Then Br^- oxidation occurs in the coordination sphere of Co(III):

$$Co(III)Br^{-} \rightleftharpoons Co(II)Br\bullet$$

The reaction with hydrocarbons is suggested to proceed without Br• atoms leaving the coordination sphere of Co(II):

$$Co(II)Br\bullet + RH \longrightarrow Co(II) + HBr + R\bullet$$

Table IX.1 gives the data on the relative rates of hydrocarbon oxidation obtained in the mixed oxidation of two hydrocarbons (usually the second hydrocarbon is ethylbenzene). For comparison, the relative reactivity for radicals and bromine atoms is also given in this table.

Table IX.1. The relative reactivity of hydrocarbons (per one reactive H atom) in the reaction with active species and in the autoxidation in the presence of Co(II) and Br^- in acetic acid.

RH	$Co(OAc)_2$ + NaBr (60 °C)	Cumylperoxy radicals (30 °C)	Br atoms (40 °C)
Cyclohexane	0.33	0.1	0.0074
p-Chlorotoluene	0.42		
Toluene	1.00	1.00	1.00
Mesitylene	1.17		
p-Xylene	1.50	1.60	
Pseudocumene	2.54		
p-Methoxytoluene	3.42		
Durene	3.83		
Ethylbenzene	8.33	9.3	17
Cumene	16.8	15.9	37
Tetralin	34.2	36.4	

Kinetic studies, combined with measurements of chemiluminescence resulting from the process of the recombination of two radicals ROO•, allow for the proposal of a scheme where cobaltous, cobaltic and bromide ions enter the chain propagation [19b, 20]:

$$(2')\ ROO\bullet + Co^{2+} + H^+ \underset{k_{-4}}{\overset{k_4}{\rightleftharpoons}} ROOH + Co^{3+}$$

$$(7)\ Co^{3+} + Br^- \xrightarrow{k_9} Co^{2+} + Br\bullet$$

$$(8)\ Br\bullet \xrightarrow{k_8} H^+ + Br^- + R\bullet$$

The other stages are similar to those of the mechanism with participation of small catalytic concentrations

$$\text{(1) } R\bullet + O_2 \longrightarrow ROO\bullet$$

$$\text{(2) } ROO\bullet + RH \xrightarrow{k_2} ROOH + R\bullet$$

$$\text{(3'') } ROOH + Co^{2+} \longrightarrow RO\bullet + Co^{3+} + HO^-$$

$$\text{(6) } 2\ ROO\bullet \xrightarrow{k_6} \text{products}$$

Thus a slow reaction of chain propagation in the absence of the catalyst, or when its concentration is small, is replaced by the more rapid reactions 2', 7 and 8 (in fact, reactions 2', 7 and 8 are a catalytic pathway of the reaction 2) with catalyst participation. In this case, the maximum rate is reached if

$$2k_{2'}\ [ROO\bullet]\ [Co^{2+}] = 2k_6\ [ROO\bullet]^2$$

Hence, in agreement with the experimental results

$$v = -\ d[O_2]dt = (2k_{2'}{}^2/k_6)\ [Co^{2+}]^2$$

Taking into account that hydroperoxide formation occurs in the reactions of ROO• not only with cobaltous ions but with hydrocarbons as well, we have, for steady-state conditions

$$2k_2\ [ROO\bullet]\ [RH] + 2k_{2'}\ [ROO\bullet]\ [Co^{2+}] = 2k_6\ [ROO\bullet]^2$$

and

$$v = (2/k_6)(k_2\ [RH] + k_{2'}\ [Co^{2+}])^2$$

In accordance with the last formula, the relation is linear, with the slope of the straight line corresponding to the ratio $k_2(2/k_6)^{½}$. The mechanism suggested is also supported by the fact that, in the case of alkylbenzenes, the dioxygen consumption rate coincides with that of the catalytic hydroperoxide decomposition. Moreover, the hydroperoxide radical recombination rate, measured by the chemiluminescence technique, is practically the same as that of dioxygen consumption.

In the suggested scheme, Br^- catalyzes an electron transfer from RH to Co^{3+}:

$$Co^{3+} + RH \longrightarrow Co^{2+} + R\bullet + H^+$$

The mechanism via bromine atoms is supported by molecular bromine formation in the interaction of Co^{3+} with Br^- in the absence of a hydrocarbon (Br_2 is apparently formed by bromine atom recombination). This mechanism is also consistent with the fact that bromide ions, while catalyzing the oxidation in the case of alkylaromatic compounds, are not particularly effective in the case of simple alkanes. This corresponds to the difference of bromine atom reactivity with respect to alkylaromatic and aliphatic hydrocarbons. The bond energy in the H–Br molecule (85 kcal $mole^{-1}$) is practically equal to the energy of the C–H bond in the α-position to the aromatic ring, so that the reaction

$$Br\bullet + HR \longrightarrow HBr + R\bullet$$

for alkylaromatic hydrocarbons is close to thermoneutrality. For aliphatic hydrocarbons, this reaction is endothermic and its activation energy is considerably higher with bromine ions ceasing to be an effective catalyst of electron transfer from RH to Co^{3+}. The question arises as to why the reaction

$$Co^{3+} + RH \longrightarrow Co^{2+} + R\bullet + H^+$$

proceeds much slower without the catalyst and why its rate is enhanced by Br^- ions. Probably the answer is that a direct electron transfer from RH to Co^{3+} requires a hydrocarbon molecule to approach Co^{3+} with H^+ being synchronously eliminated by the base. Apparently this is of rather low probability. At the same time, a negative bromine ion can easily enter the coordination sphere of a positive cobalt(III) ion and the bromine atom produced will react with the hydrocarbon molecule without serious steric hindrance. It is obvious that such catalysis reflects the characteristics of the $Br\bullet/Br^-$ pair for which the redox potential, for the optimum case, must be close to those of the Co^{3+}/Co^{2+} and $(R\bullet + H^+)/RH$ pairs.

IX.1.B. SOME NEW AUTOXIDATION PROCESSES

During the last decades some reactions between hydrocarbons and molecular oxygen have been described. These processes are often chain radical autoxidations, but in some cases the mechanisms of the reactions are not fully clear.

A weakly solvated complex containing acetonitrile, $[Co(NCMe)_4](PF_6)_2$, catalyzes the air oxidation of cyclohexane and adamantane at 75 °C and 3 atm [21]. The commercial catalyst for cyclohexane oxidation does not function under these conditions. The metal ion functions both as an initiator and as a hydroperoxide decomposition catalyst. The following steps have been proposed for the oxidation reaction:

$$[Co(NCMe)_4]^{2+} \longrightarrow Co_{Ox} \text{ (an oxidized cobalt species)}$$

$$Co_{Ox} + RH \longrightarrow R\bullet + H^+ + [Co(NCMe)_4]^{2+}$$

$$R\bullet + O_2 \longrightarrow ROO\bullet$$

$$ROO\bullet + RH \longrightarrow ROOH + R\bullet$$

$$ROOH + Co(II) \longrightarrow RO\bullet + Co(III)OH$$

$$Co(III)OH + ROOH \longrightarrow Co(III)OOR + H_2O$$

$$Co(III)OOR \longrightarrow Co(II) + ROO\bullet$$

$$Co(III)OOR \longrightarrow Co(III)OH + C_6H_{10}O$$

$$Co(III)OOR \longrightarrow CoO\bullet + RO\bullet$$

$$RO\bullet + RH \longrightarrow ROH + R\bullet$$

Cobalt chloride in diglyme is a useful catalyst for benzylic [22a] and allylic [22b] oxidation under mild conditions. The addition of sodium azide to oxidations catalyzed by transition metal acetylacetonates, heteropolyacids, phthalocyanines, bis-(pyridylimino)isoindolines, porphyrins and Schiff bases significantly enhances the rates of the low-temperature catalytic oxidation of alkanes [22c]. Ethylbenzene is slowly oxidized by air in MeCN in the presence of catalytic amount of chromium trioxide [23]. Complexes of Fe(III) and Co(II) with dipyridyl- and acetylacetone-functionalyzed polymers showed a high degree of activity and selectivity in the oxidation of ethylbenzene by oxygen [24]. Limonene can be efficiently and selectively oxidized by molecular oxygen in the presence of the $PdCl_2$–$CuCl_2$ catalytic combination [25]:

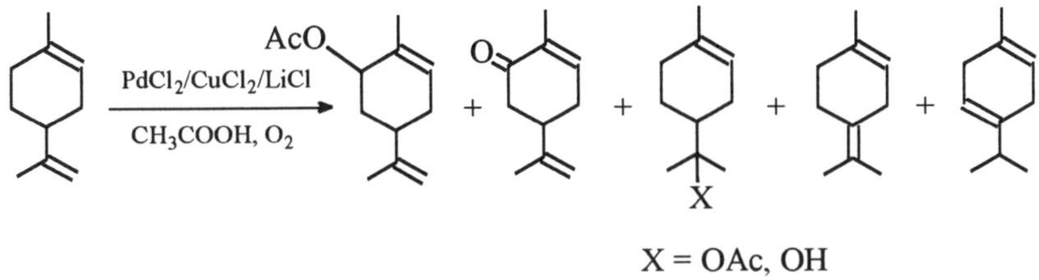

The system Ru(III)–EDTA catalyzes the oxidation of cyclohexane by molecular oxygen [26]. In the presence of manganese(II) acetate and molecular oxygen, alkenes and active methylene compounds react to yield cyclic peroxides [27]. α-Substituted cycloalkanones are oxidized to oxo acids by the copper(II) nitrate–dioxygen–acetic acid–water system [28a]:

α-Hydroxy ketones can be cleaved with a catalytic amount of dichloroethoxy-oxyvanadium under an oxygen atmosphere [28b]:

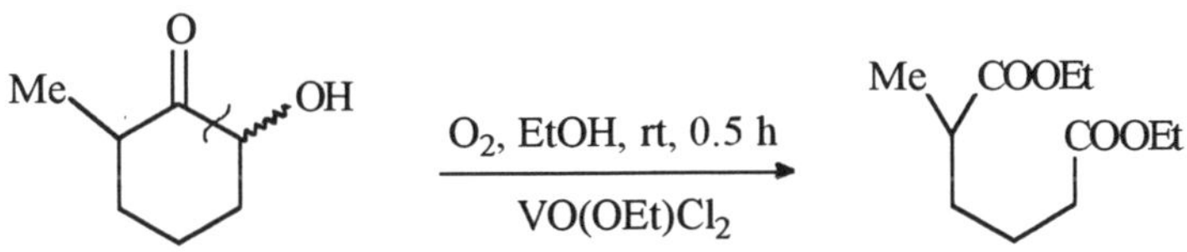

Copper(II) chloride, in combination with acetoxime, catalyzes the oxidation of a methyl group in 2,4,6-trimethylphenol in alcohols, ROH, at ambient temperature [29]:

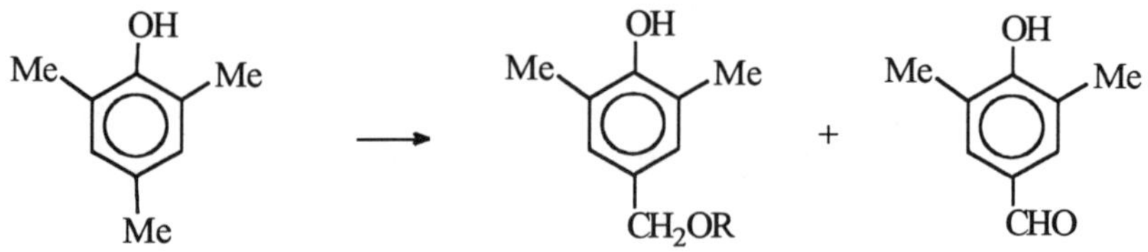

Oxygenation Catalyzed by Metalloporphyrins

Halogenated metalloporphyrins are effective catalysts for selective air oxidation of light alkanes [30] as well as of olefins [31]. The postulated mechanism of the reaction (Scheme IX.2) [30c] is similar to those proposed for biological oxidation (by cytochrome P450 and methanemonooxygenase, see Chapter XI).

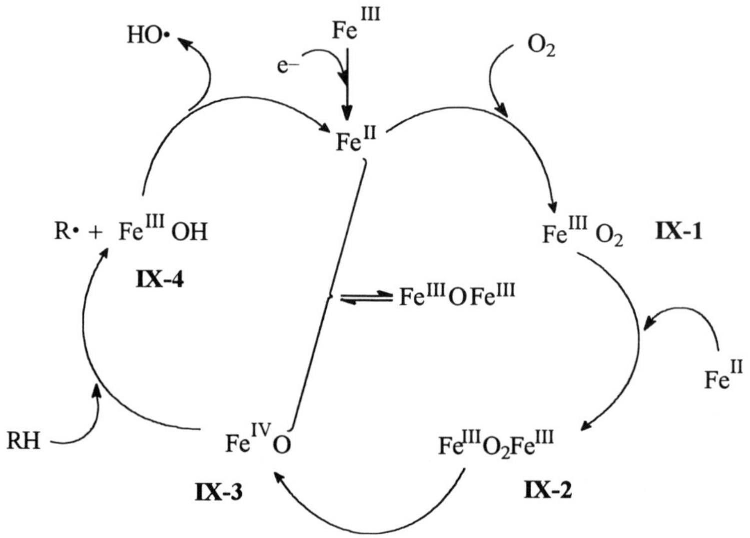

Scheme IX.2. The mechanism postulated for the air oxidation of alkanes catalyzed by halogenated metalloporphyrins.

This mechanism involves the reduction of Fe(III) followed by the addition of dioxygen to produce dioxo species **IX-1**. This species reacts with a second molecule of catalyst (intermediate **IX-2**) giving two molecules of the monooxo complex **IX-3**, which is capable of oxidizing the alkane. Species **IX-3** abstracts a hydrogen atom from the alkane to generate an alkyl radical and the hydroxy derivative **IX-4**. However, the alternative mechanism does not consider Fe(II)–O_2 species and assumes the conventional radical-chain autoxidation [30d]. Such a mechanism is depicted in Scheme IX.3.

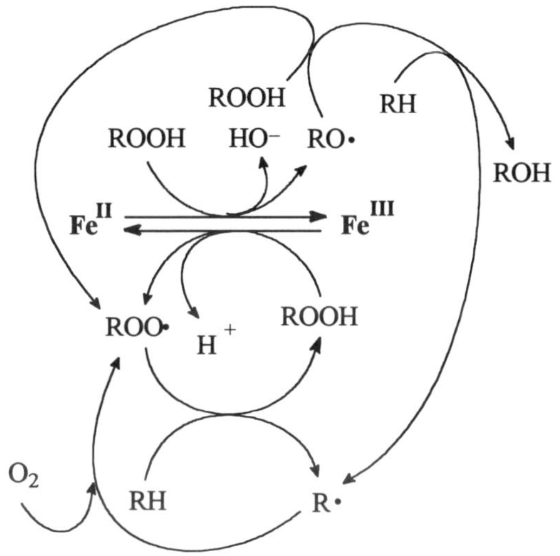

Scheme IX.3. The alternative mechanism for the air oxidation of alkanes assuming a radical–chain autoxidation.

Oxidations Catalyzed by Polymetalates

Heteropolyanions have been shown to effectively catalyze the oxidation of alkanes into the corresponding alcohols and ketones [32]. Thus, the compound $[PW_9O_{37}][Fe_2Ni(OAc)_3]^{10-}$ catalyzes [32b] the transformation of adamantane into 1-adamantanol (76%), 2-adamantanol (12%) and 2-adamantanone (12%) with a total turnover number of 25 and a 29% conversion. The mixed-addenda heteropolyanion $[PV_2Mo_{10}O_{40}]^{5-}$ catalyzes the aerobic oxidative dehydrogenation of α-terpinene to *p*-cumene [32d]. Transition metal substituted Keggin type polyoxomolybdates have been shown to be catalysts for the autoxidation of cumene or cyclohexene to afford alkylhydroperoxides which are used in the same reaction to epoxidize alkenes [32e]. Alkylbenzenes and alkanes can be oxygenated by dioxygen when ammonium molybdovanadophosphate is used as a catalyst [32f]. Finally, isophorone has been smoothly oxidized with O_2 in the presence of molybdovanadophosphate supported on active carbon [32g].

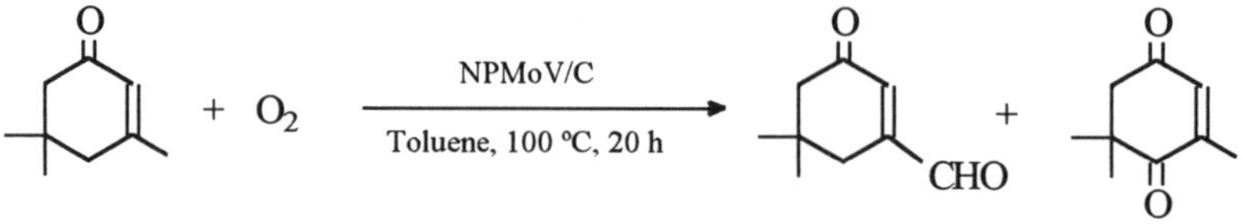

It is interesting that the regioselectivity of the oxidation is opposite to that of the conventional oxidations.

The Ishii Oxidation Reaction

An oxidation of alkanes by molecular oxygen, which is catalyzed by *N*-hydroxyphthalimide (NHPI) combined with $Co(acac)_n$ (n =2,3) or transition metal salts, has been described by Ishii and coworkers (see a review [33] and original papers [34]; in some cases NHPI also efficiently catalyzes the oxidation in the absence of metal derivatives [35]). For example, the oxidation of isobutane in the presence of catalytic amounts of NHPI and $Co(acac)_2$ under an air atmosphere in benzonitrile at 100 °C gave *tert*-butyl alcohol (yield 84%) as well as acetone (13%) [34d]. The co-catalytic effects of various metal compounds are compared in Table IX.2. The proposed mechanism of the Ishii oxidation reaction involves hydrogen abstraction from isobutane (or any other compound with reactive C–H bonds) by the phthalimidooxyl radical (PINO) (Scheme IX.4) [34d].

Direct conversion of cyclohexane into adipic acid has been achieved under the Ishii oxidation reaction conditions [34e]. The selectivity of adipic acid formation was 73% at 73% conversion in the presence of NHPI and $Mn(acac)_2$ at 100 °C for 20 h. The oxidation of other cycloalkanes is presented in Table IX.3 [34e].

Table IX.2. The aerobic oxidation of isobutane in the presence of NHPI and various metal acetylacetonates [a].

	Yield, %	
Catalyst	$(CH_3)_3COH$	CH_3COCH_3
None	3	4
TiO(acac)$_2$	6	2
V(acac)$_2$	18	9
Cr(acac)$_3$	52	9
Mn(acac)$_2$	28	8
Mn(acac)$_3$	48	12
Fe(acac)$_3$	52	9
Co(acac)$_2$	84	13
Co(acac)$_3$	58	10
Ni(acac)$_2$	25	6
Cu(acac)$_2$	47	11
Zn(acac)$_2$	2	1

[a] Reaction conditions: isobutane (5 mmol) was allowed to react in the presence of NHPI (10 mol %) and metal acetylacetonate (0.25 %) under a pressure of air (10 atm) in benzonitrile (10 mL) at 100 °C for 8 h.

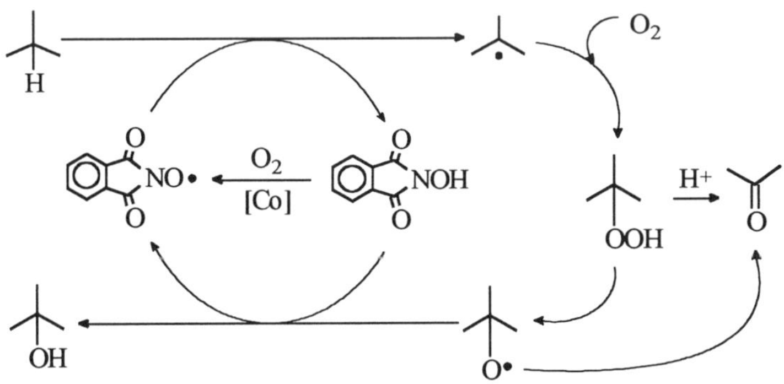

Scheme IX.4. The mechanism of *tert*-butyl alcohol formation from isobutane in the Ishii oxidation reaction.

Table IX.3. The aerobic oxidation of cycloalkanes in the presence of NHPI and $Mn(acac)_2$ at 100 °C .

		Products (%)	
Substrate	Conversion, %	Cycloalkanone	Dicarboxylic acid
Cyclohexane	70	Cyclohexanone (~0)	Adipic acid (60) [a]
Cyclooctane	83	Cyclooctanone (2) [b]	Suberic acid (53) [b]
Cyclodecane	90	Cyclodecanone (2) [b]	Sebacic acid (55) [b]
Cyclododecane [c]	81	Cyclododecanone (3) [b]	Dodecanedioic acid (68) [b]

[a] Selectivity.
[b] Yield based on substrate reacted.
[c] At 70 °C .

Oxygenation of Aromatic Hydrocarbons

Methods for hydroxylation of arenes by dioxygen to phenols have been developed in recent years. For example, the liquid-phase oxidation of benzene with molecular oxygen over the iron-heteropolyacid system gave phenol [36]. The following yields of phenol based on benzene have been obtained for various catalysts (300 K, 10 h):

Catalyst	**Yield, %**
$FeCl_2$	Trace
$H_3PW_{12}O_{40} + FeCl_2$	27.0
$H_{0.5}Fe_{1.25}PW_{12}O_{40}$	14.0
$H_{0.5}Fe_{1.25}PMo_{12}O_{40}$	4.2
$H_{0.5}Fe_{1.75}SiW_{12}O_{40}$	2.5

Copper(I) chloride promotes the oxidation of benzene to phenols with molecular oxygen [37a]. The active species is proposed to be a hydroxyl radical generated in the following manner:

$$2\,Cu(I) + O_2 + 2\,H^+ \longrightarrow 2\,Cu(II) + H_2O_2$$

$$Cu(I) + H_2O_2 + H^+ \longrightarrow Cu(II) + HO\bullet + H_2O$$

Analogously, in the presence of silica-supported palladium catalysts, benzene is oxidized under ambient conditions to give phenol, benzoquinone, hydroquinone and catechol [37b]. Palladium chloride, used for the catalyst preparation, is believed to be converted into metallic palladium. The synthesis of phenol from benzene and molecular oxygen via direct activation of a C–H bond by the catalytic system $Pd(OAc)_2$–phenanthroline in the presence of carbon monoxide has been described [38]. The proposed mechanism includes the electrophilic attack of benzene by an active palladium-containing species to to produce a σ-phenyl complex of palladium(II). Subsequent activation of dioxygen by the Pd–phen–CO complex to form a Pd–OPh complex and its reaction with acetic acid yields phenol. The oxidation of propenoidic phenols by molecular oxygen is catalyzed by [*N*,*N*′-bis(salicylidene)ethane-1,2-diaminato]cobalt(II)[Co(salen)] [39].

IX.2. Coupled Oxidation of Hydrocarbons

The oxidation of organic compounds by dioxygen, including hydrocarbons, is known to be a strongly exothermic process with a large negative free energy. For a certain mechanism to correspond to an appropriate reaction rate, the intermediate stages must be advantageous enough thermodynamically, i.e., they should not require too much energy. The probability of this will be generally higher in the case of overall strongly exothermic processes, including oxidation of hydrocarbons by molecular oxygen, than in the case of less exothermic or thermoneutral process.

There are a number of mechanisms to provide a high rate of the whole process in the case of oxidation of hydrocarbons, particularly in coupled oxidation. They are mechanisms with chemically active species, such as free radicals, readily reacting with hydrocarbons. These free radicals may be formed as intermediates in metal ion oxidation.

IX.2.A. EARLIER WORKS

The coupled oxidation was described first in the middle of the past century. Schönbein discovered the phenomenon of so-called "active oxygen" [40], i.e., oxidation of some substances by molecular oxygen only when there is a parallel oxidation of other substances present. For example, the formation of iodine from potassium iodide or the discoloring of indigo are induced by zinc or a zinc amalgam. Ferrous oxide is among the number of organic and inorganic substances whose oxidation induces the oxidation of more inert substances. A thorough review of the oxidation reactions known at that time is given in the book by N. A. Shilov, which appeared in 1905 [41]. Several theories of active oxygen already existed then. According to N. A. Shilov, "All of them come to an assumption that the coupled oxidation by molecular oxygen is conditioned by the formation of an intermediate product. Some authors consider this intermediate substance to be free oxygen atoms or ion; the others, on the contrary, suppose that the oxygen molecule in the first stage of the oxidation preserves the atomic complex –O–O–, forming this or that holoxide. This last group of theories elucidates the experimental evidence in the best way". Here holoxide is a substance of the type:

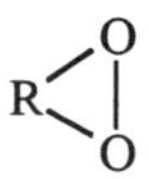

The suggestion of peroxide formation in the first stage of the oxidation by molecular oxygen was made by Schönbein, their intermediate formation being confirmed in the works of Traube, Bach, Engler and others. For the coupled oxidation of a number of substances by strong oxidizing agents (for example, H_2CrO_4 or $HMnO_4$) in the presence of ferrous oxide, Manchot [42] suggested an intermediate formation of high-valent iron oxide, Fe_2O_5.

According to Bach and Engler's peroxide theory [43], the coupling in oxidation reactions is explained in terms of intermediate formation of a peroxide. If of two substances, **A** and **B**, only one (e.g., **A**) reacts with molecular oxygen to give a peroxide (**AO**$_2$) (substance **A** is then called an "inductor"), then substance **B** ("acceptor") may be oxidized by this peroxide:

$$B + AO_2 \longrightarrow AO + BO$$

However, later it became clear that in most cases the assumption of the stable peroxide cannot by itself explain the phenomena observed. For example, though benzaldehyde oxidation initiates the coupled oxidation of indigo, specially synthesized peroxybenzoic acid can hardly oxidize indigo, and if so, the reaction proceeds far slower than it does in the process of coupled oxidation.

The development of the chain theory has produced a simple explanation of these facts in terms of the intermediate formation of radicals. The latter react with molecules much more readily than stable hydroperoxide molecules. The formation of free radicals was further established for many cases of oxidation of metal ions and complexes.

Hydroxylation of Hydrocarbons Coupled with the Oxidation of Metal Ions and Complexes

It is well-known that the biological oxidation of hydrocarbons (see Chapter XI) catalyzed by monooxygenases is coupled with the oxidation of electron donors, such as NADH or NADPH, in the presence of iron complexes. The donor in biological oxidation is believed to transfer its electrons initially to the metal ion, which is subsequently oxidized by an oxygen molecule. The fact of metal ion participation in most enzyme systems, which hydroxylate organic compounds, was the reason behind many attempts to create analogous purely chemical systems. It turned out that metal ion oxidation by molecular oxygen induces the oxidation of hydrocarbons (alkanes among them) introduced into the system. The formation of alkane oxidation products, at least with low yields, has been observed in many cases even in the oxidation of simple ions in various media.

Udenfriend suggested a simple non-enzymatic system as a possible model of monooxygenase [44]. It involves an iron(II) complex with EDTA and ascorbic acid and is capable of hydroxylating aromatic compounds. Later, Hamilton found that this system was able to hydroxylate cyclohexane (however, with very low yield) and epoxidize cyclohexene [45]. Afterwards, a number of similar systems were proposed. For example, Ullrich found two models which are more effective than the Udenfriend system. One of them includes a Sn(II) phosphate complex dissolved in water [46a], and the other one an iron(II) complex with mercaptobenzoic acid in an aqueous solution of acetone [46b]. The reactions are analogous to those catalyzed by monooxygenase: hydroxylation of alkanes and aromatic compounds, and in some cases, epoxidation of olefins.

Ascorbic acid in the Udenfriend system may be replaced by pyrimidine derivatives which resemble tetrahydropteridines; the latter may be used as donors both in the model and enzyme systems. The reaction occurs at room temperature at pH 7.

Hamilton suggested that the model systems (at least some of them) interact with the substrates via a so-called oxenoid mechanism similar to that of monooxygenase functioning. Since the reaction, in many aspects, parallels the processes with carbenes (addition to the multiple bond, insertion into the C–H bond), in the oxenoid mechanism an oxygen atom (oxene) inserts into the C–H bond without an intermediate formation of free radicals:

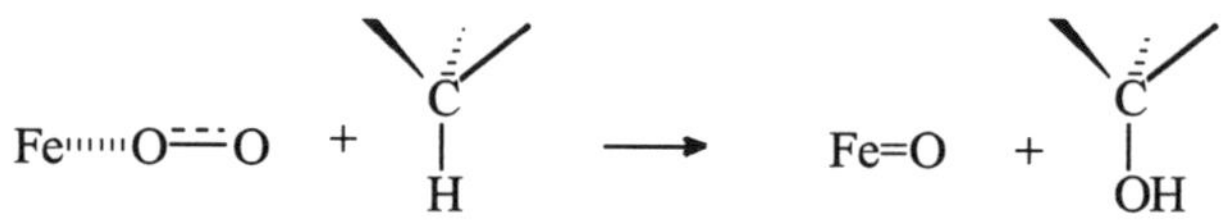

At the same time, there are some essential differences which may be noted between biological oxidation and process occuring in the model systems. The latter have no NIH-shift (migration of a hydrogen atom to the position next to the point of oxygen insertion), which is characteristic of biological systems in the hydroxylation of aromatic molecules. The model systems are mostly less selective in the reactions with aromatic compounds (a large amount of *meta*-substituted product is formed); they do not show stereospecificity in the hydroxylation of alkanes (e.g., a mixture of both possible isomers in the hydroxylation of *cis*- and *trans*-dimethylcyclohexane is formed). Evidently, the participation of reducing agents, which are analogues of biological electron donors, is not very essential. In effect, simple chlorides of some metallic ions are already active enough in organic solvents.

Thus, the coupled oxidation of C_1–C_4 alkanes and $SnCl_2$ was found to proceed in acetonitrile at room temperature [47a]. The yields of alcohols, which are the oxidation products, reach 7–15% per $SnCl_2$. Comparatively high yields of the products of cyclohexane oxidation are obtained in the presence of oxidizing tin(II) or iron(II) chlorides, when the reaction is carried out in acetonitrile as solvent [47b]. The yield, if based on the amount of metal salt present, may be further increased when the MCl_2 concentration is decreased. Under the optimum conditions of $[MCl_2] \longrightarrow 0$, the yield of products grows up to 20% with $SnCl_2$ and up to 30% with $FeCl_2$.

Table IX.4. Parameter 1° : 2° : 3° for the hydroxylation of isopentane by various systems.

System	Solvent	1° : 2° : 3°
Heme + MBA [a] + O_2	$CH_3COCH_3 + H_2O$	1 : 5.7 : 14
Fe^{2+} + MBA [a] + O_2	$CH_3COCH_3 + H_2O$	1 : 4.1 : 12.8
$SnCl_2 + O_2$	$CH_3CN + H_2O$	1 : 5.1 : 12.5
$FeCl_2 + O_2$	CH_3CN	1 : 4.3 : 13.0
$H_2O_2 + h\nu$	CH_3CN	1 : 4.5 : 11.2
CF_3COOOH	RH	1 : 60 : 720
Pyridine-*N*-oxide	CH_2Cl_2	1 : 8 : 68

[a] MBA = mercaptobenzoic acid.

Table IX.4 gives the distribution of isopentane hydroxylation products in various systems, including photochemical hydroxylation by hydrogen peroxide, where free hydroxyl radicals are known to be formed as intermediates. It is clear that the ratios of selectivities with respect to the site of attack (1° : 2° : 3°) in the systems involving Fe^{2+} and Sn^{2+} ions are very close to each other and to the selectivity observed for the attack by hydroxyl. This leads to the conclusion [47c] that the same active species (presumably hydroxyl radicals) participates in all these systems. This is also supported by a low isotope effect ($k_H/k_D \approx 1.2$–1.3, when comparing reactions of C_6H_{12} and C_6D_{12}). The yield of cyclohexane hydroxylation products depends on the nature of the solvent. The high yields in acetonitrile as compared with the other solvents are to be expected, provided the hydroxyl radicals are the intermediate active species. The rate constant of the hydroxyl radical reaction with cyclohexane is 5×10^9 dm^3 mol^{-1} s^{-1}, with acetonitrile 2.1×10^6 s^{-1}, with acetone 8.1×10^7 dm^3 mol^{-1} s^{-1}, and with methanol 7×10^8 dm^3 mol^{-1} s^{-1} [48]. In the case of cyclohexane in acetonitrile, this means that the reaction of HO• with the solvent at $[C_6H_{12}] > 10^{-2}$ mol dm^{-3} may be neglected. For the other organic solvents, it is necessary to take into account the competitive reaction of HO• with solvent, which reduces the yield of cyclohexane hydroxylation products.

In the case of iron salts, the accepted mechanism includes the formation of hydroxyl radicals:

$$Fe^{2+} + O_2 + H^+ \longrightarrow Fe^{3+} + HOO\bullet$$

$$Fe^{2+} + HOO\bullet + H^+ \longrightarrow Fe^{3+} + H_2O_2$$

$$Fe^{2+} + H_2O_2 \longrightarrow Fe^{3+} + HO\bullet + HO^-$$

$$Fe^{2+} + HO\bullet \longrightarrow Fe^{3+} + HO^-$$

$$2\,H^+ + 2\,HO^- \longrightarrow 2\,H_2O$$

$$4\,Fe^{2+} + O_2 + 4\,H^+ \longrightarrow 4\,Fe^{3+} + 2\,H_2O$$

For stannous chloride, the scheme of unbranched chain reaction without the participation of HO• radicals has been proposed [49a].

$$Sn^{2+} + O_2 \longrightarrow Sn^{3+} + HOO\bullet$$

$$Sn^{2+} + HOO\bullet \longrightarrow Sn^{3+} + H_2O_2$$

$$Sn^{3+} + O_2 \longrightarrow Sn^{4+} + HOO\bullet$$

$$Sn^{2+} + H_2O_2 \longrightarrow Sn^{4+}O^{2-} + H_2O$$

$$2\,Sn^{3+} \longrightarrow Sn^{2+} + Sn^{4+}$$

However, a detailed study of this reaction showed that it proceeds with chain branching and that it involves hydroxyl radicals (see below). The yield of oxidation products per metal ion may be increased if the reaction is carried out in the presence of specially added simple inorganic reducing agents. For example, if the coupled oxidation of $SnCl_2$ and C_6H_{12} is carried out in the presence of metallic tin, then the yield of cyclohexanol can be increased six times. For $FeCl_2$-containing systems, metallic mercury may be used as a reducing agent. This increases the yield of hydroxylation products.

In the systems involving such donors of electrons as hydrazobenzene or *o*-phenylendiamine (besides metal compounds), the yield of hydroxylation products may exceed the amount of the metal compound taken [49b]. Cyclohexane produces cyclohexanol and cyclohexanone, whereas cyclohex-1-ene-3-ol is mainly produced in the case of cyclohexene (with epoxide formation). Hydroxylation of 2-methylbutane produces isomeric alcohols, the selectivity with respect to the site of attack following the order: 3° > 2° > 1°. In toluene hydroxylation, the formation of both benzyl alcohol and isomeric cresols is observed, *o*- and *p*-isomers prevailing appreciably. An oxenoid mechanism of hydroxylation has been suggested. However, it is difficult to reconcile this mechanism with the fact that the yield of cyclohexane hydroxylation products depends so insignificantly on the nature of the metal ion (for $FeCl_2$, VCl_2, $MnCl_2$, $NiCl_2$, $CuCl_2$ the yield differs no more than 4 times; while for $FeCl_2$, $MnCl_2$, VCl_2 it is almost the same). The formation of the same intermediate active species, for example HOO•, seems to be a more natural alternative.

A particularly active system was observed when mixing copper dichloride with phenylhydrazine in various organic solvents and water [49c]. Under optimum conditions, the yield reaches up to 40% per oxidized phenylhydrazine with the copper salt acting as a catalyst. The yield of products may correspond, for example, to forty redox cycles per mole of Cu salt with one portion of the reducing agent, and even several hundred cycles, if new portions of phenyl hydrazine are added after the previous hydrazine has been oxidized.

The Mechanism of Stannous Chloride Autoxidation Coupled with the Oxidation of Cyclohexane

A great number of species participating in the autoxidation of transition metal compounds makes it difficult to select a hydroxylation mechanism. Only in a small number of cases has this mechanism and the one of coupled hydrocarbon hydroxylation been well established. Let us take as an example the coupled oxidation of cyclohexane and stannous chloride.

Tin(II) is a typical two-electron reductant since there are only two stable oxidation states for the compounds, Sn(II) and Sn(IV). The redox potential of the pair Sn(IV)/Sn(II) is +0.15 V in aqueous solutions. Thus Sn(II) is a moderately strong reducing agent. The estimation of the redox potential of the pair Sn(III)/Sn(II) gives a value +0.8 V. That means that E_0 for Sn(IV)/Sn(III) is –0.5 V, i.e., Sn(III) should be both a strong reducing agent turning into Sn(IV) and a strong oxidizing agent turning into Sn(II). The disproportionation of two Sn(III) ions is assumed to occur readily upon collision:

$$\text{Sn(III)} + \text{Sn(III)} \longrightarrow \text{Sn(IV)} + \text{Sn(II)} + \sim 30 \text{ kcal mole}^{-1}$$

The oxidation of tin(II) salts by dioxygen displays some of the features of a chain process; in particular, it is retarded by certain inhibitors and accelerated when the solution is irradiated. The kinetic behavior of the oxidation reaction of Sn(II) chloride has some peculiarities. The reaction, under certain conditions, has a marked induction period followed by a period of practically constant rate, provided the dioxygen concentration is kept constant. The study of the dependence of the reaction rate on the dioxygen concentration shows that the reaction rate under steady-state conditions can be described by a simple kinetic expression:

$$-\,d[O_2]/dt = k\,[O_2]^2$$

i.e., the reaction rate does not depend on the Sn(II) concentration. It is clear that such a kinetic expression, though very simple, cannot be explained by a simple mechanism. To explain the kinetic data obtained, a branching chain scheme has been proposed [49d,e]:

$$\text{(0) Sn(II)} + O_2 \xrightarrow{H_2O} \text{Sn(III)} + \text{HOO}\bullet \quad \textit{initiation}$$

$$\text{(1) Sn(III)} + O_2 \xrightarrow{H_2O} \text{HOO}\bullet + \text{Sn(IV)}$$

$$\text{(2) HOO}\bullet + \text{Sn(II)} \xrightarrow{\beta} H_2O_2 + \text{Sn(III)}$$

$$\phantom{\text{(2) HOO}\bullet + \text{Sn(II)}} \xrightarrow{1-\beta} \text{HO}\bullet + \text{Sn(IV)}$$

$$(3)\ H_2O_2 + Sn(II) \xrightarrow{\alpha} HO\bullet + Sn(III) \quad a$$

$$ \xrightarrow{1-\alpha} H_2O + Sn(IV) \quad b$$

$$(4)\ HO\bullet + Sn(II) \longrightarrow Sn(III)$$

$$(5)\ Sn(III) + Sn(III) \longrightarrow Sn(II) + Sn(IV)$$

Here α and β are the probabilities of one-electron reactions of Sn(II) with H_2O_2 and HOO•, respectively.

In steady-state conditions

$$\alpha\beta k_1\ [Sn(III)]\ [O_2] = k_5\ [Sn(III)]^2$$

and the reaction rate is practically equal to that of chain propagation

$$-\ d[O_2]/dt = k_1\ [Sn(III)]\ [O_2] = (\alpha\beta k_1^{\ 2}/k_5)\ [O_2]^2$$

and, in agreement with the experimental data, depends only on the dioxygen concentration.

The branching chain mechanism is confirmed, for example, by the sensitivity of the reaction to the inhibitor concentration. There exists some limiting inhibitor concentration which practically stops the reaction (the induction period increases to infinity on the addition of a definite amount of an inhibitor). Hydrogen peroxide was detected during the oxidation of Sn(II). Using the reaction scheme, it was possible to estimate the rate constant of the reaction of hydrogen peroxide with Sn(II) by the dependence of H_2O_2 concentration on time. The rate constant found coincided with the value obtained for the reaction of H_2O_2 with Sn(II) determined in a separate investigation, thus confirming the proposed mechanism. Though the probability (α) of a one-electron reaction of hydrogen peroxide with Sn(II) to give a hydroxyl radical is not very high ($\alpha \approx 5 \times 10^{-4}$ for the reaction in acetonitrile), it is quite sufficient for the reaction to ensure the high rate of radical formation in a branched chain reaction under steady-state conditions.

The rate of autoxidation is much lower in acidic aqueous solutions of Sn(II) chloride that in organic solvents (though the addition of water is necessary because the HO• and HOO• radicals, as well as hydrogen peroxide, participate in the reaction, their formation requiring protons). The rate decrease in aqueous solutions is partially associated with the decrease of dioxygen solubility, which is particularly important since the O_2 concentration enters the kinetic equation as a second-order term. However, the kinetic equation in acidic aqueous media is also different from that found for organic solvents:

$$v(O_2) = k_{\text{eff}}\,[\text{Sn(II)}]^{1/2}\,[O_2]^{3/2}$$

Detailed studies have shown that the different form of the kinetic equation for aqueous solution is due to the reaction which takes place in water:

$$(6)\ \text{Sn(III)} + H_2O_2 \longrightarrow \text{Sn(IV)} + \text{HO}\bullet + \text{HO}^-$$

and is nonessential for acetonitrile solutions. In other respects the mechanism is the same. Therefore, the autoxidation of stannous salts in a variety of solvents occurs by a branching chain mechanism with the participation of HO• and HOO• radicals. Reactions of the radicals with hydrocarbons should naturally be taken into consideration in the process of their oxidation coupled with the autoxidation of tin(II) salts.

On the introduction of cyclohexane into a Sn(II) aqueous acetonitrile solution, the rate of dioxygen consumption increases and the induction period falls. An increase of cyclohexane concentration decreases the induction period and even causes its complete disappearance. The higher the initial tin concentration, the greater is the cyclohexane concentration necessary to cause the disappearance of the induction period. The rate of hydrogen peroxide production is also increased in the presence of cyclohexane. These data can be understood on the basis of the above chain mechanism of Sn(II) oxidation, where free hydroxyl radicals HO• are intermediately formed. The rate constant of the interaction of the hydroxyl radical with cyclohexane

$$\text{HO}\bullet + C_6H_{12} \longrightarrow H_2O + C_6H_{11}\bullet$$

is 5×10^9 dm^3 mol^{-1} s^{-1}. Thus all the hydroxyl radicals, with an appropriate hydrocarbon concentration, will interact with the latter to produce hydrocarbon radicals. The radical very readily enters into an addition reaction in the presence of O_2:

$$R\bullet + O_2 \longrightarrow ROO\bullet$$

In hydroperoxide oxidation reactions, the radicals usually disappear in a bimolecular disproportionation process:

$$2\,ROO\bullet \longrightarrow R'COR'' + ROH + O_2$$

However, in the presence of large Sn(II) concentrations, practically all hydroperoxide radicals interact with the tin in a similar manner to HOO• radicals

$$ROO\bullet + Sn(II) \xrightarrow{H_2O} \begin{cases} \xrightarrow{\beta} ROOH + Sn(III) \\ \xrightarrow{1-\beta} RO\bullet + Sn(IV) \end{cases}$$

The interaction of hydroperoxide with Sn(II) is similar to that of hydrogen peroxide

$$ROOH + Sn(II) \longrightarrow \begin{cases} \xrightarrow{\alpha} RO\bullet + Sn(III) \\ \xrightarrow{1-\alpha} ROH + Sn(IV) \end{cases}$$

RO• radicals react with RH and Sn(II):

$$RO\bullet + RH \longrightarrow ROH + R\bullet$$

$$RO\bullet + Sn(II) \xrightarrow{H_2O} ROH + Sn(III)$$

causing the formation of the alcohol, which is practically the only product. Besides, in aqueous solutions, as in the case of hydroperoxide, it is necessary to take into account the following reaction:

$$ROOH + Sn(III) \longrightarrow RO\bullet + Sn(IV)$$

Thus, the participation of Sn(II) in the oxidation process ensures high selectivity of the reaction to produce cyclohexanol as virtually the sole product. It is important that the situation changes if the Sn(II) concentration is low; for example, if instead of dioxygen we add hydrogen peroxide to the Sn(II) solution. In this case Sn(II) quickly disappears, and the reaction of ROO• disproportionation becomes dominant, yielding a variety of products different from those observed for the oxidation by dioxygen. If we do not differentiate between the differences in reactions with high and low Sn(II) concentrations under the conditions of oxidation, it may lead us to an incorrect conclusion about the non-radical character of hydrocarbon oxidation coupled with Sn(II) oxidation by dioxygen.

It is interesting to note that, provided the RH concentration is sufficient for RO• radicals to disappear by the reaction with the hydrocarbon, the stoichiometry of the reaction under specified conditions corresponds to the ratio Sn(II) : O_2 : ROH = 1 : 1 : 1, which means that it will be the same as the stoichiometry of oxidation in the presence of monooxygenase. This shows that the stoichiometry of the process cannot be used as evidence for its mechanism. The example with Sn(II), treated here so extensively, is further evidence of the danger of drawing conclusions about the models of biological oxidation based merely on formal analogies.

IX.2.B. GIF SYSTEMS

Barton and co-workers have developed a family of systems for oxidation and oxidative functionalization of alkanes under mild conditions exhibiting "unusual" selectivity (see reviews [50], polemics [50d,e], the first publication [51] and some recent publications[52], as well as the papers of other authors concerning these systems[53]). These oxidations occur in pyridine in the presence of an organic acid and are catalyzed by complexes of transition metals (mainly iron). If dioxygen is used as an oxidizing regent, a reductant must also take part in the reaction. The first such a system was invented in Gif-sur-Yvette [51]. Thus their name: Gif systems. The systems with geographically based names are mentioned in Table IX.5. All Gif systems have the same chemical peculiarities:

1) the major products of the reaction are ketones; it is important that alcohols are not reaction intermediates;

Table IX.5. The nomenclature of Gif systems.

<table>
<tr><th>System</th><th>Catalyst</th><th>Electron source</th><th>Oxidant</th></tr>
<tr><td>Gif^{III}</td><td>Fe(II)</td><td>Fe (metal)</td><td>O_2</td></tr>
<tr><td>Gif^{IV}</td><td>Fe(II)</td><td>Zn (metal)</td><td>O_2</td></tr>
<tr><td>GO (Gif-Orsay)</td><td>Fe(II)</td><td>Cathode</td><td>O_2</td></tr>
<tr><td>$GoAgg^{I}$</td><td>Fe(II)</td><td colspan="2">KO_2/Ar</td></tr>
<tr><td>$GoAgg^{II}$</td><td>Fe(III)</td><td colspan="2">H_2O_2</td></tr>
<tr><td>$GoAgg^{III}$</td><td>Fe(III)/Picolinic acid</td><td colspan="2">H_2O_2</td></tr>
<tr><td>GoChAgg</td><td>Cu(II)</td><td colspan="2">H_2O_2</td></tr>
<tr><td>Cu^0/O_2</td><td>Cu(I)?</td><td>Cu(0)</td><td>O_2</td></tr>
</table>

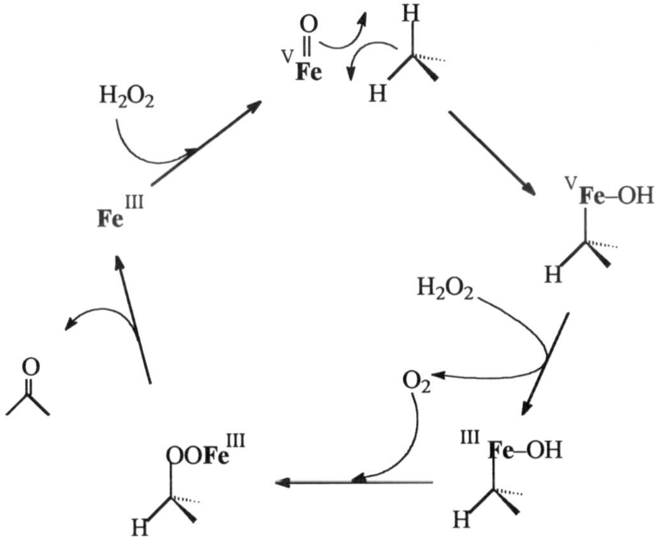

Scheme IX.5. The proposed mechanism for the Gif $GoAgg^{II}$ oxidation of alkanes.

2) the presence of an excess of some easily oxidizable compounds (e.g., alcohols, aldehydes) does not significantly suppress the alkane oxidation;

3) the selectivity of oxidation of branched hydrocarbons is secondary > tertiary > primary;

4) secondary alkyl free radicals are not reaction intermediates;

5) olefins are not epoxidized;

6) addition of different trapping reagents diverts the formation of ketones to the appropriate monosubstituted alkyl derivatives. For example, in the presence of $CBrCl_3$ alkyl bromide is formed in quantitative yield.

It should be noted that despite of numerous works devoted to Gif systems the mechanism of their action is not clear (see, for example, a discussion in [50d,e]). The proposed [52a] catalytic cycle for $GoAgg^{II}$ reactions is shown in Scheme IX.5.

IX.2.C. OTHER SYSTEMS INVOLVING O_2 AND A REDUCING REAGENT

In recent years, numerous systems for alkane oxygenation based on molecular oxygen, a reductant and a metal catalyst have been described. Some of these systems will be considered in the present section.

Systems Based on Aldehydes

Olefins can be epoxidized by dioxygen in the presence of an aldehyde and various metal complexes [54]. An analogous oxidation of alkanes gives alcohols, ketones and alkyl hydroperoxides. Examples of oxidations by dioxygen in the presence of various reducing agents catalyzed by transition metal complexes are given in Table IX.6.

A RuO_2–CH_3CHO system catalyzes the oxidative cleavage of C–C double bonds using molecular oxygen [55m]. Carbonyl compounds are formed selectively.

Systems Based on Solid Metals as Reducing Agents

Metals added to various oxidation systems can serve both as reducing reagents and catalysts. Usually transition metals (e.g., iron) are used. If non-tran-

Table IX.6. Catalytic oxidation of hydrocarbons by dioxygen in the presence of aldehydes.

Substrates	Products	Reducing agent	Catalyst	Ref.
Alkanes	Alcohol+ketone	Heptanal	$FeCl_3$, $RuCl_3$	55a
Alkanes	Alcohol+ketone	Acetaldehyde	$Cu(OH)_2$	55b
Benzene	Phenol	Crotonaldehyde	VO(1,3-diketonato)$_2$	55c
Alkanes	Alcohol+ketone	Acetaldehyde	Metalloporphyrins	55d
Alkanes	Alcohol+ketone	2-Methyl-propanal	Co(II) Schiff base complex	55e
Cyclohexane	Cyclohexanol + cyclohexanone	Aldehyde	Metalloporphyrins	55f
Alkanes	Alcohol+ketone	Acetaldehyde	Cu–Crown ether	55g
Alkanes, aryl-alkanes	Alcohol+ketone	2-Methyl-propanal	Mo,V-Heteropoly-oxometalates	55h
Alkanes	Alcohol+ketone	Aldehydes	Cu salts	55i
Arylalkanes	Ketone	2-Methyl-propanal	Co(II) cryptate	55j
Alkanes, aryl-alkanes	Alcohol+ketone	3-Methyl-butanal	M(acac)$_n$ M = Pd, Co, Fe, Ni, Cu	55k

sition metals (e.g., zinc) are employed as reducing agents, the catalysts (transition metals complexes) are necessary. In all the reactions, proton donors (carboxylic acids) take part to dissolve the solid metal. Examples of reactions using metals as reducing agents are summarized in Table IX.7.

A binuclear Mn(IV) complex with the 1,4,7-trimethyl-1,4,7-triazacyclononane macrocyclic ligand initiates oxidation of cyclohexancarboxaldehyde with atmospheric oxygen in acetonitrile under heating resulting in the formation of cyclohexanecarboxylic acid and considerable amounts of oxidative decarboxylation products, i.e., cyclohexane, cyclohexanol and cyclohexanone [55l]. In the presence of cyclooctane, the reaction gives rise also to the formation of cyclooctanol and cyclooctanone. The following stages of the reaction have been proposed (Cy = cyclohexyl; RH = cyclooctane):

$$\text{CyCHO} + \text{Mn(IV)} \longrightarrow \text{CyCO}\bullet + \text{Mn(III)} + \text{H}^+$$

$$\text{CyCO}\bullet + \text{O}_2 \longrightarrow \text{CyCOOO}\bullet$$

$$CyCOOO\bullet + CyCHO \longrightarrow CyCOOOH + CyCO\bullet$$
$$CyCOOOH + CyCHO \longrightarrow 2\ CyCOOH$$
$$CyCOOOH \longrightarrow CyCOO\bullet + HO\bullet$$
$$CyCOO\bullet \longrightarrow Cy\bullet + CO_2$$
$$Cy\bullet + O_2 \longrightarrow CyOO\bullet \longrightarrow \text{cyclohexanol} + \text{cyclohexanone}$$
$$Cy\bullet + RH \longrightarrow CyH + R\bullet$$
$$RH + CyCOO\bullet \longrightarrow R\bullet + CyCOOH$$
$$RH + CyCO\bullet \longrightarrow R\bullet + CyCHO$$
$$R\bullet + O_2 \longrightarrow ROO\bullet \longrightarrow \text{cyclooctanol} + \text{cyclooctanone}$$

Table IX.7. Catalytic oxidation of hydrocarbons by dioxygen in the presence of solid metals as reducing agents.

Substrates	Products	Reducing agent	Catalyst	Ref.
Arenes	Phenols	Fe	Fe(II)	56a
Cyclohexane	Cyclohexanol + cyclohexanone	Zn powder	Fe(II) on silica gel	56b
Alkanes, arenes	Alcohols, phenols	Zn powder	$[\{Fe(HBpz_3)(hfacac)\}_2O]$	56c
Methane	Methanol + formaldehyde	Zn powder	Fe(II) on silica gel	56d,e
Alkanes	Alcohol+ketone	Zn powder	Mn–Porphyrin	56f
Cyclohexane	Cyclohexanol + cyclohexanone	Zn powder	$SmCl_3$	56g,h
Alkanes	Alcohol+ketone	Zn powder	$EuCl_3$	56i–k
Methane	Methanol	Zn powder	$EuCl_3$	56l
Methane	CF_3COOCH_3	Zn powder	VO_3^-, $VO(acac)_2$	56m

Yamanaka *et al.* have demonstrated that aerobic hydroxylation of methane in trifluoroacetic acid at 40 °C is catalyzed by NH_4VO_3 or $VO(acac)_2$ if zinc powder is present [56m]. Methyl trifluoroacetate is formed with a turnover number ca. 10 after 1 h. The authors suggested that the active oxygen species in this system is not hydroxyl radicals and proposed that a reduced vanadium species, V(II) or V(III), that was generated by reduction with Zn(0), works as a catalyst for the activation of O_2 and hydroxylation of methane.

Systems Involving Other Reducing Agents

Various other reducing agents have been used in the aerobic coupled oxidations of hydrocarbons. Some examples are presented in Table IX.8.

Table IX.8. Catalytic oxidation of hydrocarbons by dioxygen in the presence of various reducing agents.

Substrates	Products	Reducing agent	Catalyst	Ref.
Alkanes	Alcohols	Hydrazobenzene	$Fepy_4Cl_2$	57a
Arenes	Phenols	Hydroquinones	Catecholatoiron complex	57b–c
Benzene	Phenol	H_2	Pd-Ti Silicalites	57d
Alkanes	Alcohol+ketone	H_2S	Complexes **IX-5**, **IX-6**	57e

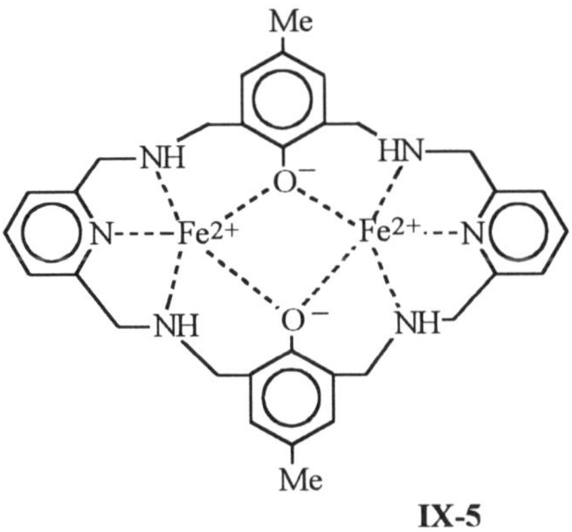

IX-5

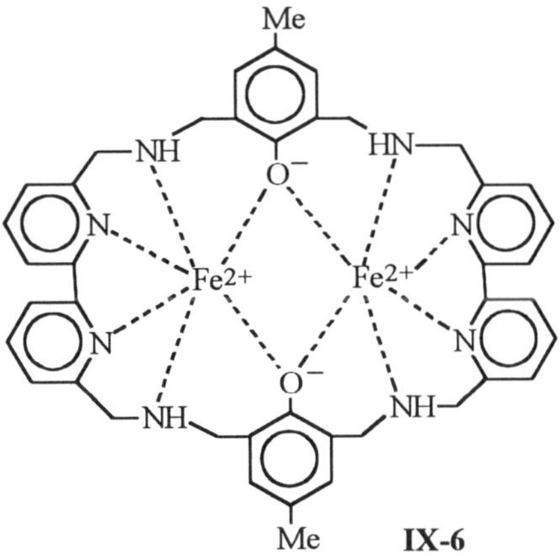

IX-6

A formal heterogeneous analog of alkane monooxygenases has been described by Lin and Sen [57f]. The system involves a metal catalyst (palladium) and a co-reductant (carbon monoxide). Ethane is transformed into acetic acid at temperatures < 100 °C. The proposed mechanism for this reaction is shown in Scheme IX.6.

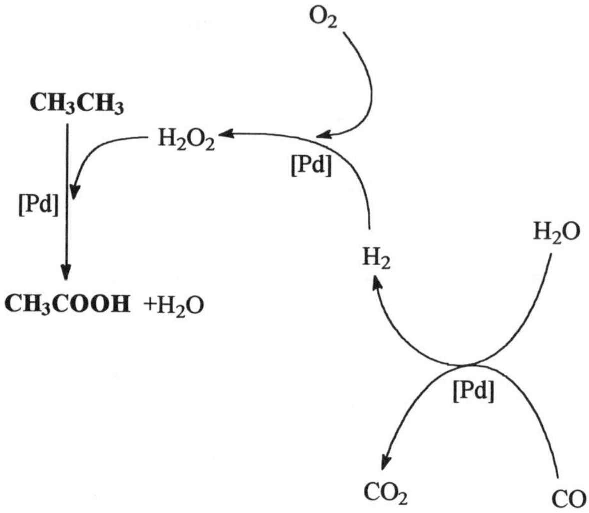

Scheme IX.6. The mechanism proposed for ethane oxidation in the presence of CO.

In some cases, the hydrocarbon oxidation in the presence of a reductant does not need a transition metal complex as catalyst. Thus, a cathode of a carbon whisker has been found to be active for the oxidation of toluene into benzaldehyde and benzyl alcohol during the H_2–O_2 fuel cell reaction [58a]. During the electrolysis of water at room temperature, the epoxidation of hex-1-ene and hydroxylation of benzene to phenol and hydroquinone occurs simultaneously on the anode and the cathode, respectively [58b]. Finally, selective oxidation of terminal isopropyl groups to the corresponding tertiary alcohols have been carried out by an electrochemical method [58c]. Using the

Tl(TFA)$_3$–hematoporphyrin–O_2–cathode system, cholesterol was converted into the corresponding alcohol:

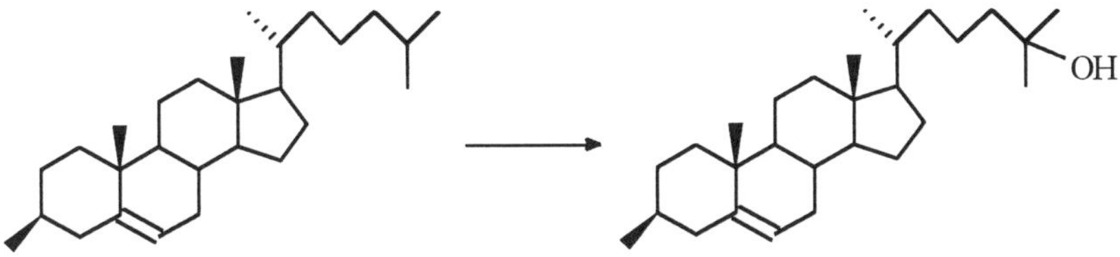

The authors assumed that Tl^{3+} binds oxygen producing an activated oxygen-like species (Scheme IX.7).

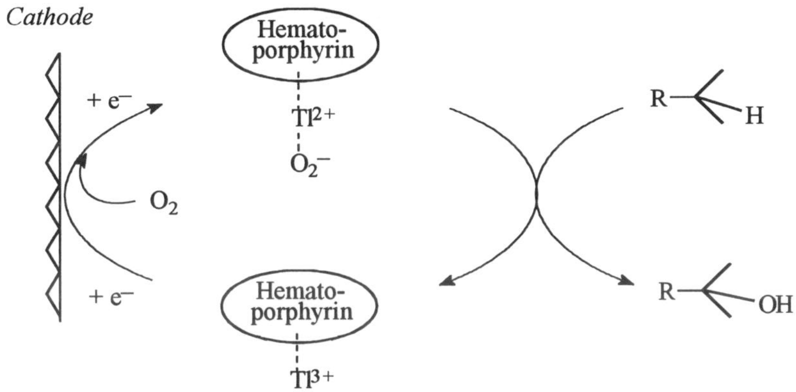

Scheme IX.7. The oxidation of tertiary C–H bonds by the Tl(TFA)$_3$–hematoporphyrin–O_2–cathode system.

IX.3. Photoinduced Metal-Catalyzed Oxidation of Hydrocarbons by Air

A new photochemical catalytic method for the transformation of alkanes and arylalkanes into hydroperoxides has been developed in the recent decade. Iron(III) chloride has been found to be an efficient photocatalyst of alkane oxidation with atmospheric oxygen [59, 60]. Kinetics of the cyclohexane photooxidation in the presence of a catalytic amount of $FeCl_3$, as well as some other compounds (see below), are shown in Figure IX.1 [60g].

Other transition metal chlorides, for example $CuCl_2$ (Figure IX.1) [60d,g, 61a,b], $AuCl_4^-$ (Figure IX.1) [60d,g, 61c,d], $PtCl_6^{2-}$ [60d, 61d,e], and $CrCl_3$ [61f] also photocatalyze the aerobic oxygenation of alkanes in acetonitrile, methylene chloride or acetic acid (Table IX.9). Methanol (0.018×10^{-2} mol dm^{-3}) and formaldehyde (0.17×10^{-2} mol dm^{-3}) were formed upon the slow bubbling of methane and air through a solution of $HAuCl_4$ in MeCN under irradiation [61c].

Table IX.9. Photocatalytic aerobic oxidation of hydrocarbons in the presence of chloride and bromide metal salts.

Photocatalyst	Substrate	Solvent	Ref.
$FeCl_3$	Toluene	CH_3COOH	59
$FeCl_3$ + phenanthroline	Toluene	CH_3COOH	59
$FeCl_3$	Alkanes, arylalkanes	CH_3CN, CH_3COCH_3	60a
$FeCl_3$	Alkanes	Alcohols	60e,f
$CuCl_2$	Alkanes	CH_3CN	60d,g,61a,b
H_2PtCl_6	Cyclohexane	CH_3CN	60d
$HAuCl_4$	Alkanes, methane	CH_3CN, CH_2Cl_2	60d,61c,d
$RhCl_3$	Cyclohexane	CH_3CN	60c
$RuCl_3$	Cyclohexane	CH_3CN	60c
$CrCl_3$ + $PhCH_2NEt_3Cl$	Alkanes	CH_3CN	61f
$FeBr_3$	Arylalkanes	CH_3CN	60c
Na_2PtBr_6	Cyclohexane	CH_3CN	60c

The first step of this process is apparently the photoexcitation of the iron chloride species to stimulate homolysis of the Fe–Cl bond [60h]. The chlorine radical (free or in the solvent cage without entering to the solvent) attacks the alkane. The iron(II) derivative that is formed can be oxidized either by molecular oxygen or the alkylperoxo radical (Scheme IX.8).

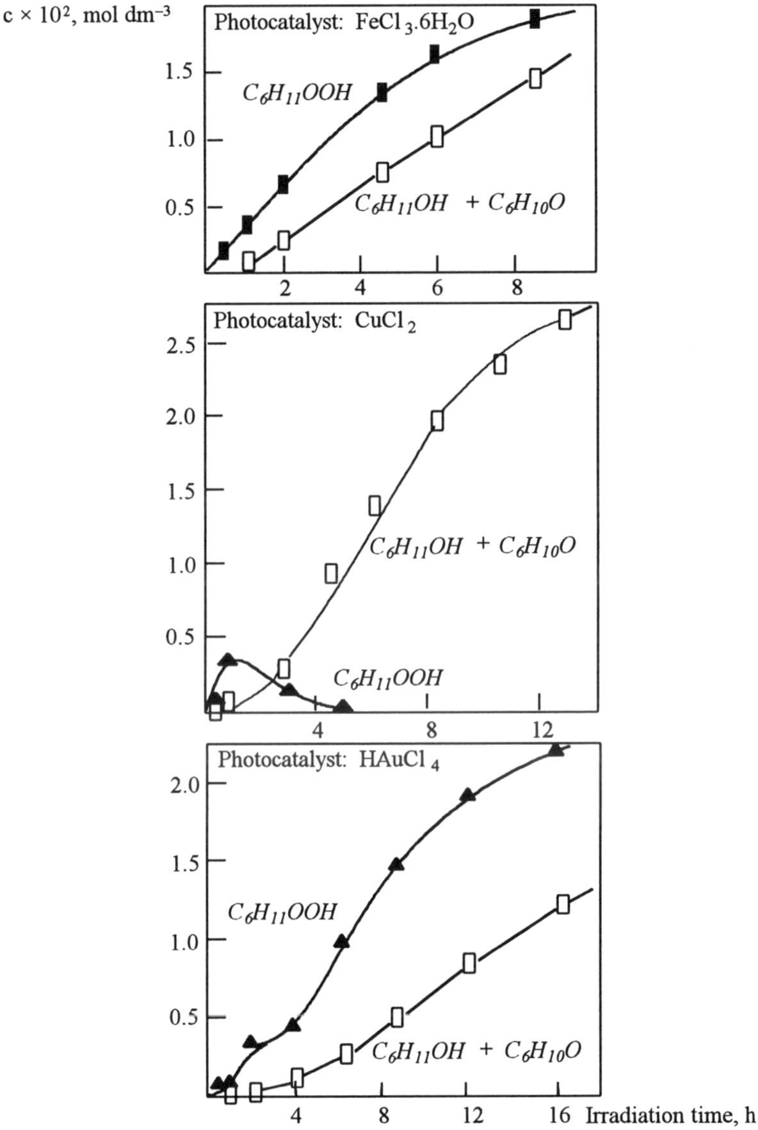

Figure IX.1. The aerobic photooxygenation ($\lambda > 300$ nm) of cyclohexane (0.46 mol dm^{-3}) catalyzed by various metal chlorides (5×10^{-4} mol dm^{-3}) in MeCN at 10 °C .

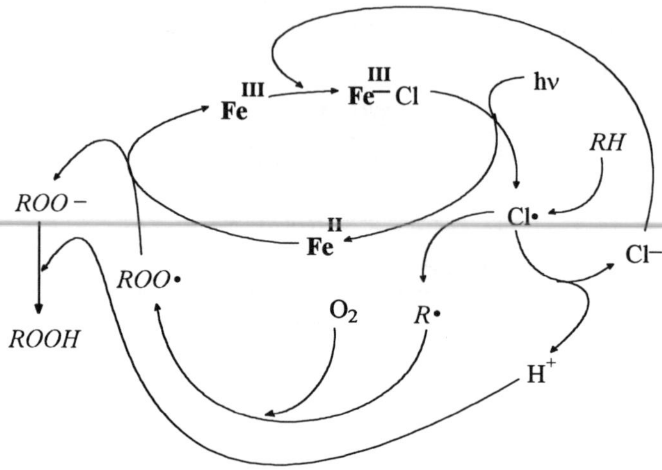

Scheme IX.8. The mechanism proposed for the $FeCl_3$-photocatalyzed oxygenation of alkanes, RH, into alkyl hydroperoxides, ROOH.

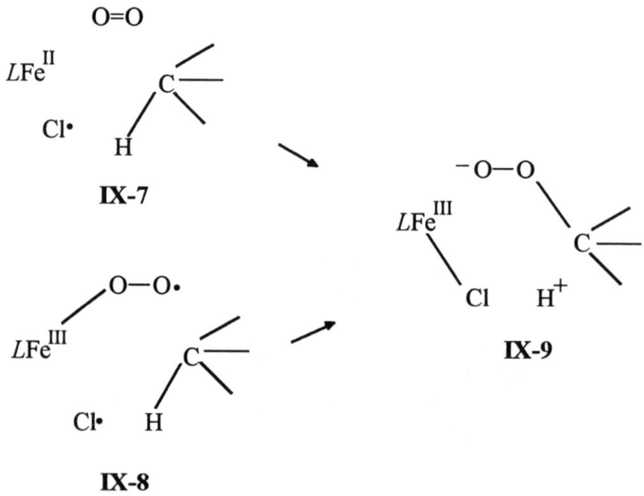

Scheme IX.9. Possible interactions between photoexcited iron chloride, C–H bond of the alkane and molecular oxygen.

It can be assumed that in one of the possible channels of the alkyl hydroperoxide formation, electron transfer Cl ⟶ Fe^{III} and reorganization of bonds occurs within a six-membered structure **IX-7** or **IX-8** (Scheme IX.9).

Bromide complexes are also efficient in the photooxidation. However, $FeBr_3$ photocatalyzes the oxygenation of only alkylbenzenes (Figure IX.2) [60c] and are not effective in the oxidation of alkanes. One can assume that the bromine radical is an active species in this reaction.

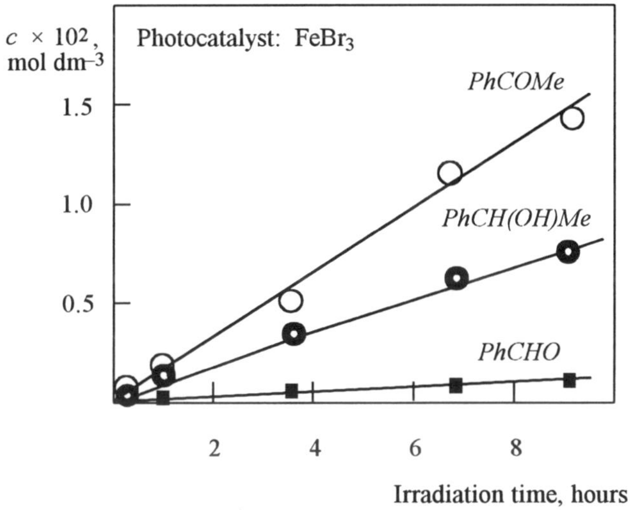

Figure IX.2. Kinetics of aerobic photooxidation (λ > 300 nm) of ethylbenzene (0.41 mol dm^{-3}) in MeCN in the presence of $FeBr_3$ (5.0×10^{-4} mol dm^{-3}). The samples were not reduced with PPh_3 before GC analysis.

Kinetics of the photooxidations catalyzed by platinum complexes are shown in Figure IX.3 [60c]. A very long induction period can be noticed in the case of catalysis by $PtBr_6^{2-}$.

Mechanisms of the photooxygenation reaction in the cases of other than iron metal seems to be different from that postulated for the $FeCl_3$-catalyzed process [60h]. Low-valent species are apparently involved in the oxidation. These species can possibly add an oxygen molecule to produce metal peroxo radicals and peroxo

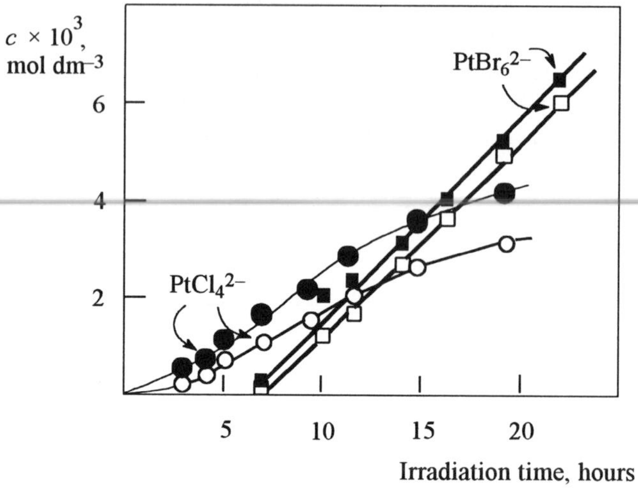

Figure IX.3. Kinetics of aerobic photooxidation (λ > 300 nm) of cyclohexane (0.46 mol dm^{-3}) in MeCN in the presence of $Na_2PtCl_4.4H_2O$ (5.0 × 10^{-4} mol dm^{-3}) (circles) or $Na_2PtBr_6.6H_2O$ (5.0 × 10^{-4} mol dm^{-3}) (squares). Concentrations of cyclohexanol (closed symbols) and cyclohexanone (open symbols) determined by GC before the reduction with PPh_3.

complexes. Such a mechanism has been proposed for the photooxidation of alkanes in acetonitrile in the presence of a catalytic amount of cyclopentadienyliron(II) complexes and bis(arene)iron(II) [62]:

$$Fe^{II} + O_2 \longrightarrow Fe^{III}\text{–}OO\bullet$$
$$Fe^{III}\text{–}OO\bullet + RH \longrightarrow Fe^{III}\text{–}OOH + R\bullet$$
$$R\bullet + O_2 \longrightarrow ROO\bullet \longrightarrow \ldots$$
$$Fe^{III}\text{–}OOH + R\bullet \longrightarrow Fe^{II} + ROOH$$

Unlike the oxygenations catalyzed by chloride complexes, the aerobic photooxidation of cyclohexane in the presence of a catalytic system $CrCl_3$–$PhCH_2NEt_3Cl$ in MeCN produces the ketone as a main product and only a small amount of the alcohol [61f]. Adding benzene, methylene chloride or ethanol to the cyclohexane solution raises the oxygenation rate and changes the ketone : alcohol ratio. In the presence of a small amount of hydroquinone, the formation rate of cyclohexanone (but not of cyclohexanol) sharply decreases. The kinetic isotope effect in the oxidation of C_6H_{12} and C_6D_{12} is *ca.* 1.0 for cyclohexanol and *ca.* 2.9 for cyclohexanone. Cyclohexanol formation has been assumed to follow a mechanism that does not involve free radicals. Free radicals can participate in the route toward the ketone [61f].

It is interesting that the irradiation of an aqueous cyclohexane emulsion in the presence of an iron(III) salt, e.g. $Fe(ClO_4)_3$, gives rise [63a] also to the formation of cyclohexanone instead of its mixture with cyclohexyl hydroperoxide and cyclohexanol. Likewise, the photooxidation of cycloalkanes by dioxygen catalyzed by the polyhalogenated porphyrinatoiron(III)-hydroxo complex, Porph-Fe–OH, has been reported to produce predominantly cycloalkanones [63b]. The proposed mechanism involves the formation of hydroxyl radicals and an Fe^{III}–σ-alkyl complex [63b]:

$$\text{PorphFe}^{\text{III}}\text{–OH} \xrightarrow{h\nu} \text{PorphFe}^{\text{II}} + \text{HO}\bullet$$

$$\text{HO}\bullet + \text{–CH}_2\text{–} \longrightarrow \text{H}_2\text{O} + \text{–CH}\bullet\text{–}$$

$$\text{–CH}\bullet\text{–} + \text{PorphFe}^{\text{II}} \longrightarrow \text{–CH(Fe}^{\text{III}}\text{Porph)–}$$

$$\text{–CH(Fe}^{\text{III}}\text{Porph)–} + \text{O}_2 \longrightarrow \text{–CH(OOFe}^{\text{III}}\text{Porph)–} \longrightarrow \text{–C(=O)–} + \text{PorphFe}^{\text{III}}\text{–OH}$$

Examples of "unusual" selectivity in alkane oxygenation (when a ketone is a predominant product) have been discussed above (Gif-type systems) and will be given below when describing alkane ketonization with peroxides and dioxygen. Another photooxidation of alkanes catalyzed by Fe(III)- or Mn(III)-porphyrin occurs in the presence of a reducing agent (triethanolamine) and photosensitizer and gives rise to the "usual" formation of a mixture of alcohol and ketone (the ratio is in the region 0.8–2.7) [63c]. The photocatalytic oxygenation of alkenes with dioxygen and porphyrinatoiron(III) complexes, which affords allylic oxygenation products and/or epoxides, has been proposed to involve the oxoiron(IV) complex, $\text{PorphFe}^{\text{IV}}\text{=O}$, as the catalytically active species. This species abstracts an allylic hydrogen atom to initiate autoxidation and "direct" oxygen-transfer reactions [63d]. For other oxygenation reactions by photoexcited iron porphyrins see in refs. [63e,f].

Some other transition metal complexes, e.g., trifluoroacetates of palladium and copper [64a], some platinum derivatives [64b], are known to promote aerobic alkane photooxidation. Mechanisms of these transformations are not clear in detail.

The irradiation of an alkane solution in acetonitrile with visible light in the presence of catalytic amounts of quinone and copper(II) acetate gives rise to the formation of almost pure alkyl hydroperoxide which is decomposed only very slowly under these conditions to produce the ketone and alcohol [64c,d]. It has been proposed [64d] that the first step of the reaction is a hydrogen atom abstraction from the alkane, RH, by a photoexcited quinone species to generate the alkyl radical R• and semiquinone. The former is rapidly transformed into ROO• and then alkyl hydroperoxide, while the latter is reoxidized by Cu(II) into the initial quinone (Scheme IX.10).

Alkanes and arylalkanes can be transformed into alkyl hydroperoxides as well as ketones and alcohols if their solutions are irradiated in the presence of catalytic amounts of metal oxo complexes. These complexes are hetero-polymetalates in methylene chloride [65a], acetic acid, alcohols, acetone, acetonitrile [65b,c], polyoxotungstate $W_{10}O_{32}^{4-}$ in acetonitrile [65d–f] or water [65g] (Table IX.10). Oxo compounds such as $K_2Cr_2O_7$ in the two-phase solvent water–1,2-dichloroethane [66a], CrO_3 [23], complexes $(n\text{-}Bu_4N)_2Cr_2O_7$, $(n\text{-}Bu_4N)_2Cr_3O_{10}$ and $(n\text{-}Bu_4N)Cr_4O_{13}$ [66b], and UO_2Cl_2 [66c] also catalyze the photooxygenation of alkanes. Photooxygenation of alkanes catalyzed by molybdenum or tungsten carbonyl begins apparently from the transformation of the carbonyl into an active oxo species [66d,e]. Some other chromium and vanadium complexes have been reported to be catalysts of alkane photooxidation [66f–j].

The mechanism of aerobic alkane photooxidation catalyzed by metal oxo complexes includes the formation of a photoexcited species which is capable of abstracting a hydrogen atom from an alkane. The alkyl radical thus formed rapidly adds a molecule of oxygen. An alkyl hydroperoxide is partially decomposed to produce a ketone and an alcohol:

$$M^{VI}{=}O \xrightarrow{h\nu} M^{V}{-}O\bullet$$

$$M^{V}\text{–O}\bullet + RH \longrightarrow M^{V}\text{–OH} + R\bullet$$

$$R\bullet + O_2 \longrightarrow ROO\bullet$$

$$ROO\bullet + M^{V}\text{–OH} \longrightarrow ROOH + M^{VI}\text{=O}$$

$$M^{V}\text{–OH} + O_2 \longrightarrow M^{VI}\text{=O} + H_2O$$

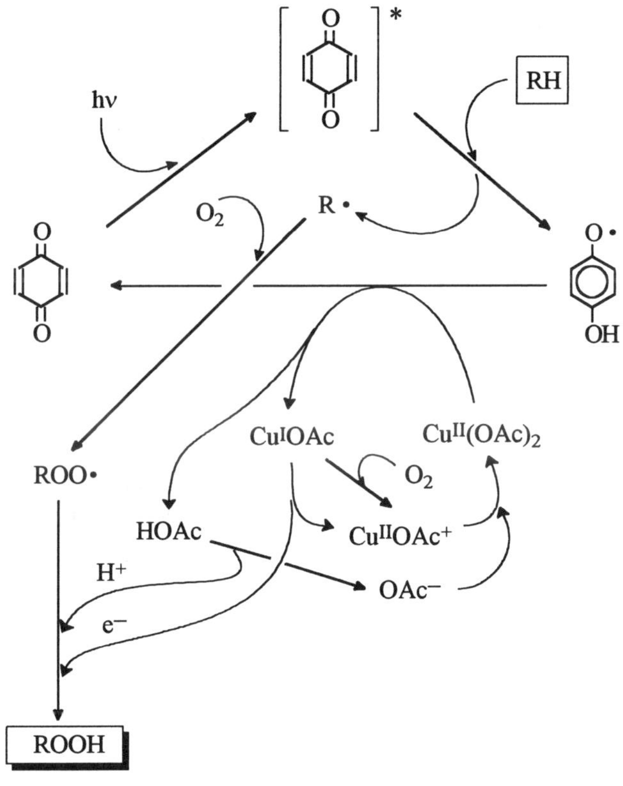

Scheme IX.10. Selective aerobic photochemical alkane hydroperoxidation catalyzed by the system "quinone–copper(II) acetate".

Table IX.10. Photocatalytic aerobic oxidation of hydrocarbons in the presence of metal oxo complexes.

Photocatalyst	Substrate	Solvent	Ref.
CrO_3 + picolinic acid	Ethylbenzene	CH_3CN	23
Oxometalate ($NaVO_3$, $H_3PW_{12}O_{40}$, $Na_6V_{10}O_{28}$, $Na_5H_7Mo_6V_5O_{39}$, $H_3PMo_{12}O_{40}$, $H_4SiMo_{12}O_{40}$) + $PhCH_2NEt_3Cl$	Cyclohexane	CH_2Cl_2	65a
$PW_{12}O_{40}^{3-}$, $PMo_{11}VO_{40}^{4-}$, $PMo_6V_5O_{39}^{12-}$	Alkylbenzenes	CH_3COOH, Me_2CHOH, CH_3OH, CH_3COCH_3, C_2H_5OH	65b
$(^nBu_4N)W_{10}O_{32}$	Cyclohexane	CH_3CN+CH_2Cl_2	65f
$K_2Cr_2O_7$ + $^nBu_4NBr$	Toluene	CH_2Cl_2–H_2O	66a
$(^nBu_4N)Cr_4O_{13}$	Cyclohexane	CH_3CN, CH_2Cl_2	66b
UO_2Cl_2	Alkylbenzenes, alkanes	CH_3COOH, CH_3CN	66c

Oxidation photocatalyzed by polyoxometalates [66k] has been applied to the functionalization of 1,8-cineole (structure **IX-10**) [66l], widely distributed in the plant kingdom. The photooxygenation of **IX-10** gave a mixture of ketones and alcohols which were transformed by the subsequent action of pyridinium chlorochromate into 5- and 6-keto derivatives in the ratio **IX-11** : **IX-12** = 2.5 : 1. A laser flash photolysis study of the mechanism has been carried out for the decatungstate anion catalyzed reaction [66m].

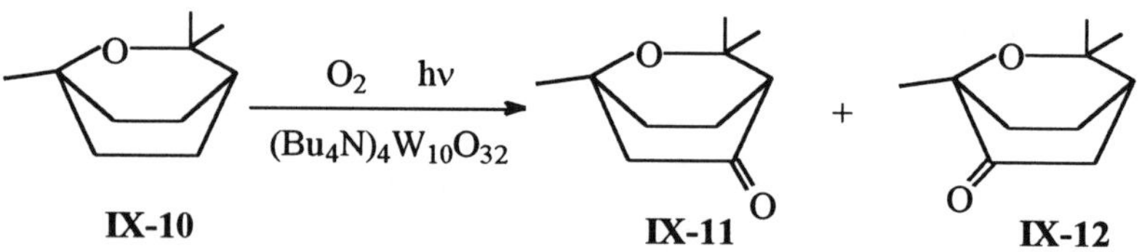

IX-10 **IX-11** **IX-12**

Cobalt(III)-alkylperoxy complexes also promote photooxidation of alkanes [67a]. Irradiation causes homolysis of the Co–O bond of the Co–O–O–Bu^t moieties in these complexes to generate ^{t}BuOO• radicals. Both ^{t}BuOO• and ^{t}BuO• are responsible for the oxidation of the C–H bonds:

$$[Co(BPI)(OAc)(OOtBu)] \longrightarrow [Co(BPI)(Oac)] + {}^tBuOO\bullet$$
$${}^tBuOO\bullet + Cy\bullet \longrightarrow {}^tBuOOCy$$
$${}^tBuO\bullet + CyH \longrightarrow {}^tBuOH + Cy\bullet$$
$${}^tBuOO\bullet \longrightarrow {}^tBuO\bullet + \tfrac{1}{2} O_2$$
$$2\,{}^tBuOO\bullet \longrightarrow {}^tBuOOOOBu^t \longrightarrow {}^tBuOOBu^t + O_2$$

Here BPI is 1,3-bis(2′-pyridylimino)isoindoline, CyH is cyclohexane. Stable final products are formed according to the equations:

$$Cy\bullet + O_2 \longrightarrow CyOO\bullet \longrightarrow CyOH + Cy{=}O + O_2$$
$$CyOO\bullet \longrightarrow CyOOH$$

Thermal aliphatic C–H bond oxygenation has been achieved by Co(II)–peroxo species [67b]. This oxygenation may proceed via homolysis of the O–O bond of the Co(II)–peroxo species.

Finally, orthometalated (arylazo)aryl(acetylacetonato)palladium(II) complexes, **IX-13**, can be oxidized by atmospheric oxygen to give the corresponding acetato derivatives, **IX-19**, upon irradiation of their acetone solutions with even long-wavelenght visible (λ > 500 nm) light [68]. The proposed mechanism (Scheme IX.11) includes the light absorbtion by the arylazo chromophore ("antenna") in complexes **IX-13**. Light energy absorbed by the transition of the azobenzene fragment is intramolecularly transferred to the metal center. The excited palladium ion is capable of promoting electron transfer from an acetylacetonate ligand to produce structures **IX-14** or **IX-15** and the radical-like acetylacetonyl fragment in **IX-15** reacts rapidly with the molecular oxygen present in the solvent to give peroxides **IX-16**, or/and **IX-17**, or/and **IX-18**. These peroxides decompose through C–C and O–O bond cleavage, affording complexes **IX-19**.

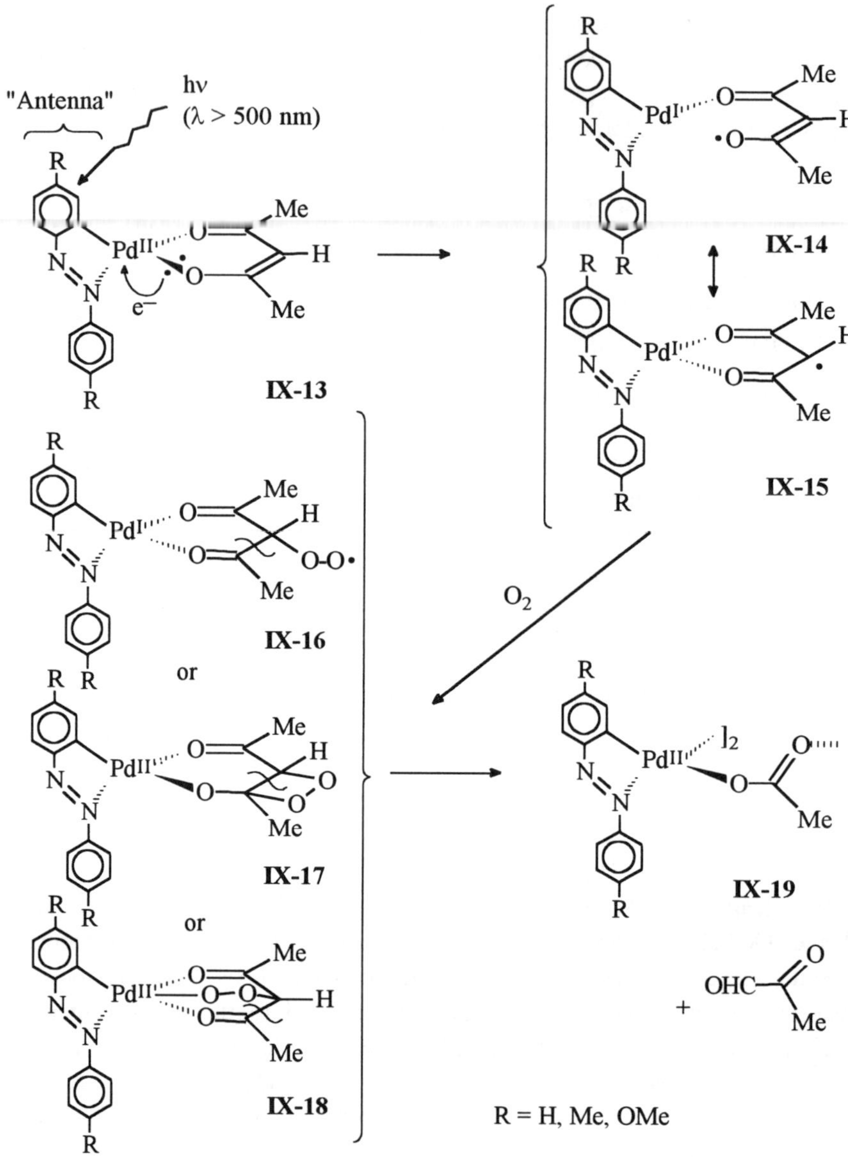

Scheme IX.11. The mechanism proposed for the photochemical C–C bond splitting under the action of dioxygen.

References for Chapter IX

1. (a) Sheldon, R. A.; Kochi, J. K. *Metal-Catalysed Oxidations of Organic Compounds*; Academic Press: New York, 1981. (b) *Oxygen Complexes and Oxygen Activation by Transition Metals*; Martell, A. E., Sawyer, D. T., Eds.; Plenum Press: New York, 1988. (c) *New Developments in Selective Oxidation*; Centi, G., Trifiro, F., Eds.; Elsevier: Amsterdam, 1990. (d) Sawyer, D. T. *Oxygen Chemistry*; Oxford University Press: Oxford, 1991. (e) *Dioxygen Activation and Homogeneous Catalytic Oxidation*; Simándi, L. I., Ed.; Elsevier: Amsterdam, 1991. (f) Drago, R. S. *Coord. Chem. Rev.* **1992**, *117*, 185. (g) *Catalytic Selective Oxidation*; Oyama, S. T.; Hightower, J. W., Eds.; ACS: Washington, 1993. (h) *The Activation of Dioxygen and Homogeneous Catalytic Oxidation*; Barton, D. H. R.; Martell, A. E., Sawyer, D. T., Eds.; Plenum Press: New York, 1993. (i) Sychev, A. Ya.; Isak, V. G. *Usp. Khim.* **1995**, *64*, 1183 (in Russian). (j) Hobbs, C. C., Jr. In *Applied Homogeneous Catalysis with Organometallic Compounds*; Cornils, B.; Herrmann, W. A., Eds.; VCH: Weinheim, 1996, p. 521. (k) Schiraldi, D. A. In *Applied Homogeneous Catalysis with Organometallic Compounds*; Cornils, B.; Herrmann, W. A., Eds.; VCH: Weinheim, 1996, p. 541. (l) *Catalytic Activation and Functionalisation of Light Alkanes*; Derouane, E. G.; Haber, J.; Lemos, F.; Ribeiro, F. R.; Guisnet, M., Eds.; Kluwer Acad. Publ.: Dordrecht, 1998. (m) Sheldon, R. A.; Arends, I. W. C. E.; Lempers, H. E. B. *Catalysis Today* **1998**, *41*, 387.
2. (a) Perkel', A. L.; Voronina, S. G.; Freidin, B. G. *Usp. Khim.* **1994**, *63*, 793 (in Russian). (b) Tanaka, K. *CHEMTECH* **1974**, 555.
3. (a) Igarashi, J.; Lusztyk, J.; Ingold, K. U. *J. Am. Chem. Soc.* **1992**, *114*, 7719. (b) Igarashi, J.; Jensen, R. K.; Lusztyk, J.; Korcek, S.; Ingold, K. U. *J. Am. Chem. Soc.* **1992**, *114*, 7727. (c) Goosen, A.; McCleland, C. W.; Morgan, D. H.; O'Connell, J. S.; Ramplin, A. *J. Chem. Soc., Chem. Commun.* **1993**, 401.
4. Denisov, E. T.; Khudyakov, I. V. *Chem. Rev.* **1987**, *87*, 1313.
5. (a) Hsu, Y. F.; Yen, M. H.; Cheng, C. P. *J. Mol. Catal. A: Chem.* **1996**, *105*, 137. (b) Hsu, Y. F.; Cheng, C. P. *J. Mol. Catal. A: Chem.* **1997**, *120*, 109.
6. (a) Emanuel, N. M.; Denisov, E. T.; Maizus, Z. K. *Liquid-Phase Oxidation of Hydrocarbons*; Plenum Press: New York, 1967. (b) Vanoppen, D. L.; De Vos, D. E.; Genet, M. J.; Rouxhet, P. G.; Jacobs, P. A. *Angew. Chem., Int. Ed. Engl.* **1995**, *34*, 560. (c) Maschmeyer, T.; Oldroyd, R. D.; Sankar, G.; Thomas, J. M.; Shannon, I. J.; Klepetko, J. A.; Masters, A. F.; Beattie, J. K.; Catlow, C. R. A.

Angew. Chem., Int. Ed. Engl. **1997**, *36*, 1639. (d) Csanyi, L. J.; Jaky, K. *J. Mol. Catal. A: Chem.* **1997**, *120*, 125. (e) Tugaut, V.; Pellegrini, S.; Castanet, Y.; Mortreux, A. *J. Mol. Catal. A: Chem.* **1997**, *127*, 25.

7. Haber, F.; Willstäter, R. *Ber.* **1931**, *64B*, 2844.
8. Maizus, Z. K.; Skibida, I. P.; Gagarina, A. B. *Zh. Fiz. Khim.* **1975**, *49*, 2491 (in Russian).
9. (a) Bawn, C. E. H.; Jolley, J. *Proc. Roy. Soc.* **1956**, *237A*, 297. (b) Bawn, C. E. H. *Disc. Faraday Soc.* **1953**, *14*, 181.
10. Barton, D. H. R.; Delanghe, N. C. *Tetrahedron Lett.* **1997**, *36*, 6351.
11. Haber, F.; Weiss, J. *Naturwiss.* **1932**, *20*, 948.
12. Kharash, M. S.; Arimoto, F. S.; Nudenberg, W. *J. Org. Chem.* **1951**, *16*, 1556.
13. Walling, C. *J. Am. Chem. Soc.* **1969**, *91*, 7590.
14. (a) Kozlova, Z. G.; Tsepalov, V. F.; Shlyapintokh, V. Ya. *Kinet. Katal.* **1964**, *5*, 868 (in Russian). (b) Zakharov, I. V.; Shlyapintokh, V. Ya. *Kinet. Katal.* **1963**, *4*, 706 (in Russian).
15. Skibida, I. P. *Usp. Khim.* **1975**, *54*, 1729 (in Russian).
16. (a) Chavez, F. A.; Rowland, J. M.; Olmstead, M. M.; Mascharak, P. K. *J. Am. Chem. Soc.* **1998**, *120*, 9015. (b) Babushkin, D. E.; Talsi, E. P. *J. Mol. Catal. A: Chem.* **1998**, *130*, 131. (c) Bakac, A. *J. Am. Chem. Soc.* **1997**, *119*, 10726. (d) Kamiya, Y.; Kashima, M. *J. Catal.* **1972**, *25*, 326.
17. (a) Onopchenko, A.; Schultz, J. G. D. *J. Org. Chem.* **1973**, *38*, 909. (b) Onopchenko, A.; Schultz, J. G. D. *J. Org. Chem.* **1973**, *38*, 3727.
18. Hanotier, J.; Camerman,, P.; Hanotier-Bridoux, M.; de Radzitzsky, P. *J. Chem. Soc., Perkin Trans. 2* **1972**, 2247.
19. (a) Fischer, R. W.; Röhrscheid, F. In *Applied Homogeneous Catalysis with Organometallic Compounds*; Cornils, B.; Herrmann, W. A., Eds.; VCH: Weinheim, 1996, p. 439. (b) Zakharov, I. V. *Kinet. Katal.* **1974**, *15*, 1457 (in Russian). (c) Felzenstein, A.; Goosen, A.; Marsh, C.; McCleland, C. W.; van Sandwyk, K. S. *S. Afr. J. Chem.* **1989**, *42*, 143. (d) Haruštiak, M.; Hronec, M.; Ilavský, J. *J. Mol. Catal.* **1989**, *53*, 209. (e) Dugmore, G. M.; Powels, G. J.; Zeelie, B. *J. Mol. Catal. A: Chem.* **1995**, *99*, 1. (f) Gomez, M. F. T.; Antunes, O. A. C. *J. Mol. Catal. A: Chem.* **1997**, *121*, 145.
20. Zakharov, I. V.; Geletii, Yu. V. *Neftekhimiya* **1978**, *18*, 615 (in Russian).
21. Goldstein, A. S.; Drago, R. S. *Inorg. Chem.* **1991**, *30*, 4506.

22. (a) Li, P.; Alper, H. *J. Mol. Catal.* **1990**, *61*, 51. (b) Alper, H.; Harustiak, M. *J. Mol. Catal.* **1993**, *84*, 87. (c) Lyons, J. E.; Ellis, P. E., Jr.; Shaikh, S. N. *Inorg. Chim. Acta* **1998**, *270*, 162.

23. Shul'pin, G. B.; Druzhinina, A. N. *Bull. Acad. Sci. USSR, Div. Chem. Sci.* **1989**, *38*, 1079.

24. Karakhanov, E. A.; Kardasheva, Yu. S.; Maksimov, A. L.; Predeina, V. V.; Runova, E. A.; Utukin, A. M. *J. Mol. Catal. A: Chem.* **1996**, *107*, 235.

25. Gusevskaya, E.; Gonsalves, J. A. *J. Mol. Catal. A: Chem.* **1997**, *121*, 131.

26. (a) Taqui Khan, M. M.; Shukla, R. S. *J. Mol. Catal.* **1988**, *44*, 85. (b) Taqui Khan, M. M.; Bajaj, H. C.; Shukla, R. S.; Mirza, S. A. *J. Mol. Catal.* **1988**, *45*, 51.

27. (a) Qian, C.-Y.; Yamada, T.; Nishino, H.; Kurosawa, K. *Bull. Chem. Soc. Jpn.* **1992**, *65*, 1371. (b) Qian, C.-Y.; Nishino, H.; Kurosawa, K.; Korp, J. D. *J. Org. Chem.* **1993**, *58*, 4448.

28. (a) Altamsani, A.; Brégeault, J.-M. *Synthesis* **1993**, 79. (b) Kirihara, M.; Takizawa, S.; Momose, T. *J. Chem. Soc., Perkin Trans. 1* **1998**, 7.

29. Shimizu, M.; Watanabe, Y.; Orita, H.; Hayakawa, T.; Takehira, K. *Tetrahedron Lett.* **1991**, *32*, 2053.

30. (a) Ellis, P. E., Jr.; Lyons, J. E. *Coord. Chem. Rev.* **1990**, *105*, 181. (b) Lyons, J. E.; Ellis, P. E., Jr. *Catalysis Lett.* **1991**, *8*, 45. (c) Chen, H. L.; Ellis, P. E., Jr.; Wijesekera, T.; Hagan, T. E.; Groh, S. E.; Lyons, J. E.; Ridge, D. P. *J. Am. Chem. Soc.* **1994**, *116*, 1086. (d) Grinstaff, M. W.; Hill, M. G.; Labinger, J. A.; Gray, H. B. *Science* **1994**, *264*, 1311. (e) Huang, J.-W.; Liu, Z.-L.; Gao, X.-R.; Yang, D.; Peng, X.-Y.; Ji, L.-N. *J. Mol. Catal. A: Chem.* **1996**, *111*, 261. (f) Ellis, P. E., Jr.; Wijesekera, T.; Lyons, J. E. *Catalysis Lett.* **1996**, *36*, 69. (g) Böttcher, A.; Birnbaum, E. R.; Day, M. W.; Gray, H. B.; Grinstaff, M. W.; Labinger, J. A. *J. Mol. Catal. A: Chem.* **1997**, *117*, 229. (h) Moore, K. T.; Horváth, I. T.; Therien, M. J. *J. Am. Chem. Soc.* **1997**, *119*, 1791. (i) Bianchini, C. *Chemtracts–Inorg. Chem.* **1997**, 588. (j) Nenoff, T. M.; Showalter, M. C.; Salaz, K. A. *J. Mol. Catal. A: Chem.* **1997**, *121*, 123.

31. (a) Birnbaum, E. R.; Grinstaff, M. W.; Labinger, J. A.; Bercaw, J. E.; Gray, H. B. *J. Mol. Catal. A: Chem.* **1995**, *104*, L119. (b) Birnbaum, E. R.; Labinger, J. A.; Bercaw, J. E.; Gray, H. B. *Inorg. Chim. Acta* **1998**, *270*, 433.

32. (a) Mizuno, N.; Nozaki, C.; Hirose, T.; Tateishi, M.; Iwamoto, M. *J. Mol. Catal. A: Chem.* **1997**, *117*, 159. (b) Mizuno, N.; Tateishi, M.; Hirose, T.;

Iwamoto, M. *Chem. Lett.* **1993**, 2137. (c) Mizuno, N.; Hirose, T.; Tateishi, M.; Iwamoto, M. *J. Mol. Catal.* **1994**, *88*, L125. (d) Neumann, R.; Levin, M. *J. Am. Chem. Soc.* **1992**, *114*, 7278. (e) Neumann, R.; Dahan, M. *J. Chem. Soc., Chem. Commun.* **1995**, 171. (f) Fujibayashi, S.; Nakayama, K.; Hamamoto, M.; Sakaguchi, S.; Nishiyama, Y. Ishii, Y. *J. Mol. Catal. A: Chem.* **1996**, *110*, 105. (g) Hanyu, A.; Sakurai, Y.; Fujibayashi, S.; Ishii, Y. *Tetrahedron Lett.* **1997**, *38*, 5659.

33. Ishii, Y. *J. Mol. Catal. A: Chem.* **1997**, *117*, 123.
34. (a) Ishii, Y.; Kato, S.; Iwahama, T.; Sakaguchi, S. *Tetrahedron Lett.* **1996**, *37*, 4993. (b) Ishii, Y.; Iwahama, T.; Sakaguchi, S.; Nakayama, K.; Nishiyama, Y. *J. Org. Chem.* **1996**, *61*, 4520. (c) Yoshino, Y.; Hayashi, Y.; Iwahama, T.; Sakaguchi, S.; Ishii, Y. *J. Org. Chem.* **1997**, *62*, 6810. (d) Sakaguchi, S.; Kato, S.; Iwahama, T.; Ishii, Y. *Bull. Chem. Soc. Jpn.* **1998**, *71*, 1237. (e) Iwahama, T.; Syojyo, K.; Sakaguchi, S.; Ishii, Y. *Org. Proc. Res. & Dev.* **1998**, *2*, 255.
35. (a) Ishii, Y.; Nakayama, K.; Takeno, M.; Sakaguchi, S.; Iwahama, T.; Nishiyama, Y. *J. Org. Chem.* **1995**, *60*, 3934. (b) Sakaguchi, S.; Eikawa, M.; Ishii, Y. *Tetrahedron Lett.* **1997**, *38*, 7075.(c) Kato, S.; Iwahama, T.; Sakaguchi, S.; Ishii, Y. *J. Org. Chem.* **1998**, *63*, 222.
36. Seo, Y.-J.; Mukai, Y.; Tagawa, T.; Goto, S. *J. Mol. Catal. A: Chem.* **1997**, *120*, 149.
37. (a) Ito, S.; Yamasaki, T.; Okada, H.; Okino, S.; Sasaki, K. *J. Chem. Soc., Perkin Trans. 2* **1988**, 285. (b) Kitano, T.; Kuroda, Y.; Itoh, A.; Li-Fen, J.; Kunai, A.; Sasaki, K. *J. Chem. Soc., Perkin Trans. 2* **1990**, 1991.
38. (a) Jintoku, T.; Takaki, K.; Fujiwara, Y.; Fuchita, Y.; Hiraki, K. *Bull. Chem. Soc. Jpn.* **1990**, *63*, 438. (b) Jintoku, T.; Nishimura, K.; Takaki, K.; Fujiwara, Y. *Chem. Lett.* **1990**, 1687.
39. Bolzacchini, E.; Canevali, C.; Morazzoni, F.; Orlandi, M.; Rindone, B.; Scotti, R. *J. Chem. Soc., Dalton Trans.* **1997**, 4695.
40. (a) Schönbein, C. F. *Ber. Natur. Ges. Bas.* **1844**, *6*, 16. (b) Schönbein, C. F. *J. Prakt. Chem.* **1852**, *55*, 1.
41. Shilov, N. A. *On the Coupled Oxidation Reactions*, Moscow, 1905 (in Russian).
42. (a) Manchot, W.; Wilhelms, O. *Ber.* **1901**, *34*, 2479. (b) Manchot, W. *Ann.* **1902**, *325*, 93.
43. (a) Bach, A. *Zh. Ross. Fiz.-Khim. Obsch.* **1897**, *29*, 25 (in Russian). (b) Engler, C.; Wild, W. *Ber.* **1897**, *12*, 1669.
44. Udenfriend, S.; Clark, C. T.; Axelrod, J.; Brodie, B. B. *J. Biol. Chem.* **1954**, *208*, 731.

45. Hamilton, G. A.; *J. Am. Chem. Soc.* **1964**, *86*, 3391.

46. (a) Ullrich, V.; Staudinger, H. *Z. Naturforsch.* **1969**, *24b*, 583. (b) Ullrich, V.. *Z. Naturforsch.* **1969**, *24b*, 699.

47. (a) Muradov, N. Z.; Shilov, A. E.; Shteinman, A. A. *Kinet. Katal.* **1972**, *13*, 1357 (in Russian). (b) Karasevich, E. I.; Muradov, N. Z.; Shteinman, A. A. *Izv. Akad. Nauk SSSR, Ser. Khim.* **1974**, 1805 (in Russian). (c) Muradov, N. Z.; Shteinman, A. A. *Izv. Akad. Nauk SSSR, Ser. Khim.* **1975**, 2294 (in Russian).

48. Denisov, E. T. *Liquid-Phase Reaction Rate Constants*, Plenum Press: New York, 1974, p. 771.

49. (a) Kutner, E. A.; Matseevskiy, B. P. *Kinet. Katal.* **1969**, *10*, 997 (in Russian). (b) Mimoun, H.; Seree de Roch, I. *Tetrahedron*, **1975**, *31*, 777. (c) Lavrushko, V. V.; Khenkin, A. M.; Shilov, A. E. *Kinet. Katal.* **1980**, *21*, 276 (in Russian). (d) Zakharov, I. V.; Karasevich, E. I.; Shilov, A. E.; Shteinman, A. A. *Kinet. Katal.* **1975**, *16*, 1151 (in Russian). (e) Geletiy, Yu. V.; Zakharov, I. V.; Karasevich, E. I.; Shteinman, A. A. *Kinet. Katal.* **1979**, *20*, 1124 (in Russian).

50. (a) Barton, D. H. R.; Doller, D. *Acc. Chem. Res.* **1992**, *25*, 504. (b) Barton, D. H. R.; Taylor, D. K. *Russ. Chem. Bull.* **1995**, *44*, 575. (c) Barton, D. H. R. *Tetrahedron* **1998**, *54*, 5805. (d) Perkins, M. J. *Chem. Soc. Rev.* **1996**, 229. (e) Barton, D. H. R. *Chem. Soc. Rev.* **1996**, 237.

51. Barton, D. H. R., Gastiger, M. J.; Motherwell, W. B. *J. Chem. Soc., Chem. Commun.* **1983**, 41.

52. (a) Barton, D. H. R.; Bévière, S. D.; Chavasiri, W.; Csuhai, E.; Doller, D.; Liu, W.-G. *J. Am. Chem. Soc.* **1992**, *114*, 2147. (b) Barton, D. H. R.; Bévière, S. D.; Chavasiri, W.; Csuhai, E.; Doller, D. *Tetrahedron* **1992**, *48*, 2895. (c) Barton, D. H. R.; Csuhai, E.; Doller, D. *Tetrahedron* **1992**, *48*, 9195. (d) Barton, D. H. R.; Bévière, S. D.; Chavasiri, W.; Doller, D.; Hu, B. *Tetrahedron Lett.* **1993**, *34*, 567. (e) Barton, D. H. R.; Chavasiri, W.; Hill, D. R.; Hu, B. *New J. Chem.* **1994**, *18*, 611. (f) Barton, D. H. R.; Hill, D. R. *Tetrahedron Lett.* **1994**, *35*, 1431. (g) Barton, D. H. R.; Chabot, B. M.; Delanghe, N. C.; Hu, B.; Le Gloahec, V. N.; Wahl, R. U. R. *Tetrahedron Lett.* **1995**, *36*, 7007. (i) Barton, D. H. R.; Hu, B.; Wahl, R. U. R.; Taylor, D. K. *New J. Chem.* **1996**, *20*, 121. (j) Barton, D. H. R.; Beck, A. H.; Delanghe, N. C. *Tetrahedron Lett.* **1996**, *37*, 1555. (k) Barton, D. H. R.; Chabot, B. M.; Hu, B. *Tetrahedron Lett.* **1996**, *37*, 1755. (l) Barton, D. H. R.; Delanghe, N. C. *Tetrahedron Lett.* **1996**, *37*, 8137. (m) Barton, D. H. R.; Hu, B.; Li, T.; MacKinnon, J. *Tetrahedron Lett.* **1996**, *37*, 8329. (n) Barton, D. H. R.; Hu, B.; Taylor, D. K.; Wahl, R. U. R. *J. Chem. Soc.,*

Perkin Trans. 2 **1996**, 1031. (o) Barton, D. H. R.; Chabot, B. M. *Tetrahedron* **1996**, *52*, 10287. (p) Barton, D. H. R.; Chabot, B. M.; Hu, B. *Tetrahedron* **1996**, *31*, 10301. (q) Barton, D. H. R.; Hu, B. *Tetrahedron* **1996**, *31*, 10313. (r) Barton, D. H. R.; Li, T.; MacKinnon, J. *Chem. Comm.* **1997**, 557. (s) Barton, D. H. R.; Delanghe, N. C. *Tetrahedron* **1998**, *54*, 4471. (t) Barton, D. H. R.; Li, T. *Chem. Comm.* **1998**, 821.

53. (a) Schuchardt, U.; Mano, V. In *New Developments in Selective Oxidation*; Centi, G., Trifiro, F., Eds.; Elsevier: Amsterdam, 1990, p. 185. (b) Schuchardt, U.; Krähembühl, C. E. Z.; Carvalho, W. A. *New J. Chem.* **1991**, *15*, 955. (c) Shul'pin, G. B.; Kitaygorodskiy, A. N. *J. Gen. Chem. USSR* **1990**, *60*, 920. (d) Knight, C.; Perkins, M. J. *J. Chem. Soc., Chem. Commun.* **1991**, 925. (e) Schuchardt, U.; Carvalho, W. A.; Spinacé, E. V. *Synlett* **1993**, 713. (f) Minisci, F.; Fontana, F.; Araneo, S.; Recupero, F. *J. Chem. Soc., Chem. Commun.* **1994**, 1823. (g) Minisci, F.; Fontana, F. *Tetrahedron Lett.* **1994**, *35*, 1427. (h) Minisci, F.; Fontana, F.; Zhao, L.; Banfi, S.; Quici, S. *Tetrahedron Lett.* **1994**, *35*, 8033. (i) Newcomb, M.; Simakov, P.A.; Park, S.-U. *Tetrahedron Lett.* **1996**, *37*, 819. (j) Sobolev, A. P.; Babushkin, D. E.; Shubin, A. A.; Talsi, E. P. *J. Mol. Catal. A: Chem.* **1996**, *112*, 253. (k) Singh, B.; Long, J. R.; Papaefthymiou, G. C.; Stavropoulos, P. *J. Am. Chem. Soc.* **1996**, *118*, 5824.

54. Mukaiyama, T.; Yamada, T. *Bull. Chem. Soc. Jpn.* **1995**, *68*, 17.

55. (a) Murahashi, S.; Oda, Y.; Naota, T. *J. Am. Chem. Soc.* **1992**, *114*, 7913. (b) Murahashi, S.; Oda, Y.; Naota, T.; Komiya, N. *J. Chem. Soc., Chem. Commun.* **1993**, 139. (c) Hata, E.; Takai, T.; Yamada, T.; Mukaiyama, T. *Chemistry Lett.* **1994**, 1849. (d) Murahashi, S.; Naota, T.; Komiya, N. *Tetrahedron Lett.* **1995**, *36*, 8059. (e) Punniyamurthy, T.; Karla, S. J. S.; Iqbal, J. *Tetrahedron Lett.* **1995**, *36*, 8497. (f) Battioni, P.; Iwanejko, R.; Mansuy, D.; Mlodnicka, T.; Poltowicz, J.; Sanches, F. *J. Mol. Catal. A: Chem.* **1996**, *109*, 91. (g) Komiya, N.; Naota, T.; Murahashi, S.; *Tetrahedron Lett.* **1996**, *37*, 1633. (h) Fujibayashi, S.; Nakayama, K.; Hamamoto, M.; Sakaguchi, S.; Nishiyama, Y.; Ishii, Y. *J. Mol. Catal. A: Chem.* **1996**, *110*, 105. (i) Komiya, N.; Naota, T.; Oda, Y.; Murahashi, S. *J. Mol. Catal. A: Chem.* **1997**, *117*, 21. (j) Chand, D. K.; Bharadwaj, P. K. *Inorg. Chem.* **1997**, *36*, 5658. (k) Dell'Anna, M. M.; Mastrorilli, P.; Nobile, C. F. *J. Mol. Catal. A: Chem.* **1998**, *130*, 65. (l) Lindsay Smith, J. R.; Shul'pin, G. B. *Russ. Chem. Bull.* **1998**, *47*, 2313. (m) Kaneda, K.; Haruna, S.; Imanaka, T.; Kawamoto, K. *J. Chem. Soc., Chem. Commun.* **1990**, 1467.

56. (a) Duprat, A. F.; Capdevielle, P.; Maumy, M. *J. Chem. Soc., Chem. Commun.* **1991**, 464. (b) Kurusu, Y.; Neckers, D. C. *J. Org. Chem.* **1991**, *56*, 1981. (c) Kitajima, N.; Ito, M.; Fukui, H.; Moro-oka, Y. *J. Chem. Soc., Chem. Commun.* **1991**, 102. (d) Belova, V. S.; Gimanova, I. M.; Stepanova, M. L.; Khenkin, A. M.; Shilov, A. E. *Dokl. Akad. Nauk SSSR* **1991**, *316*, 653 (in Russian). (e) Belova, V. S.; Khenkin, A. M.; Postnov, V. N.; Prusakov, V. E.; Shilov, A. E. Stepanova, M. L. *Mendeleev Commun.* **1992**, 7. (f) Lu, W. Y.; Bartoli, J. F.; Battioni, P.; Mansuy, D. *New J. Chem.* **1992**, *16*, 621. (g) Yamanaka, I.; Otsuka, K. *J. Mol. Catal.* **1993**, *83*, L15. (h) Yamanaka, I.; Otsuka, K. *J. Mol. Catal. A: Chem.* **1995**, *95*, 115. (i) Yamanaka, I.; Akimoto, T.; Otsuka, K. *Chemistry Lett.* **1994**, 1511. (j) Yamanaka, I.; Akimoto, T.; Nakagaki, K.; Otsuka, K. *J. Mol. Catal. A: Chem.* **1996**, *110*, 119. (k) Yamanaka, I.; Nakagaki, K.; Akimoto, T.; Otsuka, K. *J. Chem. Soc., Perkin Trans. 2* **1996**, 2511. (l) Yamanaka, I.; Soma, M.; Otsuka, K. *J. Chem. Soc., Chem. Commun.* **1995**, 2235. (m) Yamanaka, I.; Morimoto, K.; Soma, M.; Otsuka, K. *J. Mol. Catal. A: Chem.* **1998**, *133*, 251.

57. (a) Davis, R.; Durrant, J. L. A.; Khan, M. A. *Polyhedron* **1988**, *7*, 425. (b) Funabiki, T.; Tsujimoto, M.; Ozawa, S.; Yoshida, S. *Chemistry Lett.* **1989**, 1267. (c) Funabiki, T.; Toyoda, T.; Yoshida, S. *Chemistry Lett.* **1992**, 1279. (d) Tatsumi, T.; Yuasa, K.; Tominaga, H. *J. Chem. Soc., Chem. Commun.* **1992**, 1446. (e) Wang, Z.; Martell, A. E.; Motekaitis, R. J. *Chem. Comm.* **1998**, 1523. (f) Lin, M.; Sen, A. *J. Am. Chem. Soc.* **1992**, *114*, 7307.

58. (a) Otsuka, K.; Ishizuka, K.; Yamanaka, I. *Chemistry Lett.* **1992**, 773. (b) Otsuka, K.; Yamanaka, I.; Hagiwara, M. *Chemistry Lett.* **1994**, 1861. (c) Maki, S.; Konno, K.; Takayama, H. *Tetrahedron Lett.* **1997**, *38*, 7067.

59. Shul'pin, G. B.; Kats, M. M.; Lederer, P. *J. Gen. Chem. USSR* **1989**, *59*, 2450.

60. (a) Shul'pin, G. B.; Kats, M. M. *React. Kinet. Catal. Lett.* **1990**, *41*, 239. (b) Shul'pin, G. B.; Nizova, G. V.; Kats, M. M. *J. Gen. Chem. USSR* **1990**, *60*, 2447. (c) Shul'pin, G. B.; Kats, M. M. *Neftekhimiya* **1991**, *31*, 648 (in Russian); *Petrol. Chem.* **1991**, *31*, 647. (d) Lederer, P.; Nizova, G. V.; Kats, M. M.; Shul'pin, G. B. *Coll. Czech. Chem. Commun.* **1992**, *57*, 107. (e) Shul'pin, G. B.; Druzhinina, A. N. *Mendeleev Commun.* **1992**, 36. (f) Shul'pin, G. B.; Druzhinina, A. N. *Bull. Russ. Acad. Sci., Div. Chem. Sci.* **1992**, *41*, 346. (g) Shul'pin, G. B.; Nizova, G. V. *Neftekhimiya* **1993**, *33*, 118 (in Russian); *Petrol. Chem.* **1993**, *33*, 107. (h) Shul'pin, G. B.; Nizova, G. V.; Kozlov, Yu. N. *New J. Chem.* **1996**, *20*, 1243.

61. (a) Nizova, G. V.; Kats, M. M.; Shul'pin, G. B. *Bull. Acad. Sci. USSR, Div. Chem. Sci.* **1990**, *39*, 618. (b) Shul'pin, G. B.; Nizova, G. V.; Kats, M. M. *Neftekhimiya* **1991**, *31*, 658 (in Russian); *Petrol. Chem.* **1991**, *31*, 657. (c) Nizova, G. V.; Shul'pin, G. B. *Bull. Acad. Sci. USSR, Div. Chem. Sci.* **1989**, *38*, 2205. (d) Shul'pin, G. B.; Nizova, G. V. *J. Gen. Chem. USSR* **1990**, *60*, 1892. (e) Nizova, G. V.; Shul'pin, G. B. *Neftekhimiya* **1991**, *31*, 822 (in Russian); *Petrol. Chem.* **1991**, *31*, 829. (f) Nizova, G. V.; Losenkova, G. V.; Shul'pin, G. B. *React. Kinet. Catal. Lett.* **1991**, *45*, 27.

62. (a) Druzhinina, A. N.; Shul'pina, L. S.; Shul'pin, G. B. *Bull. Acad. Sci. USSR, Div. Chem. Sci.* **1991**, *40*, 1492. (b) Shul'pin, G. B.; Druzhinina, A. N.; Shul'pina, L. S. *Neftekhimiya* **1993**, *33*, 335 (in Russian); *Petrol. Chem.* **1993**, *33*, 321.

63. (a) Shul'pin, G. B.; Nizova, G. V. *Mendeleev Commun.* **1995**, 143. (b) Maldotti, A.; Bartocci, C.; Amadelli, R.; Polo, E.; Battioni, P.; Mansuy, D. *J. Chem. Soc., Chem. Commun.* **1991**, 1487. (c) Shelnutt, J. A.; Trudell, D. E. *Tetrahedron Lett.* **1989**, *30*, 5231. (d) Weber, L.; Hommel, R.; Behling, J.; Haufe, G.; Hennig, H. *J. Am. Chem. Soc.* **1994**, *116*, 2400. (e) Maldotti, A.; Bartocci, C.; Amadelli, R.; Varani, G.; Pietrogrande, M. C.; Carassiti, V. *Gazz. Chim. Ital.* **1994**, *124*, 411. (f) Maldotti, A.; Mollinari, A.; Andreotti, M.; Fogagnolo, R.; Amadelli, M. *Chem. Commun.* **1998**, 507.

64. (a) Muzart, J.; Henin, F. *C. R. Acad. Sci. Paris, Part II* **1988**, *307*, 479. (b) Monaci, A.; Dal Zotto, R.; Tarli, F. *J. Photochem. Photobiol. A: Chem.* **1990**, *52*, 327. (c) Shul'pin, G. B.; Druzhinina, A. N. *React. Kinet. Catal. Lett.* **1992**, *47*, 207. (d) Shul'pin, G. B.; Bochkova, M. M.; Nizova, G. V. *J. Chem. Soc., Perkin Trans. 2* **1995**, 1465.

65. (a) Shul'pin, G. B.; Kats, M. M. *Bull. Acad. Sci. USSR, Div. Chem. Sci.* **1989**, *38*, 2202. (b) Shul'pin, G. B.; Kats, M. M. *J. Gen. Chem. USSR* **1989**, *59*, 2447. (c) Attanasio, D.; Suber, L. *Inorg. Chem.* **1989**, *28*, 3779. (d) Chambers, R. C.; Hill, C. L. *Inorg. Chem.* **1989**, *28*, 2509. (e) Zakrzewski, J.; Chauveau, F.; Giannotti, C. *C. R. Acad. Sci. Paris, Part II* **1989**, *309*, 809. (f) Maldotti, A.; Amadelli, R.; Carassiti, V.; Molinari, A. *Inorg. Chim. Acta* **1997**, *256*, 309. (g) Muradov, N. Z.; Rustamov, M. I. *Kinet. Katal.* **1989**, *30*, 248 (in Russian).

66. (a) Shul'pin, G. B.; Kats, M. M. *J. Gen. Chem. USSR* **1989**, *59*, 2338. (b) Shul'pin, G. B.; Druzhinina, A. N.; Nizova, G. V. *Bull. Acad. Sci. USSR, Div. Chem. Sci.* **1991**, *40*, 2145. (c) Nizova, G. V.; Shul'pin, G. B. *J. Gen. Chem. USSR* **1990**, *60*, 1895. (d) Shul'pin, G. B.; Druzhinina, A. N. *React. Kinet.*

Catal. Lett. **1991**, *44*, 387. (e) Shul'pin, G. B.; Druzhinina, A. N. *Neftekhimiya* **1993**, *33*, 256 (in Russian); *Petrol. Chem.* **1993**, *33*, 247. (f) Shul'pin, G. B.; Nizova, G. V. *React. Kinet. Catal. Lett.* **1991**, *45*, 7. (g) G. V. Nizova, G. V.; Muzart, J.; Shul'pin, G. B. *React. Kinet. Catal. Lett.* **1991**, *45*, 173. (h) Muzart, J.; Druzhinina, A. N.; Shul'pin, G. B. *Neftekhimiya* **1993**, *33*, 124 (in Russian); *Petrol. Chem.* **1993**, *33*, 113. (i) Nizova, G. V.; Shul'pin, G. B. *Neftekhimiya* **1994**, *34*, 364 (in Russian); *Petrol. Chem.* **1994**, *34*, 353. (j) Shul'pin, G. B.; Druzhinina, A. N.; Nizova, G. V. *Bull. Acad. Sci. USSR, Div. Chem. Sci.* **1990**, *39*, 195. (k) Ermolenko, L. P.; Giannotti, C. *J. Chem. Soc., Perkin Trans. 2* **1996**, 1205. (l) Zakrzewski, J.; Giannotti, C. *J. Photochem. Photobiol. A: Chem.* **1992**, *63*, 173. (m) Ermolenko, L. P.; Delaire, J. A.; Giannotti, C. *J. Chem. Soc., Perkin Trans. 2* **1997**, 25.

67. (a) Farinas, E. T.; Nguyen, C. V.; Mascharak, P. K. *Inorg. Chim. Acta* **1997**, *263*, 17. (b) Hikichi, S.; Komatsuzaki, H.; Akita, M.; Moro-oka, Y. *J. Am. Chem. Soc.* **1998**, *120*, 4699.

68. Vicente, J.; Arcas, A.; Bautista, D.; Shul'pin, G. B. *J. Chem. Soc., Dalton Trans.* **1994**, 1505.

CHAPTER X

HOMOGENEOUS CATALYTIC OXIDATION OF HYDROCARBONS BY PEROXIDES AND OTHER OXYGEN ATOM DONORS

Along with oxidation with molecular oxygen, the reactions of hydrocarbons and other organic compounds with peroxides, first of all with hydrogen peroxide, are of great importance. They may be a basis for creating new technologies for direct selective transformations of alkanes and aromatics into valuable oxygen-containing products. It is noteworthy that despite numerous works of iron-promoted oxidations by hydrogen peroxide there are many questions concerning the mechanisms [1] of such processes. For example, the following scheme of the interaction of Fe(II) with H_2O_2 (well-known Fenton's reagent [2]) to produce hydroxyl radicals is adopted [1]:

$$Fe^{2+} + H_2O_2 \longrightarrow Fe^{3+} + HO^- + HO\bullet$$
$$Fe^{2+} + HO\bullet \longrightarrow Fe^{3+} + HO^-$$
$$H_2O_2 + HO\bullet \longrightarrow H_2O + HOO\bullet$$
$$Fe^{3+} + H_2O_2 \longrightarrow Fe^{2+} + HOO\bullet + H^+$$
$$Fe^{3+} + HOO\bullet \longrightarrow Fe^{2+} + O_2 + H^+$$
$$Fe^{2+} + HOO\bullet \longrightarrow Fe^{3+} + HOO^-$$
$$HOO^- + H^+ \rightleftharpoons H_2O_2$$
$$HOO\bullet + HOO\bullet \longrightarrow H_2O_2 + O_2$$

The interaction of 2,4-dimethylaniline with hydroxyl radicals generated by Fenton's reagent can be expected to produce aminophenols:

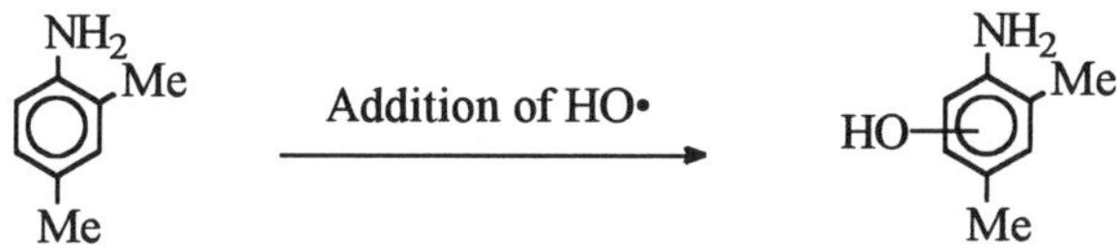

However, it has been found recently [2g] that thermal or photochemically enhanced Fenton reactions, in the presence of 2,4-dimethylaniline, yield primarily 2,4-dimethylphenol as an intermediate product, the genesis of which may only be explained by an electron transfer mechanism:

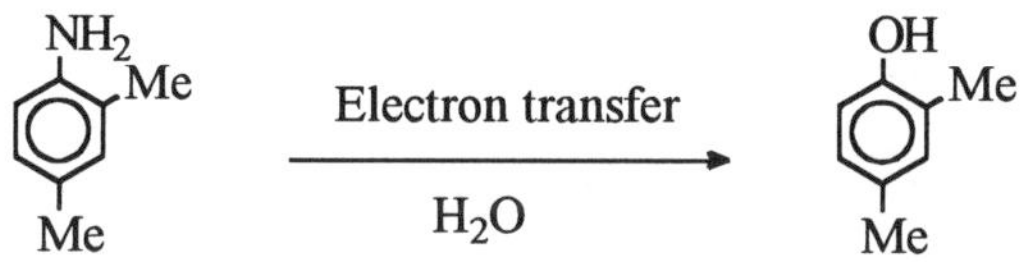

Books [3] and reviews [1, 4] have appeared in recent years on metal-catalyzed oxidations by peroxides. This chapter deals with hydrocarbon oxidations by peroxides. We will also consider briefly oxygenations by other oxygen atom donors.

X.1. Oxidation by Hydrogen Peroxide

X.1.A. ALKYL HYDROPEROXIDES AS PRODUCTS

Formation of Alkyl Hydroperoxides in H_2O_2 Oxidations of Alkanes

One can read in many publications that various metal-catalyzed oxidations of alkanes by hydrogen peroxide afford corresponding alcohols and carbonyl compounds (ketones or aldehydes) which are usually determined by GC analysis. However, it has been demonstrated [5] that H_2O_2 oxidations of alkanes catalyzed by $FeSO_4$ (in MeCN–H_2O in the presence of H_2SO_4), LMnCl [L = tetrakis(2,3,4,5,6-pentafluorophenyl)porphyrinate] (in MeCN–CH_2Cl_2 in the presence of imidazole), $Pd(OCOCF_3)_2$ (in MeCN), and $^{n}Bu_4NVO_3$–pyrazine-2-carboxylic acid (in MeCN) give substantial amounts of the corresponding alkyl hydroperoxides (in addition to usually smaller concentrations of alcohols and ketones). The method of alkyl hydroperoxide determination by GC analysis of samples *before* and *after* the reduction with PPh_3 has been used [5]. The oxidation process was monitored by withdrawing aliquots at specific intervals and analyzing them not only without any treatment but also after reduction with triphenylphosphine. One of the merits of this method is the possibility to estimate the concentration of the alkyl hydroperoxide formed from the alkane in the presence of an excess of an oxidant (hydrogen peroxide, alkyl hydroperoxide or metal peroxide). This very simple method was used in studies of hydrogen peroxide oxidations (see below) as well as in oxygenations with molecular oxygen (see, for example, [6]).

Estimation of Alkyl Hydroperoxide Content by GC Analysis of the Reaction Solution Samples Before and After Reduction with PPh_3

If an excess of solid triphenylphosphine is added to a solution of alkane (for example, cyclohexane) oxidation products 10–20 min before the GC analysis, the resulting chromatogram differs drastically from that of a sample not subjected to the reduction with triphenylphosphine. After the reduction, the cyclohexanol peak rises markedly while the intensity of the cyclohexanone peak decreases. The sum of alcohol and ketone concentrations in the reduced sample is approximately equal to the total concentration of products in the solution untreated with triphenylphosphine. These results can be explained by the fact that the mixture of products of the reaction under discussion contains cyclohexyl hydroperoxide as the main component.

Indeed, it is known that cyclohexyl hydroperoxide is usually totally decomposed in the chromatograph to produce cyclohexanol and cyclohexanone:

$$c\text{-}C_6H_{11}OOH \longrightarrow c\text{-}C_6H_{11}OH + c\text{-}C_6H_{10}O + \ldots \qquad (X.1)$$

The cyclohexyl hydroperoxide is readily and quantitatively reduced by triphenylphosphine to yield cyclohexanol:

$$c\text{-}C_6H_{11}OOH + PPh_3 \longrightarrow c\text{-}C_6H_{11}OH + OPPh_3 \qquad (X.2)$$

Thus, by comparing the data of chromatographic analysis of the reaction solution before and after reduction with triphenylphosphine, the amounts of cyclohexyl hydroperoxide, cyclohexanol and cyclohexanone present in the solution at a given moment can be estimated.

Generally, the oxidation of cyclohexane can give rise to the formation of cyclohexyl hydroperoxide [real current concentration in the reaction solution is $(C_{prxd})^{reaction}$], cyclohexan*ol* [$(C_{ol})^{reaction}$] and cyclohexan*one* [$(C_{one})^{reaction}$]. Cyclohexyl hydroperoxide decomposes in the injector of the chromatograph to produce additional amounts of cyclohexanol (the increment ΔC_{ol}) and cyclohexanone (ΔC_{one}). The ratio $\omega = \Delta C_{ol}/\Delta C_{one}$ can usually vary in the interval ~ 0.7 – 1.5 and seems to depend on the solvent, components of the solution and the chromatograph used. In this case:

$$(C_{prxd})^{reaction} = \Delta C_{ol} + \Delta C_{one} = \omega\Delta C_{one} + \Delta C_{one} = (\omega+1)\Delta C_{one} \quad \text{(X.3)}$$

The value of the coefficient ω can be determined for the chromatograph used and for the solvent in which CyOOH is dissolved. By chromatography either of the solution of pure authentic CyOOH or of the reaction solution in the initial period of the reaction when almost pure CyOOH is present in the solution and concentrations of ketone and alcohol are very low.

If it is assumed that the alkyl hydroperoxide decomposition in the injector gives the two products in approximately equal amounts, (that is $\Delta C_{ol} \approx \Delta C_{one}$), the concentrations are

$$(C_{prxd})^{reaction} = \Delta C_{ol} + \Delta C_{one} \approx 2\,\Delta C_{one}. \quad \text{(X.4)}$$

Total concentrations of detectable stable products (cyclohexanol and cyclohexanone) in the chromatogram of the sample unreduced with PPh_3 will be $(C_{ol} + \Delta C_{ol})^{unred}$ and $(C_{one} + \Delta C_{one})^{unred}$, respectively. Triphenylphosphine reduces the alkyl hydroperoxide to yield, quantitatively, cyclohexanol. So the concentrations of *ol* and *one* determined by GC in the sample treated with PPh_3 will be $(C_{ol} + C_{prxd})^{red}$ and $(C_{one})^{red}$, respectively. The real amount of ketone present in the reaction mixture may thus be determined by measuring the concentration $(C_{one})^{red}$ of this product in the reduced sample.

$$(C_{one})^{reaction} = (C_{one})^{red} \quad \text{(X.5)}$$

Then, since $(C_{one})^{reaction} = (C_{ol})^{reaction}/\omega$, we can calculate the concentration of the alkyl hydroperoxide present in the reaction solution:

$$(C_{prxd})^{reaction} = (C_{ol})^{red} - (C_{one})^{red}/\omega \qquad (X.6)$$

The second way to determine the real concentrations of the products in the solution is based on measuring values ΔC_{ol} and ΔC_{one}.

$$\Delta C_{one} = (C_{one})^{unred} - (C_{one})^{red} \qquad (X.7)$$

$$\Delta C_{ol} = (C_{ol})^{unred} - (C_{ol})^{red} \qquad (X.8)$$

Using equations (X.7), (X.8) and (X.3), it is possible to calculate $(C_{prxd})^{reaction}$ as well as $(C_{one})^{reaction}$ and $(C_{ol})^{reaction}$. If the values determined by the two ways are similar, it testifies that these concentrations are the real concentrations of the products in the solution.

However, if there is only one channel of the *ol* and *one* formation in the course of the oxidation, i.e., gradual decomposition of the alkyl hydroperoxide to produce both products again in approximately equal amounts, the concentrations of *ol* and *one* in the chromatogram of the untreated sample should be equal. This situation has been noticed in many cases of alkane oxidation. If the concentrations of *ol* and *one*, determined by GC, in the untreated sample are different, it testifies that the real amounts of products are not equal and either the alkyl hydroperoxide decomposes to produce predominantly the alcohol (or, on the contrary, the ketone) or there is an additional channel leading to cyclohexanol or cyclohexanone. This third simplified (without determination of ω) method can give relatively precise values of the real concentrations of the alkyl hydroperoxide (as well as cyclohexanol and cyclohexanone) only if all conditions mentioned above are valid. For example, if the chromatogram of a solution before the reduction exhibits two peaks of approximately equal area for cyclohexanol and cyclohexanone, and after the reduction only cyclohexanol is determined by the GC, these data testify that only cyclohexyl hydroperoxide is present in the solution and its concentration can be determined precisely.

It should be noted, that the approximate equality of alcohol and ketone yields determined in the untreated sample generally by no means can be interpreted as evidence for the presence of the alkyl hydroperoxide in the reaction solution. The alkyl hydroperoxide may be decomposed in the course of a reaction to produce approximately equal amounts of alcohol and ketone; this equality could, in certain cases, arise through the coincidence of competitive reactions. Only on the basis of the comparison of the chromatograms of the initial and treated with PPh_3 samples, one may testify to the formation of an alkyl hydroperoxide as well as estimate its concentration. In any case, the difference between the chromatogams of the reaction solution samples before and after the reduction with PPh_3 can unambiguously certify the formation of an alkyl hydroperoxide in the course of the reaction. In order to determine the concentrations of all components of the reaction mixture in the oxidation of branched alkanes, it is necessary to measure the ratio ketone:alcohol (coefficient ω) separately for the decomposition of each isomer of the alkyl hydroperoxide formed. Principally, it is possible to do this if the reaction time is short enough and only isomeric hydroperoxides are present in the solution.

In some cases, using a quartz-lined injector and quartz columns in the GC, it is possible to find peaks due to alkyl hydroperoxides. These peaks disappear completely after the treatment of the solution with PPh_3, while peaks of the corresponding alcohols grow [7].

X.1.B. METAL-CATALYZED OXIDATIONS WITH H_2O_2

Various metal complexes catalyze efficient oxygenations of organic compounds with hydrogen peroxide (see, e.g., [8]). Tables X.1 and X.2 summarize some examples of hydrocarbon oxidations with hydrogen peroxide catalyzed by transition metal complexes.

Mechanisms of metal-catalyzed oxidations by hydrogen peroxide may be different for alkanes and arenes as well as for different metal complexes. For example, for the oxidation of alkanes by the complex $[Ru(dmp)_2(S)_2](PF_6)_2$ (where S = MeCN or H_2O) a mechanism analogous to the "oxygen rebound" radical mechanism, assumed for cytochrome P450 and its models (see Chapter XI), has been proposed (Scheme X.1) [13b].

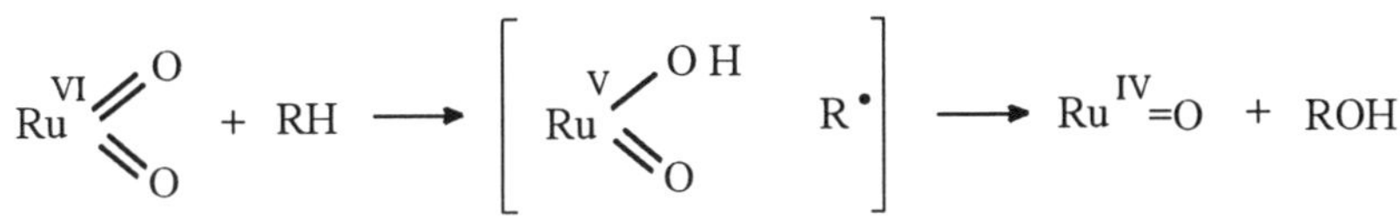

Table X.1. Oxidation of saturated hydrocarbons or fragments with hydrogen peroxide catalyzed by metal complexes.

Substrates	Catalyst	Ref.
Alkanes, arylalkanes	Mn-porphyrin + imidazole	9
Alkanes	Mn-porphyrin + carboxylic acid + heterocyclic base	10a
Adamantane	Mn-porphyrin	10b
Cyclohexane, ethylbenzene	CrO_3, $(^nBu_4N)_2CrO_4$, $(^nBu_4N)_2Cr_2O_7$, $(^nBu_4N)_2Cr_4O_{13}$	11
Cycloalkanes	Iron salts	12
Methane	*cis*-$[Ru(dmp)_2S_2](PF_6)_2$[a]	13a
Alkanes	*cis*-$[Ru(dmp)_2S_2](PF_6)_2$[a]	13b
Alkylaromatics	Polyoxometalates	14a
Alkylaromatics	$H_5PV_2Mo_{10}O_{40}$	14b
Alkanes	$[\gamma\text{-}SiW_{10}\{Fe(OH_2)\}_2O_{38}]^{6-}$	14c
Arylalkanes	$PZnMo_2W_9O_{39}{}^{5-}$	14d
Alkanes	$(Bu_3SnO)_2CrO_2$	14e
Alkanes	$FeSO_4$, $Pd(OCOCF_3)_2$, CrO_3, LMnCl[b] + imidazole	5
Alkanes	V derivative + pyrazine-2-carboxylic acid	5, 7, 15
Cyclohexane	$VO(PA)_2$[c] encapsulated into NaY zeolite	16a
Cyclohexane	Cu(II) complexes	17
Cyclohexane, ethylbenzene	μ-Oxo diferric complexes	18a
Alkanes	Iron(II) complexes	18b,c
Cyclohexane	Peroxide adduct of Fe(III)	18d
Cyclohexane	Fe-porphyrin	18e
n-Hexane	Zeolite TS-1	19a,b
Cyclohexane	Ti silicate	19c
Toluenes	Ce(III) + Br^-	20a
Toluenes	Co(II) acetate + bromide source	20b
Arylalkanes	Cu(II) nitrate on clay	21
Alkanes	$[L_2Mn_2O_3](PF_6)_2$ (**X-1**)[d] + carboxylic acid	16p,q, 22

[a] dmp = 2,9-dimethyl-1,10-phenanthroline, S = H_2O or MeCN. [b] L = tetrakis(2,3,4,5,6-pentafluorophenyl)porphyrinate. [c] PA = picolinic acid anion. [d] L = 1,4,7-trimethyl-1,4-7-triazacyclononane.

Table X.2. Oxidation of arenes with hydrogen peroxide catalyzed by metal complexes.

Substrates	Catalyst	Ref.
Benzene	α-Pyrrolidonate-bridged tetranuclear Pt complexes	23a
Arenes	$(P\text{–}P)Pt(CF_3)X$ [a]	23b
Benzene	Iron salts	24
Methylnaphthalene	CH_3ReO_3	25a
Arenes	CH_3ReO_3 + pyrazine-2-carboxylic acid	25b
Arenes	CH_3ReO_3	25c
Arenes	$VO(O_2)(PA)(H_2O)_2$ [b]	26a
Phenol	Ti silicalite	27
Phenol	Metallosilicalite xerogels	28
Arenes	Ti silicate molecular sieve	29
Phenol	Fe(II)–1,10-phenanthroline	30
Benzene	Hexagonal silicas	31
Benzene	$[PMo_{11}VO_{40}]^{4-}$	32

[a] P-P = tetraaryldiphosphine, X = OH, OPh. [b] PA = picolinic acid anion.

Analogously, the oxidation of alkanes catalyzed by manganese(III) tetra-arylporphyrins in the presence of carboxylic acids and heterocyclic bases can involve the formation of a Mn(V) derivative and proceed via the oxygen rebound mechanism (Scheme X.2) [10a].

Meanwhile, the mechanism proposed for the hydroxylation of aromatics catalyzed by cationic complexes of platinum(II) involves an electrophilic metalation of the aromatic ring to yield platinum–aryl intermediates followed by oxygen transfer from a platinum-hydroperoxy species (Scheme X.3) [23b]. Finally, the oxidation [26a] of aromatic compounds by hydrogen peroxide catalyzed by the peroxovanadium complex $VO(O_2)(Picolinate)(H_2O)_2$ is proposed to occur via oxygenation of the arene by this complex, which is restored under the action of H_2O_2.

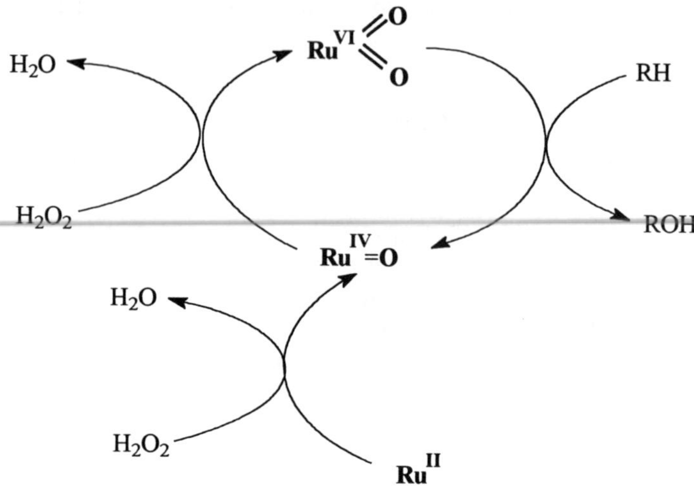

Scheme X.1. The mechanism proposed for the H_2O_2 oxidation of alkanes, RH, catalyzed by a Ru complex.

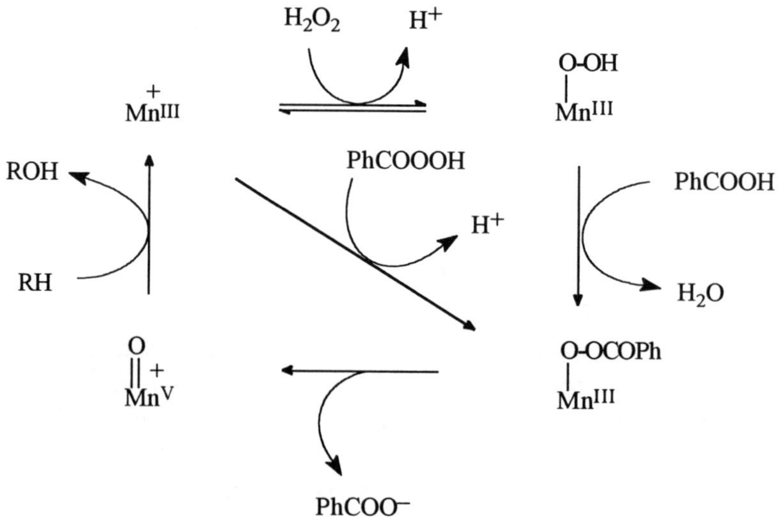

Scheme X.2. The mechanism proposed for the H_2O_2 oxidation of alkanes, RH, catalyzed by Mn(III) porphyrins.

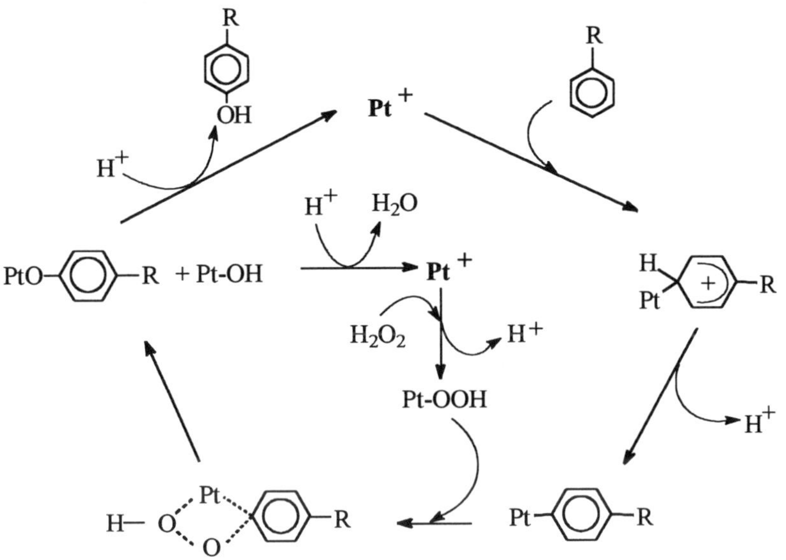

Scheme X.3. The mechanism proposed for the H_2O_2 oxidation of aromatic compounds catalyzed by cationic Pt complexes.

An efficient oxidation of alkanes by the reagent "O_2 – H_2O_2 – vanadium complex – pyrazine-2-carboxylic acid" has been described [15] (the kinetics of the cyclohexane oxidation are shown in Figure X.1). Any soluble vanadium derivative [vanadate-anion in the form ^{n}Bu$_4$NVO$_3$, or $VOSO_4$, VCl_3, VO(acac)$_2$] can be used as a catalyst. The reagent also oxygenates arenes to phenols, alcohols to ketones and hydroperoxidizes the allylic position in olefins. At low temperatures in acetonitrile, the predominant product of alkane oxidation is the corresponding alkyl hydroperoxide (alcohols and ketones or aldehydes are formed simultaneously in smaller amounts). This alkyl hydroperoxide then slowly decomposes to produce the corresponding ketone and alcohol. The amounts of alkyl hydroperoxide, alkanol and alkanone were estimated by comparing the data of chromatographic analysis of the solution before and after reduction with triphenylphosphine (see above), as well as by directly measuring the intensities of peaks corresponding to ROOH [15]. It has been demonstrated that atmospheric

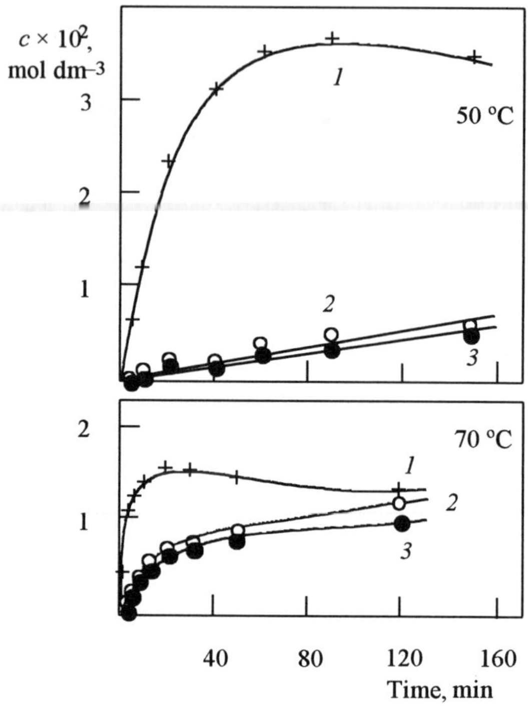

Figure X.1. Kinetic curves of cyclohexyl hydroperoxide (*1*) formation and its decomposition to yield cyclohexanol (*2*) and cyclohexanone (*3*) in the oxidation of cyclohexane (0.46 mol dm^{-3}) by the reagent 'O_2 – H_2O_2 – vanadium complex – pyrazine-2-carboxylic acid' in MeCN at 50 (top) and 70 °C (bottom).

oxygen takes part in this reaction; in the absence of air the oxygenation reaction does not proceed. The cyclohexane oxidation under an $^{18}O_2$ atmosphere unambiguously showed a high degree of ^{18}O incorporation into the oxygenated products. Thus it may be concluded that in alkane oxidation, hydrogen peroxide plays the role of a promoter while atmospheric oxygen is the true oxidant. The oxidation of *n*-heptane by the reagent under consideration exhibits low selectivity, C(1) : C(2) : C(3) : C(4) ≈ 1 : 4 : 4 : 4. This parameter is close to that found for the oxidation of *n*-heptane by H_2O_2 in MeCN under UV irradiation (1.0

: 3.4 : 3.2 : 3.0). It is interesting that the oxidation of higher linear alkanes yields some amounts of C–C bond splitting (lower alcohols, especially methanol). The selectivities of the reactions with branched alkanes (2- and 3-methylhexane, *cis*- and *trans*-decalin) are also very similar to those observed with hydroxylation of the alkanes with hydrogen peroxide under UV irradiation. Methane, ethane, propane, *n*-butane and isobutane can be also readily oxidized in acetonitrile by the same reagent. In addition to the primary oxidation products (alkyl hydroperoxides), alcohols, aldehydes or ketones, and carboxylic acids are obtained with high total turnover numbers (at 75 °C after 4 h: 420 for methane and 2130 for ethane) and H_2O_2 efficiency. Methane can also be oxidized in aqueous solution, giving in this case methanol as the product (after 20 h at 20 °C the turnover number equals 250).

It has been proposed that the crucial step of the oxidation by the reagent "O_2 – H_2O_2 – VO_3^- – pyrazine-2-carboxylic acid" is the very efficient generation of HO• radicals [15]. These radicals abstract a hydrogen atom from the alkane, RH, to generate the alkyl radical, R•. The latter reacts rapidly with an O_2 molecule affording the peroxo radical, ROO•. This radical is then transformed simultaneously into three products: alkyl hydroperoxide, ketone, and alcohol. The relative content of the last two products is increased if the reaction temperature is higher.

The proposed mechanism involves the reduction of V(V) by the first molecule of H_2O_2 to give a V(IV) derivative. The possible role of pyrazine-2-carboxylic acid is its participation (in a zwitter-ionic form) in the proton transfer which gives the hydroperoxy derivative of vanadium (Scheme X.4) [15p,q]. No oxidation occurs in the absence of pyrazine-2-carboxylic acid.

Picolinic acid also accelerates the H_2O_2 oxidations but less efficiently than pyrazine-2-carboxylic acid. It has been demonstrated recently that the vanadium complex with picolinic acid, $VO(PA)_2$, encapsulated into the NaY zeolite retains solution-like activity in the liquid-phase oxidation of hydrocarbons [16a]. It is noteworthy that pyrazine-2-carboxylic acid accelerates the hydrocarbon oxidation catalyzed by CH_3ReO_3 [25b]. Employing a (+)-camphor derived pyrazine-2-carboxylic acid as a potential co-catalyst in the CH_3ReO_3-catalyzed oxidation of methyl phenyl sulfide with urea-H_2O_2 adduct, the corresponding sulfoxide was obtained with an e.e. of 15% [16b].

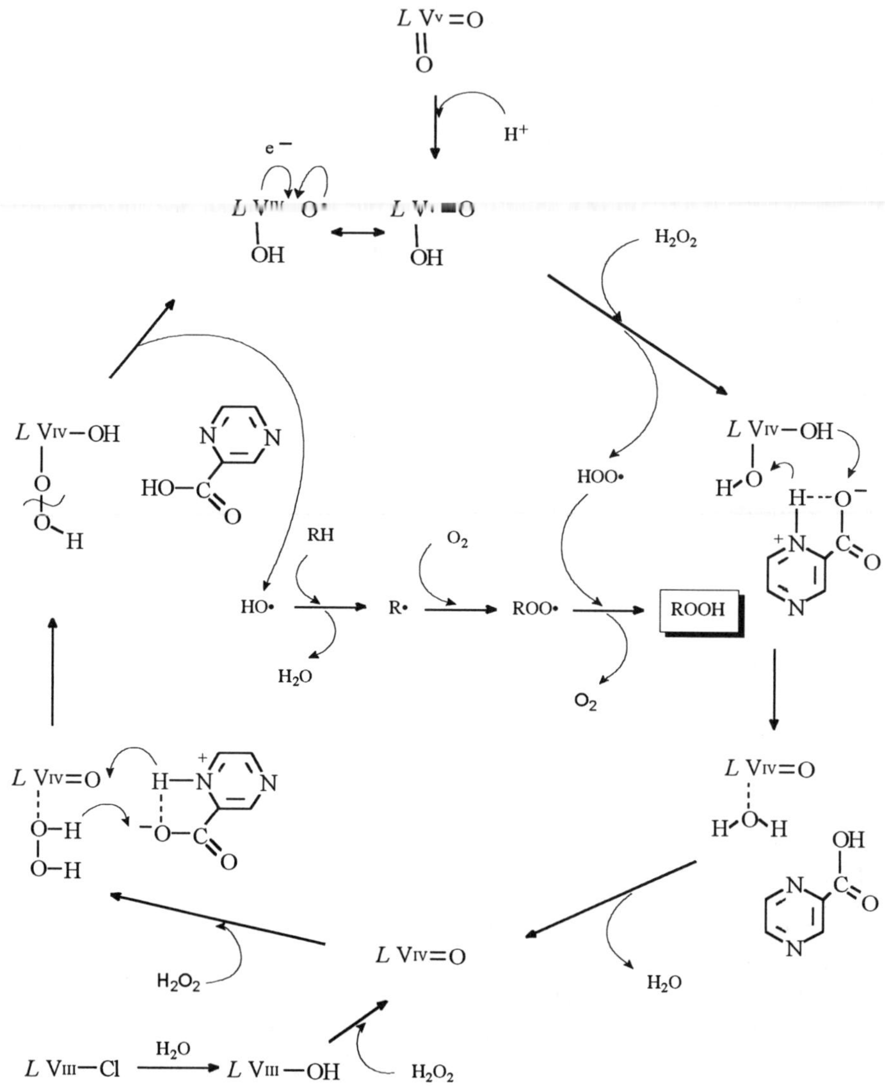

Scheme X.4. The simplified mechanism proposed for the oxidation of alkanes, RH, by the reagent "O_2 – H_2O_2 – vanadium complex – pyrazine-2-carboxylic acid".

It is interesting that oxidations of alkanes by Gif systems (see Chapter IX, as well as certain recent publications [33]) can afford in pyridine–acetic acid mixture, not only products of ketonization but also carboxylic acids (PA = picolinate):

$$RH + Fe(PA)_3 + P(OMe)_3 + CO + H_2O_2 \longrightarrow RCOOH \qquad [34a]$$

$$RH + Fe(CO)_5 + H_2O_2 \longrightarrow RCOOH \qquad [34b]$$

and products of dehydrogenation and peroxidation (Scheme X.5) [35].

$$Fe^{III} + H_2O_2 \longrightarrow Fe^{III}\text{–O–OH} \xrightarrow{H^+} Fe^{V}{=}O + H_2O \xrightarrow{H_2O_2} HO\text{–}Fe^{V}\text{–O–OH}$$

$$HO\text{–}Fe^{V}\text{–O–OH} \longrightarrow Fe^{III}\text{–OH} + O_2 + H^+$$

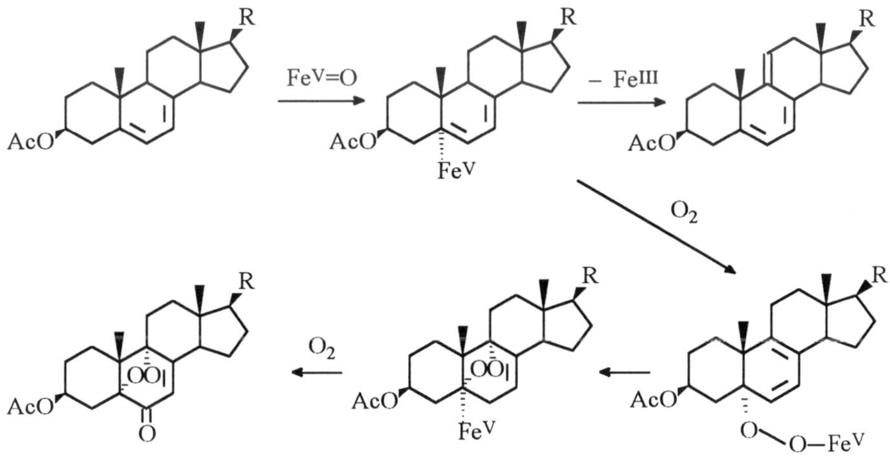

Scheme X.5. The mechanism proposed for the peroxidation of ergosterol acetate by H_2O_2–$FeCl_3$ in pyridine.

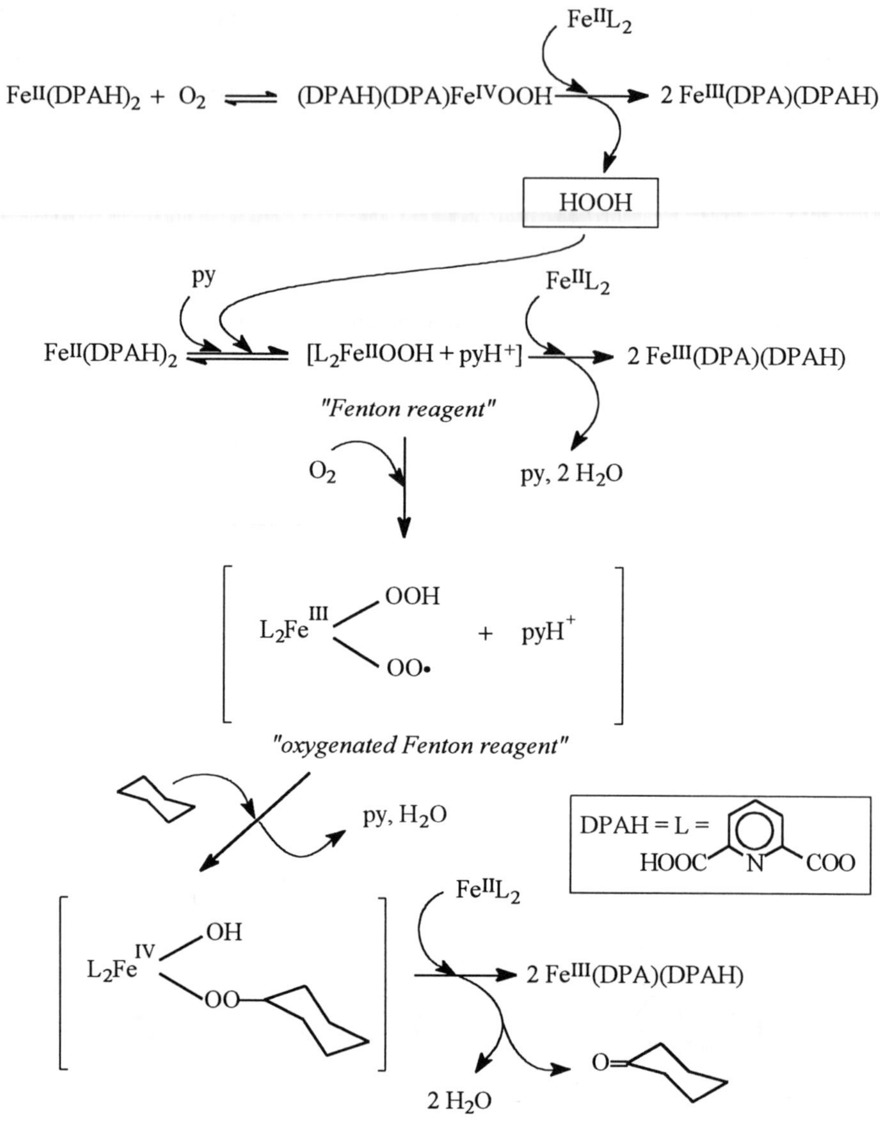

Scheme X.6. The mechanism proposed for the ketonization of alkanes, RH, catalyzed by iron complexes ("oxygenated Fenton chemistry").

Sawyer *et al.* have developed "oxygenated Fenton chemistry": the iron-, cobalt- and copper-induced activation of hydrogen peroxide (along with dioxygen) for the oxidation of organic substances, especially alkanes [36]. The main feature of these reactions is the predominant formation of ketones from cycloalkanes. The ketonization is carried out in a pyridine–acetic acid solution and so is relevant to the oxidation by Gif-type systems. For example, it has been found that the combination of $Fe(DPAH)_2$ ($DPAH_2$ = 2,6-dicarboxypyridine) and dioxygen in a pyridine–acetic acid (2 : 1) mixture results in rapid autoxidation to produce hydrogen peroxide and the complex Fe(DPA)(DPAH). The hydrogen peroxide thus formed reacts with an excess of the starting iron(II) compound to yield "a Fenton reagent", $[(DPAH)_2–FeOOH + pyH^+]$. This species adds dioxygen and then attacks a cyclohexane molecule to produce cyclohexanone as shown in Scheme X.6. It should be noted, however, that according to the results obtained by Ingold *et al.* [37a] and Walling [37b], "oxygenated Fenton chemistry" in organic solvents (and by implication in water) involves simple *free*-radical-mediated chemistry. It is not radical-free as Sawyer has suggested [37a].

Recently, a highly efficient alkane oxidation has been described [15p,q, 22] for the system consisting of hydrogen peroxide, the manganese(IV) salt $[L_2Mn_2O_3](PF_6)_2$ (**X-1**) (L = 1,4,7-trimethyl-1,4-7-triazacyclononane) as the catalyst, and a carboxylic acid as an obligatory component of the reaction mixture (Scheme X.7).

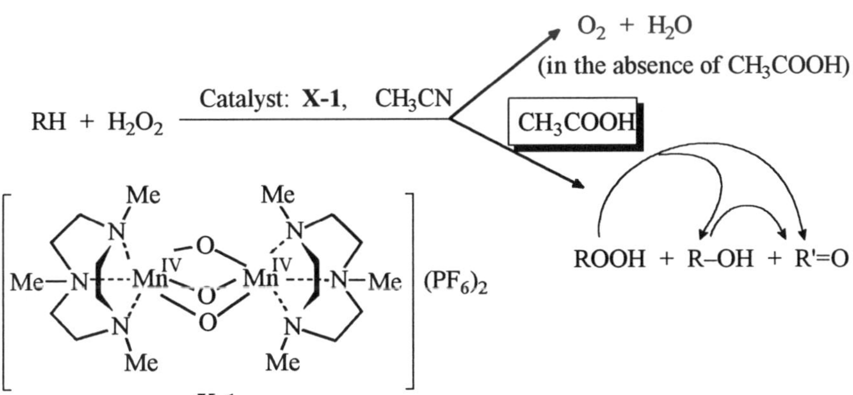

X-1

Scheme X.7. Different routes of the alkane oxidation by H_2O_2 catalyzed by complex **X-1** in the absence and presence of acetic acid.

Higher and light (methane, ethane, propane, normal butane and isobutane) alkanes can be easily oxidized by this system at room temperature, at 0 °C and even at –22 °C if acetonitrile (or nitromethane) is used as a solvent. Turnover numbers of 3300 have been attained and the yield of oxygenated products is 46% based on the alkane. The oxidation initially affords the corresponding alkyl hydroperoxide as the predominant product, however this compound decomposes later to produce the corresponding ketones and alcohols (see Figure X.2 [15p,q]).

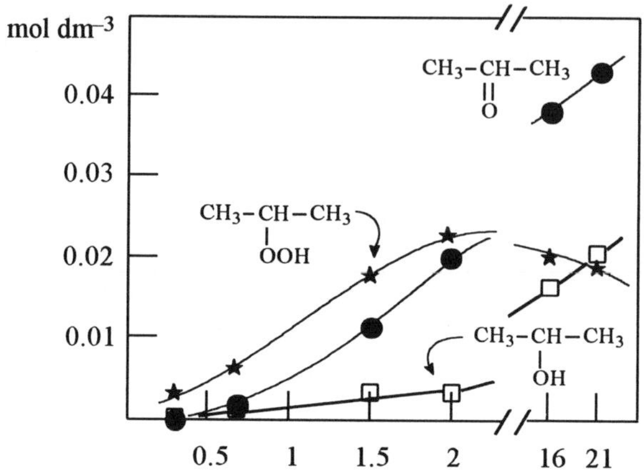

Figure X.2. Kinetic curves of the formation of 2-propyl hydroperoxide, 2-propanol, and acetone from the oxidation of propane (1 bar) by the system "H_2O_2 (0.5 mol dm^{-3}) – complex **X-1** (1.0×10^{-4} mol dm^{-3}) – acetic acid (0.25 mol dm^{-3})" in MeCN at 0 °C.

Regio and bond selectivities of the reaction are high: C(1) : C(2) : C(3) : C(4) ≈ 1 : 40 : 35 : 35 and 1° : 2° : 3° is 1 : (15–40) : (180–300). The reaction with *cis*- or *trans*- isomers of decalin gives (after treatment with PPh_3) alcohols hydroxylated in the tertiary position with a *cis*/*trans* ratio of ~2 in the case of *cis*-decalin, and of ~30 in the case of *trans*-decalin (i.e., in the latter case the reaction is stereospecific). The authors [15p,q, 22] believe that the alkane oxidation by the system under consideration begins with hydrogen atom abstraction from the alkane by an oxygen-centered radical or radical-like species.

The active oxidant is probably a manganese complex, possibly an oxomanganese species, and the reaction occurs in a cage or via an "oxygen-rebound mechanism", both these routes would afford products with retention of stereochemistry. Further, alkyl radicals (R•) that escape from the solvent cage react with dioxygen to generate ROO• and subsequently ROOH with some loss of stereochemistry. An alternative route, abstraction of the alkane H atom by a species Mn–OO• to produce R• and Mn–OOH, followed by the interaction of the two latter particles to give ROOH with retention of configuration cannot be excluded.

X.2. OXYGENATIONS BY ALKYL HYDROPEROXIDES

The oxidation of saturated hydrocarbons by alkyl hydroperoxides catalyzed by various metal complexes yields alcohols, carbonyl compounds (ketones and aldehydes) and peroxides. Examples of recent works devoted to these reactions are given in Table X.3. Usually *tert*-butyl hydroperoxide is employed for these oxidations. Other hydroperoxides, e.g., cumyl hydroperoxide [38a, 42a], have been also reported to oxygenate alkanes.

Scheme X.8 illustrates the proposed [53] mode of alkane oxidation with both peroxo and oxo groups bound to chromium(VI). This scheme also shows a possible way to regenerate an active Cr(VI) species starting from Cr(IV).

It has been shown [54] that the complex $VO(acac)_2$, under the action of alkyl hydroperoxide, ROOH, is initially transformed into the alkylperoxo complex $VO(acac)_2OOR$ and the alkoxo complex $VO(acac)_2OR$, which further decomposes to yield $VO(OR)_3$ and two other vanadium(V) species (an alkylperoxo complex **X-2** and an alkoxo complex **X-3**). Complex **X-2** does not react directly with cyclohexane, but produces a free radical:

$$V^{V}O(OOR) \rightleftharpoons V^{IV}O + ROO\bullet$$

The peroxo radical thus formed initiates the oxidation of cyclohexane.

The complex $[Fe(PMA)]^{2+}$ also efficiently catalyzes the oxidation of alkanes with *tert*-butyl hydroperoxide in acetonitrile at room temperature [55]. The mechanism proposed by the authors for this reaction involves the formation of free alkyl radicals and their subsequent reaction with O_2 (Scheme X.9). The

CyOO• radical may be involved in a Russell-type termination which produces cyclohexanol and cyclohexanone in equal amounts, along with molecular oxygen:

$$2\ R_2CHOO\bullet \rightleftharpoons R_2CHOOOOHCR_2 \longrightarrow R_2CHOH + R_2C{=}O + O_2$$

Table X.3. Hydrocarbon oxidation with *tert*-butyl hydroperoxide catalyzed by metal complexes.

Substrates	Catalyst	Ref.
Alkanes	$RuCl_2(PPh_3)_3$	38a
Cyclohexane	$[RuLO_2(CF_3CO_2)]ClO_4$ [a]	38b
Alkanes	$K_5[Ru(H_2O)PW_{11}O_{39}]$	39a
Cyclohexane, adamantane	Transition-metal-containing heteropolycomplexes	39b
Benzylic C-H	CrO_3	40
Cyclohexane	Co(III) complexes	41a
Nitriles	$VO(OAc)_2$	41b
Cyclohexane	Fe(TPA) [b] complexes	42
Cyclohexane	Fe biomimetic complexes	43
Alkanes, arylalkanes	Gif conditions	44
Cyclohexane	Binuclear Fe and Mn oxo-bridged complexes	45
Cyclohexane	Fe(III) and Cu(II) complexes	46a
Δ^5-Steroids	Cu(II) and Cu(I) salts and Cu metal	46b
Cyclooctane	Cu and Co complexes of tetraazacyclotetradecane bearing highly fluorinated tails	46c
Cyclooctane, adamantane	Mn carboxylates	47a
Alkanes	$[Mn_3O_4(dipy)_4(OH_2)_2](ClO_4)_4$	47b
Cyclohexane etc.	$[Mn(bipy)_3](BF_4)_2$	47c
Alkanes	Mn clusters	47d
Ethane	$Mn_4O_2(O_2CPh)_7(bipy)_2$	47e
Alkanes, arylalkanes	Polynuclear Mn Schiff base complexes	48
Cyclohexane	Binuclear Mn and Fe complexes	49
Cycloalkanes	Mn^{2+}-exchanged clay catalyst	50
Benzylic C-H	VAPO-5	51
Cyclooctane	Ru colloids	52

[a] L = *N,N',N''*-trimethyl-1,4,7-triazacyclononane. [b] TPA = tris(2-pyridylmethyl)amine.

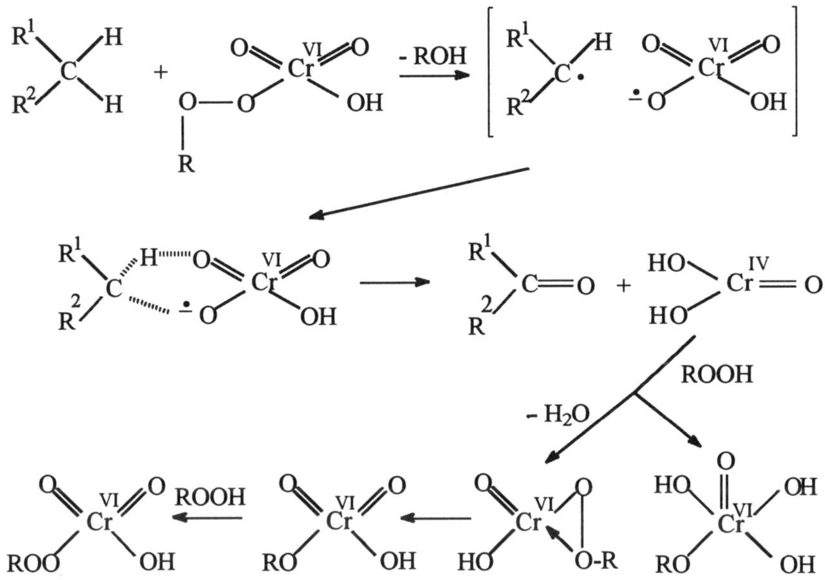

Scheme X.8. The alkane oxidation by alkyl hydroperoxide catalyzed by chromium oxo complexes.

In conclusion, it is useful to cite the papers by Meunier [56]:

> "...many hydroxylation reactions with alkyl hydroperoxides in the presence of transition-metal complexes are not due to a metal-centered active species, but to a free-radical process initiated by RO•"

and by Ingold *et al.* [37a]:

> "We urge all investigators who would like to claim that *tert*-butyl hydroperoxide-derived high-valent metal-oxo species are the effective oxidizing agents in their systems to check that the mechanistic probe hydroperoxides we have described yield the same results as *tert*-butyl hydroperoxide before they draw any mechanistic conclusions".

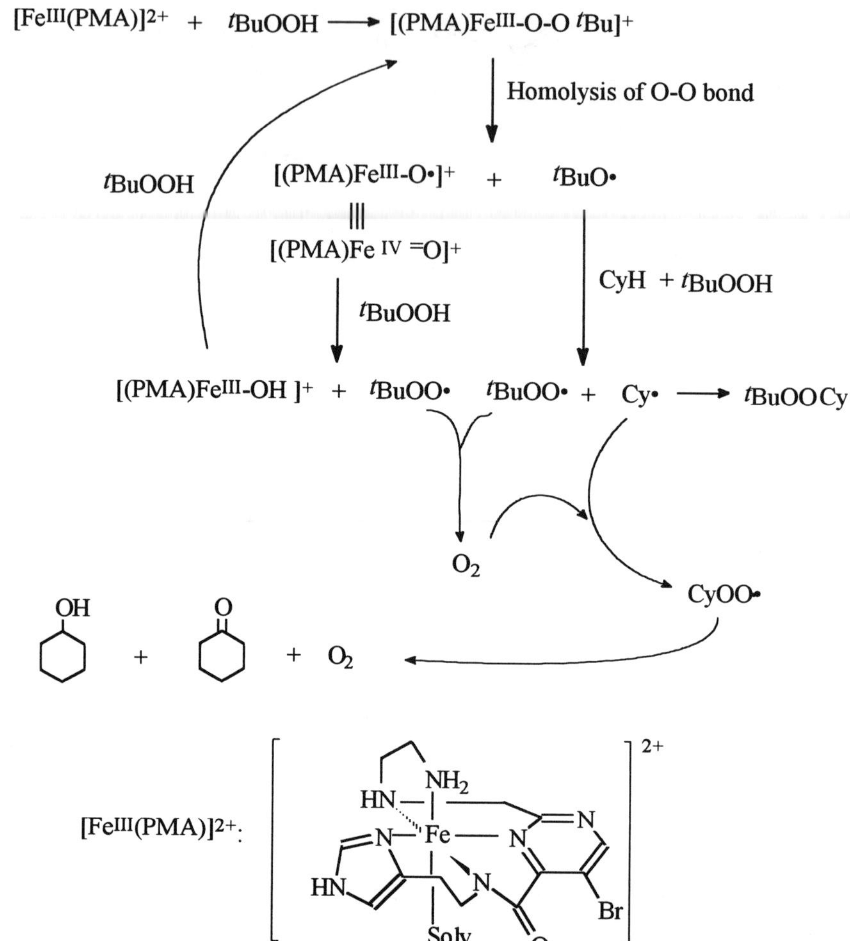

Scheme X.9. The alkane oxidation by alkyl hydroperoxide catalyzed by the iron complex $[Fe(PMA)]^{2+}$.

The formation of free radicals in alkane oxidations by ^{t}BuOOH, for example, via a route:

$$Fe(II) + {}^{t}BuOOH \longrightarrow Fe(III) + {}^{t}BuO\bullet + HO^{-}$$

which are responsible for the free radical initiation process, has also been proposed in the papers by Minisci [44] (for the "Gif catalysis") and Fish [43] (for the oxidations in the presence of "biomimetic methane monooxygenase enzyme complexes").

X.3. OXYGENATIONS BY PEROXYACIDS

Soluble derivatives of various metals catalyze the alkane oxidation by peroxy-acids (or by H_2O_2 in strong acids as solvents). Examples of these reactions are summarized in Table X.4.

The mechanism proposed for the rhodium-catalyzed oxidation of alkanes with H_2O_2 in trifluoroacetic acid is demonstrated in Scheme X.10 [56]. The rhodium oxo derivative is assumed to be a key intermediate in this process.

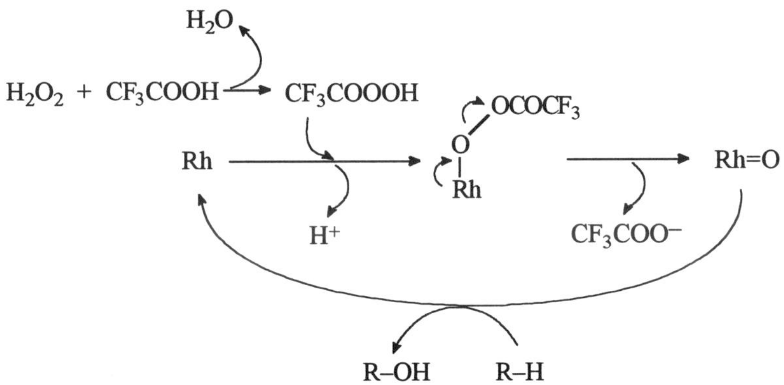

Scheme X.10. Rhodium-catalyzed oxidation of an alkane by H_2O_2 in CF_3COOH.

X.4. OXIDATIONS BY OTHER OXYGEN ATOM DONORS

Some oxygen-containing oxidizing reagents, A=O, can functionalize hydrocarbons in the presence of transition metal complexes (see, for example, [62]). Usually the oxygen atom transfer from an oxidizing reagent to a metal-complex catalyst, M, gives rise to the formation of a high-valent oxo complex.

Table X.4. Hydrocarbon oxidation by peroxyacids catalyzed by metal complexes.

Substrates	Oxidant	Catalyst	Ref.
Alkanes	H_2O_2–CF_3COOH	$[Rh_3O(OAc)_6(H_2O)_3]OAc$ $Rh_2(OAc)_4$, $RhCl_3$, $Rh(acac)_3$, $RhCl(PPh_3)_3$, $Pd(OAc)_2$, $RuCl_3$, H_2PtCl_6, $FeCl_3$	56
Alkanes	Peroxyacetic acid	Ru derivatives	57
Adamantane, cyclooctane, ethylbenzene	*m*-Chloroperoxybenzoic acid	Ru complexes	58
Alkanes	H_2O_2–CF_3COOH	Vanadium derivatives	59
Cyclohexane, methylcyclo-hexane, adamantane	*m*-Chloroperoxybenzoic acid	Fe μ-oxo-di-μ-acetate complex	60
Cyclohexane	Peroxy acids	Binuclear Mn and Fe complexes	49
Alkanes	Peroxyacetic acid, *m*-chloroperoxybenzoic acid	$[L_2Mn_2O_3](PF_6)_2$ (**X-1**) [a] + carboxylic acid	16p,q, 22
Alkanes	H_2O_2–CF_3COOH	Cu complexes	61

[a] L = 1,4,7-trimethyl-1,4-7-triazacyclononane.

This complex then is capable of oxidizing the alkane, RH, for example, via the "oxygen rebound" radical mechanism:

$$A{=}O + M \longrightarrow A + M{=}O$$

$$M{=}O + RH \longrightarrow [M{-}OH \;\; R\bullet] \longrightarrow M + ROH$$

Table X.5. Oxidation of hydrocarbons with iodosylbenzene catalyzed by metal complexes.

Catalyst	Substrate	Ref.
Manganese porphyrins as insoluble zinc phosphonates	Alkanes	63a
Metalloporphyrins; soluble and on imidazole propyl gel	Cyclohexane	63b–d
Iron porphyrin	α-Alkylbenzyl alcohols (to form C–H and C–C bond cleavage products)	64a
Iron porphyrins	Alkanes	64b
Iron porphyrins	Cyclohexane	64c
Iron porphyrin	Cyclohexane	64d
Manganese porphyrin	Cyclohexane	65a
$[Mn_2L_2(\mu\text{-OAc})_2]^{2+}$ (L: *N,N'*-dimethyl-*N,N'*-bis(2-pyridilmethyl)ethane-1,2-diamine	Cyclohexane, adamantane 2,3-dimethylbutane	65b
Mn-Salen complex	Benzylic C–H bond	66a
Chiral Mn-Salen complexes	*Meso*-Pyrrolidine derivatives	66b
Chiral Mn-Salen complexes	Benzylic C–H bond	66c
Chiral Mn-Salen complexes	Alkanes	66d
Racemic Mn-Salen complexes	Benzylic C–H bond	66e
Ru(III) diphosphino complexes	Adamantane, hexane	67a
$P_2W_{17}O_{61}(M^{n+}Br)^{(n-11)}$ (M^{n+} = Mn^{3+}, Fe^{3+}, Co^{2+}, Ni^{2+}, Cu^{2+})	Cyclohexane, heptane, adamantane	67b
CrO_3, $VOCl_2$	Cyclohexane, ethylbenzene	67c

Table X.6. Oxidation of hydrocarbons with monopersulfate (HSO_5^-) catalyzed by metal complexes.

Catalyst	Substrate	Ref.
Mn porphyrins	Ethylbenzene	68a
Metalloporphyrins	4-(1-Hydroxy-1-methylethyl)benzoic acid	68b
Mn porphyrins	Alkanes	68c
Mn, Fe porphyrins	2-Methylnaphthalene	68d
$SiRu(H_2O)W_{11}O_{39}^{5-}$	Adamantane, cyclohexane	69a
Ru complexes	Ethylbenzene	69b

Examples of hydrocarbon oxidations by some oxygen atom donors which are catalyzed by metal complexes are given in Tables X.5, X.6, and X.7.

Highly efficient oxygenation reactions have been described by Higuchi, Hirobe and Nagano [74]. Alkanes and other C–H compounds can be oxidized by aromatic *N*-oxides (usually by 2,6-dichloropyridine *N*-oxide) under catalysis with ruthenium porphyrin complexes in the presence of strong mineral acids. The oxidation of adamantane under mild conditions gives a turnover number of up to 120000. The reaction occurs via the mechanism depicted in Scheme X.11 [74b] and allows for the hydroxylation of steroids with retention of configuration at the asymmetric center, giving novel steroids (Scheme X.12) [74d].

It is important to note that alkane oxidation reactions by various oxygen atom donors in cases when a metal-complex catalyst is a metalloporphyrin (or even any other complex) can be considered to be a model for biological hydrocarbon oxidation. The next chapter will be fully devoted to the oxidations of alkanes and arenes in living cells and modeling these processes using metal complexes.

Table X.7. Oxidation of hydrocarbons with various oxygen atom donors catalyzed by metal complexes.

Oxygen atom donor	Catalyst	Substrate	Ref.
Sodium percarbonate ($Na_2CO_3 \cdot 1.5H_2O_2$)	(n-$Bu_3SnO)_2CrO_2$	Benzylic methylene groups	70a
Magnesium monoperoxyphthalate	Mn porphyrins	Alkanes	70b
Magnesium monoperoxyphthalate	Fe porphyrin	1,2-Dimethoxyarenes	70c
$NaIO_4$	$SiRu(H_2O)W_{11}O_{39}^{5-}$	Adamantane, cyclohexane	71a
$LiClO_4$	Ru(II) complexes	Adamantane, decane, cyclohexane	71b
$HClO_4$	Polyphenylferrosiloxane	1,4-Dimethylcyclohexane	71c
Hypochlorite	$RuCl_3$	Substituted xylenes	72a
$KBrO_3$	Ce salts	Alkyl benzenes	72b
Me_3NO	L_2Cu [a]	Aromatics	73
2,6-Dichloropyridine *N*-oxide	Ru porphyrins	C–H Compounds	74
2,6-Dichloropyridine *N*-oxide	Ru pentafluoro phenylporphyrin	Alkanes	75
(Ad=Ad + $^1O_2$) [b]	metalloporphyrins	Adamantane, cyclohexane	76a
XeO_3	CrO_3; Mn, Fe porphyrins	Benzene, alkylbenzenes	76b
Ozone	Fe porphyrins	Ethylbenzene	76c

[a] L: 2-(*N*-amido)-4-nitrophenolate. [b] Ad=Ad: adamantalydeneadamantane; $^1O_2$: singlet oxygen.

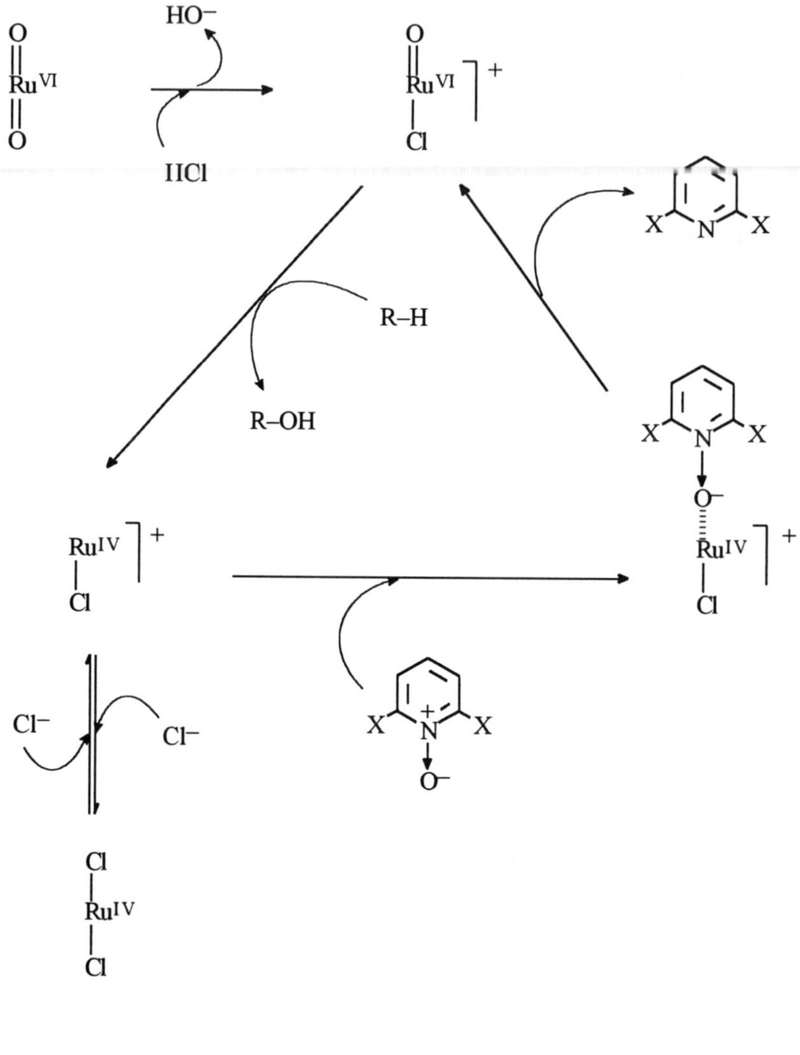

Scheme X.11. The mechanism proposed for the hydroxylation of alkanes, RH, by aromatic *N*-oxides under catalysis with ruthenium porphyrin complexes in the presence of strong mineral acids.

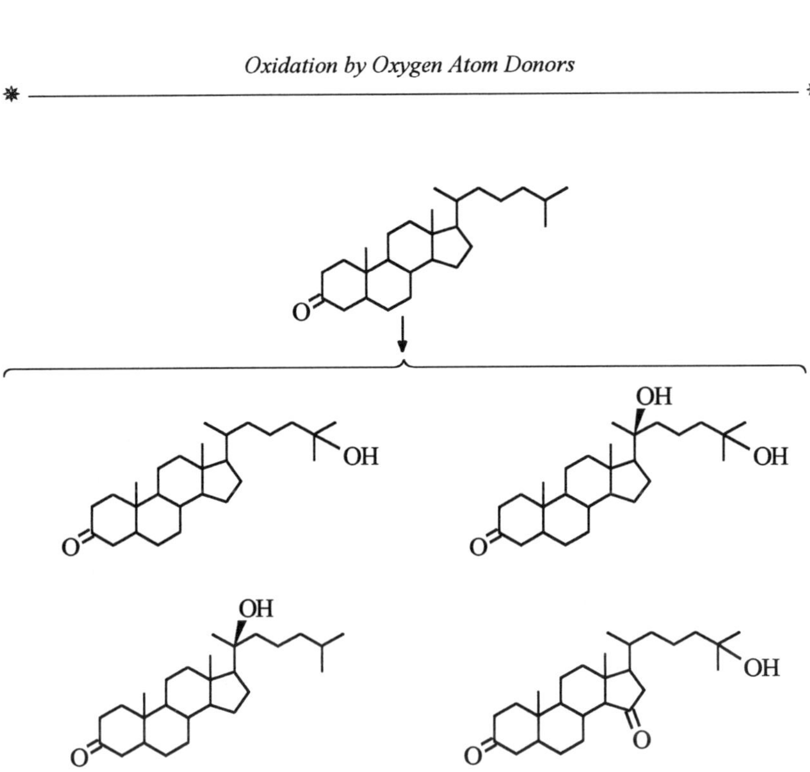

Scheme X.12. The hydroxylation of 5α-cholestan-3-one by 2,6-dichloropyridine *N*-oxide catalyzed with ruthenium porphyrin.

References for Chapter X

1. Karakhanov, E. A.; Narin, S. Yu.; Dedov, A. G. *Appl. Organomet. Chem.* **1991**, *5*, 445.

2. (a) Kurata, F.; Watanabe, Y.; Katoh, M.; Sawaki, Y. *J. Am. Chem. Soc,* **1988**, *110*, 7472. (b) Maskos, Z.; Rush, J. D.; Koppenol, W. H. *Free Radical Biol. Med.* **1990**, *8*, 153. (c) Zakharov, I. V.; Kumpan, Yu. V. *Kinet. Katal.* **1996**, *37*, 190 (in Russian). (d) Litvintsev, I. Yu.; Timofeev, S. V.; Sapunov, V. N. *Kinet. Katal.* **1996**, *37*, 209 (in Russian). (e) Van Dyke, B. R.; Clopton, D. A.; Saltman, P. *Inorg. Chim. Acta* **1996**, *242*, 57. (f) Park, J. S. B.; Wood, P. M.; Davies, M. J.; Gilbert, B. C.; Whitwood, A. C. *Free Rad. Res.* **1997**, *27*, 447. (g) Bossmann, S. H.; Oliveros, E.; Göb, S.; Siegwart, S.; Dahlen, E. P.; Payawan, L., Jr.; Straub, M.; Wörner, M.; Braun, A. M. *J. Phys. Chem. A* **1998**, *102*, 5542.

3. (a) *New Developments in Selective Oxidation*; Centi, G., Trifiro, F., Eds.; Elsevier: Amsterdam, 1990. (b) Sawyer, D. T. *Oxygen Chemistry*; Oxford University Press: Oxford, 1991. (c) *Dioxygen Activation and Homogeneous Catalytic Oxidation*; Simándi, L. I., Ed.; Elsevier: Amsterdam, 1991. (d) *Catalytic Oxidations with Hydrogen Peroxide as Oxidant*; Strukul, G., Ed.; Kluwer: Dordrecht, 1992. (e) *Catalytic Selective Oxidation*; Oyama, S. T.; Hightower, J. W., Eds.; ACS: Washington, 1993. (f) *The Activation of Dioxygen and Homogeneous Catalytic Oxidation*; Barton, D. H. R.; Martell, A. E., Sawyer, D. T., Eds.; Plenum Press: New York, 1993. (g) Simándi, L. I. *Catalytic Activation of Dioxygen by Metal Complexes*; Kluwer: Dordrecht, 1992.

4. (a) Luňák, S.; Sedlák, P.; Lederer, P. *Chem. Listy* **1989**, *83*, 785 (in Czech). (b) Kislenko, V. N.; Berlin, A. A. *Usp. Khim.* **1991**, *60*, 949 (in Russian). (c) Luňák, S.; Sedlák, P. *J. Photochem. Photobiol. A: Chem.* **1992**, *68*, 1. (d) Sobkowiak, A.; Tung, H.-C.; Sawyer, D. T. *Progr. Inorg. Chem.* **1992**, *40*, 291. (e) Drago, R. S.; *Coord. Chem. Rev.* **1992**, *117*, 185. (f) Sychev, A. Ya.; Isak, V. G. *Usp. Khim.* **1995**, *64*, 1183 (in Russian).

5. Shul'pin, G. B.; Nizova, G. V. *React. Kinet. Catal. Lett.* **1992**, *48*, 333.

6. (a) Shul'pin, G. B.; Druzhinina, A. N. *React. Kinet. Catal. Lett.* **1992**, *47*, 207. (b) Shul'pin, G. B.; Druzhinina, A. N.; Shul'pina, L. S. *Neftekhimiya* **1993**, *33*, 335 (in Russian); *Petrol. Chem.* **1993**, *33*, 321. (c) Shul'pin, G. B.; Bochkova,

M. M.; Nizova, G. V. *J. Chem. Soc., Perkin Trans. 2* **1995**, 1465. (d) Shul'pin, G. B.; Nizova, G. V.; Kozlov, Yu. N. *New J. Chem.* **1996**, *20*, 1243.

7. Shul'pin, G. B.; Guerreiro, M. C.; Schuchardt, U. *Tetrahedron* **1996**, *52*, 13051.
8. (a) Ishii, Y. Ogawa, M. In *Reviews on Heteroatom Chemistry*; Oae, S., Ed.; MYU: Tokyo, 1990, p. 121. (b) Kodera, M.; Shimakoshi, H.; Tachi, Y.; Katayama, K.; Kano, K. *Chemistry Lett.* **1998**, 441.
9. (a) Battioni, P.; Renaud, J.-P.; Bartoli, J. F.; Mansuy, D. *J. Chem. Soc., Chem. Commun.* **1986**, 341. (b) Battioni, P.; Renaud, J.-P.; Bartoli, J. F.; Reina-Artiles, M.; Fort, M.; Mansuy, D. *J. Am. Chem. Soc.* **1988**, *110*, 8462.
10. (a) Banfi, S.; Maiocchi, A.; Moggi, A.; Montanari, F.; Quici, S. *J. Chem. Soc., Chem. Commun.* **1990**, 1794. (b) Baciocchi, E.; Boschi, T.; Cassioli, L.; Galli, C.; Lapi, A.; Tagliatesta, P. *Tetrahedron Lett.* **1997**, *38*, 7283.
11. Druzhinina, A. N.; Nizova, G. V.; Shul'pin, G. B. *Bull. Acad. Sci. USSR, Div. Chem. Sci.* **1990**, *39*, 194.
12. Briffaud, T.; Larpent, C.; Patin, H. *J. Chem. Soc., Chem. Commun.* **1990**, 1193.
13. (a) Goldstein, A. S.; Drago, R. S. *J. Chem. Soc., Chem. Commun.* **1991**, 21. (b) Goldstein, A. S.; Beer, R. H.; Drago, R. S. *J. Am. Chem. Soc.* **1994**, *116*, 2424.
14. (a) Attanasio, D.; Orru', D.; Suber, L. *J. Mol. Catal.* **1989**, *57*, L1. (b) Neumann, R.; de la Vega, M. *J. Mol. Catal.* **1993**, *84*, 93. (c) Mizuno, N.; Nozaki, C.; Kiyoto, I.; Misono, M. *J. Am. Chem. Soc.* **1998**, *120*, 9267. (d) Tangestaninejad, S.; Yadollahi, B. *Chemistry Lett.* **1998**, 511. (e) Muzart, J.; N'Ait Ajjou, A.; Nizova, G. V.; Shul'pin, G. B. *Bull. Acad. Sci. USSR, Div. Chem. Sci.* **1991**, *40*, 129.
15. (a) Shul'pin, G. B.; Attanasio, D.; Suber, L. *J. Catal.* **1993**, *142*, 147. (b) Shul'pin, G. B.; Attanasio, D.; Suber, L. *Russ. Chem. Bull.* **1993**, *42*, 55. (c) Shul'pin, G. B.; Druzhinina, A. N.; Nizova, G. V. *Russ. Chem. Bull.* **1993**, *42*, 1327. (d) Nizova, G. V.; Shul'pin, G. B. *Russ. Chem. Bull.* **1994**, *43*, 1146. (e) Shul'pin, G. B.; Süss-Fink, G. *J. Chem. Soc., Perkin Trans. 2* **1995**, 1459. (f) Shul'pin, G. B.; Drago, R. S.; Gonzalez, M. *Russ. Chem. Bull.* **1996**, *45*, 2386. (g) Guerreiro, M. C.; Schuchardt, U.; Shul'pin, G. B. *Russ. Chem. Bull.* **1997**, *46*, 749. (h) Schuchardt, U.; Guerreiro, M. C.; Shul'pin, G. B. *Russ. Chem. Bull.* **1998**, *47*, 247. (i) Nizova, G. V.; Süss-Fink, G.; Shul'pin, G. B. *Chem. Comm.* **1997**, 397. (j) Nizova, G. V.; Süss-Fink, G.; Shul'pin, G. B. *Tetrahedron* **1997**, *53*, 3603. (k) Süss-Fink, G.; Hong Yan, Nizova, G. V.; Stanislas, S.; Shul'pin, G. B. *Russ. Chem. Bull.* **1997**, *46*, 1801. (l) Süss-Fink, G.; Nizova, G. V.; Stanislas, S.; Shul'pin, G. B. *J. Mol. Catal. A: Chem.* **1998**, *130*, 163.

(m) Shul'pin, G. B.; Ishii, Y.; Sakaguchi, S.; Iwahama, T. *Russ. Chem. Bull.* **1999**, *48*, 887. (n) Nizova, G. V.; Süss-Fink, G.; Stanislas, S.; Shul'pin, G. B. *Chem. Comm.* **1998**, 1885. (o) Hoehn, A.; Suess-Fink, G.; Shulpin, G. B.; Nizova, G. V. *Pat. Ger. Offen. DE 19,720,344* (to BASF A.-G.) [*Chem. Abstr.* **1998**, *128*: 49802p]. (p) Shul'pin, G. B. Presented at the *217th Am. Chem. Soc. Natl. Meeting*; Anaheim, March **1999**; paper INOR No. 353. (q) Shul'pin, G. B. Presented at the *7th Int. Symp. on Dioxygen Activation and Homogeneous Catalytic Oxidation*; York, UK, July **1999**.

16. (a) Kozlov, A.; Asakura, K.; Iwasawa, Y. *J. Chem. Soc., Faraday Trans.* **1998**, *94*, 809. (b) Gunaratne, H. Q. N.; McKervey, M. A.; Feutren, S.; Finlay, J.; Boyd, J. *Tetrahedron Lett.* **1998**, *39*, 5655.
17. Okuno, T.; Ohba, S.; Nishida, Y. *Polyhedron* **1997**, *16*, 3765.
18. (a) Duboc-Toia, C.; Menage, S.; Lambeaux, C.; Fontecave, M. *Tetrahedron Lett.* **1997**, *38*, 3727. (b) Kim, C.; Chen, K.; Kim, J.; Que, L., Jr. *J. Am. Chem. Soc.* **1997**, *119*, 5964. (c) Okuno, T.; Ito, S.; Ohba, S.; Nishida, Y. *J. Chem. Soc., Dalton Trans.* **1997**, 3547. (d) Ito, S.; Suzuki, M.; Kobayashi, T.; Itoh, H.; Harada, A.; Ohba, S.; Nishida, Y. *J. Chem. Soc., Dalton Trans.* **1996**, 2579. (e) Lee, K. A.; Nam, W. *J. Am. Chem. Soc.* **1997**, *119*, 1916.
19. (a) Kooyman, P. J.; Luijikx, G. C. A.; Arafat, A.; van Bekkum, H. *J. Mol. Catal. A: Chem.* **1996**, *111*, 167. (b) Vankelecom, I.; Vercruysse, K.; Moens, N.; Parton, R.; Reddy, J. S.; Jacobs, P. *Chem. Comm.* **1997**, 137. (c) Reddy, J. S.; Sivasanker, S. *Catalysis Lett.* **1991**, *11*, 241.
20. (a) Auty, K.; Gilbert, B. C.; Thomas, C. B.; Brown, S. W.; Jones, C. W.; Sanderson, W. R. *J. Mol. Catal. A: Chem.* **1997**, *117*, 279. (b) Jones, C. W.; Hackett, A.; Pattinson, I.; Johnstone, A.; Wilson, S. L. *J. Chem. Res. (S)* **1996**, 438.
21. Varma, R. S.; Dahiya, R. *Tetrahedron Lett.* **1998**, *39*, 1307.
22. (a) Shul'pin, G. B.; Lindsay Smith, J. R. *Russ. Chem. Bull.* **1998**, *47*, 2379. (b) Shul'pin, G. B.; Süss-Fink, G.; Lindsay Smith, J. R. *Tetrahedron* **1999**, *55*, 5345. (c) Lindsay Smith, J. R.; Shul'pin, G. B. *Tetrahedron Lett.* **1998**, *39*, 4909.
23. (a) Sakai, K.; Matsumoto, K. *J. Mol. Catal.* **1991**, *67*, 7. (b) Marsella, A.; Agapakis, S.; Pinna, F.; Strukul, G. *Organometallics* **1992**, *11*, 3578.
24. Tagawa, T.; Seo, Y.-J.; Goto, S. *J. Mol. Catal.* **1993**, *78*, 201.
25. (a) Adam, W.; Herrmann, W. A.; Lin, J.; Saha-Möller, C. R.; Fischer, R. W.; Correia, J. D. G. *Angew. Chem.* **1994**, *106*, 2545. (b) Schuchardt, U.; Mandelli,

D.; Shul'pin, G. B. *Tetrahedron Lett.* **1996**, *37*, 6487. (c) Jacob, J.; Espenson, J. H. *Inorg. Chim. Acta* **1998**, *270*, 55.

26. Bianki, M.; Bonchio, M.; Conte, V.; Coppa, F.; Di Furia, F.; Modena, G.; Moro, S.; Standen, S. *J. Mol. Catal.* **1993**, *83*, 107.

27. Romano, U.; Esposito, A.; Maspero, F.; Neri, C.; Clerici, M.G. In *New Developments in Selective Oxidation*; Centi, G., Trifiro, F., Eds.; Elsevier: Amsterdam, 1990, p. 33.

28. Neumann, R.; Chava, M.; Levin, M. *J. Chem. Soc., Chem. Commun.* **1993**, 1685.

29. Bhaumik, A.; Kumar, R. *J. Chem. Soc., Chem. Commun.* **1995**, 349.

30. Liu, C.; Ye, X.; Zhan, R.; Wu, Y. *J. Mol. Catal. A: Chem.* **1996**, *112*, 15.

31. Zhang, W.; Wang, J.; Tanev, P. T.; Pinnavaia, T. J. *Chem. Comm.* **1996**, 979.

32. Nomiya, K.; Yagishita, K.; Nemoto, Y.; Kamataki, T. *J. Mol. Catal. A: Chem.* **1997**, *126*, 43.

33. (a) Barton, D. H. R.; Bardin, C. O.; Doller, D. *Bioorg. and Med. Chem.* **1994**, *2*, 259. (b) Barton, D. H. R.; Beck, A. H.; Taylor, D. K. *Tetrahedron* **1995**, *51*, 5245. (c) Sobolev, A. P.; Babushkin, D. E.; Shubin, A. A.; Talsi, E. P. *J. Mol. Catal. A: Chem.* **1996**, *112*, 253. (d) Barton, D. H. R. *Synlett* **1997**, 229. (e) Barton, D. H. R.; Launay, F.; Le Gloahec, V. N.; Li, T.; Smith, F. *Tetrahedron Lett.* **1997**, *38*, 8491. (f) Singh, B.; Long, J. R.; de Biani, F. F.; Gatteschi, D.; Stavropoulos, P. *J. Am. Chem. Soc.* **1997**, *119*, 7030. (g) Barton, D. H. R.; Li, T. *Tetrahedron* **1998**, *54*, 1735. (h) Barton, D. H. R.; Le Gloahec, V. N. *Tetrahedron Lett.* **1998**, *39*, 4413. (i) Barton, D. H. R.; Launay, F. *Tetrahedron* **1998**, *54*, 12699. (j) Newcomb, M.; Simakov, P. A. *Tetrahedron Lett.* **1998**, *39*, 965.

34. (a) Barton, D. H. R.; Beck, A. H.; Delanghe, N. C. *Tetrahedron Lett.* **1996**, *37*, 1555. (b) Barton, D. H. R.; Delanghe, N. C. *Tetrahedron Lett.* **1996**, *37*, 8137.

35. Barton, D. H. R.; Choi, S.-Y.; Hu, B.; Smith, J. A. *Tetrahedron* **1998**, *54*, 3367.

36. (a) Sawer, D. T.; Sobkowiak, A.; Matsushita, T. *Acc. Chem. Res.* **1996**, *29*, 409. (b) Sawer, D. T. *Coord. Chem. Rev.* **1997**, *165*, 297. (c) Sheu, C.; Sobkowiak, A.; Jeon, S.; Sawyer, D. T. *J. Am. Chem. Soc.* **1990**, *112*, 879. (d) Sheu, C.; Richert, S. A.; Cofré, P.; Ross, B., Jr.; Sobkowiak, A.; Sawyer, D. T.; Kanofsky, J. R. *J. Am. Chem. Soc.* **1990**, *112*, 1936. (e) Tung, H.-C.; Kang, C.; Sawyer, D. T. *J. Am. Chem. Soc.* **1992**, *114*, 3445. (f) Sobkowiak, A.; Qui, A.; Liu, X.;

Llobet, A.; Sawyer, D. T. *J. Am. Chem. Soc.* **1993**, *115*, 609. (g) Kang, C.; Sobkowiak, A.; Sawyer, D. T. *Inorg. Chem.* **1994**, *33*, 79.

37. (a) MacFaul, P. A.; Wayner, D. D. M.; Ingold, K. U. *Acc. Chem. Res.* **1998**, *31*, 159. (b) Walling, C. *Acc. Chem. Res.* **1998**, *31*, 155.

38. (a) Murahashi, S.-I.; Oda, Y.; Naota, T.; Kuwabara, T. *Tetrahedron Lett.* **1993**, *34*, 1299. (b) Cheng, W.-C.; Yu, W.-Y.; Cheung, K.-K.; Che, C.-M. *J. Chem. Soc., Chem. Commun.* **1994**, 1063.

39. (a) Bressan, M.; Morvillo, A.; Romanello, G. *J. Mol. Catal.* **1992**, *77*, 283. (b) Cramarossa, M. R.; Forti, L.; Fedotov, M. A.; Detusheva, L. G.; Likholobov, V. A.; Kuznetsova, L. I.; Semin, G. L.; Cavani, F.; Trifirò, F. *J. Mol. Catal. A: Chem.* **1997**, *127*, 85.

40. (a) Muzart, J. *Chem. Rev.* **1992**, *92*, 113. (b) Muzart, J.; N'Ait Ajjou, A. *J. Mol. Catal.* **1991**, *66*, 155.

41. (a) Chavez, F. A.; Nguyen, C. V.; Olmstead, M. M.; Marscharak, P. K. *Inorg. Chem.* **1996**, *35*, 6282. (b) Choudary, B. M.; Reddy, P. N. *J. Mol. Catal. A: Chem.* **1996**, *112*, 385.

42. (a) Kim, J.; Harrison, R.G.; Kim, C.; Que, L., Jr. *J. Am. Chem. Soc.* **1996**, *118*, 4373. (b) Kim, J.; Kim, C.; Harrison, R. G.; Wilkinson, E. C.; Que, L., Jr. *J. Mol. Catal. A: Chem.* **1997**, *117*, 83. (c) MacFaul, P. A.; Arends, I. W. C. E.; Ingold, K. U.; Wayner, D. D. M. *J. Chem. Soc., Perkin Trans. 2* **1997**, 135.

43. Rabion, A.; Buchanan, R. M.; Seris, J.-L.; Fish, R. H. *J. Mol. Catal. A: Chem.* **1997**, *116*, 43.

44. Minisci, F.; Fontana, F.; Araneo, S.; Recupero, F.; Zhao, L. *Synlett* **1996**, 119.

45. Vincent, J. M.; Ménage, S.; Lambeaux, C.; Fontecave, M. *Tetrahedron Lett.* **1994**, *35*, 6287.

46. (a) Schuchardt, U.; Pereira, R.; Rufo, M. *J. Mol. Catal. A: Chem.* **1998**, *135*, 257. (b) Salvador, J. A. R.; Sá e Melo, M. L.; Neves, A. S. C. *Tetrahedron Lett.* **1997**, *38*, 119. (c) Pozzi, G.; Cavazzini, M.; Quici, S.; Fontana, S. *Tetrahedron Lett.* **1997**, *38*, 7605.

47. (a) Taft, K. L.; Kulawiec, R. J.; Sarneski, J. E.; Crabtree, R. H. *Tetrahedron Lett.* **1989**, *30*, 5689. (b) Sarneski, J. E.; Michos, D.; Thorp, H. H.; Didiuk, M.; Poon, T.; Blewitt, J.; Brudwig, G. W.; Crabtree, R. H. *Tetrahedron Lett.* **1991**, *32*, 1153. (c) Ménage, S.; Collomb-Dunand-Sauthier, M.-N.; Lambeaux, C.; Fontecave, M. *J. Chem. Soc., Chem. Commun.* **1994**, 1885. (d) Fish, R. H.; Fong, R. H.; Vincent, J. B.; Christou, G. *J. Chem. Soc., Chem. Commun.* **1988**,

1504. (e)) Fish, R. H.; Fong, R. H.; Oberhausen, K. J.; Konings, M. S.; Vega, M. C.; Christou, G.; Vincent, J. B.; Buchanan, R. M. *New J. Chem.* **1992**, *16*, 727.

48. (a) Ganeshpure, P. A.; Satish, S.; Tembe, G. L. *Tetrahedron Lett.* **1995**, *36*, 8861. (b) Ganeshpure, P. A.; Tembe, G. L.; Satish, S. *J. Mol. Catal. A: Chem.* **1996**, *113*, L423. (c) Tembe, G. L.; Ganeshpure, P. A.; Satish, S. *J. Mol. Catal. A: Chem.* **1997**, *121*, 17.

49. Tetard, D.; Verlhac, J.-B. *J. Mol. Catal. A: Chem.* **1996**, *113*, 223.

50. Tateiwa, J.; Horiuchi, H.; Uemura, S. *J. Chem. Soc., Chem. Commun.* **1994**, 2567.

51. Rigutto, M. S.; van Bekkum, H. *J. Mol. Catal.* **1993**, *81*, 77.

52. Launay, F.; Roucoux, A.; Patin, H. *Tetrahedron Lett.* **1998**, *39*, 1353.

53. Glushakova, V. N.; Skorodumova, N. A.; Gryzunova, O. Yu. *et al. Kinet. Katal.* **1989**, *30*, 656 (in Russian).

54. Talsi, E. P.; Chinakov, V. D.; Babenko, V. P.; Zamaraev, K. I. *J. Mol. Catal.* **1993**, *81*, 235.

55. Nguyen, C.; Guajardo, R. J.; Marscharak, P. K. *Inorg. Chem.* **1996**, *35*, 6273.

56. Nomura, K.; Uemura, S. *J. Chem. Soc., Chem. Commun.* **1994**, 129.

57. Murahashi, S.-I.; Oda, Y.; Komiya, N.; Naota, T. *Tetrahedron Lett.* **1994**, *35*, 7953.

58. (a) Yamaguchi, M.; Iida, T.; Yamagishi, T. *Inorg. Chem. Commun.* **1998**, *1*, 299. (b) Yamaguchi, M.; Kousaka, H.; Yamagishi, T. *Chemistry Lett.* **1997**, 769.

59. Moiseeva, N. I.; Gekhman, A. E.; Moiseev, I. I. *J. Mol. Catal. A: Chem.* **1997**, *117*, 39.

60. Kodera, M.; Shimakoshi, H.; Kano, K. *Chem. Comm.* **1996**, 1737.

61. Camaioni, D. M.; Lilga, A.; Bays, J. T.; Birnbaum, J. C.; Linehan, J. C. Presented at the *217th Am. Chem. Soc. Natl. Meeting*; Anaheim, March **1999**; paper INOR No. 436.

62. Holm, R. H. *Chem. Rev.* **1987**, *87*, 1401.

63. (a) Deniaud, D.; Spyroulias, G. A.; Bartoli, J.-F.; Battioni, P.; Mansuy, D.; Pinel, C.; Odobel, F.; Bujoli, B. *New J. Chem.* **1998**, 901. (b) Iamamoto, Y.; Ciuffi, K. J.; Sacco, H. C.; Prado, C. M. C.; de Moraes, M.; Nascimento, O. R. *J. Mol. Catal.* **1994**, *88*, 167. (c) Iamamoto, Y.; Assis, M. D.; Ciuffi, K. J.;

Prado, C. M. C.; Prellwitz, B. Z.; Moraes, M.; Nascimento, O. R.; Sacco. H. C. *J. Mol. Catal. A: Chem.* **1997**, *116*, 365. (d) Iamamoto, Y.; Ciuffi, K. J.; Sacco, H. C.; Iwamoto, L. S.; Nascimento, O. R.; Prado, C. M. C. *J. Mol. Catal. A: Chem.* **1997**, *116*, 405.

64. (a) Baciocchi, E.; Belvedere, S. *Tetrahedron Lett.* **1998**, *39*, 4711. (b) Gross, Z.; Simkhovich, L. *Tetrahedron Lett.* **1998**, *39*, 8171. (c) Porhiel, E.; Bondon, A.; Leroy, J. *Tetrahedron Lett.* **1998**, *39*, 4029. (d) Iamamoto, Y.; Assis, M. D., Ciuffi, K. J.; Sacco, H. C.; Iwamoto, L.; Melo, A. J. B.; Nascimento, O. R.; Prado, C. M. C. *J. Mol. Catal. A: Chem.* **1996**, *109*, 189.

65. (a) Jin, R.; Cho, C. S.; Jiang, L. H.; Shim, S. C. *J. Mol. Catal. A: Chem.* **1997**, *116*, 343. (b) Che, C.-M.; Tang, W.-T.; Wong, K.-Y.; Wong, W.-T.; Lai, T.-F. *J. Chem. Res. (S)* **1991**, 30.

66. (a) Hamachi, K.; Irie, R.; Katsuki, T. *Tetrahedron Lett.* **1996**, *37*, 4979. (b) Punniyamurthy, T.; Miyafuji, A.; Katsuki, T. *Tetrahedron Lett.* **1998**, *39*, 8295. (c) Hamada, T.; Irie, R.; Mihara, J.; Hamachi, K.; Katsuki, T. *Tetrahedron*, **1998**, *54*, 10017. (d) Komiya, N.; Noji, S.; Murahashi, S.-I. *Tetrahedron Lett.* **1998**, *39*, 7921. (e) Lee, N. H.; Lee, C.-S.; Jung, D.-S. *Tetrahedron Lett.* **1998**, *39*, 1385.

67. (a) Bressan, M.; Forti, L.; Morvillo, A. *Inorg. Chim. Acta* **1993**, *211*, 217. (b) Mansuy, D.; Bartoli, J.-F.; Battioni, P.; Lyon, D. K.; Finke, R. G. *J. Am. Chem. Soc.* **1991**, *113*, 7222. (c) Shul'pin, G. B.; Druzhinina, A. N.; Kats, M. M. *Neftekhimiya* **1989**, *29*, 697 (in Russian).

68. (a) Cagnina, A.; Campestrini, S.; Di Furia, F.; Ghiotti, P. *J. Mol. Catal. A: Chem.* **1998**, *130*, 221. (b) Balahura, R. J.; Sorokin, A.; Bernadou, J.; Meunier, B. *Inorg. Chem.* **1997**, *36*, 3488. (c) Robert, A.; Meunier, B. *New J. Chem.* **1988**, *12*, 885. (d) Song, R.; Sorokin, A.; Bernadou, J.; Meunier, B. *J. Org. Chem.* **1997**, *62*, 673.

69. (a) Neumann, R.; Abu-Gnim, C. *J. Chem. Soc., Chem. Commun.* **1989**, 1324. (b) Bressan, M.; Forti, L.; Morvillo, A. *J. Mol. Catal.* **1993**, *84*, 59.

70. (a) Muzart, J.; Ait-Mohand, S. *Tetrahedron Lett.* **1995**, *36*, 5735. (b) Querci, C.; Ricci, M. *Tetrahedron Lett.* **1990**, *31*, 1779. (c) Artaud, I.; Grennberg, H.; Mansuy, D. *J. Chem. Soc., Chem. Commun.* **1992**, 1036.

71. (a) Neumann, R.; Abu-Gnim, C. *J. Chem. Soc., Chem. Commun.* **1989**, 1324. (b) Bressan, M.; Morvillo, A. *J. Chem. Soc., Chem. Commun.* **1989**, 421. (c) Kulikova, V. S.; Levitsky, M. M.; Shestakov, A. F.; Shilov, A. E. *Izv. Akad. Nauk, Ser. Khim.* **1998**, 450 (in Russian).

72. (a) Sasson, Y.; Al Quntar, A. E.-A.; Zoran, A. *Chem. Comm.* **1998**, 73. (b) Ganin, E.; Amer, I. *J. Mol. Catal. A: Chem.* **1997**, *116*, 323.

73. Reinaud, O.; Capdevielle, P.; Maumy, M. *J. Mol. Catal.* **1991**, *68*, L13.

74. (a) Ohtake, H.; Higuchi, T.; Hirobe, M. *J. Am. Chem. Soc.* **1992**, *114*, 10660. (b) Ohtake, H.; Higuchi, T.; Hirobe, M. *Heterocycles* **1995**, *40*, 867. (c) Higuchi, T.; Satake, C.; Hirobe, M. *J. Am. Chem. Soc.* **1995**, *117*, 8879. (d) Shingaki, T.; Miura, K.; Higuchi, T.; Hirobe, M.; Nagano, T. *Chem. Comm.* **1997**, 861.

75. (a) Groves, J. T.; Shalyaev, K. V.; Bonchio, M.; Carofiglio, T. *3rd World Congress on Oxidation Catalysis*; Grasselli, R. K.; Oyama, S. T.; Gaffney, A. M.; Lyons, J. E., Eds.; Elsevier: Amsterdam, 1997, p. 865. (b) Groves, J.T.; Bonchio, M.; Carofiglio, T.; Shalyaev, K. *J. Am. Chem. Soc.* **1996**, *118*, 8961.

76. (a) Akasaka, T.; Haranaka, M.; Ando, W. *J. Am. Chem. Soc.* **1993**, *115*, 7005. (b) Shul'pin, G. B.; Kats, M. M.; Kozlov, Yu. N. *Bull. Acad. Sci. USSR, Div. Chem. Sci.* **1988**, *37*, 2396. (c) Gross, Z.; Simkhovich, L. *J. Mol. Catal. A: Chem.* **1997**, *117*, 243.

CHAPTER XI

OXIDATION IN LIVING CELLS AND ITS CHEMICAL MODELS

One of the main pathways of alkane functionalization is their oxidation. In the living nature, this problem has been successfully solved in two ways: through aerobic oxidation, which involves molecular oxygen, and anaerobic oxidation. The corresponding natural enzymic systems still remain unsurpassed in regards to the rate and the selectivity of alkane oxidation. Alkanes, arenes and other organic compounds can be surprisingly easily oxidized by dioxygen in the cells of bacteria, fungi, plants, insects, fish, and mammals, including man. Some recent examples of various biotransformations (see, e.g., references [1–10]) including biodegradations of poisonous compounds [11–13] are shown in Schemes XI.I and XI.2.

Therefore, the biomimetic approach, that is the creation of chemical analogues of enzymes, seems to be especially promising in this respect. The creation of chemical models of the enzymatic oxidation of alkanes and arenes makes it possible not only to understand its mechanisms better, but also to develop what are likely to be fundamentally new processes for the conversion of the hydrocarbon raw material.

The current views about aerobic biological oxidation of alkanes and arenes involving various oxygenases. Its chemical simulations are discussed in this chapter, which gives only a brief survey of the most recent data for biological C–H activation and hydrocarbon oxidation. It is noteworthy that, somewhat unexpectedly, important and profound analogies exist between chemical activation by metal complexes and biological C–H oxidation. It is necessary to note that many books [14] and reviews [15] have been devoted to enzymatic oxidations and processes that more or less closely model these oxidations.

The term “monooxygenases” is generally applied to the group of enzymes catalyzing the hydroxylation of C–H compounds by molecular oxygen. Monooxy-

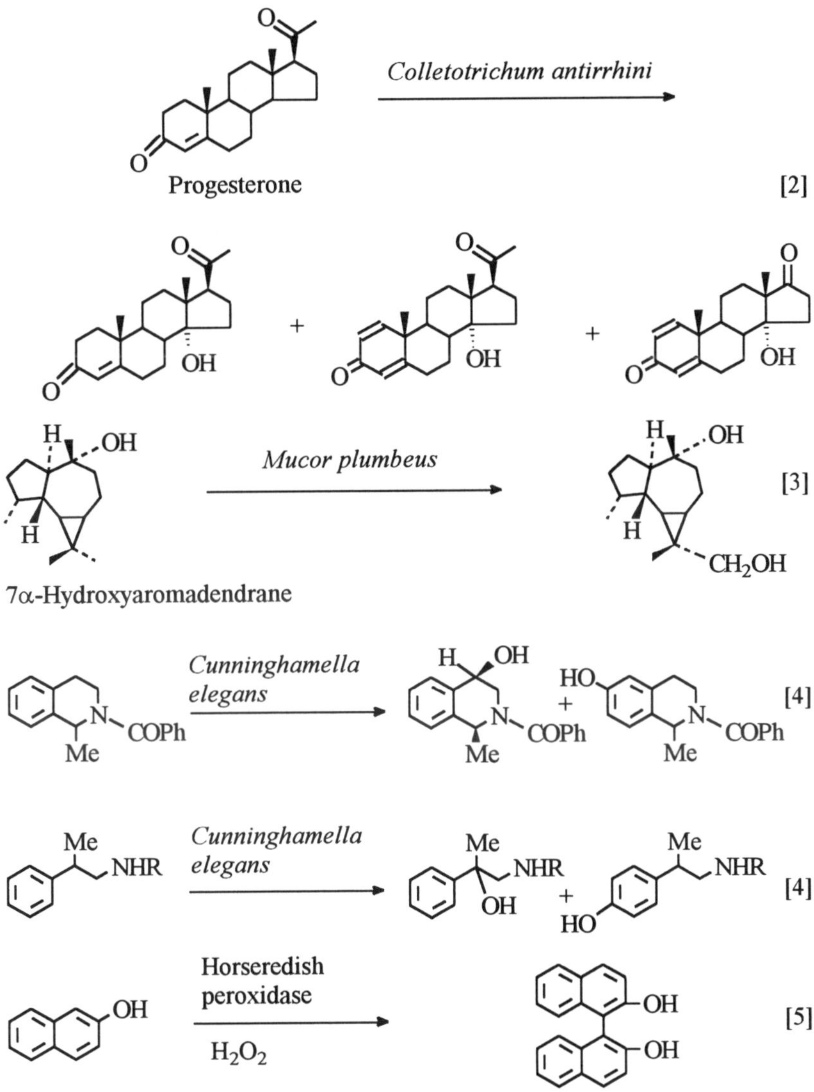

Scheme XI.1 (part 1). Examples of hydrocarbon biotransformations which occur in the living cells of various organisms as well as under the action of some enzymes. Reference numbers are in brackets.

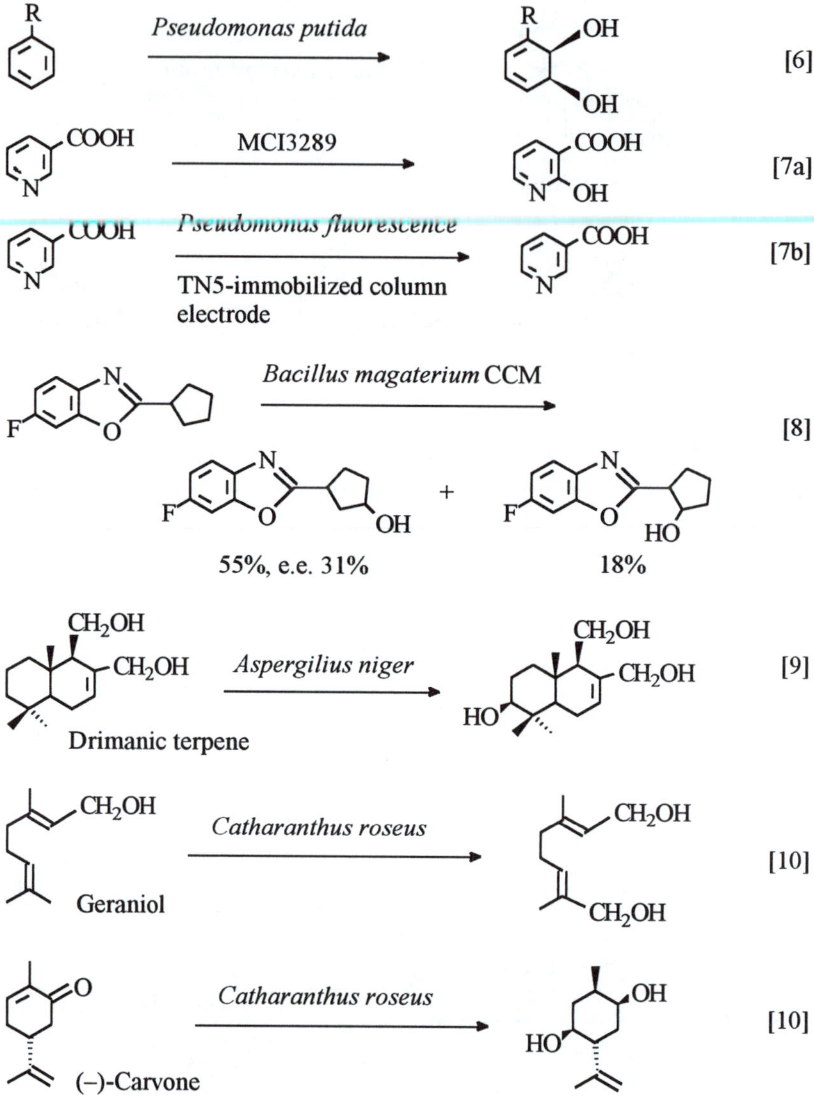

Scheme XI.1 (part 2). Examples of hydrocarbon biotransformations which occur in the living cells of various organisms as well as under the action of some enzymes. Reference numbers are in brackets.

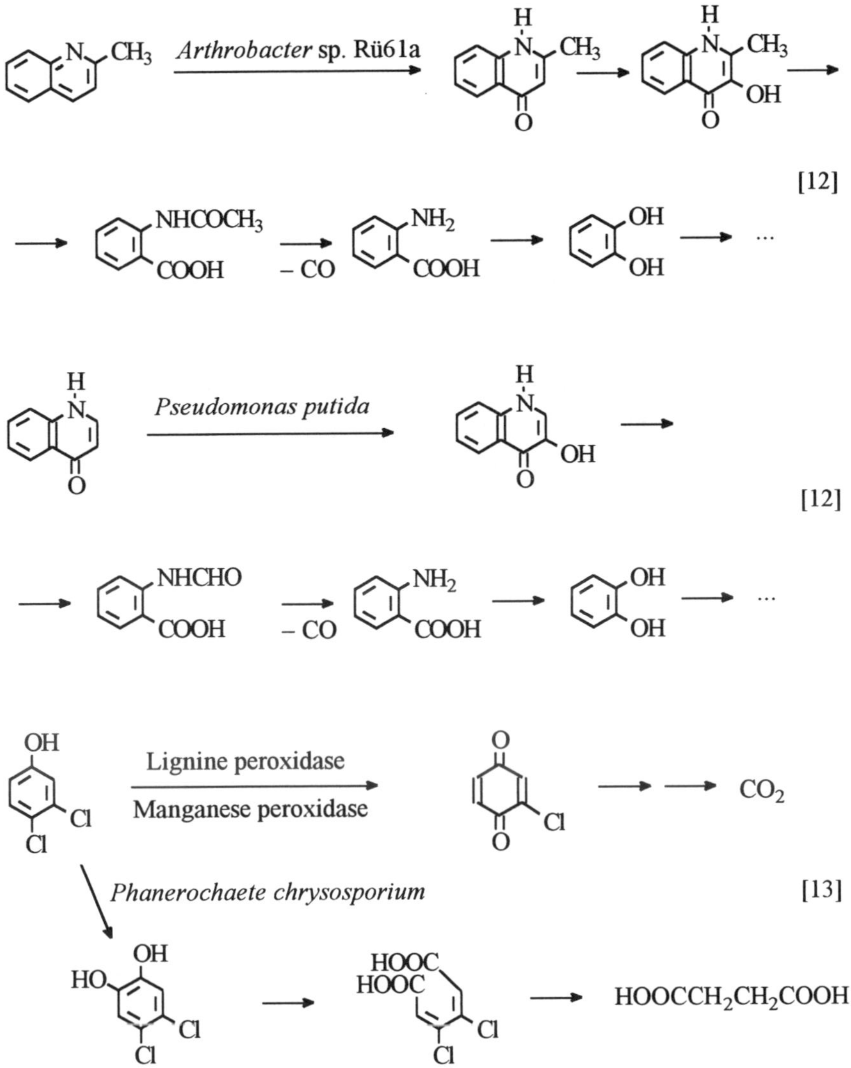

Scheme XI.2. Biodegradation of poisonous compounds. Reference numbers are in brackets.

genases can induce the insertion of only one oxygen atom from O_2 into the C–H bond, while the second oxygen atom is reduced with formation of water:

$$-\overset{|}{\underset{|}{C}}-H + \mathbf{O}_2 + AH_2 \longrightarrow -\overset{|}{\underset{|}{C}}-\mathbf{O}H + H_2\mathbf{O} + A$$

In this case the enzymes are referred to as monooxygenases. The hydrogen donors AH_2 are NADH, NADPH, the ascorbate anion, and other biological reductants. In virtually all the biological oxidations of hydrocarbons, hydroxylation is preceded by the activation of oxygen on the metalloenzyme. Many monooxygenases contain metals as constituents, namely iron and copper. Two equivalents of a reducing agent, which is needed for the monooxygenase cycle to be accomplished, come from an organic component that must always be present in the system. The role of this compound is to transfer electrons to the metal atom of the enzyme, which passes into the reduced state, and thereby acquires the ability to react with molecular oxygen and activate it for further interactions with alkanes. Hence, according to the current concepts, activation of molecular oxygen resulting in the formation of metal peroxo- and oxo-complexes is the basic mechanism of the catalytic effect of monooxygenases. The chemistry of these complexes and their reactions with alkanes have attracted the keen attention of investigators in the past decade.

The group of enzymes called dioxygenases can oxidize substrates, XH, according to the equation:

$$XH + \mathbf{O}_2 \longrightarrow X(\mathbf{O}_2)H$$

Dioxygenases can incorporate the two oxygen atoms into separate substrates (intermolecular dioxygenases) or into a single molecule (intramolecular dioxygenases) [16a].

The monooxygenases of the bacterium *Pseudomonas putida* hydroxylate camphor, while the monooxygenases from mitochondria and microsomes in the adrenal cortex cause the side chain of cholesterol to be split off with subsequent hydroxylation of the saturated C–H bonds. In the presence of liver microsomes, alkanes and aliphatic acids are hydroxylated predominantly at the terminal (ω) carbon atom. The C–H bonds at the neighboring (ω–1) carbon atom of the aliphatic chain are much less reactive. Thus the oxidation of decanoic acid results in the formation of 92% of 10-hydroxydecanoic and 8% of 9-hydroxydecanoic acid. Since the oxidation in the ω- and (ω–1)-positions is inhibited by carbon monoxide to different extents, presumably the C–H bond activation processes in these positions are catalyzed by different monooxygenases. At the same time, monooxygenases are known (for example those isolated from rat liver microsomes) which hydroxylate alkanes with the "usual" selectivity (thus 1° : 2° : 3° = 1 : 15.4 : 113 for isopentane). In many cases, the characteristic feature of the oxidation in the presence of monooxygenases is the retention of the configuration of the substrate. The hydroxylation of aromatic compounds proceeds as an electrophilic substitution (the ratio *o* : *m* : *p* = 59 : 13 : 28 for toluene) and is accompanied by the so-called NIH shift. This phenomenon, consisting of the transfer of a hydrogen (deuterium) atom to a neighboring position relative to that at which hydroxylation takes place, is depicted below:

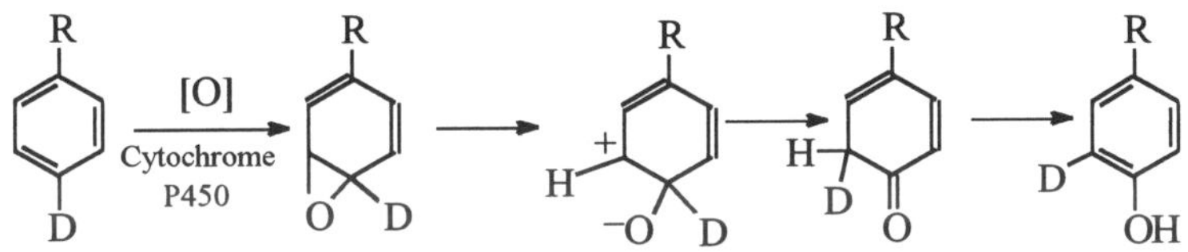

XI.1. HEME-CONTAINING MONOOXYGENASES

Many oxygenases contain the heme prosthetic group which usually is protoheme IX (Scheme XI.3). For example, similarly to other porphyrin metal complexes, four coordination sites in the iron involved in the active center of cytochrome P450 are coupled by nitrogen atoms from the porphyrin molecule. The fifth position is ocuupied by a sulfur atom from the cyctein molecule. This distinguish cytochrome P450 from hemoproteins involved in other enzymes, such as catalases and peroxidases (where the fifth position is occupied by a nitrogen atom of the histidine molecule).

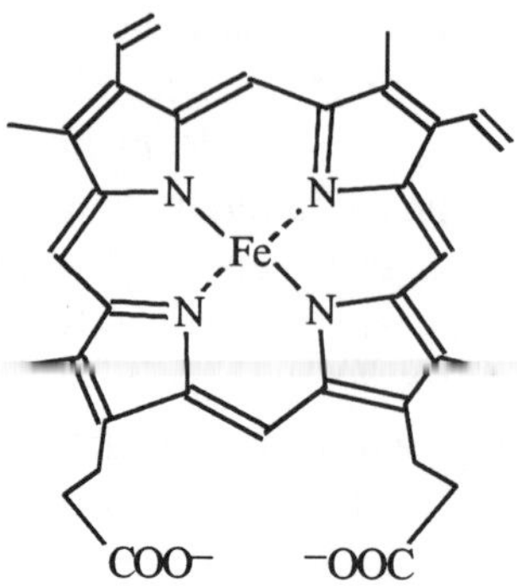

Scheme XI.3. Structure of protoheme IX.

XI.1.A. CYTOCHROME P450

Among monooxygenases, the suprafamily of cytochrome P450 has been known for the longest time. Cytochrome P450 is a widely distributed enzyme in living nature. It is capable, in particular, of oxidizing alkanes, arenes, and their derivatives (for example, prostaglandins and aliphatic acids). Some examples of oxidations in the presence of this enzyme are shown in Scheme XI.4. Cytochrome P450 also catalyzes the epoxidation of alkenes, *O*- and *N*-dealkylation, and other reactions. About 500 representatives of the cytochrome P450 suprafamily are currently known. They differ from one another in the structure of their globular protein in the region of the active center, but have identical prosthetic groups. Cytochrome P450 is the component of a multienzyme system which also contains electron transfer enzymes. In mitochondria and most bacteria, electrons are transferred from NADH and NADPH to the enzyme by FAD-dependent reductase via iron-sulfur ferredoxin. In systems of the microsomal endoplasmic reticulum devoid of Fe_2S_2-ferredoxin, the electrons are transferred from NADPH by cytochrome P450 oxidoreductase, a flavoprotein that contains FAD and FMN.

A number of books [17] and reviews [18] provide a description of cytochrome P450. Also see recent papers devoted to its reactions [19], theoretical studies [20], and mechanism of action [21], as well as certain biochemical investigations [22].

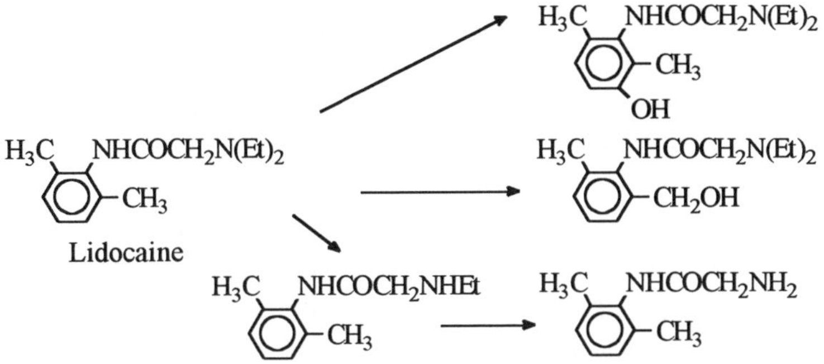

[16b]

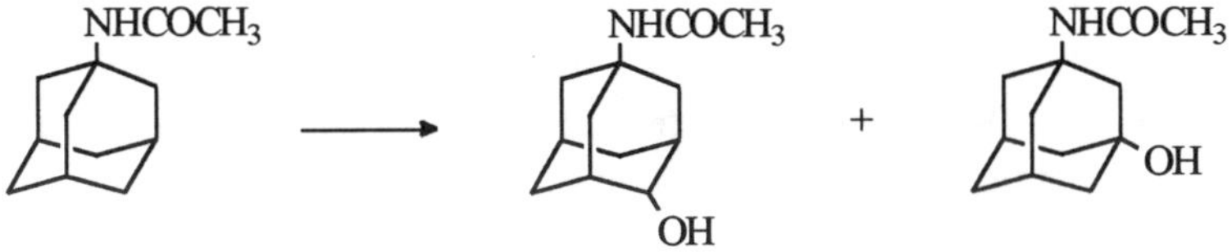

[16c]

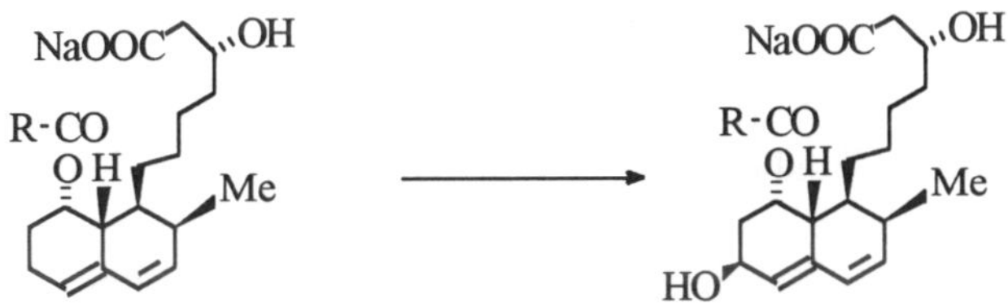

[16d]

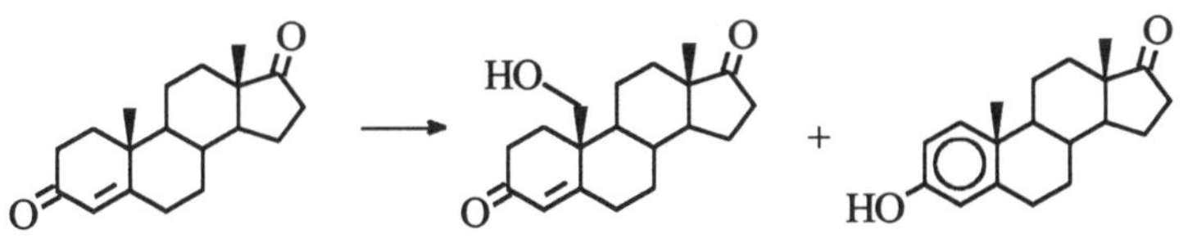

[16e,f]

[16g]

Scheme XI.4. Examples of oxidation by cytochrome P450. Reference numbers are in brackets.

The catalytic cycle proposed for alkane oxidation by dioxygen in the presence of cytochrome P450 is shown in Scheme XI.5. The oxidation of an alkane, RH, includes at least eight steps. The formation of the complex **XI-2** in the first step converts cytochrome P450 from the low-spin form into the high-spin form. Reduction takes place in the second step of the process, after which a dioxygen molecule is coordinated to iron(II) atom.

The X-ray structure of cytochrome P450 is known for camphor hydroxylating $P450_{cam}$ with and without the substrate. The structure confirms that the iron atom of the porphyrin ring is open to O_2 addition and activation and the substrate molecule is in close vicinity to the iron core to be directly involved in C–H hydroxylation. An X-ray diffraction analysis of the complex of myoglobin with dioxygen showed that oxygen was coordinated in such a way that the molecular axis of O_2 forms an angle of 120° with the plane of the porphyrin ring (Pauling's structure **XI-4**). The fourth and fifth steps terminate with the reduction of the complex and the elimination of a water molecule. This results in the formation of the oxenoid **XI-6**. The sixth step consists in the cleavage of a C–H bond in the substrate molecule. Transfer of the second electron to oxycytochrome P450 results in the formation of an $Fe^{2+}O_2^{2-}$ or $Fe^{3+}O_2^{2-}$ species. In the absence of a substrate, decomposition of these species yields a superoxide, the rate constant being 29 min^{-1}. Binding of a substrate with the enzyme inhibits this process; the rate constant for the formation of oxidation products is about 30 min^{-1} and it changes depending on the nature of the substrate and the enzyme. The intermediate steps following the oxycytochrome reduction occur too fast for the intermediates to be detected by spectral methods. The mechanism of transformation of the substrate–reduced oxycytochrome complex is the least understood in the catalytic cycle of cytochrome P450. It is usually assumed that the known oxidation products are formed through heterolytic or homolytic cleavage of the O–O bond in the iron peroxo-complex. Different versions of this mechanism agree only with certain parts of the experimental data available. Presumably, the operation of a particular mechanism depends on the nature of the substrate and the enzyme.

Most of investigators postulate the pentavalent iron oxene complex, P450–Fe^{V}=O, as the direct oxidant of a substrate by analogy with the reactive intermediates of the catalytic cycle of peroxidase, which have the oxoferryl group, Fe^{IV}=O, in their active center. However, the Mössbauer and Raman spec-

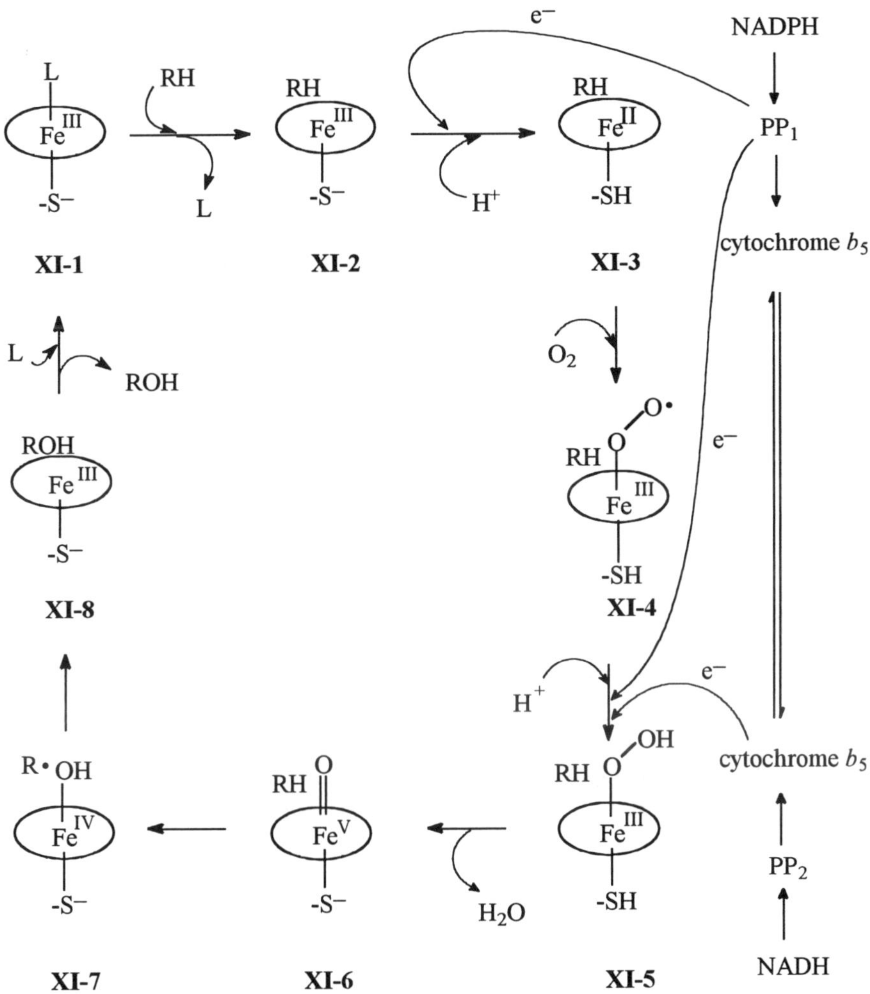

Scheme XI.5. The catalytic cycle proposed for alkane oxidation by O_2 promoted by cytochrome P450 (PP_1: NADPH–cytochrome P450 reducatase; PP_2: NADH–cytochrome b_5 reductase; S: cysteine residue).

tra of cytochrome P450 recorded under catalytic conditions suggest the presence of iron-superoxide structures similar to oxyhemoglobin HbO_2^-, but different from the ferryl species of peroxidase. Probably the superoxidecomplexes are formed as intermediate products in the activation of oxygen on cytochrome P450. Unfortunately, the iron oxene species postulated for the catalytic cycle of cytochrome P450 has not been identified yet by any of the existing physical methods. At the same time, the oxene species in peroxidase systems are relatively stable well-characterized intermediates. The formation of reactive species in the conversion of H_2O_2 by peroxidase occurs under conditions of acid-alkaline catalysis in the imidazole fragment of histidine localized in the distal region of the enzyme heme. In contrast to peroxidase, thiolate is the proximal ligand of the cytochrome P450 heme. Presumably, the effect of the thiolate ligand and the presence of a substrate in close proximity to the reactive oxygen of cytochrome P450 decrease the lifetime of the active species and prevent their registration.

According to calculations, the P450–Fe=O complexes must be electrophilic. In practice, the reactions catalyzed by this enzymic system suggest either the electrophilic or the nucleophilic character of the reactive species. The role of the nucleophile in the catalytic cycle of cytochrome P450 can be played by iron peroxide, whereas iron oxene and iron hydroperoxide (or hydroxide) are the possible candidates for the electrophilic reactive intermediates. Electron density calculations demonstrate that the oxygen atom that is directly bound to the heme in peroxo-ferryl complexes with thiolate ligands, bears a partial positive charge, whereas the external oxygen atom bears a partial negative charge. It may thus be assumed that peroxo-complexes function as monooxygenating species. The catalytic activity of ferrocytochrome $P450_{cam}$ with superoxide is additional evidence in favor of this hypothesis. The peroxo intermediate is regarded as a direct oxidant in the oxidation of lauric acid and in the oxidative deformylation of aldehydes by aromatase. However, the formation of epoxides in these reactions is inconsistent with reactive peroxide or superoxide intermediates and attests to the involvement of oxene species. Thus the nature of monooxygenating reactive species in the catalytic cycle of cytochrome P450 is still open to discussion.

XI.1.B. OTHER MONOOXYGENASES

Secondary amine monooxygenase (SAMO) from *Pseudomonas asinovorans* catalyzes the oxidative *N*-dealkylation of secondary aimines [16a]:

$$(CH_3)_2NH_2 + NADPH + O_2 + H^+ \longrightarrow CH_3NH_2 + NADP^+ + HCHO + H_2O$$

Like peroxidase, and unlike cytochrome P450, SAMO contains the imidazole moiety of hystidine in the fifth axial (proximal) position of the heme; however, similarly to cytochrome P450, it oxygenates the C–H bonds of methylamines. The intramolecular kinetic isotope effect (KIE) observed for the methyl group of 1,1,1′,1′-tetradeuteriodimethylamine (1.7) is close to the KIE for the analogous reactions of cytochrome P450 and chloroperoxidase. Heme-containing chloroperoxidase and dehaloperoxidase catalyze the halogenation of phenols and dehalogenation of halophenols, respectively [23a], as well as the halogenation of activated C–H bonds [23b].

XI.2. None-Heme Iron-Containing Monooxygenases

Non-heme iron-containing monooxygenases, such as methane monooxygenase and ω-hydroxylase, analogously to cytochrome P450, can transfer the oxygen atom from molecular oxygen to the C–H bonds of alkanes [24].

XI.2.A. METHANE MONOOXYGENASE

The enzymes isolated from methane oxidizing bacteria are known under the general name of methane monooxygenases (MMO) [25, 26]. The oxidation takes place in accordance with the equation:

$$CH_4 + O_2 + NAD(P)H + H^+ \xrightarrow{\text{MMO}} CH_3OH + NAD(P)^+ + H_2O$$

Depending on the growth conditions, the enzyme can exist both in the membrane-bound and the cytoplasmic (soluble) form. Water-soluble MMO from *Methylococcus capsulatus* has been studied in the most detail. Methane monooxygenases catalyze the oxidation of alkanes, methane showing the highest activity among them in this reaction [27]. The rate of oxidation falls on going from methane to butane. Specific activity (milliunits/mg) [28a] for higher alkanes is the following: pentane (22), hexane (16), heptane (12), neopentane (8), 2-methylpropane (33), 2,3-dimethylpentane (20), and adamantane (3). Aromatic hydrocarbons are not oxidized. It has been shown [28b] that MMO isolated from *Methylococcus capsulatus* contains 4 atoms of non-heme iron and 1 atom of copper per molecule with a molecular weight of 240000. The enzymes also include cytochrome *c*, but its presence does not influence the activity of MMO. The active center of MMO contains a binuclear iron complex [28c–f]. X-ray structural analysis has confirmed the binuclear structure of the MMO active center. One of the methane oxygenation mechanisms [29] proposed for this binuclear MMO center is similar to that of cytochrome P450 hydroxylation and involves hydrogen abstraction preceding hydroxylation (Scheme XI.6).

A theoretical study of the three possible oxygen coordination modes, μ-η^1:η^1-O_2^{2-}, μ-O_2^{2-} and μ-η^2:η^2-O_2^{2-} of $Fe_2(\mu\text{-OH})(\mu\text{-OOCH})(OH)_6$ has been carried out [29g]. The calculations suggested that the μ-η^1:η^1-O_2^{2-} mode is the most favorable.

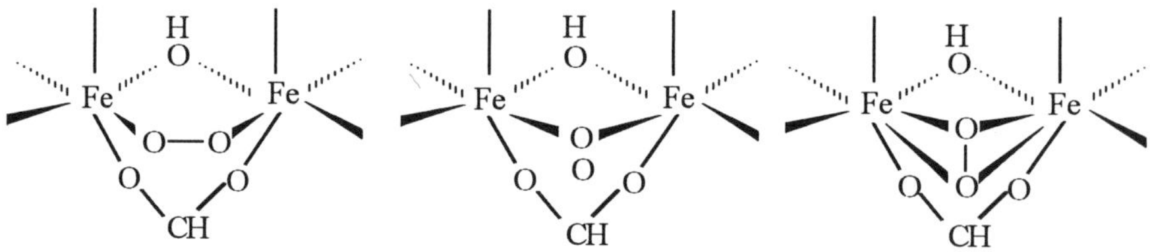

μ-$\eta1$:$\eta1$-$O_2{}^{2-}$ μ-$O_2{}^{2-}$ μ-$\eta2$:$\eta2$-$O_2{}^{2-}$

According to Shteinman's concept [30], activation of dioxygen occurs with the formation of μ-η^2:η^2-peroxide as a result of laterial binding of the O_2 molecule. This intermediate is further converted into the bis-μ-oxo-$Fe^{IV}{}_2$ intermediate, which reacts with methane and is an alternative to ferryl. Unlike ferryl, the oxygen atom in this intermediate occupies the bridging position between two iron atoms and thus ensures its more efficient stabilization. The interaction of the oxygen atom of the bis-μ-oxo-$Fe^{IV}{}_2$ intermediate with the C–H

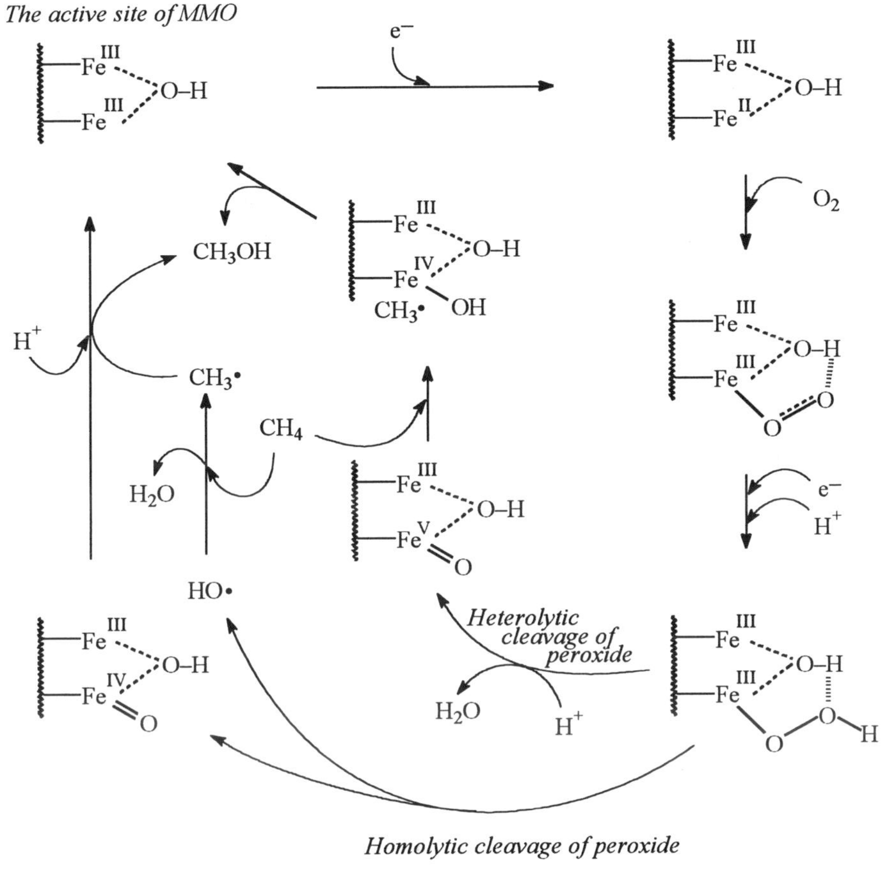

Scheme XI.6. The catalytic cycle proposed for alkane oxidation by O_2 promoted by the binuclear methane monooxygenase center.

bond of methane, which has the smallest length, allows an additional nucleophilic assistance of the second O-bridge.

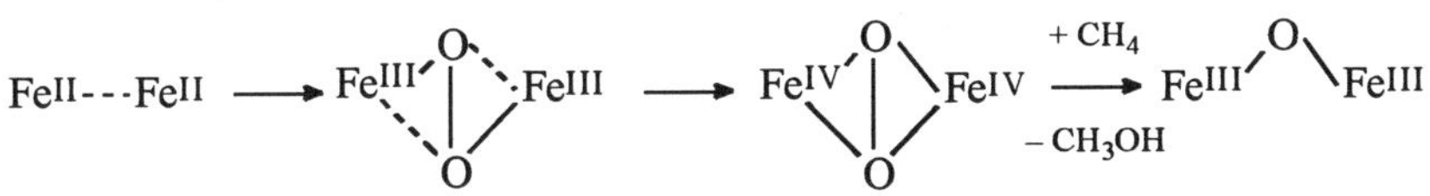

The transient kinetics of the reduced $Fe^{II}Fe^{II}$ hydroxylase reaction with dioxygen was studied by Lipscomb *et al.* and by Lippard *et al.* Three consecutive intermediates, termed compounds P, Q and T, were identified during one catalytic cycle of the MMO in the work for *Methylosinus trichosporium* OB3B. The colorless intermediate P is formed immediately after mixing of reduced hydroxylase with O_2 and then slowly transforms into the colored intermediate Q. Substrates were found to have little effect on the rate of compound Q formation, but they greatly accelerate their decay. The decay rate depends upon both the concentration and the type of substrate present. Therefore compound Q may be the activated form of the enzyme that leads directly to substrate hydroxylation or its immediate precursor. The compound Q was trapped by the rapid freeze quench technique and characterized by Mössbauer spectroscopy. The results indicate that the iron atoms of compound Q are in Fe^{IV} oxidation level. The spectra recorded at 4.2 and 50 K show that Fe^{IV} sites of compound Q are diamagnetic and the Fe^{IV} spectrum originates from clusters containing two indistinguishable coupled iron atoms. A huge kinetic isotope effect (50–100) was reported [31] for the reaction of Q with CH_4 as compared with CD_4 and proton tunneling was proposed for the explanation. It is interesting that k_H/k_D is only ca. 4 for ethane (C_2H_6 vs. C_2D_6).

Studies by Chan and Floss *et al.* [32a] with another kind of MMO, the so-called particulate form pMMO from *Methylococcus capsulatus* (Bath) (which is membrane bound and contains copper), confirmed that the rate controlling step of methane hydroxylation has a very high KIE, while ethane produces only a moderate KIE. The data obtained in this work "point to a mechanism in which C–H bond cleavage is preceded by bond formation at the alkyl carbon, i.e., one proceeding through a pentacoordinated carbon species" [32a].

Floss and Lippard *et al.* [32b] studied the oxidation of tritiated chiral alkanes by soluble MMO from *Methylococcus capsulatus* (Bath). The product alcohol displayed only a 72% retention of stereochemistry at the labeled carbon for (*S*)-[1-$^2H_1$,1-^3H]ethane and (*R*)-[1-$^2H_1$,1-^3H]ethane. These results are best accounted for by a nonsynchronous concerted process, which is shown conventionally in Scheme XI.7 with intermediate Q depicted as a diiron(IV) dioxo species. The alkane molecule can be doubly activated by one iron and its bound oxygen atom, thus leading to retention.

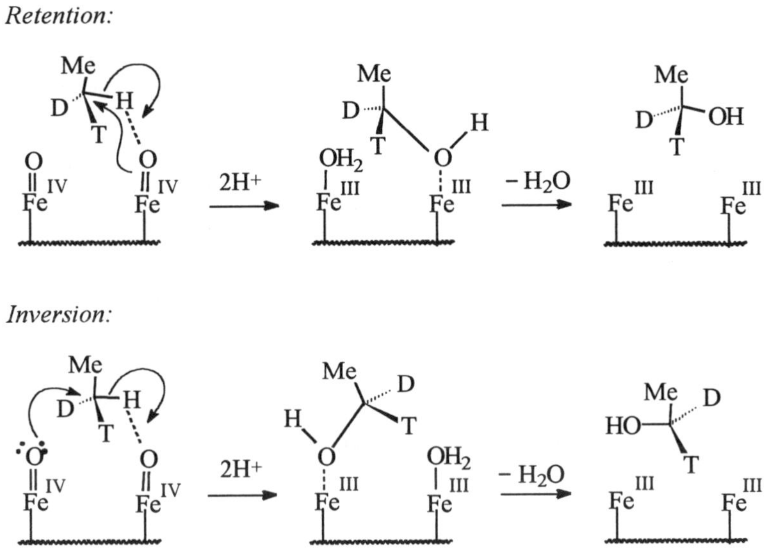

Scheme XI.7. The oxidation of tritiated chiral alkanes by soluble MMO.

Many other recent works have been devoted to the mechanisms of oxidations catalyzed by non-heme monooxygenases [33]. See also a special issue of *Journal of Biological Inorganic Chemistry (JBIC)* containing papers on the mechanism of methane monooxygenase [30d, 34].

XI.2.B. OTHER NONE-HEME IRON-CONTAINING OXYGENASES

Toluene monooxygenases catalyze the transformation of toluene into *o*-, *m*-, and *p*-cresoles. These enzymes can also oxidize other hydrocarbons. The active center of toluene monooxygenase is comprised of two irom atoms linked together by a hydroxyl bridge. Toluene monooxygenase catalyzes the oxidation of substrates by either molecular oxygen in the presence of a reducing agent or H_2O_2 [35].

ω-Hydroxylases, which contain one iron atom in their active center, mediate targeted hydroxylation of the methyl group of *n*-alkyl chains of various compounds including alkanes. Non-heme ω-hydroxylase from *Pseudomonas aeruginosa* oxidizes *n*-alkanes into the corresponding acids via the primary alcohols [36]. The efficiency of oxidation is lower for branched alkanes. The methyl group in 1,4-dimethylcyclohexane is oxidized only when it occupies the equatorial position. It was concluded that a hydrophobic cavity exists in the vicinity of the enzyme active center. The presence of such a cavity in this particular enzymic system, as well as in other monooxygenases, may account for the unusual substrate specificity and regioselectivity observed in the oxidation of hydrocarbons.

XI.3. MECHANISMS OF MONOOXYGENATIONS

The electronic structure of atomic oxygen is similar to that of the bivalent carbon compounds (carbenes) and by analogy it is termed oxene. Like carbenes, oxene in the triplet state reacts with alkanes as a free radical and abstracts hydrogen. In the singlet state, it can also be inserted into the C–H bond to yield an alcohol. Studies of the gas-phase reactions of singlet oxene with alkanes resulting in the generation of the hydroxyl radicals showed that this reaction can occur by two different routes, which are interpreted as the insertion of the oxygen atom into the C–H bond and the abstraction of the hydrogen atom. The insertion mechanism predominates if small molecules with strong C–H bonds are subject to attack, whereas hydrogen abstraction takes place predominantly in the reaction with higher hydrocarbons that are either sterically hindered and/or contain weaker C–H bonds [37a]. The oxygen atom in the active center of monooxygenase may be bound with the metal complex and its capacity to be inserted into the C–H bond or to detach the hydrogen atom is maintained. Such compounds, oxenoids, are related to the oxo-salts of transition metals. These compounds are also thought to be the intermediates possessing the superoxide properties.

Two alternative mechanisms have been proposed for the crucial step of the oxidation process: the oxenoid mechanism of direct oxygen atom insertion and the radical mechanism of C–H bond cleavage with subsequent recombination in the cage ("oxygen rebound" radical mechanism) (Scheme XI.8).

The oxenoid mechanism:

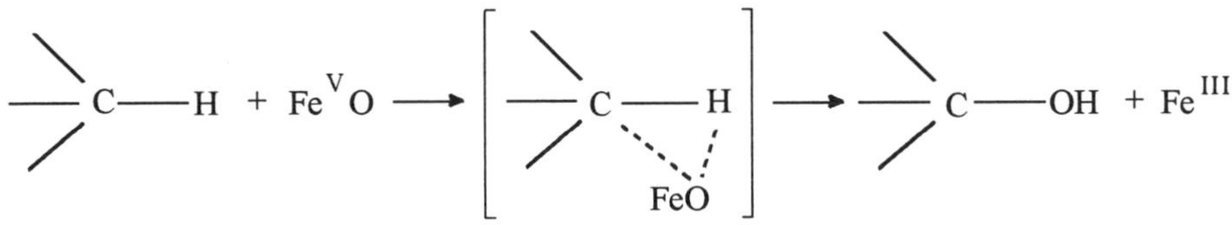

The "rebound" radical mechanism:

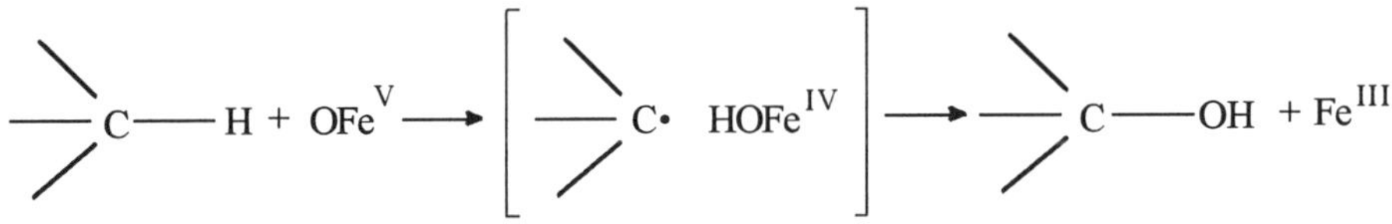

Scheme XI.8. Two alternative mechanisms proposed for C–H bond activation by cytochrome P450.

The synchronous insertion mechanism is supported by the following experimental data. In the monooxygenase reaction, alkanes yield alcohols and retention of the original configuration of the asymmetric carbon atom, where the oxidation involves the C–H bond, is a characteristic feature in many cases. It may therefore be assumed that this reaction proceeds without the intermediate formation of free radicals, which might lead to racemization. Migration of a group bound to the aromatic ring (the NIH shift) is consistent with this conclusion. However, the oxygen rebound mechanism for heme and non-heme monooxygenases has recently received the most ample recognition. This mechanism is usually supported by the following facts: (a) the large primary kinetic isotope effect in the cytochrome P450-catalyzed hydroxylation is attributed to the linear structure of the transition state ($Fe\cdots O\cdots H\cdots R$); it is assumed that the insertion of the oxygen atom into the C–H bond must be characterized by a small KIE; (b) partial or complete isomerization of the original compound occurs in some cases; presumably, this takes place in the intermediate alkyl radical; (c) in the case of methane oxidation by MMO, the use of radical traps allowed for the identification of the methyl radical [37b]. There are other facts in favor of the oxygen-rebound mechanism. Allylic oxidation in a system with cytochrome P450 resulted in the formation of isomeric allylic alcohols, which can be accounted for by the generation of the intermediate radical [37c]: C=C–CH• $\longleftrightarrow$ •C–C=CH.

However, in both biological and model chemical systems, the oxygen rebound mechanism, which is proposed as an alternative to the traditional free radical mechanism, at present gives rise to serious doubts. First, the evidence usually given for this mechanism no longer looks very convincing. A very high KIE was explained as due to involvement of proton tunneling which does not necessarily require a linear transition state. When the KIE values become smaller (e.g., in more polar media) and can be attributed to the classic picture with a linear transition state, actually the tunneling can still be involved, since it is more likely that its contribution is decreased gradually than it suddenly completely disappears.

There are other data which are in conflict with the oxygen rebound mechanism. For example, hydroxylation of the stereoisomers of deuterium-substituted camphor by cytochrome $P450_{cam}$ is accompanied by partial isomerization and transfer of deuterium atoms from the *exo*- to the *endo*-position and *vice versa* [37d]. This phenomenon is considered as a proof of the radical mechanism. However, the camphor molecule at the active center of the enzyme is fixed by a hydrogen bond and thus cannot rotate, which follows from the selectivity of this reaction (only 5-*exo*-hydroxycamphor is formed). This means that isomerization owing to the rotation of the molecule is impossible. It was suggested that a certain radical exists which can migrate freely and attack the camphor molecule at both the *exo*- and the *endo*-positions of the ring. It must be noted that the presence of this radical in the enzymic reaction is hardly probable. The isomerization observed in biological and model chemical systems, while demonstrating the stepwise character of the process, is not necessarily due to free radicals as intermediates. Epoxidation of olefins as well as hydroxylation of aromatics is often accompanied by isomerization, but the origin of the isomerization is not formation of free alkyl radicals. Rather, the oxygen atom bound to metal adds to a double bond or to an aromatic molecule, giving intermediates capable of isomerizing. If a similar addition of the oxygen atom takes place also in the case of a C–H single bond, then the intermediate formed might be able to isomerize without free radical formation. Different from a free radical (which is a three-coordinate carbon compound), the product of oxygen atom addition to the C–H bond is a five-coordinate carbon intermediate, and A. F. Shestakov and A. E. Shilov proposed to call this mechanism a "five-coordinate carbon mechanism" [37e,f] shown below:

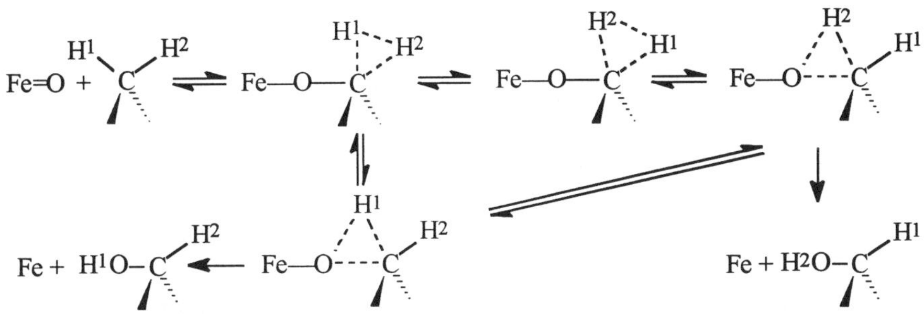

Within the framework of the pentacoordinated carbon mechanism [37g], the unusual stereoselectivity of the hydroxylation of the isotopically substituted camphor in a system with cytochrome $P450_{cam}$ has received natural interpretation without any recourse to improbable speculations [37g,h]. The existence of an intermediate with pentacoordinated carbon has also been postulated in reference [32a]. The analysis of the isomerization results published in the literature from the point of view of the oxygen rebound mechanism and from the mechanism of five-coordinate carbon often witnesses in favor of the latter, for both biological and model systems. A remarkable case of ethylbenzene hydroxylation by iodosylbenzene with chiral binaphthyl iron porphyrin has been described by Groves and Viski [37i]. Ethylbenzene afforded 1-phenylethanol. Both (*R*)- and (*S*)-(1-deuterioethyl)benzenes give all possible isomers of 1-phenylethanol, (*R*)- and (*S*)-enantiomers, each containing either deuterium or protium next to the hydroxyl group. The major product in both cases (with retention of configuration) is virtually pure monodeuteriophenylethanol, whereas the second (minor) product, with the inverse configuration, contains nearly equal amounts of hydrogen and deuterium. This result leads to a natural explanation in the framework of the five-coordinate carbon mechanism. If the insertion of the oxygen atom is accompanied by a considerable H–D isotope effect, then in the case of the C–H bond the insertion can be much faster than the isomerization, while in the case of the C–D bond insertion and isomerization can occur in parallel with comparable rates. Therefore, in the case of C–H hydroxylation, there will be retention of configuration at the carbon atom, while for C–D hydroxylation there will be partial inversion, leading to racemization. The ratios of the products obtained for both (*R*)-

and (*S*)-isomers make it possible to calculate the rate constant ratios for elementary steps in the hydroxylation. The obtained values of the parameters from the results of, for example, the (*R*)-enantiomer allow us to predict the distribution of the products for the (*S*)-enantiomer without any additional postulate. The results obtained in this way are in very good agreement with those from experiment [37g]. This example helps us to understand how, even for similar systems, one can observe both full retention of configuration and a considerable extent of isomerization. Thus, whereas for soluble sMMO, considerable isomerization is observed for chiral ethane substituted by D and T atoms in one of the methyl groups, for particular MMO (pMMO) there is 100% retention of configuration. It is more likely that the insertion-to-isomerisation rate constant ratio changes with some structural and polar effects changes than that the mechanisms are completely different for pMMO as compared with sMMO.

A paradoxal situation has been observed for *N*-dealkylation. The isotope effect (8–10) was high for peroxidases, the mechanism of which involved an electron transfer, but it was small (1–3) for various forms of cytochrome P450, chloroperoxidase, and *N*-methylaminomonooxygenase, where the activated oxygen atom directly reacts with the C–H bond. This paradox can also be overcome by assuming a two-step mechanism involving intermediates with pentacoordinated carbon (structures **XI-9** and **XI-10)** for cytochrome P450 and its analogues. The first step results in the formation of the intermediate **XI-9**, which is further converted into the intermediate **XI-10**. If the rate of this reaction is limited by the first stage, the isotope effect will be small. The formation of an intermediate complex of the metal-bound oxygen atom with the carbon atom of an unsaturated molecule is beyond doubt for the epoxidation of alkenes and the hydroxylation of aromatic compounds. At the same time, the absence of saturation in the case of alkane hydroxylation apparently rules out the possibility of the formation of an analogous intermediate complex. This hypothesis is attractive in that it unifies the mechanisms of atomic oxygen transfer to the C–H bonds of saturated compounds as well as to alkene and aromatic C=C bonds.

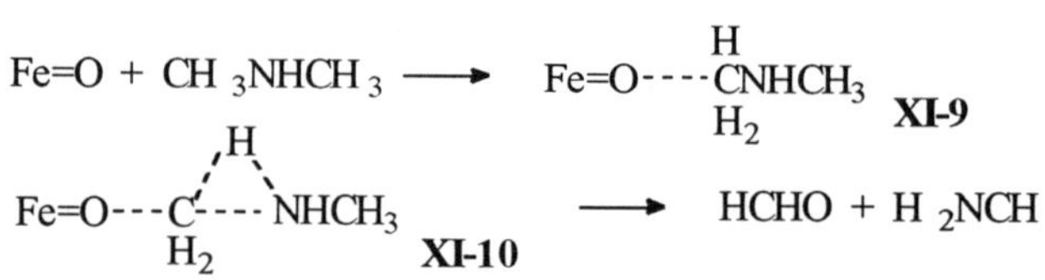

Analysis of the results described here as well as those published in the literature for monooxygenases and their chemical analogues shows that, besides the traditional radical mechanism of C–H hydroxylation in the presence of metal complexes, another, five-coordinate, carbon mechanism exists which involves addition of an oxygen atom to a C–H bond, followed by insertion into the C–H bond to form an alcohol. However, the exact nature of the latter mechanism remains to be elucidated and may be somewhat different for different systems.

XI.4. OTHER OXYGENASES CONTAINING IRON

Some of other mono- and dioxygenases [38] which are capable of activating the C–H bond also contain non-heme iron sites. Examples are discussed below.

Phenylalanine hydroxylase [39] catalyzes the first step of phenylalanine degradation in mammals. Phenylalanine is converted into tyrosine (Scheme XI.9). The enzyme has one tightly bound, non-heme iron atom per subunit. Tyrosine hydroxylase catalyzes [40] the hydroxylation of tyrosine to produce dihydroxyphenylalanine (DOPA), the first step in the biosynthesis of catechol-amine neurotransmitters (Scheme XI.9). This enzyme also contains one ferrous iron atom per subunit. These two enzymes, together with tryptophane hydroxylase (Scheme XI.9) [41], constitute a family of tetrahydropterin-dependent aromatic acid hydroxylases (monooxygenases) [42]. Other dioxygenases catalyze the hydroxylation of arenes [43].

Non-heme iron dioxygenases, catechol-2,3-dioxygenase [44], gentisate 1,2-dioxygenase [45], and phthalate dioxygenase [46] from *Pseudomonas*, catalyze the oxidative cleavage of catechol, gentisic acid (Scheme XI.10) and phthalate respectively. *Comamonas testosteroni* T-2 oxidizes toluene-*p*-sulphonate by monooxygenation of the methyl group to the alcohol and then 4-sulfobenzoate 3,4-dioxygenase induces the following degradation of *p*-sulfobenzoate [47]:

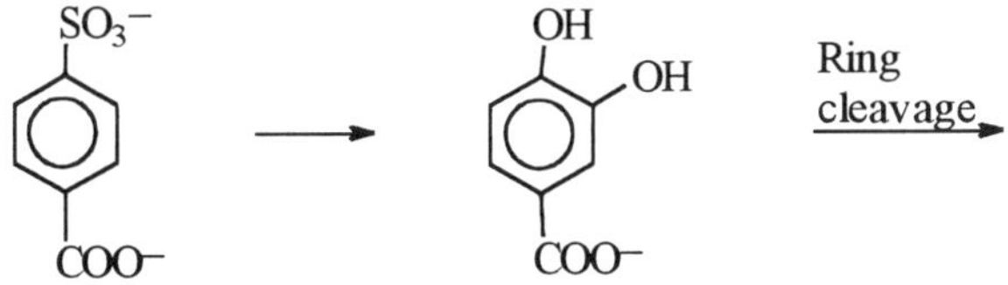

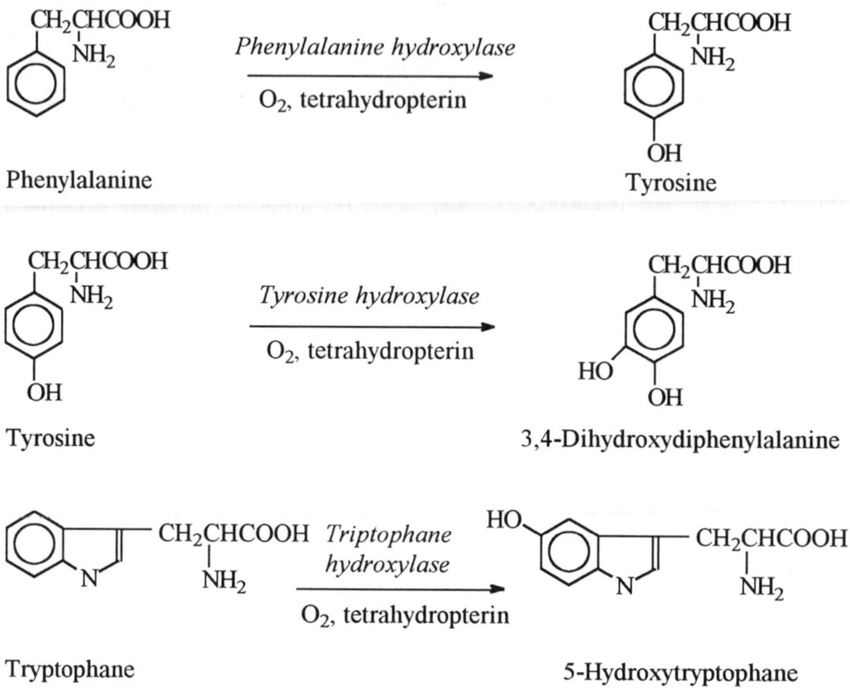

Scheme XI.9. Oxygenases which are capable of activating a C–H bond.

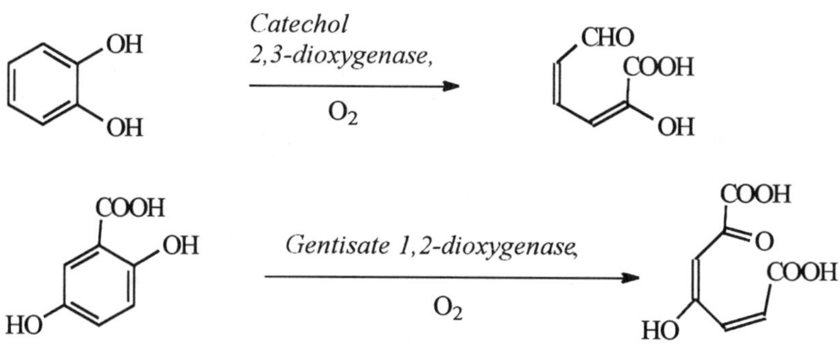

Scheme XI.10. Non-heme iron-containing dioxygenases.

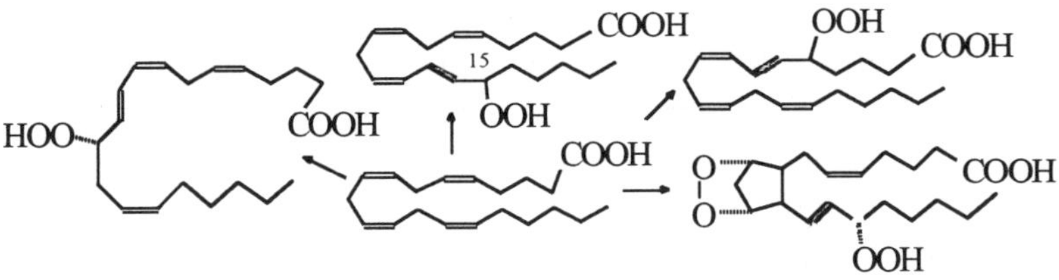

Scheme XI.11. Products formed from arachidonic acid under the action of different lypoxygenases.

Lipoxygenase is a dioxygenase which incorporates one molecule of oxygen at a certain position of an unsaturated fatty acid [48], for example arachidonate 15-lipoxygenase peroxidizes the carbon-15 of arachidonic acid as depicted in Scheme XI.11 (which shows a variety of products formed under the action of different lypoxygenases). Lypoxygenases contain non-heme iron. Thus the purified 12-lipoxygenase of porcine leukocytes contains 0.45 atoms of iron per mole of enzyme, the rabbit reticulocyte 15-lipoxygenase contains about one atom of iron.

α-Ketoglutarate-dependent dioxygenases [49] catalyze various oxidation reactions of non-activated C–H bonds, α-ketoglutarate being consumed simultaneously with the substrate and molecular oxygen. The substrate is transformed into the hydroxylated derivative; α-ketoglutarate gives rise to the formation of succinate and carbon dioxide.

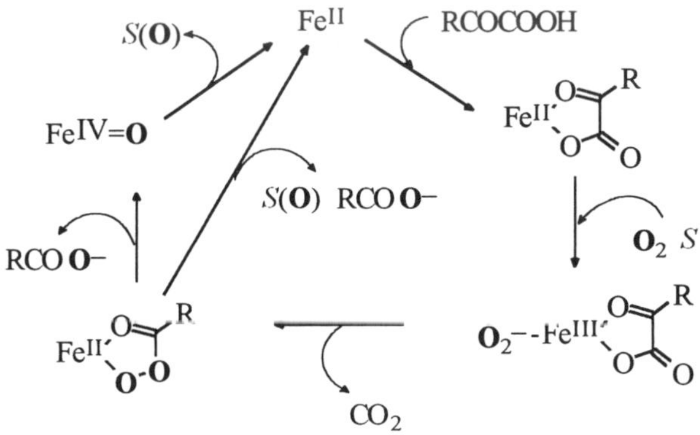

Scheme XI.12. The mechanism proposed for the α-keto acid-dependent enzymes.

In this manner, γ-butyrobetaine hydroxylase catalyzes the terminal step in carnitine biosynthesis, the hydroxylation of 4-*N*-trimethyl-aminobutyrate. α-Ketoglutarate-dependent dioxygenases act as oxygenation catalysts only in the presence of iron ions. Formation of thermodynamically stable CO_2 helps to produce a high-valent $Fe^{IV}O$ center (Scheme XI.12) [38a].

The importance of high-valent Fe=O in the molecular mechanism of action of antimalarial trioxane analogs of artemisinin has been demonstrated [50]. It is interesting that chloroperoxidase is able to catalyze not only the chlorination of organic compounds, but it is an oxygenase which can also oxidize indoles [51].

XI.5. COPPER-CONTAINING ENZYMES

Some enzymes capable of oxidizing C–H bonds contain copper ions [52]. For example, tyrosinase [53a] contains a coupled, binuclear copper active site which reversibly binds dioxygen as a peroxide that bridges between the two copper ions. This enzyme catalyzes the orthohydroxylation of phenols with further oxidation of catechol to an *ortho*-quinone [53b–d]. The mechanism proposed for phenolase activity of tyrosinase is shown in Scheme XI.13 [52l].

In bacterial (*Chromobacterium violaceum*) pterin-dependent phenylalanine hydroxylase, a cupric ion is in a tetragonal coordination environment with several nitrogen donors [54]. Dopamine β-monooxygenase takes part in the biosynthetic pathway for the production of epinephrine from tyrosine [55]:

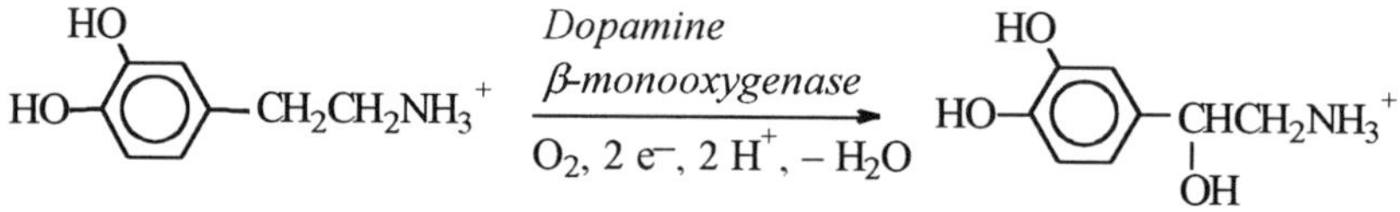

It is assumed that the oxidation by dopamine β-hydroxylase, containing a mononuclear active copper site, occurs via an intermediate copper peroxo complex Cu–OOH which generates a hydroxyl radical [52l]:

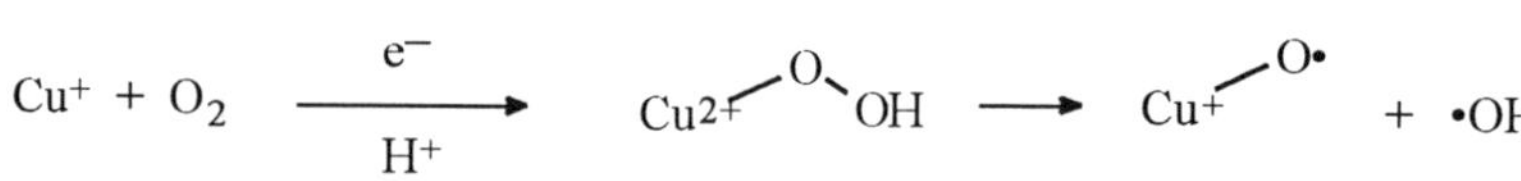

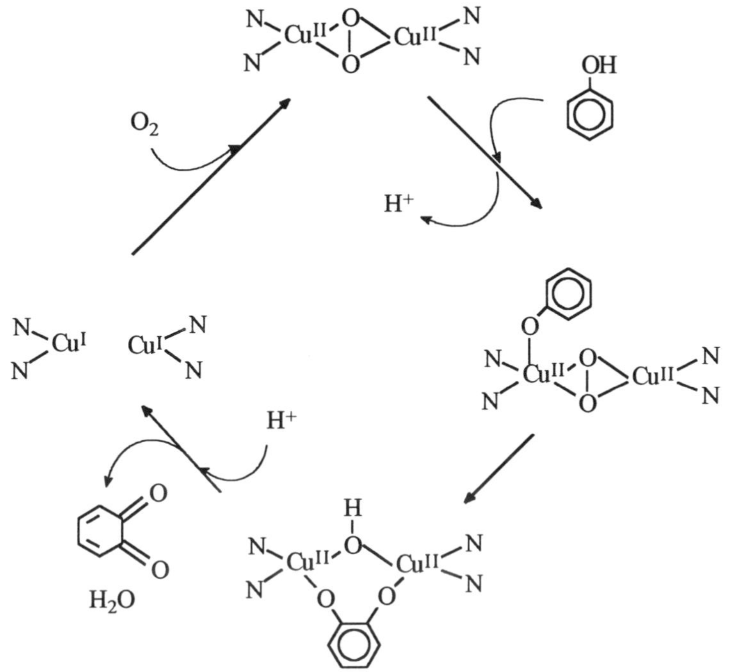

Scheme XI.13. The mechanism proposed for tyrosinase.

XI.6. MOLYBDENUM-CONTAINING ENZYMES

The pterin-containing molybdenum enzymes catalyze particularly the oxidation of C–H compounds [56, 57]. For example, xanthine oxidase transforms xanthine into uric acid according to the equation:

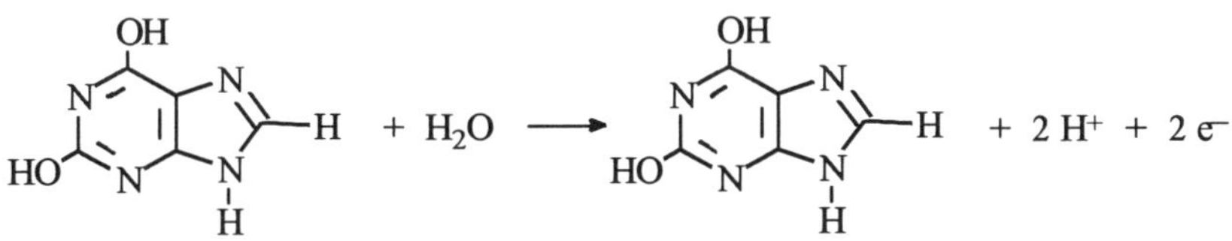

Mechanism *A*

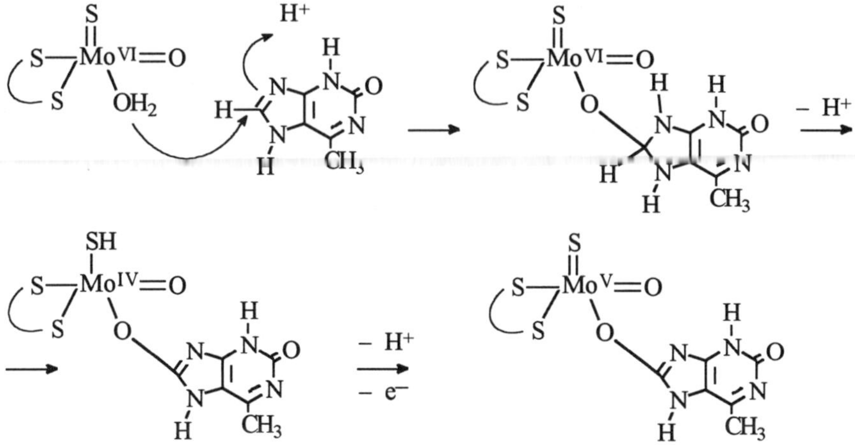

Mechanism *B*

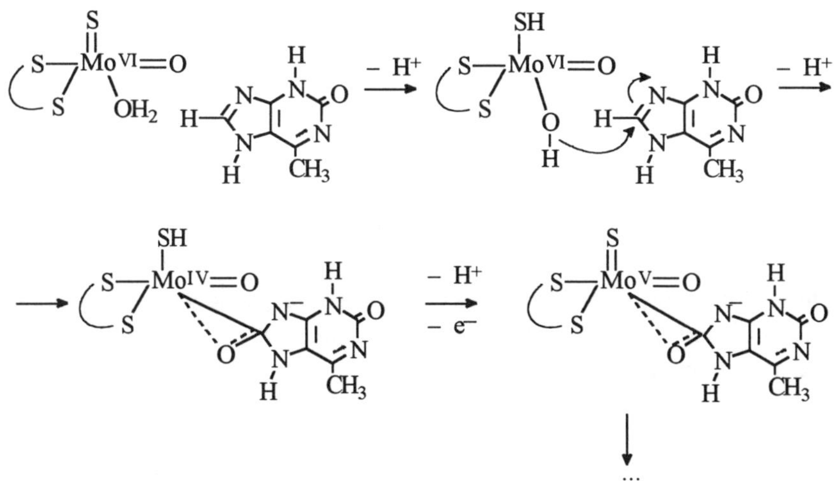

Scheme XI.14. Two mechanisms proposed for the molybdenum hydroxylases.

Two alternative reaction mechanisms (*A* and *B*) proposed for the molybdenum hydroxylases [58] are shown in Scheme XI.14. The first mechanism is based on the assumption that the catalytically labile oxygen might be a hydroxide/water rather than an oxo group.

XI.7. MANGANESE-CONTAINING ENZYMES

Manganese is a constituent of some enzymes capable of oxygenating C–H compounds (see [59]). For example, chlorocatechol 1,2-dioxygenase from *Rhodococcus erythropolis* 1CP, capable of oxidizing different catecholes, is a protein which contains one atom each of iron and manganese per homodimer [60]. The biodegradation of lignine, one of the main polymeric constituent of plants, occurs with the participation of manganese peroxidase [61].

XI.8. VANADIUM-CONTAINING ENZYMES

Vanadium ions take part in some an important processes occurring in biological systems [62]. One vanadium-containing enzyme, bromoperoxidase, has been isolated from several species of marine brown algae. This enzyme catalyzes the oxidation of the bromide anion by hydrogen peroxide, resulting in the bromination of organic compounds:

$$RH + H_2O_2 + Br^- \xrightarrow{\textit{Bromoperoxidase}} RBr + H_2O + HO^-$$

For example, the bromination of monochlorodimedone proceeds according to the following equation:

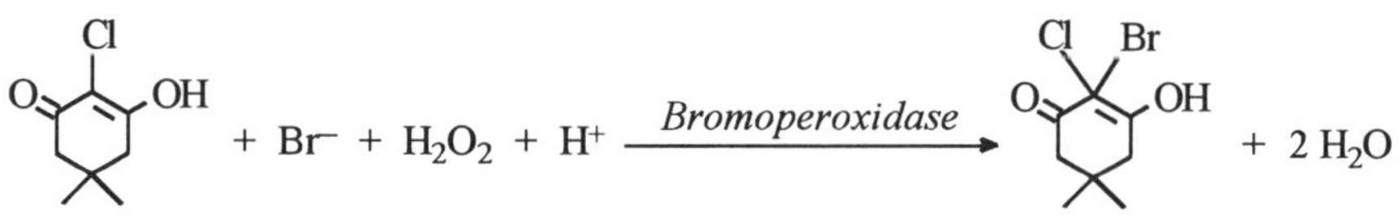

In the catalytic cycle proposed for the oxidation by bromoperoxidase, hydrogen peroxide coordinates first, followed by bromide oxidation (Scheme XI.15) [62b].

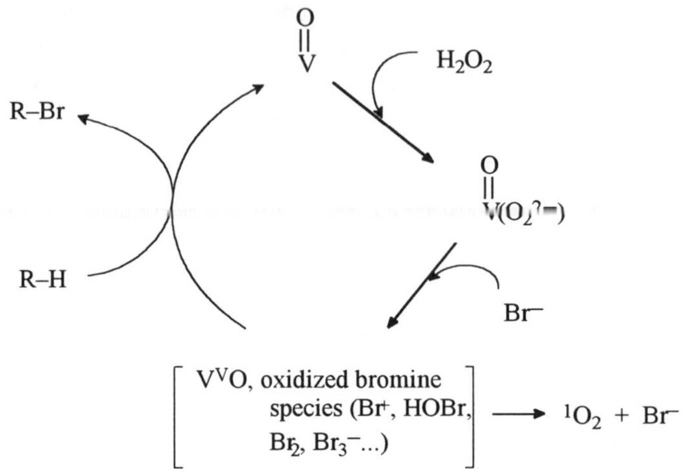

Scheme XI.15. The bromination of an organic substrate, RH, by H_2O_2 and Br^- and catalyzed by bromoperoxidase.

XI.9. CHEMICAL MODELS OF ENZYMES

There is the increasing interest of investigators not only in the enzymatic oxidation of alkanes but also in the construction of chemical models of these reactions. The progress in the chemical simulation of enzymes demonstrates that novel efficient, highly selective catalytic systems for alkane oxidation will hopefully be created based on them in the future.

XI.9.A. MODELS OF CYTOCHROME P450

The iron porphyrin complex is present not only in monooxygenases, but also in the dioxygen carriers (hemoglobin, myoglobin) and enzymes (catalase, peroxidase). One of the reasons for such a wide occurrence of the porphyrin system in living nature is its stability. In addition, the porphyrin ring can play the role of an electron donor owing to its ability to generate the radical cation and thus become an active participant of the catalytic process. Therefore, it was natural to employ metal porphyrins as catalysts in the simulation of heme monooxygenases. In modeling cytochrome P450 (see reviews [15s] and [63]), porphyrin complexes of iron(III) as well as manganese(III), chromium(III), and ruthenium(III) are usually employed as models of active centers of enzymes (see [64]). For example, it has been shown that the oxidative aromatization of aldehydes by a peroxo iron(III) porphyrin complex, $[PorphFeOO]^-$, suggests that direct nucleophilic attack by a ferric peroxo heme enzimatic intermediate on an aldehyde substrate is feasible [65]. Thus, a mechanism proposed previously for both cytochrome P450 2B4 and aromatase [66] has been supported.

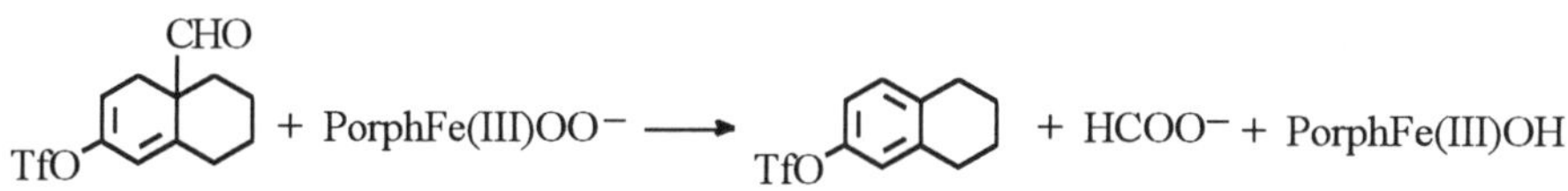

In this case, a ferric porphyrin peroxo complex plays the role of the analogue of the reactive intermediate implicated in the third oxidative step of aromatase, the enzyme responsible for the conversion of androgens to estrogens (Scheme XI.16).

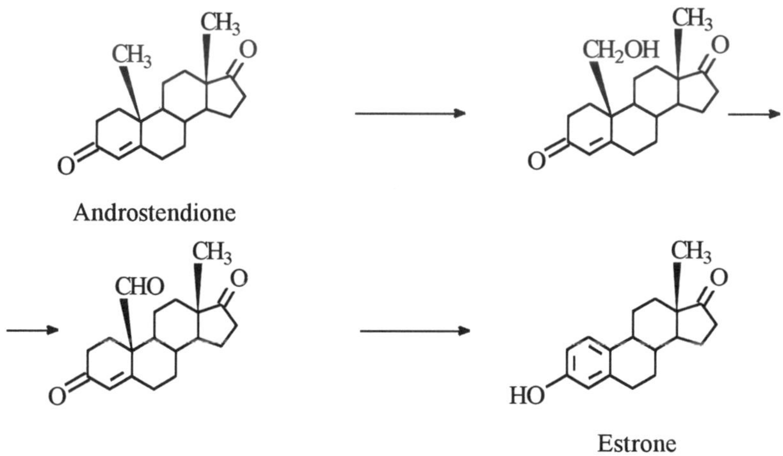

Scheme XI.16. The oxidative aromatization of androstendione into estrone under the action of aromatase.

If molecular oxygen is the oxidant in model reactions, a reducing agent is required whose role can be played by an electric current [67a,b]. Thiosalicylic acid [67c], molecular hydrogen [67d], metallic zinc [67e] and ferrocene [67f] (also see [67g]) can also serve as the reductant. Alkane oxidation in the presence of molecular oxygen and an electron (and proton) source occurs at normal temperature and pressure. The selectivity and the rate of the reaction depend on the nature of the metalloporphyrin as well as on the electron and proton donors. The product yield with respect to the reducing agent is usually low. The best yield (37%) was obtained in the oxidation of hydrocarbons having weak C–H bonds (indane and tetralin) with zinc as the reducing agent [67h].

In non-enzymic systems, it is difficult to obtain high yields of the reaction products with respect to the reducing agent, since the reactive oxidant formed in this process reacts with the substrate and the reducing agent with equal ease. Model systems based on molecular oxygen and the reducing agent are usually less selective than the systems in which active oxygen donors play the role of the oxidant. Nevertheless, the obvious advantage of such systems is a high oxidation rate, which sometimes exceeds that of enzymic reactions by more than one order of magnitude. The experimental results lead to a conclusion about a single mechanism of the generation of reactive species in the systems based on iron and manganese porpyrins with molecular oxygen and proton and electron donors. The candidates for the reactive intermediates in these systems are high-valent oxo-complexes of metalloporphyrins formed by reductive activation of molecular oxygen in the presence of a proton donor via a catalytic cycle that is similar to that of cytochrome P450. The reactive species in these systems, as well as in the systems with different active oxygen donors, may differ from each other depending on the starting reagents.

Instead of the dioxygen–reductant pair, one can employ oxo componds containing an oxygen atom which is already partly reduced: H_2O_2 [68], ROOH [69], PhIO [70], NaOCl [71], $KHSO_5$ [72], amine *N*-oxides [73], and magnesium monoperoxyphtalate [74] (see also Chapter X). One of the most efficient (in terms of reaction rate and turnover number) systems is the combination of ruthenium porphyrin and 2,6-dichloropyridine *N*-oxide [73]. A simplified mechanism of alkane oxidation with iodosylbenzene catalyzed by iron porphyrinate is demonstrated in Scheme XI.17.

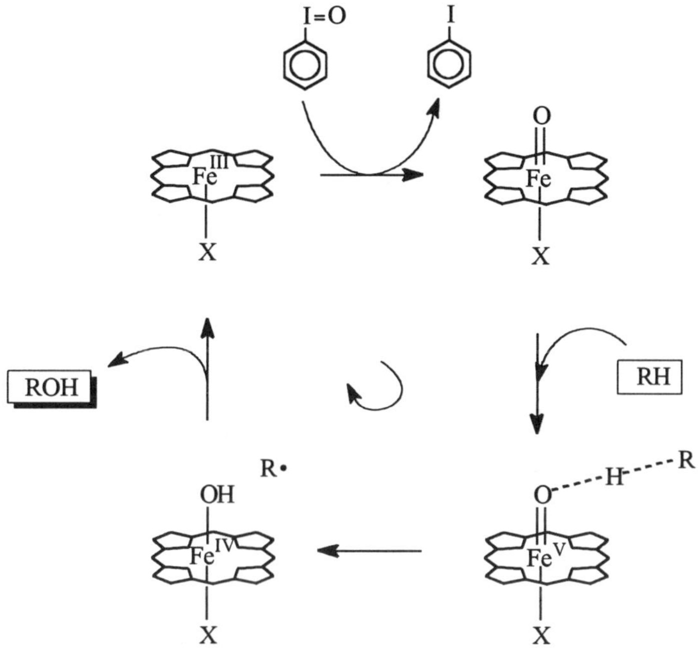

Scheme XI.17. The mechanism proposed for the alkane oxidation with iodosylbenzene catalyzed by iron porphyrinate.

In many cases, other metal complexes can be considered as models of an enzyme active center (e.g., [75]). Free radicals are definitely the intermediates in some reactions [76], particularly for hydrocarbons with weak C–H bonds, however, similarly to their biological analogs there is certainly a mechanism of direct oxygen atom insertion, possibly preceded by coordination of the alkane to form a five-coordinate carbon intermediate. In the case of chemical models with active oxygen atom donors, the formation of the (Porph^{+}• Fe^{IV}=O) intermediates was recorded by low-temperature spectroscopy for the interaction of iron tetramesitylporphyrinate with *m*-chloroperoxybenzoic acid [77a] and for a $PorphFe^{III}–O_2–Ac_2O$ system [77b,c]. The stable Mn^{V}=O complexes known so far contain macrocyclic tetraamide, porphyrin or *N,N′*-bis(salicylidene)ethylenediamine dianion as the ligands (see [77d–f]). It is interesting that the high-valent iron–oxo intermediate of thiolate-ligated porphyrin has a stronger hydrogen-atom-abstracting ability than that of chloride anion- or imidazole-ligated porphyrins [78]. Thus, the preferred structure of the active oxidizing intermediate of the thiolate-ligated porphyrins is of the oxygen radical type.

One of the ways of increasing the selectivity and the stability of catalysts is the simulation of the microenvironment of the active center of enzymes through the construction of organized molecular systems. This term refers to multicomponent catalytic systems with an optimum spatial arrangement and energetic and orbital compatibility between the components of the catalytic system and the substrates. some general principles for the design of organized molecular species have been formulated in order to achive optimum rates and selectivity of the monooxygenase reaction [79a]. The selectivity of cyclohexane oxidation in an iron tetrakis(*p*-hexadecyloxyphenyl)porphyrinate–cyclodextrin–PhIO microheterogeneous system is higher than that of the homogeneous analogue but the rate of the reaction decreases [79b]. The selectivity of microheterogeneous systems is close to that of the most regioselective isoform of cytochrome $P450_{LM2}$ (CYP2B4). Apparently, the molecules of hydrophobic iron porphirynate and alkane in these organized molecular systems are specifically arranged relative to one another in such a way that the reaction at the terminal methyl group of normal alkanes becomes preferential. Breslow *et al.* have shown recently that a metallophorphyrin carrying a β-cyclodextrin group on tetrafluorophenyl rings **XI-11** catalyzes the selective hydroxylation of an androstandiol derivative with complete positional selectivity (Scheme XI.18) [79c].

Another approach to the simulation of the catalyst's microenvironment is its immobilization on a solid support, i.e., the formation of heterogeneous oxygenase systems. Immobilization of a metalloporphyrin on porous glass [80] or polyvinyl-pyrrolidone [81] markedly increases the selectivity of alcohol formation. Immobilization of an iron–porphyrin complex on zeolites [82], SiO_2 [83] and especially on graphite, strongly increases the steric effects, which are observed in hexane oxidation in model systems with active oxygen species [84]. The effect of the matrix is not confined to the increase in the steric hindrances around the active oxidant. It can also be accompanied by a sieve effect, which is well-known for zeolites and accounts for the differences in the substrate specificity.

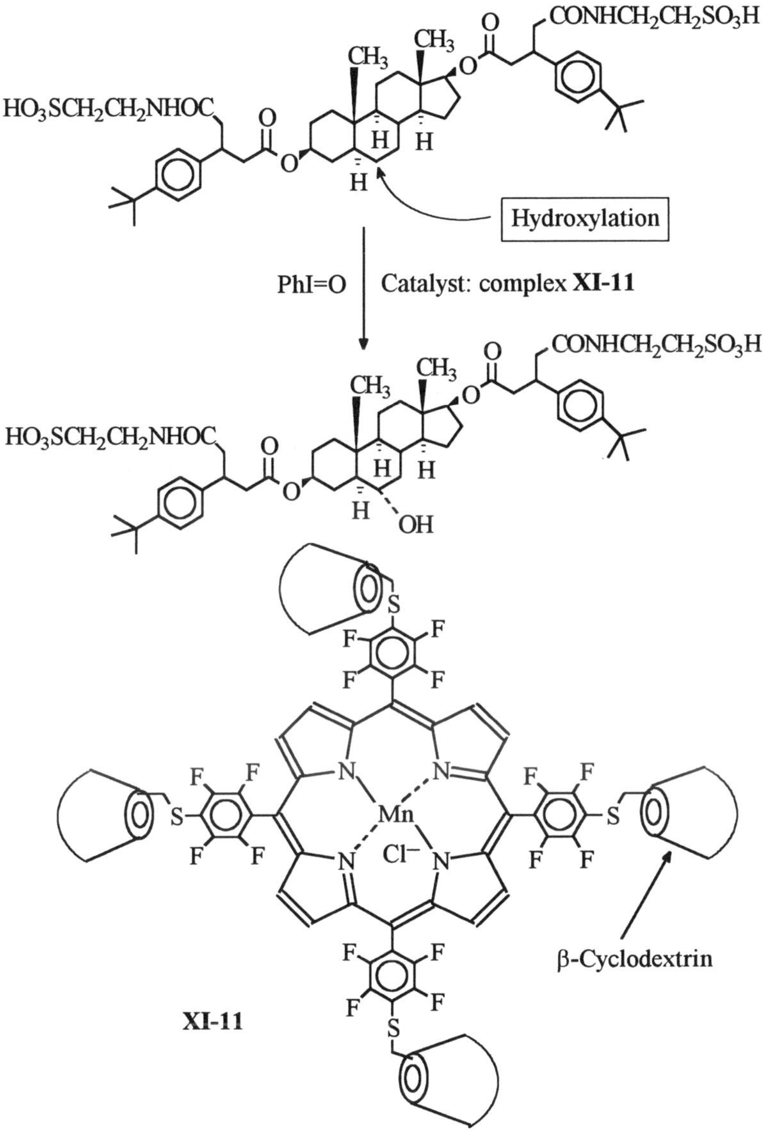

Scheme XI.18. The selective hydroxylation of an androstandiol derivative by PhIO catalyzed by β-cyclodextrin-containing Mn porphyrin complex **XI-11**.

As expected, the heterogenation of metalloporphyrin catalysts is accompanied by a decrease in the rate of hydrocarbon oxidation. At the same time, heterogeneous systems have significant advantages over their homogeneous analogues. Immobilization of metalloporhyrins on various supports not only enhances the selectivity of the reactions catalyzed by them but also increases their stability in the majority of cases. For example, immobilization of iron porphyrinate on porous glass increased the number of reaction cycles nearly 200-fold [80]. It is generally accepted that the reactive species in systems with metalloporphyrins immobilized on supports are similar to those in solution. Apparently, in most cases this conclusion reflects the real situation; however, it should be kept in mind that binding of the catalyst with the support can change its coordination properties and the ability to produce reactive intermediates.

Heterogeneous systems of alkane photooxygenation by molecular oxygen in the presence of iron and manganese porphyrins immobilized on TiO_2-modified montmorillonite have been developed by Bagrii and Nekhaev [85]. Finaly, Inoue *et al.* [86] reported the homogeneous photochemical oxygenation of cyclohexene coupled with two-electron oxidative activation of water as an axial porphyrin ligand.

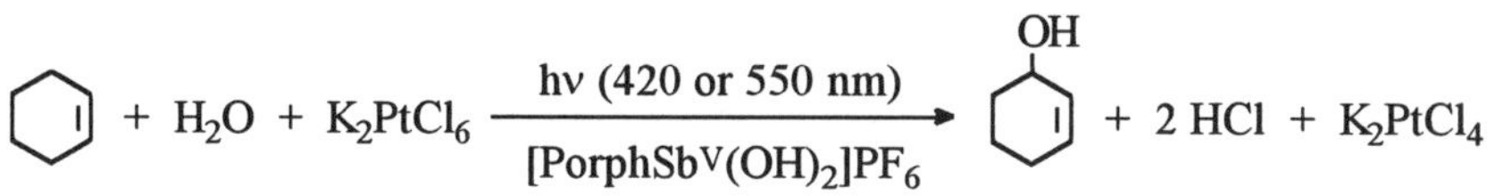

A metal-oxo type complex is supposed to be a key intermediate in the oxygenation reaction which could be called a "photochemical P450 reaction".

XI.9.B. MODELS OF IRON-CONTAINING NON-HEME OXYGENASES

Various systems oxidizing hydrocarbons and containing iron ions have been described which can be considered as models of non-heme mono- and dioxygenases [87] (see also Chapter X). Mononuclear iron derivatives have been used as catalysts in oxidations modeling the action of methane monooxygenase [88]. For example, a mononuclear iron carboxylate complex immobilized on a modified silica surface catalyzes oxidation of hexane in the presence of mercaptane, acetic acid, triphenylphosphine and molecular oxygen at ambient conditions [88c]. However, usually binuclear iron complexes were employed as MMO active site model [30, 89a–af]. One of the most effective functional models of MMO was discovered recently by Panov *et al.* They have shown that so-called α-oxygen, formed in the high-temperature reaction of nitrous oxide, N_2O, with iron compounds in the ZSM-5 zeolite, readily reacts with methane at room and lower temperatures (even at –50 °C) producing methanol in nearly quantitative yield [89ag]. So far, the reaction is not catalytic since the product (methanol) is very tightly bound to the zeolite and can be removed from it only by extraction with acetonitrile; which destroys the system. However, the fact of selective alcohol formation speaks for a non-radical mechanism presumably similar to that of methane reaction in the presence of MMO.

Complexes of iron ions with various ligands [90] as well as ruthenium(III) derivatives [91] were investigated as catalysts in oxidation reactions which mimic other iron-containing monooxygenase and dioxygenase oxidations. Iron(III) perchlorate or bis(salycilidene ethylenediamine) iron(III) complex in the presence of dioxygen, reductant (zinc amalgam), mediator (methylviologen) and effector (pyruvic acid) have been found to be a model of α-ketoglutarate-dependent dioxygenases [92]. A manganese porphyrin bound to bovine serum albumin modified with poly(ethylene glycol) has been reported to exhibit enzymatic tryptophan 2,3-dioxygenase-like activity [93]. Metal complexes and their reactions relevant to catechol dioxygenases [94], lipoxygenases [95], tyrosine hydroxylase [96] and other enzymes [97] have been described recently.

XI.9.C. MODELS OF OTHER ENZYMES

Numerous papers have been published in recent years which are devoted to the synthesis and reactions [98], structural [99], spectroscopic [100], theoretical [101], and mechanistic [102] studies of copper complexes, especially multinuclear copper complexes which are related to aromatic and aliphatic hydroxylation enzymes (also see reviews [103]). The reactions between molecular oxygen and such complexes give rise usually to the formation of hydroxylated compounds [104]. Two examples of intramolecular hydroxylation are shown in Scheme XI.19. Copper ions and their complexes are known to cata-

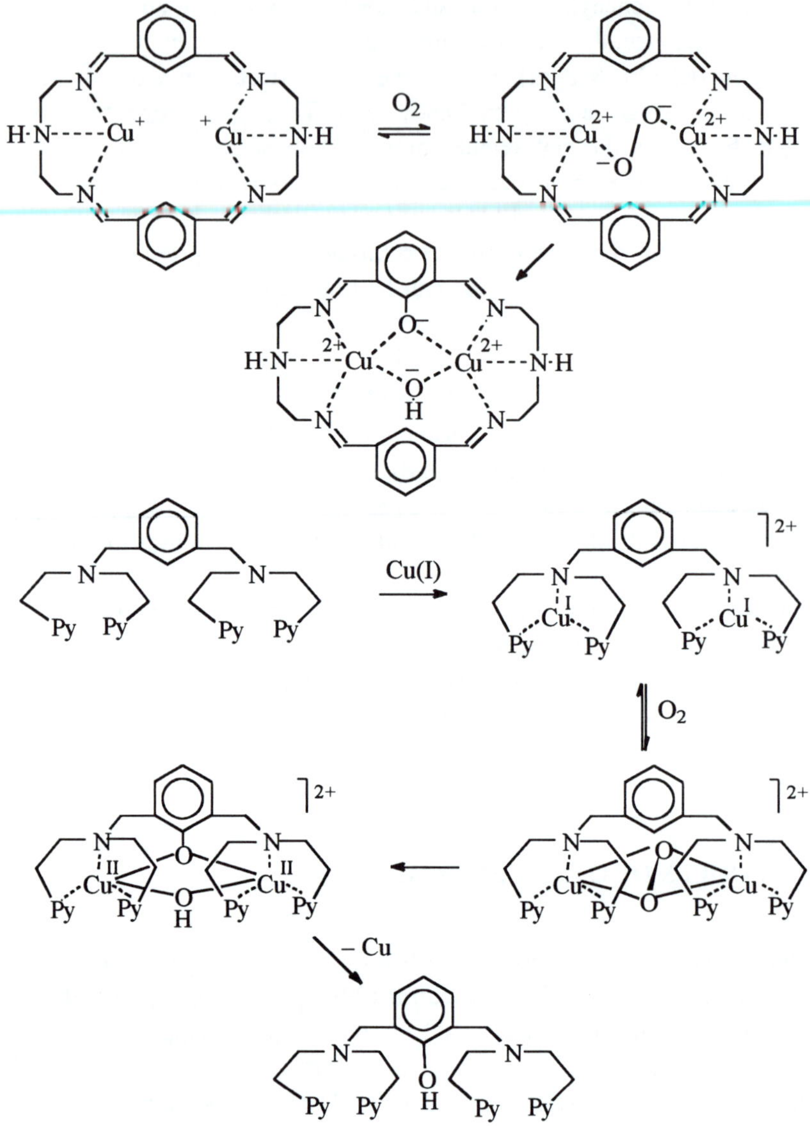

Scheme XI.19. The intramolecular aromatic hydroxylation in binuclear copper complexes modeling the hydroxylation by copper-containing enzymes.

lyze the oxygenation and even oxidative rupture of the aromatic ring in phenols [105].

Recent papers describe models for molybdenum-containing enzymes [106]. Certain vanadium complexes have been described to mimic the binding site reactions of vanadium haloperoxidases [62, 107]. Scheme XI.20 demonstrates the mechanism of active species formation in bromoperoxidase proposed on the basis of the investigation of model reactions [108].

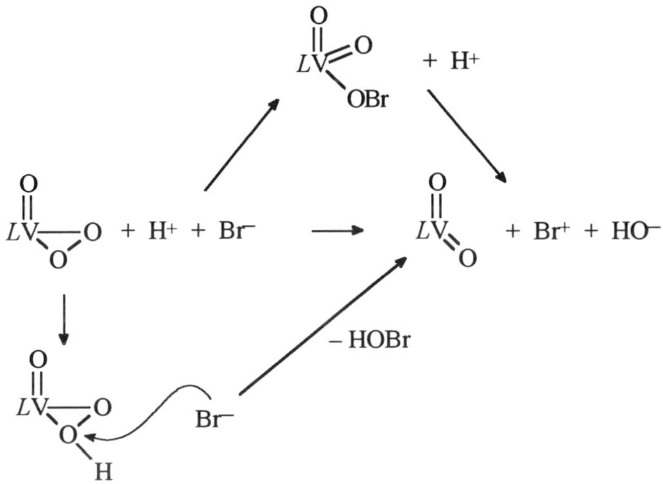

Scheme XI.20. The formation of the active species in the bromoperoxidase oxidation.

Oxidation with H_2O_2 catalyzed by MoO_4^{2-} also has been shown to mimic the action of vanadium-containing enzymes [109].

XI.10. Anaerobic Oxidation of Alkanes

Aerobic oxidation of alkanes requires dioxygen, hence it was originated only after plant photosynthesis thereby producing most part of the dioxygen on earth. Meanwhile methane is produced in anaerobic methanogenesis, therefore apparently this biological process existed before when life on earth was essentially anaerobic. Does it mean that in anaerobic conditions methane was only produced and not biologically consumed?

It turned out that the anaerobic oxidation of alkanes exists in living nature as well, although until recently biochemists were in doubt about its existence and importance. Now anaerobic biological oxidation of alkanes, including methane, was confirmed and the scale of this process is probably larger than it was expected. According to the recent findings, the oxidant in the anaerobic alkane oxidation can be sulfate which is of course unable to react directly with alkanes. Recently some evidence was found that the last steps of biological methanogenesis are reversible, therefore methane can be activated at the same complex on which it is produced in methanogenesis. This complex is factor-430 and contains nickel included in the porphynoid ring [110]. Thus a similar complex, perhaps at more positive redox potential, could be a catalyst for methane oxidation.

Nickel is an analogue of platinum and its oxidation state at which methane is produced is Ni(II) with a d^8 electronic configuration, that is the same as that of Pt(II) which activates methane and other alkanes (see Chapter VII). Evidently the activation of methane on platinum complexes may be considered a conditional model for biological anaerobic oxidation of alkanes. It is of importance that Ni(II) as well as Pt(II) is a so-called "soft" acid and could prefer to react with methane ("soft" base) rather than with such strong "hard" base as water, therefore surrounding water does not prevent this reaction.

References for Chapter XI

1. (a) Dorovska-Taran, V.; Posthumus, M. A.; Boeren, S.; Boersma, M. G.; Teunis, C. J.; Rietjens, I. M. C. M.; Veeger, C. *Eur. J. Biochem.* **1998**, *253*, 659. (b) Holland, H. L.; Allen, L. J.; Chernishenko, M. J.; Diez, M.; Kohl, A.; Ozog, J.; Gu, J.-X. *J. Mol. Catal. B: Enzymatic* **1997**, *3*, 311. (c) Held, M.; Suske, W.; Schmid, A.; Engesser, K.-H.; Kohler, H.-P.; Witholt, B.; Wubbolts, M. G. *J. Mol. Catal. B: Enzymatic* **1998**, *5*, 87. (d) Ridder, L.; Mulholland, A. J.; Vervoort, J.; Rietjens I. M. C. M. *J. Am. Chem. Soc.* **1998**, *120*, 7641.
2. (a) Njar, V. C. O.; Shapiro, S.; Aruchalam, T.; Caspi, E. *J. Steroid. Biochem.* **1985**, *22*, 339. (b) Shapiro, S.; Caspi, E. *Tetrahedron* **1998**, *54*, 5005.
3. Gullillermo, R.; Hanson, J. R.; Truneh, A. *J. Chem. Res. (S)* **1997**, 28.
4. Canfield, L. M.; Crabb, T. A. *J. Chem. Res. (S)* **1997**, 381.
5. Schmidt, M. M.; Schüler, E.; Braun, M.; Häring, D.; Schreier, P. *Tetrahedron Lett.* **1998**, *39*, 2945.
6. Boyd, D. R.; Sharma, N. D.; Byrne, B.; Hand, M. V.; Malone, J. F.; Sheldrake, G. N.; Blacker, J.; Dalton, H. *J. Chem. Soc., Perkin Trans. 2* **1998**, 1935.
7. (a) Ueda, M.; Sashida, R. *J. Mol. Catal. B: Enzymatic* **1998**, *4*, 199. (b) Torimura, M.; Yoshida, H.; Kano, K.; Ikeda, T.; Nagasawa, T.; Ueda, T. *Chemistry Lett.* **1998**, 295.
8. Weber, H.; Braunegg, G.; de Raadt, A.; Feichtenhofer, S.; Gringl, H.; Lübke, K.; Klinger, M. F.; Kreiner, M.; Lehmann, A. *J. Mol. Catal. B: Enzymatic* **1998**, *5*, 191.
9. Aranda, G.; Bertranne-Delahaye, M.; Lallemand, J.-Y.; Azerad, R.; Maurs, M.; Cortes, M.; Lopez, J. *J. Mol. Catal. B: Enzymatic* **1998**, *5*, 203.
10. Hamada, H.; Miyamoto, Y.; Nakajima, N.; Furuya, T. *J. Mol. Catal. B: Enzymatic* **1998**, *5*, 187.
11. (a) Tay, S. T.-L.; Hemond, H. F.; Polz, M. F.; Cavanaugh, C. M.; Dejesus, I.; Krumholz, L. R. *Appl. Environ. Micobiol.* **1998**, *64*, 1715. (b) Weiner, J. M.; Loveley, D. R. *Appl. Environ. Micobiol.* **1998**, *64*, 1937. (c) Barton, J. W.; Klasson, K. T.; Koran, L. J., Jr.; Davison, B. H. *Biotechnol. Progr.* **1997**, *13*, 814. (d) Schmid, A.; Kohler, H.-P. E.; Engesser, K.-H. *J. Mol. Catal. B: Enzymatic* **1998**, *5*, 311. (e) Bidaud, C.; Tran-Minh, C. *J. Mol. Catal. B: Enzymatic* **1998**, *5*, 417.

12. Fetzner, S.; Tshisuaka, B.; Lingens, F.; Kappl, R.; Hüttermann, J. *Angew. Chem., Int. Ed. Engl.* **1998**, *37*, 577.

13. Deschler, C.; Duran, R.; Junqua, M.; Landou, C.; Salvado, J.-C.; Goulas, P. *J. Mol. Catal. B: Enzymatic* **1998**, *5*, 423.

14. (a) *Molecular Mechanisms of Oxygen Activation*; Hayaishi, O., Ed.; Academic Press: New York, 1974. (b) *Microbal Degradation of Organic Compounds*; Gibson, D. T., Ed.; Marcell Dekker: New York, 1984. (c) Metelitsa, D. I. *The Modelling of Oxidation-Reduction Enzymes*; Nauka i Tekhnika: Minsk, 1984 (in Russian). (d) Large, P. J.; Bamforth, C. W. *Methylotrophy and Biotechnology*; Longman Sci. Techn.: Essex, 1988. (e) *Oxidases and Related Redox Systems*; A. R. Liss, Inc.: New York, 1988. (f) *The Role of Oxygen in Chemistry and Biochemistry*; Ando, W., Moro-oka, Y., Eds.; Elsevier: Amsterdam, 1988. (g) *Oxygen Complexes and Oxygen Activation by Transition Metals*; Martell, A. E., Sawyer, D. T., Eds.; Plenum: New York, 1988. (h) *Free Radicals in Synthesis and Biology*; Minisci, F., Ed.; Kluwer: Dordrecht, 1989. (i) *Biological Oxidation Systems*; Reddy, C., Hamilton, G. A., Madyastha, K. M., Eds.; Academic Press: New York, 1990. (j) *Catalytic Activation of Dioxygen by Metal Complexes*; Simándi, L. I., Ed.; Kluwer: Dordrecht, 1992. (k) *Metal Ions in Biological Systems*; Sigel, H., Sigel, A., Eds.; Marcel Dekker: New York, 1992. (l) *Bioinorganic Catalysis*; Reedijk, J., Ed.; Marcel Dekker: New York, 1993. (m) Lippard, S. J.; Berg, J. M. *Principles of Bioinorganic Chemistry*; University Science Books: Mill Valley, CA, 1994. (n) *Metalloporphyrins Catalyzed Oxidations*; Montanari, F., Casella, L, Eds.; Kluwer: Dordrecht, 1994. (o) *Metalloporphyrins in Catalytic Oxidations*; Sheldon, R., Ed.; Marcel Dekker: New York, 1994. (p) *Microbal Growth on C_1 Compounds*; Lindstrom, M. E., Tabita, F. R., Eds.; Kluwer: Dordrecht, 1996. (q) *Molecular Modeling and Dynamics of Bioinorganic Systems*; Banci, L., Comba, P., Eds.; Kluwer: Dordrecht, 1997. (r) *Oxygenases and Model Systems*; Funabiki, T., Ed.; Kluwer: Dordrecht, 1997. (s) Shilov, A. E. *Metal Complexes in Biomimetic Chemical Reactions*; CRC Press: Boca Raton, 1997.

15. (a) Kanofsky, J. R. *Chem.-Biol. Interactions* **1989**, *70*, 1. (b) Kozlova, N. B.; Skurlatov, Yu. I. *Usp. Khim.* **1989**, *58*, 234 (in Russian). (c) Khenkin, A. M.; Shteinman, A. A. *Kinet. Katal.* **1989**, *30*, 7 (in Russian). (d) Minotti, G.; Aust, S. D. *Chem.-Biol. Interactions* **1989**, *71*, 1. (e) Frey, P. A. *Chem. Rev.* **1990**, *90*, 1343. (f) Mansuy, D. *Pure Appl. Chem.* **1990**, *62*, 741. (g) Miller, D. M.; Buettner, G. R.; Aust, S. D. *Free Radical Biol. Med.* **1990**, *8*, 95. (h)

Ramasarma, T. *Ind. J. Biochem. Biophys.* **1990**, *27*, 269. (i) Ramasarma, T. *Ind. J. Biochem. Biophys.* **1990**, *27*, 269. (j) Adams, J. D., Jr.; Odunze, I. N. *Free Rad. Biol. Med.* **1991**, *10*, 161. (k) Nogatsu, T.; Ichinose, H. *Comp. Biochem. Physiol.* **1991**, *98C*, 203. (l) Bagriy, E. I.; Plate, N. A.; Neverova, I. N.; Krentsel', B. A. *Neftekhimiya* **1991**, *31*, 190 (in Russian). (m) Bertini, I.; Messori, L.; Viezzoli, M. S. *Coord. Chem. Rev.* **1992**, *120*, 163. (n) Mansuy, D. *Coord. Chem. Rev.* **1993**, *125*, 129. (o) Maldotti, A.; Amadelli, R.; Bartocci, C.; Carassiti, V.; Polo, E.; Varani, G. *Coord. Chem. Rev.* **1993**, *125*, 143. (p) Bertini, I.; Briganti, F.; Scozzafava, A.; Luchinat, C. *New J. Chem.* **1996**, *20*, 187. (q) Kovalenko, G. A. *Usp. Khim.* **1996**, *65*, 676 (in Russian). (r) Holm, R. H.; Kennepohl, P.; Solomon, E. I. *Chem. Rev.* **1996**, *96*, 2239. (s) Karasevich, E. I.; Kulikova, V. S.; Shilov, A. E.; Shteinman, A. A. *Usp. Khim.* **1998**, *67*, 376 (in Russian).

16. (a) Sono, M.; Roach, M. P.; Coulter, E. D.; Dawson, J. H. *Chem. Rev.* **1996**, *96*, 2841. (b) Hanna, I. H.; Roberts, E. S.; Hollenberg, P. F. *Biochemistry* **1998**, *37*, 311. (c) De Voss, J. J.; Sibbesen, O.; Zhang, Z.; Ortiz de Montellano, P. R. *J. Am. Chem. Soc.* **1997**, *119*, 5489. (d) Matsuoka, T.; Miyakoshi, S.; Tanzawa, K. *et al. Eur. J. Biochem.* **1989**, *184*, 707. (e) Thompson, E. A.; Siiteri, P. K. *J. Biol. Chem.* **1974**, *249*, 5364, 5373. (f) Wright, J. N.; Van Leersum, P. T.; Chamberlin, S. G.; Athtar, M. *J. Chem. Soc., Perkin Trans. 1* **1989**, 1647. (g) Fretz, H.; Woggen, W.-D.; Voges, R. *Helv. Chim. Acta.* **1989**, *184*, 707.

17. (a) *Basis and Mechanism of Regulation of Cytochrome P-450*; Ruckpaul, K., Rein, H., Eds.; Akademie-Verlag: Berlin, 1989. (b) Archakov, A. I.; Bachmanova, G. I. *Cytochrome P-450 and Active Oxygen*, Taylor & Francis: London, 1990. (c) *Cytochrome P-450: Structure, Mechanism and Biochemistry*; Ortiz de Montellano, P. Ed.; Plenum Press: New York, 1995.

18. (a) Metelitsa, D. I. *Usp. Khim.* **1981**, *50*, 2019 (in Russian). (b) Guengerich, F. P.; Macdonald, T. L. *Acc. Chem. Res.* **1984**, *17*, 9.

19. (a) Boddupalli, S. S., Estabrook, R. W.; Peterson, J. A. *J. Biol. Chem.* **1990**, *265*, 4233. (b) Yoshimura, R.; Kusunose, E.; Yokotani, N.; Yamamoto, S.; Kubota, I.; Kusunose, M. *J. Biochem.* **1990**, *108*, 544. (c) Lehmann, W. D.; Fürstenberger, G. *Angew. Chem., Int. Ed. Engl.* **1993**, *32*, 1027. (d) Robichaud, P.; Wright, J. N.; Akhtar, M. *J. Chem. Soc., Chem. Commun.* **1994**, 1501. (e) England, P. A.; Rouch, D. A.; Westlake, C. G.; Bell, S. G.; Nickerson, D. P.; Webberley, M.; Flitsch, S. L.; Wong, L.-L. *Chem. Comm.*

1996, 357. (f) Jones, N. E.; England, P. A.; Rouch, D. A.; Wong, L.-L. *Chem. Comm.* **1996**, 2413. (g) Stevenson, J.-A.; Bearpark, J. K.; Wong. L.-L. *New J. Chem.* **1998**, 551. (h) Green, S. P.; Whiting, D. A. *J. Chem. Soc., Perkin Trans. 1* **1998**, 193. (i) Tschirret-Guth, R. A.; Hoa, G. H. B.; Ortiz de Montellano, P. R. *J. Am. Chem. Soc.* **1998**, *120*, 3590. (j) Lvov, Yu. M.; Lu, Z.; Schenkman, J. B.; Zu, X.; Rusling, J. F. *J. Am. Chem. Soc.* **1998**, *120*, 4073. (k) Racha, J. K.; Kunze, K. L. *J. Am. Chem. Soc.* **1998**, *120*, 5337.

20. (a) Eberson, L. *Acta Chem. Scand.* **1990**, *44*, 733. (b) Korzekwa, K. R.; Jones, J. P.; Gillette, J. R. *J. Am. Chem. Soc.* **1990**, *112*, 7042. (c) Korzekwa, K. R.; Trager, W. F.; Smith, S. J.; Osawa, Y.; Gillette, J. R. *Biochemistry* **1991**, *30*, 6155. (d) Harris, D.; Loew, G. *J. Am. Chem. Soc.* **1995**, *117*, 2738. (e) De Voss, J. J.; Ortiz de Montellano, P. R. *J. Am. Chem. Soc.* **1995**, *117*, 4185. (f) Harris, D. L.; Loew, G. H. *J. Am. Chem. Soc.* **1996**, *118*, 6377. (g) Zakharieva, O.; Grodzicki, M.; Trautwein, A. X.; Veeger, C.; Rietjens, I. M. C. M. *JBIC*, **1996**, *1*, 192. (h) Torrens, F. *J. Mol. Catal. A: Chem.* **199x**, *119*, 393. (i) Yoshizawa, K.; Shiota, Y.; Yamabe, T. *Chem Eur. J.* **1997**, *3*, 1160. (j) Yoshizawa, K.; Shiota, Y.; Yamabe, T. *J. Am. Chem. Soc.* **1998**, *120*, 564. (k) Yoshizawa, K.; Shiota, Y.; Yamabe, T. *Organometallics* **1998**, *17*, 2825. (l) Harris, D.; Loew, G.; Waskell, L. *J. Am. Chem. Soc.* **1998**, *120*, 4308. (m) Roitberg, A. E.; Holden, M. J.; Mayhew, M. P.; Kurnikov, I. V.; Beratan, D. N.; Vilker, V. L. *J. Am. Chem. Soc.* **1998**, *120*, 8927. (n) Harris, D. L.; Loew, G. H. *J. Am. Chem. Soc.* **1998**, *120*, 8941. (o) Maseras, F. *New J. Chem.* **1998**, 327.

21. (a) Hanzlik, R. P.; Ling, K.-H. J. *J. Org. Chem.* **1990**, *55*, 3992. (b) Amodeo, R.; Baciocchi, E.; Crescenzi, M.; Lanzalunga, O. *Tetrahedron Lett.* **1990**, *31*, 3477. (c) Cole, P. A.; Robinson, C. H. *J. Med. Chem.* **1990**, *33*, 2933. (d) Baciocchi, E.; Crescenzi, M.; Lanzalunga, O. *J. Chem. Soc., Chem. Commun.* **1990**, 687. (e) Bowry, V. W.; Lusztyk, J.; Ingold, K. U. *J. Am. Chem. Soc.* **1991**, *113*, 5687. (f) Bowry, V. W.; Ingold, K. U. *J. Am. Chem. Soc.* **1991**, *113*, 5699. (g) Wright, J. N.; Slatcher, G.; Akhtar, M. *Biochem. J.* **1991**, *273*, 533. (h) Hanzlik, R. P.; Ling, K.-H. J. *J. Am. Chem. Soc.* **1993**, *115*, 9363. (i) Newcomb, M.; Le Tadic, M.-H.; Putt, D. A.; Hollenberg, P. F. *J. Am. Chem. Soc.* **1995**, *117*, 3312. (j) Bach, R. D.; Mintcheva, I.; Estévez, C. M.; Schlegel, H. B. *J. Am. Chem. Soc.* **1995**, *117*, 10121. (k) Newcomb, M.; Le Tadic-Biadatti, M.-H.; Chestney, D. L.; Roberts, E. S.; Hollenberg, P. F. *J. Am. Chem. Soc.* **1995**, *117*, 12085. (l) Choi, S.-Y.; Eaton, P. E.; Hollenberg, P. F.; Liu, K. E.; Lippard, S. J.; Newcomb, M.; Putt, D. A.; Upadhyaya, S. P.;

Xiong, Y. *J. Am. Chem. Soc.* **1996**, *118*, 6547. (m) Urano, Y.; Higuchi, T.; Hirobe, M. *J. Chem. Soc., Perkin Trans. 2* **1996**, 1169. (n) Green, S. P.; Whiting, D. A. *J. Chem. Soc., Perkin Trans. 1* **1996**, 1027. (o) Stevenson, J.-A.; Westlake, A. C. G.; Whittock, C.; Wong, L.-L. *J. Am. Chem. Soc.* **1996**, *118*, 12846. (p) Toy, P. H.; Dhanabalasingam, B.; Newcomb, M.; Hanna, I. H.; Hollenberg, P. F. *J. Org. Chem.* **1997**, *62*, 9114. (q) Gross, Z.; Nimri, S.; Barzilay, C. M.; Simkhovich, L. *JBIC*, **1997**, *2*, 492. (r) Manchester, J. I.; Dinnocenzo, J. P.; Higgins, L., Jones, J. P. *J. Am. Chem. Soc.* **1997**, *119*, 5069. (s) Shaik, S.; Filatov, M.; Schröder, D.; Schwarz, H. *Chem Eur. J.* **1998**, *4*, 193. (t) Toy, P. H.; Newcomb, M.; Coon, M. J.; Vaz, A. D. N. *J. Am. Chem. Soc.* **1998**, *120*, 9718. (u) Toy, P. H.; Newcomb, M.; Hollenberg, P. F. *J. Am. Chem. Soc.* **1998**, *120*, 7719. (v) Murakami, T,; Yamaguchi, K.; Watanabe, Y.; Morichima, I. *Bull. Chem. Soc. Jpn.* **1998**, *71*, 1343. (w) Higgins, L.; Bennett, G. A.; Shimoji, M.; Jones, J. P. *Biochemistry* **1998**, *37*, 7039. (x) Collman, J. P.; Chien, A. S.; Eberspacher, T. A.; Brauman, J. I. *J. Am. Chem. Soc.* **1998**, *120*, 425. (y) Di Gleria, K.; Nickerson, D. P.; Hill, H. A. O.; Wong, L.-L.; Fülöp, V. *J. Am. Chem. Soc.* **1998**, *120*, 46.

22. (a) Bell, S. G.; Rouch, D. A.; Wong, L.-L. *J. Mol. Catal. B: Enzymatic* **1997**, *3*, 293. (b) Smith, G.; Modi, S.; Lillai, I.; Lian, L.-Y.; Sutcliffe, M. J.; Pritchard, M. P.; Friedberg, T.; Roberts, G. C. K.; Wolf, C. R. *Biochem. J.* **1998**, *331*, 783. (c) Cane, D. E. Graziani, E. I. *J. Am. Chem. Soc.* **1998**, *120*, 2682. (d) Yamazaki, T.; Ohno, T.; Sakaki, T.; Akiyoshi-Shibata, M.; Yabusaki, Y.; Imai, T.; Kominami, S. *Biochemistry* **1998**, *37*, 2800. (e) Kobayashi, Y.; Fang, X.; Szklarz, G. D.; Halpert, J. R. *Biochemistry* **1998**, *37*, 6679. (f) Lee-Robichaud, P.; Akhtar, M. E.; Akhtar, M. *Biochem. J.* **1998**, *330*, 967. (g) Numazawa, M.; Yoshimura, A.; Oshibe, M. *Biochem. J.* **1998**, *329*, 151. (h) Schmid, M.; Löffler, G. *Eur. J. Biochem.* **1998**, *252*, 147. (i) Sopkova-De Oliveira Santos, J.; Smith, J. C.; Delaforge, M.; Virelizier, H.; Jankowski, C. K. *Eur. J. Biochem.* **1998**, *251*, 398. (j) Korzekwa, K. R.; Krishnamachary, N.; Shou, M.; Ogai, A.; Parise, R. A.; Rettie, A. E.; Gonzalez, F. J.; Tracy, T. S. *Biochemistry* **1998**, *37*, 4137. (k) Racha, J. K.; Rettie, A. E.; Kunze, K. L. *Biochemistry* **1998**, *37*, 7407. (l) Contzen, J.; Jung, C. *Biochemistry* **1998**, *37*, 4317.

23. (a) Franzen, S.; Roach, M. P.; Chen, Y.-P.; Dyer, R. B.; Woodruff, W. H.; Dawson, J. H. *J. Am. Chem. Soc.* **1998**, *120*, 4658. (b) Wagenknecht, H.-A.; Woggon, W.-D. *Angew. Chem., Int. Ed. Engl.* **1997**, *36*, 390.

24. (a) Wallar, B. J.; Lipscomb, J. D. *Chem. Rev.* **1996**, *96*, 2625. (b) Kurtz, D. M., Jr. *IBIC*, **1997**, *2*, 159. (c) van den Beuken, E. K.; Feringa, B. L. *Tetrahedron*, **1998**, *54*, 12985.

25. (a) Vincent, J. B.; Olivier-Lilley, G. L.; Averill, B. A. *Chem. Rev.* **1990**, *90*, 1447. (b) Wilkins, R. G. *Chem. Soc. Rev.* **1992**, 171. (c) Shteinman, A. A. *Russ. Chem. Bull.* **1995**, *44*, 975. (d) Liu, K. E.; Lippard, S. J. *Adv. Inorg. Chem.* **1995**, *42*, 263. (e) Que, L., Jr.; Dong, Y. *Acc. Chem. Res.* **1996**, *29*, 190. (f) Valentine, A. M.; Lippard, S. J. *J. Chem. Soc., Dalton Trans.* **1997**, 3925.

26. (a) Takeguchi, M.; Miyakawa, K.; Okura, I. *J. Mol. Catal. A: Chem.* **1998**, *132*, 145. (b) Tellez, C. M.; Gaus, K. P.; Graham, D. W.; Arnold, R. G.; Guzman, R. Z. *Appl. Eniron. Microbiol.* **1998**, *64*, 1115. (c) Shinohara, Y.; Uchiyama, H.; Yagi, O.; Kusakabe, I. *J. Fermentation and Bioengineering* **1998**, *85*, 37. (d) Nguyen, H.-H.; Nakagawa, K. H.; Hedman, B.; Elliott, S. J.; Lindstrom, M. E.; Hodgson, K. O.; Chan, S. I. *J. Am. Chem. Soc.* **1996**, *118*, 12766. (e) Elliott, S. J.; Randall, D. W.; Britt, R. D.; Chan, S. I. *J. Am. Chem. Soc.* **1998**, *120*, 3247. (f) Valentine, A. M.; Tavares, P.; Pereira, A. S.; Davydov, R.; Krebs, C.; Hoffman, B. M.; Edmodson, D. E.; Huynh, B. H.; Lippard, S. J. *J. Am. Chem. Soc.* **1998**, *120*, 2190. (g) Bollinger, J. M., Jr.; Krebs, C.; Vicol, A.; Chen, S.; Ley, B. A.; Edmondson, D. E.; Huynh, B. H. *J. Am. Chem. Soc.* **1998**, *120*, 1094.

27. (a) Shimoda, M.; Okura, I. *J. Chem. Soc., Chem. Commun.* **1990**, 533. (b) Nemoto, S.; Shimoda, M.; Okura, I. *J. Mol. Catal.* **1991**, *64*, L35. (c) Shimoda, M.; Okura, I. *J. Mol. Catal.* **1991**, *64*, L39. (d) Shimoda, M.; Nemoto, S.; Okura, I. *J. Mol. Catal.* **1991**, *64*, 373. (e) Shimoda, M.; Seki, Y.; Okura, I. *J. Mol. Catal.* **1993**, *83*, L5. (f) Sugimori, D.; Okura, I. *J. Mol. Catal. A: Chem.* **1995**, *97*, L135. (g) Liu, K. E.; Valentine, A. M.; Wang, D.; Huynh, B. H.; Edmondson, D. E.; Salifoglou, A.; Lippard, S. J. *J. Am. Chem. Soc.* **1995**, *117*, 10174. (h) Liu, A. E.; Valentine, A. M.; Qiu, D.; Edmondson, D. E.; Appelman, E. H.; Spiro, T. G.; Lippard, S. J. *J. Am. Chem. Soc.* **1995**, *117*, 4997. (i) Newman, L. M.; Wackett, L. P. *Biochemistry*, **1995**, *34*, 14066. (j) Elliott, S. J.; Zhu, M.; Tso, L.; Nguenh, H.-H. T.; Yip, J. H.-K.; Chan, S. I. *J. Am. Chem. Soc.* **1997**, *119*, 9949. (k) Shinohara, Y.; Uchiyama, H.; Kusakabe, I. *J. Fermentation and Bioengineering* **1998**, *85*, 266.

28. (a) Green, J.; Dalton, H. *J. Biol. Chem.* **1989**, *264*, 17698. (b) Akent'eva, N. P.; Gvozdev, R. I. *Biokhimiya*, **1988**, *53*, 91 (in Russian). (c) Fox, B. G.; Hendrich, M. P.; Surerus, K. K.; Andersson, K. K.; Froland, W. A.;

Lipscomb, J. D.; Münck, E. *J. Am. Chem. Soc.* **1993**, *115*, 3688. (d) Lee, S.-K.; Fox, B. G.; Froland, W. A.; Lipscomb, J. D.; Münck, E. *J. Am. Chem. Soc.* **1993**, *115*, 6450. (e) Thomann, H.; Bernardo, M.; McCormick, J. M.; Pulver, S.; Andersson, K. K.; Lipscomb, J. D.; Solomon, E. I. *J. Am. Chem. Soc.* **1993**, *115*, 8881. (f) Pulver, S.; Froland, W. A.; Fox, B. G.; Lipscomb, J. D.; Solomon, E. I. *J. Am. Chem. Soc.* **1993**, *115*, 12409.

29. (a) Liu, K. E.; Johnson, C. C.; Newcomb, M.; Lippard, S. J. *J. Am. Chem. Soc.* **1993**, *115*, 939. (b) Liu, K. E.; Wang, D.; Huynh, B. H.; Edmondson, D. E.; Salifoglou, A.; Lippard, S. J. *J. Am. Chem. Soc.* **1994**, *116*, 7465. (c) Feig, A. L.; Lippard, S. J. *Chem. Rev.* **1994**, *94*, 759. (d) DeRose, V. J.; Liu, K. E.; Lippard, S. J.; Hoffman, B. M. *J. Am. Chem. Soc.* **1996**, *118*, 121. (e) Nesheim, J. C.; Lipscomb, J. D. *Biochemistry*, **1996**, *35*, 10240. (f) Yoshizawa, K.; Hoffmann, R. *Inorg. Chem.* **1996**, *35*, 2409.

30. (a) Shteinman, A. A. *FEBS Lett.* **1995**, *362*, 5 (b) Shteinman, A. A. *Izv. Akad. Nauk, Ser. Khim.* **1995**, 1011 (in Russian). (c) Shteinman, A. A. *Izv. Akad. Nauk, Ser. Khim.* **1997**, 1676 (in Russian). (d) Shteinman, A. A. *JBIC*, **1998**, *3*, 325.

31. (a) Nesheim, J. C.; Lipscomb, J. D. *J. Inorg. Biochem.* **1995**, *59*, 269. (b) Korshunova, L. A.; Gvozdev, R. I.; Dolton, G. *Dokl. Russ. Akad. Nauk* **1997**, *352*, 545 (in Russian).

32. (a) Wilkinson, B.; Zhu, M.; Priestley, N. D.; Nguyen, H.-H. T.; Morimoto, H.; Williams, P. G.; Chan, S. I.; Floss, H. G. *J. Am. Chem. Soc.* **1996**, *118*, 921. (b) Valentine, A. M.; Wilkinson, B.; Liu, K. E.; Komar-Panicucci, S.; Priestley, N. D.; Williams, P. G.; Morimoto, H.; Floss, H. G.; Lippard, S. J. *J. Am. Chem. Soc.* **1997**, *119*, 1818.

33. (a) Yoshizawa, K.; Ohta, T.; Shiota, Y.; Yamabe, T. *Chemistry Lett.* **1997**, 1213. (b) Yoshizawa, K.; Ohta, T.; Yamabe, T.; Hoffmann, R. *J. Am. Chem. Soc.* **1997**, *119*, 12311. (c) Siegbahn, P. E. M.; Crabtree, R. H. *J. Am. Chem. Soc.* **1997**, *119*, 3103. (d) Ortiz de Montellano, P. R. *Acc. Chem. Res.* **1998**, *31*, 543. (e) Willems, J.-P.; Valentine, A. M.; Gurbiel, R.; Lippard, S. J.; Hoffman, B. M. *J. Am. Chem. Soc.* **1998**, *120*, 9410.

34. (a) Nordlander, E.; Andersson, K. K. *JBIC*, **1998**, *3*, 300. (b) Deeth, R. J.; Dalton, H. *JBIC*, **1998**, *3*, 302. (c) Whittington, D. A.; Valentine, A. M.; Lippard, S. J. *JBIC*, **1998**, *3*, 307. (d) Siegbahn, P. E. M.; Crabtree, R. H.; Nordlund, P. *JBIC*, **1998**, *3*, 314. (e) Yoshizawa, K. *JBIC*, **1998**, *3*, 318. (f) Lipscomb, J. D.; Que, L., Jr. *JBIC*, **1998**, *3*, 331.

35. (a) Newman, L. M.; Wackett, L. P. *Biochemistry* **1995**, *34*, 14066. (b) Pikus, J. D.; Studts, J. M.; Achim, C.; Kauffmann, K. E.; Munck, E.; Steffan, R. J.; McClay, K.; Fox, B. G. *Biochemistry* **1996**, *36*, 9106.

36. van Beilen J. B.; Kingma, J.; Witchold, B. *Enzyme Microbiol. Technol.* **1994**, *106*, 7928.

37. (a) Lunts, A. C. *J. Chem. Phys.* **1980**, *73*, 1143. (b) Dalton, H.; Wilkins, P. C.; Deighton, N.; Podmore, I. D.; Symous, M. C. R. *Faraday Discuss. Chem. Soc.* **1992**, *93*, 163. (c) Groves, J. T.; Subramanian, D. V. *J. Am. Chem. Soc.* **1984**, *106*, 2177. (d) Gelb, M. H.; Heimbrook, D. C.; Malkonen, P.; Sligar, S. G. *Biochemistry* **1982**, *21*, 370. (e) Shestakov, A. F.; Shilov, A. E. *Zh. Obsch. Khim.* **1995**, *65*, 622 (in Russian). (f) Shestakov, A. F.; Shilov, A. E. *J. Mol. Catal. A: Chem.* **1996**, *105*, 1. (g) Shilov, A. E.; Shteinman, A. A. *Acc. Chem. Res.* **1999**, *32*, 763. (h) Karasevich, E. I.; Shestakov, A. F.; Shilov, A. E. *Kinet. Katal.* **1997**, *38*, 852 (in Russian). (i) Groves, J. T.; Viski, P. *J. Am. Chem. Soc.* **1989**, *111*, 8537.

38. (a) Que, L., Jr.; Ho, R. Y. N. *Chem. Rev.* **1996**, *96*, 2607. (b) Boyd, D. R.; Sharma, N. D.; Carroll, J. G.; Malone, J. F.; Mackerracher, D. G.; Allen, C. C. R. *Chem. Comm.* **1998**, 683. (c) Eppink, M. H. M.; Schreuder, H. A.; van Berkel, W. J. H. *Eur. J. Biochem.* **1998**, *253*, 194. (d) Burton, S. G.; Boshoff, A.; Edwards, W.; Rose, P. D. *J. Mol. Catal. B: Enzymatic* **1998**, *5*, 411. (e) Frazee, R. W.; Orville, A. M.; Dolbeare, K. B.; Yu, H.; Ohlendorf, D. H.; Lipscomb, J. D. *Biochemisry* **1998**, *37*, 2131. (f) Ueda, H.; Kitayama, A.; Suzuki, E.; Nagamune, T. *Kagaku Kogaku Ronbunshu* **1998**, *24*, 169 (in Japaneese). (g) Mitrochkine, A. A.; Eydoux, F.; Gil, G.; Relier, M. *Eur. J. Org. Chem.* **1998**, 1171.

39. Shiman, R.; Gray, D. W.; Hill, M. A. *J. Biol. Chem.* **1994**, *269*, 24637.

40. (a) Fitzpatrick, P. F. *Biochemistry* **1991**, *30*, 3658. (b) Francisco, W. A.; Tian, G.; Fitzpatrick, P. F.; Klinman, J. P. *J. Am. Chem. Soc.* **1998**, *120*, 4057. (c) Alterio, J.; Ravassard, P.; Haavik, J.; Le Caer, J.-P.; Biguet, N. F.; Waksman, G.; Mallet, J. *J. Biol. Chem.* **1998**, *273*, 10196.

41. Moran, G. R.; Daubner, S. C.; Fitzpatrick, P. F. *J. Biol. Chem.* **1998**, *273*, 12259.

42. Fitzpatrick, P. F. *J. Am. Chem. Soc.* **1994**, *116*, 1133.

43. (a) Bertini, I.; Cremonini, M.A.; Ferretti, S.; Lozzi, I.; Luchinat, C.; Viezzoli, M. S. *Coord. Chem. Rev.* **1996**, *151*, 145. (b) Boyd, D. R.; Sharma, N. D.; Kerley, N. A.; McMordie, R. A. S.; Sheldrake, G. N.; Williams, P.; Dalton, H. *J. Chem. Soc., Perkin Trans. 1* **1996**, 67. (c) Hudlicky, T.; Tian, X.;

Königsberger, K.; Maurya, R.; Rouden, J.; Fan, B. *J. Am. Chem. Soc.* **1996**, *118*, 10752. (d) Hudlicky, T.; Abboud, K. A.; Entwistle, D. A.; Fan, R.; Maurya, R.; Thorpe, A. J.; Bolonick, J.; Myers, B. *Synthesis* **1996**, 897.

44. (a) Wallis, M. G.; Chapman, S. K. *Biochem. J.* **1990**, *266*, 605. (b) Murakami, S.; Nakanishi, Y.; Kodama, N.; Takenaka, S.; Shinke, R.; Aoki, K. *Biosci. Biotechnol. Biochem.* **1998**, *62*, 747.

45. Harpel, M. R.; Lipscomb, J. D. *J. Biol. Chem.* **1990**, *265*, 22187.

46. Tsang, H.-T.; Batie, C. J.; Ballou, D. P.; Penner-Hahn, J. E. *JBIC* **1996**, *1*, 24.

47. Locher, H. H.; Leisinger, T.; Cook, A. M. *Biochem. J.* **1991**, *274*, 833.

48. (a) Nelson, M. J.; Batt, D. G.; Thompson, J. S.; Wright, S. W. *J. Biol. Chem.* **1991**, *266*, 8225. ((b) Yamamoto, S. *Free Radical Biol. Med.* **1991**, *10*, 149. (c) Glickman, M. H.; Wiseman, J. S.; Klinman, J. P. *J. Am. Chem. Soc.* **1994**, *116*, 793. (d) Hwang, C.-C.; Grissom, C. B. *J. Am. Chem. Soc.* **1994**, *116*, 795. (e) Clapp, C. H.; Grandizio, A. M.; McAskill, S. A.; Moser, R. *Biochemistry* **1995**, *34*, 264. (f) Nieuwenhuizen, W. F.; Schilstra, M. J.; van der Kerk-van Hoof, A.; Brandsma, L.; Veldink, G. A.; Vliegenthart, J. F. G. *Biochemistry* **1995**, *34*, 10538. (g) Glickman, M. H.; Klinman, J. P. *Biochemistry* **1995**, *34*, 14077. (h) Jonsson, T.; Glickman, M. H.; Sun, S.; Klinman, J. P. *J. Am. Chem. Soc.* **1996**, *118*, 10319. (i) Guajardo, R. J.; Marscharak, P. *Inorg. Chem.* **1995**, *34*, 802. (j) Jonas, R. T.; Stack, T. D. P. *J. Am. Chem. Soc.* **1997**, *119*, 8566. (k) Moiseyev, N.; Rucker, J.; Glickman, M. H. *J. Am. Chem. Soc.* **1997**, *119*, No. 17. (l) Glickman, M. H. Cliff, S.; Thiemens, M.; Klinman, J. P. *J. Am. Chem. Soc.* **1997**, *119*, 11357. (m) Vörös, K.; Feussner, I.; Kühn, H.; Lee, J.; Graner, A.; Löbler, M.; Parthier, B.; Wasternack, C. *Eur. J. Biochem.* **1998**, *251*, 36. (n) Husson, F.; Couturier, A.; Kermasha, S.; Belin, J. M. *J. Mol. Catal. B: Enzymatic* **1998**, *5*, 159. (o) Hubatsch, I.; Ridderström, M.; Mannervik, B. *Biochem. J.* **1998**, *330*, 175. (p) Kuban, R. J.; Wiesner, R.; Rathman, J.; Veldink, G.; Nolting, H.; Sole, V. A.; Kühn, H. *Biochem. J.* **1998**, *332*, 237.

49. (a) Han, H.; Pascal, R. A., Jr. *J. Org. Chem.* **1990**, *55*, 5173. (b) Ng, S.-F.; Hanauske-Abel, H. M.; Englard, S. *J. Biol. Chem.* **1991**, *266*, 1526. (c) Baldwin, J. E.; Adlington, R. M.; Crouch, N. P.; Pereira, I. A. C.; Aplin, R. T.; Robinson, C. *J. Chem. Soc., Chem. Commun.* **1993**, 105.

50. Posner, G. H.; Park, S. B.; Gonzalez, L.; Wang, D.; Cumming, J. N.; Klinedinst, D.; Shapiro, T. A.; Bachi, M. D. *J. Am. Chem. Soc.* **1996**, *118*, 3537.

51. van Deurzen, M. P. J.; van Rantwijk, F.; Sheldon, R. A. *J. Mol. Catal. B: Enzym*atic **1996**, *2*, 33.

52. (a) Tyeklár, Z.; Karlin, K. D. *Acc. Chem. Res.* **1989**, *22*, 241. (b) Vigato, P. A.; Tamburini, S.; Fenton, D. E. *Coord. Chem. Rev.* **1990**, *106*, 25. (c) Spodine, E.; Manzur, J. *Coord. Chem. Rev.* **1992**, *119*, 171. (d) Solomon, E. I.; Baldwin, M. J.; Lowery, M. D. *Chem. Rev.* **1992**, *92*, 521. (e) Kitajima, N. *Adv. Inorg. Chem.* **1992**, *39*, 1. (f) Karlin, K. D.; Tyeklár, Z. *Adv. Inorg. Biochem.* **1993**, *9*, 123. (g) *Bioinorganic Chemistry of Copper*; Karlin, K. D., Tyeklár, Z., Eds.; Chapman & Hall: New York, 1993. (h) Kitajima, N.; Moro-oka, Y. *Chem. Rev.* **1994**, *94*, 737. (i) Klinman, J. P. *Chem. Rev.* **1996**, *96*, 2541. (j) Solomon, E. I.; Sundaram, U. M.; Machonkin, T. E. *Chem. Rev.* **1996**, *96*, 2563. (k) Ferguson-Miller, S.; Babcock, G. T. *Chem. Rev.* **1996**, *96*, 2889. (l) Fontecave, M.; Pierre, J.-L. *Coord. Chem. Rev.* **1998**, *170*, 125.

53. (a) Espin, J. C.; Garcia-Ruiz, P. A.; Tudela, J.; Garcia-Canovas, F. *Biochem. J.* **1998**, *331*, 547. (b) Rodríguez, P. S.; López, J. N. R.; Tudela, J.; Varón, R.; Cánovas, F. G. *Biochem. J.* **1990**, *272*, 459. (c) Winder, A.J.; Harris, H. *Eur. J. Biochem.* **1991**, *198*, 317. (d) Gahn, L.G.; Roskoski, R., Jr. *Biochemistry* **1995**, *34*, 252.

54. McCracken, J.; Pember, S.; Benkovic, S. J.; Villafranca, J. J.; Miller, R. J.; Peisach, J. *J. Am. Chem. Soc.* **1988**, *110*, 1069.

55. (a) Stewart, L. C.; Klinman, J. P. *Annu. Rev. Biochem.* **1988**, *57*, 551. (b) Bossard, M. J.; Klinman, J. P. *J. Biol. Chem.* **1990**, *265*, 5640. (c) Wang, N.; Southan, C.; DeWolf, W. E.; Wells, T. N. C.; Kruse, L. I.; Leatherbarrow, R. J. *Biochemistry* **1990**, *29*, 6466. (d) Wimalasena, K.; Alliston, K. R. *J. Am. Chem. Soc.* **1995**, *117*, 1220. (e) Ghosh, D.; Lal, T. K.; Ghosh, S.; Mukherjee, R. *Chem. Commun.* **1996**, 13.

56. (a) *Molybdenum Enzymes*; Spiro, T. G., Ed.; John Wiley: New York, 1985. (b) Enemark, J. H.; Young, C. G. *Adv. Inorg. Chem.* **1993**, *40*, 1. (c) *Molybdenum Enzymes, Cofactors and Model Systems*; Stiefel, E. I., Coucouvanis, D.; Newton, W. E., Eds.; ACS Symposium Series 535; ACS: Washington, DC., 1993. (d) Hille, R. *Chem. Rev.* **1996**, *96*, 2757.

57. (a) Greenwood, R. J.; Wilson, G. L.; Pilbrow, J. R.; Wedd, A. G. *J. Am. Chem. Soc.* **1993**, 115, 5385. (b) Howes, B. D.; Bennett, B.; Bray, R. C.; Richards, R. L.; Lowe, D. J. *J. Am. Chem. Soc.* **1994**, *116*, 11624. (c) Carducci, M. D.; Brown, C.; Solomon, E. I.; Enemark, J. H. *J. Am. Chem. Soc.* **1994**, 116, 11856. (d) Hille, R. *JBIC*, **1996**, *1*, 397. (e) Ishikawa, T.; Maeda, K.; Hayakawa, K.; Kojima, T. *J. Mol. Catal. B: Enzym*atic **1996**, *1*, 173. (f) Bray, M. R.; Deeth, R. J. *J. Chem. Soc., Dalton Trans.* **1997**, 1267.

(g) Bray, M. R.; Deeth, R. J. *J. Chem. Soc., Dalton Trans.* **1997**, 4005. (h) Conrads, T.; Hemann, C.; Hille, R. *Biochemistry* **1998**, *37*, 7787. (i) Voityuk, A. A.; Albert, K.; Romão, M. J.; Huber, R.; Rösch, N. *Inorg. Chem.* **1998**, *37*, 176.

58. (a) Hille, R. *JBIC* **1997**, *2*, 804. (b) Lowe, D. J.; Richards, R. L.; Bray, R. C. *JBIC* **1998**, *3*, 557. (c) Hille, R. *JBIC* **1998**, *3*, 559.

59. (a) Bono, J.-J.; Goulas, P.; Boe, J.-F.; Portet, N.; Seris, J.-L. *Eur. J. Biochem.* **1990**, *192*, 189. (b) Tittmann, K.; Proske, D.; Spinka, M.; Ghisla, Rudolph, R.; Hübner, G.; Kern, G. *J. Biol. Chem.* **1998**, *273*, 12929. (c) Su, C.; Oliw, E. H. *J. Biol. Chem.* **1998**, *273*, 13072. (d) Hamberg, M.; Su, C.; Oliw, E. H. *J. Biol. Chem.* **1998**, *273*, 13080.

60. Maltseva, O. V.; Solyanikova, I. P.; Golovleva, L. A. *Eur. J. Biochem.* **1994**, *226*, 1053.

61. (a) Labat, G.; Meunier, B. *Bull. Soc. Chim. Fr.* **1990**, *127*, 553. (b) Sarkanen, S.; Razal, R. A.; Piccariello, T.; Yamamoto, E.; Lewis, N. G. *J. Biol. Chem.* **1991**, *266*, 3636.

62. (a) *Vanadium in Bilogical Systems*; Chasteen, N. D., Ed.; Kluwer: Dordrecht, 1990. (b) Butler, A.; Carrano, C. J. *Coord. Chem. Rev.* **1991**, *109*, 61. (c) Butler, A.; Walker, J. V. *Chem. Rev.* **1993**, *93*, 1937. (d) Kaim, W.; Schwederski, B. *Bioinorganic Chemistry: Inorganic Elements in the Chemistry of Life*; John Wiley: New York, 1994. (e) Butler, A.; Clague, M. J.; Meister, G. E. *Chem. Rev.* **1994**, *94*, 625. (f) Messerschmidt, A.; Wever, R. *Proc. Natl. Acad. Sci.* **1996**, *93*, 392. (g) Messerschmidt, A.; Wever, R. *Inorg. Chim. Acta* **1998**, *273*, 160.

63. (a) Tabushi, I. *Coord. Chem. Rev.* **1988**, *86*, 1. (b) Mansuy, D.; Battioni, P.; Battioni, J.-P. *Eur. J. Biochem.* **1989**, *184*, 267. (c) Gunter, M. J.; Turner, P. *Coord. Chem. Rev.* **1991**, *108*, 115. (d) Bruice, T. C. *Acc. Chem. Res.* **1991**, *24*, 243. (e) Meunier, B. *Chem. Rev.* **1992**, *92*, 1411. (f) Quici, S.; Banfi, S.; Pozzi, G. *Gazz. Chim. Ital.* **1993**, *123*, 597. (g) Corbin, D. R.; Herron, N. *J. Mol. Catal.* **1994**, *86*, 343.

64. (a) Tajima, K.; Tada, K.; Shigematsu, M.; Kanaori, K.; Azuma, N.; Makino, K. *Chem. Comm.* **1997**, 1069. (b) Volz, H.; Müller, W. *Chem. Ber./Recueil* **1997**, *130*, 1099. (c) Gold, A.; Jayaraj, K.; Ball, L. M.; Brust, K. *J. Mol. Catal. A: Chem.* **1997**, *125*, 23. (d) Selke, M.; Sisemore, M. F.; Ho, R. Y. N.; Wertz, D. L.; Valentine, J. S. *J. Mol. Catal. A: Chem.* **1997**, *117*, 71. (e) Sisemore, M. F.; Selke, M.; Burstyn, J. N.; Valentine, J. S. *Inorg. Chem.*

1997, *36*, 979. (f) Wolowiec, S.; Latos-Grazynski, L. *Inorg. Chem.* **1998**, *37*, 2984. (g) Selke, M.; Valentine, S. *J. Am. Chem. Soc.* **1998**, *120*, 2652. (h) Traylor, T. G.; Kim, C.; Fann, W.-P.; Perrin, C. L. *Tetrahedron*, **1998**, *54*, 7977.

65. Wertz, D. L.; Sisemore, M. F.; Selke, M.; Driscoll, J.; Valentine, J. S. *J. Am. Chem. Soc.* **1998**, *120*, 5331.

66. Akhtar, M.; Calder, M. R.; Corina, D. L.; Wright, J. N. *Biochem. J.* **1982**, *201*, 569. (b) Vaz, A. D. N.; Chakraborty, S.; Massey, V. *Biochemistry* **1995**, *34*, 4246.

67. (a) Khenkin, A. M.; Shilov, A. E. *React. Kinet. Catal. Lett.* **1987**, *33*, 125. (b) Leduc, P.; Battioni, P.; Bartoli, J.F.; Mansuy, D. *Tetrahedron Lett.* **1988**, *29*, 205. (c) Belova, V. S.; Nikonova, L. A.; Raihkman, L. M.; Borukaeva, M. R. *Dokl. Akad. Nauk SSSR* **1972**, **204**, 897 (in Russian). (d) Tabushi, I.; Yazaki, A. *J. Am. Chem. Soc.* **1981**, *103*, 7371. (e) Karasevich, E. I.; Khenkin, A. M.; Shilov, A. E. *Dokl. Akad. Nauk SSSR* **1987**, *295*, 639 (in Russian). (b) Shul'pin, G. B.; Druzhinina, A. N. *Bull. Acad. Sci. USSR, Div. Chem. Sci.* **1991**, *40*, 2385. (g) Borovkov, V. V.; Solovieva, A. B.; Cheremenskaya, O. V.; Belkina, N. V. *J. Mol. Catal. A: Chem.* **1997**, *120*, L1. (h) Lu, W. Y.; Bartoli, J. F.; Battioni, P.; Mansuy, D. *New J. Chem.* **1992**, *16*, 341.

68. (a) Banfi, S.; Maiocchi, A.; Moggi, A.; Montanari, F.; Quici, S. *J. Chem. Soc., Chem. Commun.* **1990**, 1794. (b) Yamaki, S.; Kobayashi, S.; Kotani, E.; Tobinaga, S. *Chem. Pharm. Bull.* **1990**, *38*, 1501. (c) Artaud, I.; Ben-Aziza, K.; Mansuy, D. *J. Org. Chem.* **1993**, *58*, 3373.

69. (a) Mansuy, D.; Battioni, P.; Renaud, J.-P. *J. Chem. Soc., Chem. Commun.* **1984**, 1255. (b) Higuchi, T.; Shimada, K.; Maruyama, N.; Hirobe, M. *J. Am. Chem. Soc.* **1993**, *115*, 7551.

70. (a) Lindsay Smith, J. R.; Sleath, P. S. *J. Chem. Soc., Perkin Trans. 2* **1982**, 1009. (b) Bartoli, J. F.; Brigaud, O.; Battioni, P.; Mansuy, D. *J. Chem. Soc., Chem. Commun.* **1991**, 440. (c) Khanna, R. K.; Pauling, T. M.; Vajpayee, D. *Tetrahedron Lett.* **1991**, *31*, 3759. (d) Battioni, P.; Bartoli, J. F.; Mansuy, D.; Byun, Y. S.; Traylor, T. G. *J. Chem. Soc., Chem. Commun.* **1992**, 1050. (e) Bartoli, J. F.; Battioni, P.; De Foor, W. R.; Mansuy, D. *J. Chem. Soc., Chem. Commun.* **1994**, 23. (f) Nam, W.; Valentine, J. S. *J. Am. Chem. Soc.* **1993**, *115*, 1772.

71. (a) De Poorter, B.; Ricci, M.; Bortolini, O.; Meunier, B. *J. Mol. Catal.* **1985**, *31*, 221. (b) Sorokin, A. B.; Khenkin, A. M. *New J. Chem.* **1990**, *14*, 63.

72. (a) Robert, A.; Meunier, B. *New J. Chem.* **1988**, *12*, 885. (b) Meunier, B. *New J. Chem.* **1992**, 16, 203. (c) Hoffmann, P.; Robert, A.; Meunier, B. *C. R. Acad. Sci., Ser. 2* **1992**, *314*, 51.

73. Ohtake, H.; Higuchi, T.; Hirobe, M. *J. Am. Chem. Soc.* **1992**, *114*, 10660.

74. (a) Querci, C.; Ricci, M. *Tetrahedron Lett.* **1990**, *31*, 1779. (b) Artaud, I.; Grennberg, H.; Mansuy, D. *J. Chem. Soc., Chem. Commun.* **1992**, 1036.

75. (a) Moro-oka, Y.; Akita, M. *Catalysis Today* **1998**, *41*, 327. (b) Murahashi, S.-I.; Komiya, N. *Catalysis Today* **1998**, *41*, 339. (c) Kulikova, V. S.; Shestakov, A. F.; Shilov, A. E. *Kinet. Katal.* **1999**, *40*, No. 2 (in Russian).

76. MacFaul, P. A.; Ingold, K. U.; Wayner, D. D. M.; Que, L., Jr. *J. Am. Chem. Soc.* **1997**, *119*, 10594.

77. (a) Groves, J. T.; Haushalter, R. C.; Nakamura, M.; Nemo, T. E.; Evans, B. I. *J. Am. Chem. Soc.* **1981**, *103*, 2884. (b) Khenkin, A. M.; Shteinman, A. A. *Kinet. Katal.* **1982**, *23*, 219 (in Russian). (c) Khenkin, A. M.; Shteinman, A. A. *Izv. Akad. Nauk SSSR, Ser. Khim.* **1982**, 1668 (in Russian). (d) Groves, J. T.; Bonchio, M.; Carofiglio, T.; Shalyaev, K. *J. Am. Chem. Soc.* **1996**, *118*, 8961. (e) Groves, J. T.; Lee, J.; Marla, S. S. *J. Am. Chem. Soc.* **1997**, *119*, 6269. (f) Feichtinger, D.; Plattner, D. A. *Angew. Chem., Int. Ed. Engl.* **1997**, *36*, 1718.

78. Urano, Y.; Higuchi, T.; Hirobe, M.; Nagano, T. *J. Am. Chem. Soc.* **1997**, *117*, 12008.

79. (a) Shilov, A. E. *Izv. Akad. Nauk SSSR, Ser. Khim.* **1990**, 2407 (in Russian). (b) Sorokin, A. B.; Marakushev, S. A.; Khenkin, A. M.; Shilov, A. E.; Shteinman, A. A. *Dokl. Akad. Nauk SSSR* **1984**, *279*, 939 (in Russian). (c) Breslow, R.; Gabriele, B.; Yang, J. *Tetrahedron Lett.* **1998**, *39*, 2887.

80. Karasevich, E. I.; Anisimova, B. L.; Rubailo, V. L.; Shilov, A. E. *Kinet. Katal.* **1993**, *34*, 650 (in Russian).

81. Campestrini, S.; Meunier, B. *Inorg. Chem.* **1992**, *31*, 1999.

82. *Inclusion Chemistry with Zeolites: Nanoscale Materials by Design*; Topics in Inclusion Science No 6.; Herron, N., Corbin, D. R., Eds.; Kluwer: Dordrecht, 1995.

83. (a) Barloy, L.; Lallier, J. P.; Battioni, P.; Mansuy, D. *New J. Chem.* **1992**, *16*, 71. (b) Battioni, P.; Ivanejko, R.; Mansuy, D.; Mlodnicka, T.; Poltowicz, J.; Sanchez, *J. Mol. Catal. A: Chem.* **1996**, *109*, 91.

84. Khenkin, A. M.; Shteinman, A. A. *Kinet. Katal.* **1989**, *30*, 7 (in Russian).

85. Bagrii, E. I.; Nekhaev, A. I. *Neftekhimiya* **1996**, *36*, 483 (in Russian).

86. Takagi, S.; Suzuki, M.; Shiragami, T.; Inoue, H. *J. Am. Chem. Soc.* **1997**, *119*, 8712.

87. (a) Que, L.; Jr.; Dong, Y. *Acc. Chem. Res.* **1996**, *29*, 190. (b) Mandal, S. K.; Que, L.; Jr. *Inorg. Chem.* **1997**, *36*, 5424.

88. (a) Sheu, C.; Sawyer, D. T. *J. Am. Chem. Soc.* **1990**, *112*, 8212. (b) Nivorozhkin, A. L.; Girord, J. J. *Angew. Chem. Intern. Ed. Engl.* **1996**, *35*, 609. (c) Miki, K.; Furuya, T. *Chem. Comm.* **1998**, 97.

89. (a) Stassinopoulos, A.; Schulte, G.; Papaefthymiou, G. C.; Caradonna, J. P. *J. Am. Chem. Soc.* **1991**, *113*, 8686. (b) Belova, V. S.; Gimanova, I. M.; Stepanova, M. L.; Khenkin, A. M.; Shilov, A. E. *Dokl. Akad. Nauk SSSR* **1991**, 3*16*, 653 (in Russian). (c) Shteinman, A. A. *Dokl. Akad. Nauk SSSR* **1992**, *324*, 617 (in Russian). (d) Belova, V. S.; Khenkin, A. M.; Postnov, V. N.; Prusakov, V. E.; Shilov, A. E.; Stepanova, M. L. *Mendeleev Commun.* **1992**, 7. (e) Dong, Y.; Ménage, S.; Brennan, B. A.; Elgren, T E.; Jang, H. G.; Pearce, L. L.; Que, L., Jr. *J. Am. Chem. Soc.* **1993**, *115*, 1851. (f) Ménage, S.; Vincent, J. M.; Lambeaux, C.; Chottard, G.; Grand, A.; Fontecave, M. *Inorg. Chem.* **1993**, *32*, 4766. (g) Leising, R .A.; Kim, J.; Pérez, M. A.; Que, L., Jr. *J. Am. Chem. Soc.* **1993**, *115*, 9524. (h) Kojima, T.; Leising, R. A.; Yan, S.; Que, L., Jr. *J. Am. Chem. Soc.* **1993**, *115*, 11328. (i) Buchanan, R. M.; Chen, S.; Richardson, J. F.; Bressan, M.; Forti, L.; Morvillo, A.; Fish, R. H. *Inorg. Chem.* **1994**, *33*, 3208. (j) Zang, Y.; Dong, Y.; Que, L., Jr.; Kauffmann, K.; Münck, E. *J. Am. Chem. Soc.* **1995**, *117*, 1169. (k) Hayashi, Y.; Kayatani, T.; Sugimoto, H.; Suzuki, M.; Inomata, K.; Uehara, A.; Mizutani, Y.; Kitagawa, T.; Maeda, Y. *J. Am. Chem. Soc.* **1995**, *117*, 11220. (l) Tétard, D.; Rabion, A.; Verlhac, J.-B.; Guilhem, J. *J. Chem. Soc., Chem. Commun.* **1995**, 531. (m) Ménage, S.; Galey, J.-B.; Hussler, G.; Seité, M.; Fontecave, M. *Angew. Chem.* **1996**, *108*, 2535. (n) Kodera, M.; Shimakoshi, H.; Kano, K. *Chem. Commun.* **1996**, 1737. (o) Feig, A. L.; Becker, M.; Schindler, S.; van Eldik, R.; Lippard, S. J. *Inorg. Chem.* **1996**, *35*, 2590. (p) Kim, K.; Lippard, S. J. *J. Am. Chem. Soc.* **1996**, *118*, 4914. (q) Dong, Y.; Yan, S.; Young, V. G., Jr.; Que, L., Jr. *Angew. Chem.* **1996**, *108*, 673. (r) Davydov, R. M.; Ménage, S.; Fontecave, M.; Gräslund, A.; Ehrenberg, A. *JBIC* **1997**, *2*, 242. (s) Yoshizawa, K.; Ohta, T.; Yamabe, T. *Bull. Chem. Soc. Jpn.* **1998**, *71*, 1899. (t) Müller, M.; Weyhermüller, T.; Bill, E.; Wieghardt, K. *JBIC* **1998**, *3*, 96. (u) LeCloux, D. D.; Barrios, A. M.; Mizoguchi, T. J.; Lippard, S. J. *J. Am. Chem. Soc.* **1998**, *120*, 9001. (v) Kauffmann, K. E.; Popescu, C. V.; Dong, Y.;

Lipscomb, J. D.; Que, L., Jr.; Münck, E. *J. Am. Chem. Soc.* **1998**, *120*, 8739. (w) Cai, L.; Han, Y.; Mahmpud, H.; Xie, W.; Segal, B. M. *Inorg. Chem. Commun.* **1998**, *1*, 71. (x) Poussereau, S.; Blondin, G.; Casario, M.; Guilheim, J.; Chottard, G.; Gonnet, F.; Girerd, J.-J. *Inorg. Chem.* **1998**, *37*, 3127. (y) Zhang, X.-X.; Fuhrmann, P.; Lippard, S. J. *J. Am. Chem. Soc.* **1998**, *120*, 10260. (z) Kim, C.; Dong, Y.; Que, L., Jr. *J. Am. Chem. Soc.* **1997**, *119*, 3635. (aa) Ito, S.; Okuno, T.; Matsushima, H.; Tokii, T.; Nishida, Y. *J. Chem. Soc., Dalton Trans.* **1996**, 4037, 4479. (ab) Que, L., Jr. *J. Chem. Soc., Dalton Trans.* **1997**, 3933. (ac) Dong, Y.; Zang, Y.; Shu, L.; Wilkinson, E. C.; Que, L., Jr. *J. Am. Chem. Soc.* **1997**, *119*, 12683. (ad) Kulikova, V. S.; Gritsenko, O. N.; Shteinman, A. A. *Mendeleev Commun.* **1996**, 119. (ae) Eulering, B.; Schmidt, M.; Pinkernell, U.; Karst, U.; Krebs, B. *Angew. Chem., Int. Ed. Engl.* **1996**, *35*, 1973. (af) Knops-Gerrits, P. P.; Dick, S.; Weiss, A.; Genet, M.; Roughet, P.; Li, X. Y.; Jacobs, P. A. *3rd World Congress on Oxidation Catalysis*; Grasselli, R. K.; Oyama, S. T.; Gaffney, A. M.; Lyons, J. E., Eds.; Elsevier: Amsterdam, 1997, p. 1061. (ag) Panov, G. I.; Sobolev, V. I.; Dubkov, K. A.; Parmon, V. N.; Ovanesyan, N. S.; Shilov, A. E.; Shteinman, A. A. *React. Kinet. Catal. Lett.* **1997**, *61*, 251.

90. (a) Funabiki, T.; Toyoda, T.; Ishida, H.; Tsujimoto, M.; Ozawa, S.; Yoshida, S. *J. Mol. Catal.* **1990**, *61*, 235. (b) Ohkubo, K.; Ishida, H.; Sagawa, T.; Miyata, K.; Yoshinaga, K. *J. Mol. Catal.* **1990**, *62*, 107. (c) Tajima, K.; Yoshino, M.; Mikami, K.; Edo, T.; Ishizu, K.; Ohya-Nishiguchi, H. *Inorg. Chim. Acta* **1990**, *172*, 83. (d) Fujii, S.; Ohya-Nishiguchi, H.; Hirota, N.; Nishinaga, A. *Bull. Chem. Soc. Jpn.* **1993**, *66*, 1408. (e) Kitajima, N.; Ito, M.; Fukui, H.; Moro-oka, Y. *J. Am. Chem. Soc.* **1993**, *115*, 9335.

91. Taqui Khan, M. M.; Shukla, R. S. *J. Mol. Catal.* **1992**, *77*, 221.

92. Kulikova, V. S.; Khenkin, A. M.; Shilov, A. E. *Kinet. Katal.* **1988**, *29*, 1278 (in Russian).

93. Sagawa, T.; Ishida, H.; Urabe, K.; Yoshinaga, K.; Ohkubo, K. *J. Chem. Soc., Perkin Trans. 2* **1993**, 1.

94. (a) Mialane, P.; Anxolabehere-Mallart, E.; Blondin, G.; Nivorojkine, A.; Guilhem, J.; Tchertanova, L.; Casario, M.; Ravi, N.; Bominaar, E.; Girerd, J.-J.; Münck, E. *Inorg. Chim. Acta* **1997**, *263*, 367. (b) Duda, M.; Pascaly, M.; Krebs, B. *Chem. Comm.* **1997**, 835. (c) Funabiki, T.; Yamazaki, T.; Fukui, A.; Tanaka, T.; Yoshida, S. *Angew. Chem., Int. Ed. Engl.* **1998**, *37*, 513. (d) Simándi, L. I.; Simándi, T. L. *J. Chem. Soc., Dalton Trans.* **1998**, 3275. (e)

Viswanathan, R.; Palaniandavar, M.; Balasumbramanian, T.; Muthiah, T. P. *Inorg. Chem.* **1998**, *37*, 2943.

95. (a) Jonas, R. T.; Stack, T. D. P. *J. Am. Chem. Soc.* **1997**, *117*, 8566. (b) Ogo, S.; Wada, S.; Watanabe, Y.; Iwase, M.; Wada, A.; Harata, M.; Jitsukawa, K.; Masuda, H.; Einaga, H. *Angew. Chem., Int. Ed. Engl.* **1998**, *37*, 2102.

96. Funabiki, T.; Yokomizo, T.; Suzuki, S.; Yoshida, S. *Chem. Comm.* **1997**, 151.

97. Shu, L.; Broadwater, J. A.; Achim, C.; Fox, B. G.; Münck, E.; Que, L., Jr. *JBIC* **1998**, *3*, 392.

98. (a) Nasir, M. S.; Karlin, K. D.; Chen, Q.; Zubieta, J. *J. Am. Chem. Soc.* **1992**, *114*, 2264. (b) Alilou, E. H.; Amadei, E.; Giorgi, M.; Pierrot, M.; Réglier, M. *J. Chem. Soc., Dalton Trans.* **1993**, 549. (c) Fujisama, K.; Tanaka, M.; Moro-oka, Y.; Kitajima, N. *J. Am. Chem. Soc.* **1994**, *116*, 12079. (d) Malachowski, M. R.; Dorsey, B.; Sackett, J. G.; Kelly, R. S.; Ferko, A. L.; Hardin, R. N. *Inorg. Chim. Acta* **1996**, *249*, 85. (e) Allen, W. E. *Inorg. Chem.* **1997**, *36*, 1732. (f) Alzuet, G.; Casella, L.; Villa, M. L.; Carugo, O.; Gullotti, M. *J. Chem. Soc., Dalton Trans.* **1997**, 4789. (g) Zheng, H.; Que, L., Jr. *Inorg. Chim. Acta* **1997**, *263*, 301. (h) Mahapatra, S.; Kaderli, S.; Llobet, A.; Neuhold, Y.-M.; Palanche, T.; Halfen, J. A.; Young, V. G., Jr.; Kaden, T. A.; Que, L., Jr.; Zuberbühler, A. D.; Tolman, W. B. *Inorg. Chem.* **1997**, *36*, 6343. (i) Martell, A. E.; Motekaitis, R. J.; Menif, R.; Rockcliffe, D. A.; Llobet, A. *J. Mol. Catal. A: Chem.* **1997**, *117*, 205. (j) Fahrni, C. J. *Tetrahedron*, **1998**, *54*, 5465. (k) Pidkock, E.; Obias, H. V.; Zhang, C. X.; Karlin, K. D.; Solomon, E. I. *J. Am. Chem. Soc.* **1998**, *120*, 7841. (l) Eswaramoorthy, M.; Neeraj, Rao, C. N. R. *Chem. Comm.* **1998**, 615. (m) Monzani, E.; Quinti, L.; Perotti, A.; Casella, L.; Gullotti, M.; Randaccio, L.; Geremia, S.; Nardin, G.; Faleschini, P.; Tabbi, G. *Inorg. Chem.* **1998**, *37*, 553.

99. (a) Perkinson, J.; Brodie, S.; Yoon, K.; Mosny, K.; Carroll, P. J.; Morgan, T. V.; Burgmayer, S. J. N. *Inorg. Chem.* **1991**, 30, 719. (b) Maddaluno, J.; Giessner-Prettre, C. *Inorg. Chem.* **1991**, *30*, 3439. (c) Ross, P. K.; Solomon, E. I. *J. Am. Chem. Soc.* **1991**, *113*, 3246. (d) Kitajima, N.; Fujisawa, K.; Fujimoto, C.; Moro-oka, Y.; Hasimoto, S.; Kitagawa, T.; Toriumi, K.; Tatsumi, K.; Nakamura, A. *J. Am. Chem. Soc.* **1992**, *114*, 1277. (e) Chaundhuri, P.; Karpenstein, I.; Winter, M.; Butzlaff, C.; Bill, E.; Trautwein, A. X.; Flörke, U.; Haupt, H.-J. *J. Chem. Soc., Chem. Commun.* **1992**, 321. (f) Mahroof-Tahir, M.; Karlin, K. D. *J. Am. Chem. Soc.* **1992**, *114*, 7599. (g) Tyeklár, Z.; Jacobson, R. R.; Wei, N.; Murthy, N.; Zubieta, J.; Karlin, K. D. *J. Am. Chem. Soc.* **1993**, *115*, 2677. (h) Wada, A.; Harata, M.; Hasegawa, K.;

Jitsukawa, K.; Masuda, H.; Mukai, M.; Kitagawa, T.; Einaga, H. *Angew. Chem., Int. Ed. Engl.* **1998**, *37*, 798. (i) Champloy, F.; Benali-Cherif, N.; Bruno, P.; Blain, I.; Pierrot, M.; Réglier, M. *Inorg. Chem.* **1998**, *37*, 3910. (j) Neves, A.; Rossi, L. M.; Vencato, I.; Drago, V.; Haase, W.; Werner, R. *Inorg. Chim. Acta* **1998**, *281*, 111.

100. (a) Cole, J. L.; Ballou, D. P.; Solomon, E. I. *J. Am. Chem. Soc.* **1991**, *113*, 8544. (b) Baldwin, M. J.; Root, D. E.; Pate, J. E.; Fujisawa, K.; Kitajima, N.; Solomon, E. I. *J. Am. Chem. Soc.* **1992**, *114*, 10421. (c) Reedy, B. J.; Blackburn, N. J. *J. Am. Chem. Soc.* **1994**, *116*, 1924.

101. (a) Yoshizawa, K.; Ohta, T.; Yamabe, T. *Bull. Chem. Soc. Jpn.* **1997**, *70*, 1911. (b) Bérces, A. *Chem Eur. J.* **1998**, *4*, 1297.

102. (a) Casella, L.; Monzani, E.; Gullotti, M.; Cavagnino, D.; Cerina, G.; Santagostini, L.; Ugo, R. *Inorg. Chem.* **1996**, *35*, 7517. (b) Karlin, K. D.; Tolman, W. B.; Kaderli, S.; Zuberbühler, A. D. *J. Mol. Catal. A: Chem.* **1997**, *117*, 215. (c) Halfen, J. A.; Young, V. G, Jr.; Tolman, W. B. *Inorg. Chem.* **1998**, *37*, 2102. (d) Ryan, S.; Adams, H.; Fenton, D. E.; Becker, M.; Schindler, S. *Inorg. Chem.* **1998**, *37*, 2134. (e) Itoh, S.; Nakano, H.; Berreau, L. M.; Kondo, T.; Komatsu, M.; Fukuzumi, S. *J. Am. Chem. Soc.* **1998**, *120*, 2890.

103. (a) Karlin, K. D.; Kaderli, S.; Zuberbühler, A. D. *Acc. Chem. Res.* **1997**, *30*, 139. (b) Tolman, W. B. *Acc. Chem. Res.* **1997**, *30*, 227.

104. (a) Trocha-Grimshaw, J.; McKillop, K. P. *Polyhedron* **1989**, *8*, 2513. (b) Réglier, M.; Jordan, C.; Waegell, B. *J. Chem. Soc., Chem. Commun.* **1990**, 1752. (c) Casella, L.; Gullotti, M.; Bartosek, M.; Pallanza, G.; Laurenti, E. *J. Chem. Soc., Chem. Commun.* **1991**, 1235. (d) Casella, L.; Gullotti, M.; Radaelli, R.; Di Gennaro, P. *J. Chem. Soc., Chem. Commun.* **1991**, 1611. (e) Paul, P. P.; Tyeklár, Z.; Jacobson, R. R.; Karlin, K. D. *J. Am. Chem. Soc.* **1991**, *113*, 5322. (f) Nasir, M. S.; Cohen, B. I.; Karlin, K. D. *J. Am. Chem. Soc.* **1992**, *114*, 2482. (g) Amadéi, E.; Alilou, E. H.; Eydoux, F.; Pierrot, M.; Réglier, M.; Waegell, B. *J. Chem. Soc., Chem. Commun.* **1992**, 1782. (h) Kitajima, N.; Katayama, T.; Fujisawa, K.; Iwata, Y.; Moro-oka, Y. *J. Am. Chem. Soc.* **1993**, *115*, 7872. (i) Capdevielle, P.; Maumy, M. *Tetrahedron Lett.* **1993**, *34*, 2953. (j) Karlin, K. D.; Wei, N.; Jung, B.; Kaderli, S.; Niklaus, P.; Zuberbühler, A. D. *J. Am. Chem. Soc.* **1993**, *115*, 9506. (k) Sanyal, I.; Murthy, N.; Karlin, K. D. *Inorg. Chem.* **1993**, *32*, 5330. (l) Mahapatra, S.; Halfen, J. A.; Wilkinson, E. C.; Que, L., Jr.; Tolman, W. B. *J. Am. Chem. Soc.* **1994**, *116*, 9785. (m) Sayre, L. M.; Nadkarni, D. V. *J. Am. Chem. Soc.*

1994, *116*, 3157. (n) Karlin, K. D.; Nasir, M. S.; Cohen, B. I.; Cruse, R. W.; Kaderli, S.; Zuberbühler, A. D. *J. Am. Chem. Soc.* **1994**, *116*, 1324.

105. (a) Lim, Y. Y.; Tan, E. H. L. *J. Mol. Catal.* **1993**, *81*, L1. (b) Shimizu, M.; Watanabe, Y.; Orita, H.; Hayakawa, T.; Takehira, K. *Bull. Chem. Soc. Jpn.* **1993**, *66*, 251. (c) Speier, G. *New J. Chem.* **1994**, *18*, 143. (d) Malachowski, M. R.; Huynh, H. B.; Tomlinson, L. J.; Kelly, R. S.; Furbe, J. W., Jr. *J. Chem. Soc., Dalton Trans.* **1995**, 31.

106. (a) Young, C. G. *JBIC* **1997**, *2*, 810. (b) Rastelli, G.; Costantino, L.; Albasini, A. *J. Am. Chem. Soc.* **1997**, *119*, 3007. (c) Young, C. G.; Wedd, A. G. *Chem. Comm.* **1997**, 1251.

107. (a) Carrano, C. J.; Mohan, M.; Holmes, S. M.; De la Rosa, R.; Butler, A.; Charnock, J. M.; Garner, C. D. *Inorg. Chem.* **1994**, *33*, 646. (b) Conte, V.; Di Furia, F.; Moro, S. *Tetrahedron Lett.* **1994**, *35*, 7429. (c) Dinesh, C. U.; Kumar, R.; Pandey, B.; Kumar, P. *J. Chem. Soc., Chem. Commun.* **1995**, 611. (d) Butler, A.; Clague, M. J. In *Mechanistic Bioinorganic Chemistry*; Adv. Chemistry Series *246*; Thorp, H. H., Pecoraro, V. L., Eds; A.C.S.: Washington DC, 1995, p. 330. (e) Anderson, M.; Conte, V.; Di Furia, F.; Moro, S. *Tetrahedron Lett.* **1995**, *36*, 2675. (f) Plass, W. *Inorg. Chim. Acta* **1996**, *244*, 221. (g) Hamstra, B. J.; Houseman, A. L. P.; Coplas, G. J.; Kampf, J. W.; LoBrutto, R.; Frasch, W. D.; Pecoraro, V. L. *Inorg. Chem.* **1997**, *37*, 4866. (h) Arends, I. W. C. E.; Birelli, M. P.; Sheldon, R. A. *3rd World Congress on Oxidation Catalysis*; Grasselli, R. K.; Oyama, S. T.; Gaffney, A. M.; Lyons, J. E., Eds.; Elsevier: Amsterdam, 1997, p. 1031. (i) Bashirpoor, M.; Schmidt, H.; Schulzke, C.; Rehder, D. *Chem. Ber./Recueil* **1997**, *130*, 651. (j) Fukui, K.; Ohya-Nishiguchi, H.; Kamada, H. *Inorg. Chem.* **1998**, *37*, 2326.

108. Coplas, G. J.; Hamstra, B. J.; Kampf, J. W.; Pecoraro, V. L. *J. Am. Chem. Soc.* **1996**, *118*, 3469.

109. Conte, V.; Di Furia, F.; Moro, S. *Tetrahedron Lett.* **1996**, *37*, 8609.

110. Lin, S.-K.; Jaun, B. *Helv. Chim. Acta* **1991**, *74*, 1725.

CONCLUSION

During the three decades which have elapsed since the discovery of the first reactions of alkanes in solutions of metal complexes, this field of chemistry has developed vigorously. There is no doubt that one may expect the discovery of new, especially catalytic, processes with the participation of methane and its homologues, which will find practical applications. In turn, this will intensify the interest in alkanes from the standpoint of their involvement in selective chemical processes. The new chemistry of alkanes thus established will be in line with the ever more urgent need for economy in the consumption of the existing resources of this chemical raw material on our planet.

The biomimetic approach for the search and development of new important reactions is a fruitful method at present and promising in the future. The existence in living nature of catalytic reactions unknown in chemistry encourages the search for their non-enzymatic analogues. More and more complicated enzymatic systems functioning in living organisms will be understood using new methods of investigation, and we will learn how to mimic these complicated systems in chemical laboratories. Eventually, when the origin of life and the chemical evolution of the biological systems will be understood sufficiently well, self developing and self perfecting catalysts can be visualized.

ABBREVIATIONS

AIBN = azobis(isobutyronitrile)

BDE's = bond dissociation enthalpies.

Cp = cyclopentadienyl. Cp* = pentamethylcyclopentadienyl.

KIE = kinetic isotope effect.

M = multiple exchange factor denoting the average number of hydrogen atoms exchanged within one contact.

MMO = methane monooxygenases.

NIH shift = transfer of a hydrogen (deuterium) atom to a neighboring position relative to that at which hydroxylation takes place (NIH = National Institute of Health).

NHPI = *N*-hydroxyphthalimide.

Organyl group = alkyl, aryl, vinyl, acyl, etc. group which is bound to M via a carbon atom.

Ox = oxidant. Red = reductant.

PINO = phthalimide *N*-oxyl.

Porph = porphyrinate.

Rad, R = radical. Alk = alkyl radical. Ar = aryl radical.

SET = single-electron transfer.

Tp = hydridotris(pyrazolyl)borate, $HB(Pz)_3$. Pz = pyrazolate. Tp* = hydridotris(3,5-dimethyl-pyrazolyl)borate, $HB(3,5\text{-}Me_2Pz)_3$,

C(1) : C(2) : C(3) : C(4) = the relative normalized (taking into account the number of hydrogen atoms at each of the carbon atoms) reactivities of hydrogen atoms at carbon atoms 1, 2, 3 and 4 of the linear hydrocarbon chain.

1° : 2° : 3° = the relative normalized (taking into account the number of hydrogen atoms at each of the carbon atoms) reactivities of hydrogen atoms at primary, secondary, and tertiary carbon atoms of the branched alkane

5/6 effect = the normalized ratio of the rate constants for cyclopentane and cyclohexane.

" ——→ " = agostic bond.

* = a vacant site.

INDEX

Catalysis by Metal Complexes

Series Editors:
R. Ugo, *University of Milan, Milan, Italy*
B. R. James, *University of British Columbia, Vancouver, Canada*

1.* F. J. McQuillin: *Homogeneous Hydrogenation in Organic Chemistry.* 1976 ISBN 90-277-0646-8
2. P. M. Henry: *Palladium Catalyzed Oxidation of Hydrocarbons.* 1980 ISBN 90-277-0986-6
3. R. A. Sheldon: *Chemicals from Synthesis Gas.* Catalytic Reactions of CO and H_2. 1983 ISBN 90-277-1489-4
4. W. Keim (ed.): *Catalysis in* C_1 *Chemistry.* 1983 ISBN 90-277-1527-0
5. A. E. Shilov: *Activation of Saturated Hydrocarbons by Transition Metal Complexes.* 1984 ISBN 90-277-1628-5
6. F. R. Hartley: *Supported Metal Complexes.* A New Generation of Catalysts. 1985 ISBN 90-277-1855-5
7. Y. Iwasawa (ed.): *Tailored Metal Catalysts.* 1986 ISBN 90-277-1866-0
8. R. S. Dickson: *Homogeneous Catalysis with Compounds of Rhodium and Iridium.* 1985 ISBN 90-277-1880-6
9. G. Strukul (ed.): *Catalytic Oxidations with Hydrogen Peroxide as Oxidant.* 1993 ISBN 0-7923-1771-8
10. A. Mortreux and F. Petit (eds.): *Industrial Applications of Homogeneous Catalysis.* 1988 ISBN 90-277-2520-9
11. N. Farrell: *Transition Metal Complexes as Drugs and Chemotherapeutic Agents.* 1989 ISBN 90-277-2828-3
12. A. F. Noels, M. Graziani and A. J. Hubert (eds.): *Metal Promoted Selectivity in Organic Synthesis.* 1991 ISBN 0-7923-1184-1
13. L. I. Simndi: *Catalytic Activation of Dioxygen by Metal Complexes.* 1992 ISBN 0-7923-1896-X
14. K. Kalyanasundaram and M. GrÎtzel (eds.), *Photosensitization and Photocatalysis Using Inorganic and Organometallic Compounds.* 1993 ISBN 0-7923-2261-4
15. P. A. Chaloner, M. A. Esteruelas, F. Jo¢ and L. A. Oro: *Homogeneous Hydrogenation.* 1994 ISBN 0-7923-2474-9

Catalysis by Metal Complexes

16. G. Braca (ed.): *Oxygenates by Homologation or CO Hydrogenation with Metal Complexes*. 1994 ISBN 0-7923-2628-8
17. F. Montanari and L. Casella (eds.): *Metalloporphyrins Catalyzed Oxidations*. 1994 ISBN 0-7923-2657-1
18. P.W.N.M. van Leeuwen, K. Morokuma and J.H. van Lenthe (eds.): *Theoretical Aspects of Homogeneous Catalysis*. Applications of *Ab Initio* Molecular Orbital Theory. 1995 ISBN 0-7923-3107-9
19. T. Funabiki (ed.): *Oxygenases and Model Systems*. 1997 ISBN 0-7923-4240-2
20. S. Cenini and F. Ragaini: *Catalytic Reductive Carbonylation of Organic Nitro Compounds*. 1997 ISBN 0-7923-4307-7
21. A.E. Shilov and G.B. Shul'pin: *Activation and Catalytic Reactions of Saturated Hydrocarbons in the Presence of Metal Complexes*. 2000 ISBN 0-7923-6101-6

KLUWER ACADEMIC PUBLISHERS – DORDRECHT / BOSTON / LONDON

**Volume 1 is previously published under the Series Title:*
Homogeneous Catalysis in Organic and Inorganic Chemistry.